# Inherited Disorders of the Skeleton

# GENETICS IN MEDICINE AND SURGERY

*General Editor*

**Alan E. H. Emery** MD PhD DSc FRCP FLS FRS(E)
Emeritus Professor of Human Genetics and University Fellow, Edinburgh.
Visiting Fellow, Green College, Oxford

The series is designed to provide trainee and practising doctors with easily accessible and authoritative information about genetic aspects of their specialties.
Volumes describe what is known of the hereditary nature of various groups of disease and indicate how this knowledge can be applied in genetic counselling.

*Two general volumes of relevance to the series*

**The Kidney in Genetic Disease**
Amin Y. Barakat, Vazken M. der Kaloustian, Amjad A. Mufarrji and Adel E. Birbari 1986

**Genetics and Neurology**
Sarah Bundey 1985

# Inherited Disorders of the Skeleton

**Peter Beighton**
M.D., Ph. D., F.R.C.P., D.C.H., F.R.S.(S.A.)

Professor of Human Genetics, Medical School, University of Cape Town.
Consultant in Medical Genetics, Groote Schuur Hospital.
Director, South African Medical Research Council Unit for
Inherited Skeletal Disorders, Cape Town, South Africa.

SECOND EDITION

FOREWORD BY
Alan E. H. Emery
M.D., Ph.D., D.Sc., F.R.C.P., F.L.S., F.R.S.(E).
Emeritus Professor of Human Genetics and University
Fellow, Edinburgh.
Visiting Fellow, Green College, Oxford.

CHURCHILL LIVINGSTONE
EDINBURGH LONDON MELBOURNE AND NEW YORK 1988

CHURCHILL LIVINGSTONE
Medical Division of Longman Group UK Limited

Distributed in the United States of America by
Churchill Livingstone Inc., 1560 Broadway, New York,
N.Y. 10036, and by associated companies, branches
and representatives throughout the world.

First edition 1978
Second edition 1988

ISBN 0 443 03506 7

British Library Cataloguing in Publication Data
Beighton, Peter
    Inherited disorders of the skeleton.
    [2nd ed.] — (Genetics in medicine
    and surgery).
    1. Bones — Diseases — Genetic aspects
    I. Title  II. Series
    616.7'1042    RC930

Library of Congress Cataloging in Publication Data
Beighton, Peter.
    Inherited disorders of the skeleton/Peter Beighton:
foreword by Alan E. H. Emery. — 2nd ed.
        p.     cm. — (Genetics in medicine and surgery)
    Includes bibliographies and index.
    ISBN 0-443-03506-7
    1. Bones — Diseases — Genetic
aspects.  2. Connective tissues — Diseases — Genetic
aspects.  3. Human skeleton. I. Title.  II. Series.
    [DNLM: 1. Bone Diseases — familial &
genetic.    WE 250 B422i] RC930.4.B45   1988
616.7'1042 — dc19      87-27621

Printed in Great Britain at The Bath Press, Avon

TO THE WOMEN IN MY LIFE
Greta, Victoria and Mary Violet

# Foreword

'Evolution . . . is a change from an indefinite, incoherent homogeneity, to a definite, coherent heterogeneity.'
  (Herbert Spencer's *First Principles*, 1862)

Since the first edition of this book appeared ten years ago, the subject of inherited skeletal disorders has advanced considerably, in no small measure due to the author's own researches in the field. Many new syndromes have been identified and the problems of resolving genetic heterogeneity in this group of disorders are now yielding to clinical, genetic and biochemical studies. Furthermore, developments in recombinant DNA technology are opening up novel approaches to prenatal diagnosis. The hope is that it may not be too long before these new developments lead to effective treatments for many of these distressing conditions which Professor Beighton has so interestingly reviewed in this scholarly monograph.

Edinburgh, 1988                                         Alan E. H. Emery

# Preface

The dramatic manifestations of the inherited disorders of the skeleton have always attracted the attention of connoisseurs of rare syndromes. More than 300 genetic skeletal dysplasias have been delineated and there is little doubt that many others await recognition. Equally, it can confidently be anticipated that a number of conditions which are at present regarded as distinct entities will ultimately turn out to be heterogeneous. Although individually rare, these dysplasias are collectively quite common. In the period 1971–1987 more than 3000 patients with disorders of this type have been investigated by the author in genetic clinics in Southern Africa. Their case details, which have been recorded in the registry of the Medical Research Council Unit for Inherited Skeletal Disorders, Department of Human Genetics, Medical School, University of Cape Town, have formed the basis for this book. The illustrations are derived from the same source.

These disorders have been discussed from the point of view of the practice of medical genetics. Pertinent clinical and radiological features are described and illustrated in order to provide a balanced perspective. However, their detailed consideration is outside the scope of this book, and no attempt has been made to review them in depth. Similarly, management and pathogenesis are mentioned only when strictly relevant.

The terminology of the inherited disorders of the skeleton is complex and some have been given a different label in each publication in which they have been reported. It is not without significance that the philosopher Immanuel Kant commented: 'physicians often think that they do a lot for a patient when they give his disease a name'. For the sake of simplicity and clarity, the designations used in this book are those which have gained general acceptance. Technical terms which are applicable to the intrinsic diseases of bones are defined in the text and listed in the glossary.

It is assumed that the reader has a working knowledge of medical genetics and no attempt has been made to explain established concepts such as the various Mendelian modes of inheritance. Similarly, it is taken for granted that the consequences of any particular form of genetic transmission, with regard to counselling and antenatal diagnosis, are understood. In this respect, the 'burden of disease', as it influences the quality of life of the individual with any particular condition, is of paramount importance.

The references have been chosen to include key reviews and papers with a genetic slant. Every effort has been made to ensure that these are up-to-date. However, it is axiomatic that the period of time necessary for the preparation of a book always exceeds the author's initial forecast. The fact that a number of the children portrayed in the illustrations have now reached adulthood is ample testimony to the truth of this adage!

# Acknowledgements

It is a source of considerable pleasure that the original acknowledgements are still generally applicable to this new edition, that long established collaboration continues and that old friendships have been sustained.

I am grateful to numerous friends and colleagues for assistance in many ways:

to Dr C. O. Carter of London and Professor V. A. McKusick of Baltimore for fostering my early interest in clinical genetics and to Professor Alan E. H. Emery of Edinburgh for his continued guidance and encouragement.

to Professor J. Spranger of Germany for facilitating the development of the Cape Town Skeletal Dysplasia Registry by consistently reaching the correct diagnosis in the difficult cases which I have referred to him.

to Mr F. Horan, orthopaedic surgeon, my friend and collaborator over many years, for his amicable criticisms and his valuable comments.

to orthopaedic surgeons for allowing access to their patients: Professor C. E. L. Allen and Mr B. Jones, formerly of Princess Alice Orthopaedic Hospital and Groote Schuur Hospital, Cape Town: Professor L. Solomon and Mr J. Handelsman, University of the Witwatersrand, and Mr C. A. Bathfield, Baragwanath Hospital, Johannesburg.

to Professor B. Cremin for access to radiographic material

to the late Dr S. Goldberg for his good natured advice from his vast knowledge of bone disorders and to Professor H. Hamersma for patient referrals.

to clinical photographers for the illustrations — Messrs R. A. de Méneaud, C. Russ, R. C. Clow, E. Norman and Mrs S. Henderson.

to genetic nursing sisters for their assistance with patients — Miss Lecia Durr, Mrs Elizabeth Napier, Mrs Rosemary Duggan, and Mrs Lorraine Groeneveldt of Cape Town; Mrs Pam Otto, Miss Ann Williams and Mrs Judith Mathee of Johannesburg.

to the South African Medical Research Council and the University of Cape Town Staff Research Fund for generous financial support for the skeletal dysplasia clinics.

to Dr M. Nelson and Dr. W. S. Winship for their helpful suggestions on the contents of the text.

Most of all to Greta, who prepared the manuscript with enthusiasm and dedication which far exceeded her professional and marital obligations.

I offer my thanks for assistance with the second edition:

to the genetic nursing sisters, Meriel Macrae, Sue Dunstan, Lesley Merckel, Gail Christy, Janis Schapera and Luzanne McAllister.

to the orthopaedic surgeons, Professor A. W. B. Heywood and Professor George Dall, for access to the patients.

to Mrs E. M. Petersen, biochemist, Department of Human Genetics, for her kind assistance with Chapter 11.

to Drs Denis Viljoen and Colin Wallis for checking the manuscript.

to Gillian Shapley for preparing the manuscript with good humoured efficiency and diligence in the face of many forms of adversity and to Diane Cannel for assistance in the final stages during the race to meet the deadline.

to Solly Yach for his interest and encouragement and to the Mauerberger Fund and the Harry Crossley Foundation for their support of my investigations in the inherited skeletal disorders.

The Medical Research Council of South Africa have established a research unit for inherited skeletal disorders in the Department of Human Genetics, University of Cape Town. This contribution is acknowledged with gratitude.

P. B.

# Contents

# 1

# Dwarfing skeletal dysplasias without significant spinal involvement

For practical purposes, the osteochondrodysplasias of unknown pathogenesis which cause dwarfism can be divided into two groups: those in which spinal changes are minimal or absent and those in which the spine is significantly involved. The following conditions fall into the former category.

1. Achondroplasia
2. Hypochondroplasia
3. Asphyxiating thoracic dysplasia (Jeune)
4. Chondroectodermal dysplasia (Ellis-van Creveld)
5. Multiple epiphyseal dysplasia
   a. Other forms of MED
6. Chondrodysplasia punctata
   a. Autosomal dominant type (Conradi-Hünerman)
   b. X-linked dominant type
   c. Other forms of chondrodysplasia punctata
7. Metaphyseal chondrodysplasia
   a. Jansen type
   b. Schmid type
   c. McKusick type
   d. Shwachman type
   e. Davis type
   f. Other types of metaphyseal chondrodysplasia
8. Mesomelic dysplasia
   a. Nievergelt
   b. Langer
   c. Other types of mesomelic dysplasia
9. Acromesomelic dysplasia
10. Rhizomelic dysplasia
11. Grebe chondrodysplasia

Delineation of the osteochondrodystrophies was initially undertaken on a clinical basis. Nelson (1970) described the development of the concept of heterogeneity of these conditions and Dorst et al (1972), Kozlowski (1976) and Kaufman (1976) reviewed their radiographic features. The specific

1

ultrastructural changes which have now been identified in many of these disorders have been discussed by Rimoin et al (1976) and Stanescu et al (1984). It is likely that histopathological investigations of this type will reveal further heterogeneity. In the future it is conceivable that nomenclature and classification will be based upon abnormalities which are recognisable at the histological, biochemical or molecular levels.

The use of the word 'dwarfism' as a component of a name for a disorder has proved to be offensive to affected persons and their families and it is now conventionally replaced by 'dysplasia'. For example, the condition

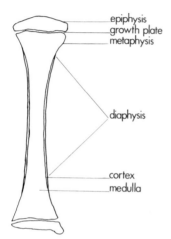

epiphysis
growth plate
metaphysis

diaphysis

cortex
medulla

**Fig. 1.1** (left) The components of a tubular bone.

**Fig. 1.2** (below left) Genu varum or bowlegs; a feature of many skeletal dysplasias.

**Fig. 1.3** (below right) Genu valgum or knock-knees, affected twins.

formerly known as 'mesomelic dwarfism' is now termed 'mesomelic dysplasia'. In the descriptive and anatomical sense 'dwarfism' has been retained and it is still perfectly correct to speak of 'disorders which cause dwarfism' or to allude to 'dwarfing skeletal dysplasias'.

Radiologically the epiphyses appear to be small and fragmented in many of the osteochondrodysplasias. However, hip arthrograms in various conditions in this category have revealed that the radiolucent cartilage of the femoral head is surprisingly normal in size and configuration. Lachman et al (1974), who undertook arthrographic studies, concluded that defective epiphyseal ossification must be a significant pathogenic factor. These findings are of importance from many points of view, ranging from the investigation of the basic defect to the formulation of regimes of management.

Orthopaedic problems predominate in the osteochondrodysplasias. The management of these complications has been reviewed by Kopits (1976) and formed the subject of a monograph by Horan & Beighton (1982). Prosthetic replacement of the hip and knee joints has been successfully accomplished in many patients and procedures of this type hold great promise for the future.

The term 'dwarfism' implies an abnormal degree of short stature. However, there is marked racial variation in normal height and it is impossible to give a precise definition of this designation. For this reason an individual's stature must be evaluated on a basis of his ethnic background. The last word on this ambiguous situation undoubtedly remains with the dwarfed fairground exhibitionist who advertised himself as 'the smallest giant in the world'!

## ACHONDROPLASIA

Achondroplasia is by far the commonest and best-known form of short-limbed dwarfism and it has attracted attention since ancient times. For instance, achondroplasts can be recognised amongst the dwarfed goldsmiths portrayed in a frieze on an Egyptian tomb. Similarly, the Egyptian deity Bes was conventionally depicted with the features of achondroplasia. Other classical portrayals include a statue of an achondroplast gladiator dating from the time of the Roman Emperor Domitian (AD 51–96) and Velasquez's portrait of Don Sebastian de Morro, an affected nobleman at the court of Philip V of Spain. It has even been suggested that Attila the Hun, the 'Scourge of God', might have had achondroplasia.

In the past the term 'achondroplast' was used indiscriminately for any individual with short-limbed dwarfism. However, with increasing diagnostic sophistication, many forms of dwarfism have been delineated as distinct disorders in their own right. Achondroplasia now remains as a specific and well-defined entity.

Warkany (1971) calculated that there are about 5000 achondroplasts in the USA and 65 000 on Earth. A measure of the relatively high frequency

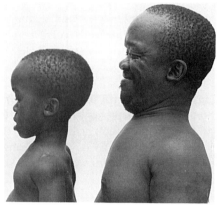

**Fig. 1.4** (left) Achondroplasia in a father and his son.

**Fig. 1.5** (above) Achondroplasia; the forehead is prominent and the nasal bridge is depressed.

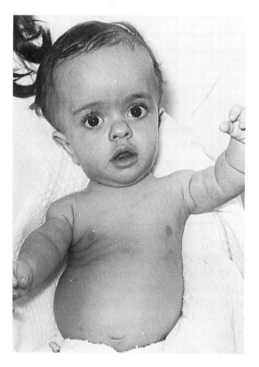

**Fig. 1.6** Achondroplasia: shortening of the arms is maximal in the proximal regions (rhizomelia).

of the condition can be gained from the fact that investigators at the Moore Clinic, Johns Hopkins Hospital, Baltimore, USA, were able to collect data concerning 393 cases (Todorov & Bolling 1974).

## Clinical and radiographic features

The characteristic facies, habitus and stance of the achondroplast are unmistakable. The limbs are disproportionately short and the knees are often bowed, while the lumbar spine is lordotic. The forehead is bossed and the nasal bridge is depressed. Inability to approximate the third and fourth fingers produces a 'trident' configuration of the hand.

The skull is relatively large, with frontal prominence and a small foramen magnum. The interpeduncular distances in the lower lumbar spine become progressively narrowed, vertebral pedicles are short and the disc spaces are wide. The iliac bones have a 'tombstone' configuration, the sacroiliac notch is narrow and the acetabular portions of the iliac bones are horizontal. Limb shortening is predominantly rhizomelic (proximal). The tubular bones have widened shafts and flared metaphyses. The radiographic features of achondroplasia have been comprehensively reviewed by Langer et al (1967).

General health is good and the life span is not reduced. Restriction of the calibre of the nasal airways predisposes to upper respiratory tract infection. Mentality is usually normal but intellectual ability is compromised in a minority of achondroplasts, probably due to mild internal hydrocephaly. Infrequently, severe hydrocephalus warrants a shunt procedure.

A high proportion of achondroplasts experience orthopaedic complications, notably degenerative osteoarthropathy of the knee joints. Spinal malalignment is generally uncommon in achondroplasia but in Africa many infants with the condition develop a significant thoraco-lumbar gibbus. This deformity, which poses a threat to the spinal cord, is probably the result of mechanical forces generated by the traditional custom of carrying the child on the mother's back, with the vulnerable spine in flexion (Beighton & Bathfield 1981).

Following an investigation of 48 affected persons in Britain, Wynne-Davies et al (1981) pointed out that there is a high risk of neurological damage when spinal canal narrowing and spinal malalignment co-exist. These observations were confirmed when the clinical problems caused by lumbar spine involvement in 47 adult achondroplasts in the USA were reviewed in detail by Kahanovitz et al (1982).

Compression of the spinal cord in the cervical region is an uncommon but life-threatening complication which results from anatomical abnormalities at the base of the skull (Blondeau et al 1984, Hecht et al 1984). The dimensions of the foramen magnum were measured using computerised tomography in 63 achondroplasts by Hecht et al (1985). These researchers demonstrated that this aperture was invariably small and that there was a relationship between size and liability to neurological complications.

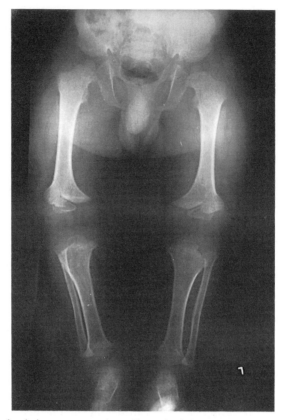

**Fig. 1.7** Achondroplasia: anteroposterior radiograph of the knees and shins. The metaphyses are flared and the diaphyses are short. The fibula is disproportionately long.

In a massive study involving 400 achondroplasts, Horton et al (1978) established standard curves for height, growth velocity, upper and lower body segments and head circumference. These charts are invaluable for predictory adult stature in affected children and they will serve as a baseline for the assessment of therapy for promoting growth, if this ever becomes available.

*Genetics*

Achondroplasia provides an excellent example of autosomal dominant inheritance as the gene is invariably penetrant, with consistent clinical expression. The prevalence per million in the newborn has been estimated at 23 in Denmark (Mørch 1941), 15 in the USA (Potter & Coverstone 1948) and 28 in Northern Ireland (Stevenson 1957). It is likely that dwarfing conditions other than achondroplasia were included in these series and these figures are probably inflated. Until recently it was assumed that a significant proportion of achondroplasts were stillborn or died in the neonatal period and this fact was taken into account when mutation rates were calculated. However, it is now known that achondroplasia is not usually lethal and the majority of stillborn infants with short-limbed dwarfism do not have achondroplasia (see Ch. 3). Using pooled data from four centres, Gardner (1977) identified 7 sporadic achondroplasts in 242 257 births and estimated that the mutation rate was of the order of 1:4 × 10⁻⁵. As the diagnostic criteria for achondroplasia are now well established there is little doubt that this figure reflects the true situation. Comparable figures were obtained from a survey of 22 063 consecutive newborns in Tokyo, in which Higurashi et al (1985) detected 2 achondroplasts. In a population survey of skeletal dysplasias in Britain, Wynne-Davies & Gormley (1985) estimated the minimum frequency of achondroplasia as 4.3 per million.

More than 80% of achondroplasts have normal parents (Scott 1976). These 'sporadic' patients are assumed to be the result of new mutation of the particular gene, which has taken place before conception. Murdoch et al (1970) and Collipp et al (1976) have shown that there is a significant increase in the average age of the unaffected fathers of sporadic achondroplasts. In a similar study of more than 100 achondroplast members of the French association for persons of small stature Stoll et al (1983) calculated a mean paternal age which was 3.9 years in excess of that of a control group. Achondroplasia therefore ranks as one of the disorders in which paternal age effect in the genesis of a new mutation has been demonstrated.

Anomalous kindreds have been reported in which achondroplasia was present in two cousins (Opitz 1969), in two sisters (Bowen 1974) and in three siblings (Fryns et al 1983). The other members of these families were normal. It was proposed that gonadal mosaicism was the most likely explanation for this latter situation. Six families in which achondroplasia had recurred in a manner which was not in keeping with autosomal dominant inheritance were reviewed by Reiser et al (1985) and an additional kindred containing affected second cousins was described by Fitzsimmons (1985). The problem of anomalous genetic transmission of achondroplasia and the development of concepts of unstable premutation have been discussed by Opitz (1985). Genetic counselling in achondroplasia is usually a straight forward matter but in view of the accumulating atypical pedigrees, a measure of caution is warranted.

In terms of genetic counselling, unaffected parents of an achondroplastic infant can be reassured that the risk of recurrence in further offspring is very low. Conversely, there is an even chance that any child born to an achondroplast of either sex, who is married to a normal individual, will have achondroplasia. Theoretically, 25% of infants born to parents who are both achondroplasts will be homozygous for the abnormal gene, 25% will be homozygous normal individuals and 50% will be heterozygous achondroplasts. Hall et al (1969) reported two probable homozygotes, while

Murdoch et al (1970) described further cases. These 'double affected' individuals had severe respiratory distress and died soon after birth. Their external features resembled classical achondroplasia, although the radiographic changes were of greater severity (Langer et al 1969).

The molecular basis of achondroplasia is unknown but Strom (1984) and Eng et al (1985) claimed to have identified a DNA insertion in the structural gene which encodes type II collagen in an affected person. This article was subsequently withdrawn. Thereafter Wordsworth et al (1986) undertook similar studies but were unable to detect any rearrangement of the gene at this locus in 12 persons with achondroplasia, using a variety of restriction endonucleases.

Type II collagen plays an important role in endochondral bone formation, which is defective in achondroplasia and it seems possible that the mutation is present in this gene or in the mechanism which regulates its function. As achondroplasia is relatively common, biological material is available for investigation and it can be anticipated that the molecular pathology will be elucidated in the foreseeable future.

*Antenatal diagnosis*

A female with achondroplasia married to a normal male, or a male achondroplast with an unaffected wife, has an even chance of transmitting the faulty gene at any episode of procreation. Early attempts at antenatal diagnosis in potentially affected pregnancies by conventional radiography were unsuccessful. Subsequently, radiography was supplemented with ultrasound, but achondroplasia still defied early prenatal diagnosis (Hall et al 1979). The difficulty revolved around the fact that the limb bones in achondroplasia are not grossly shortened in early pregnancy. ⦁

When tables of precise sequential anthropometric measurements of the skull and long bone dimensions became available, deviations from the norm could be recognised and antenatal diagnosis of achondroplasia by ultrasound became possible. In this way Filly et al (1981) were able to recognise achondroplasia and two instances of homozygous achondroplasia in at-risk pregnancies by comparison of femoral and biparietal measurements. Elejalde et al (1983) reviewed the diagnostic situation and published brief details of successful ultrasonic monitoring of two normal pregnancies of women with achondroplasia. In this context it is important that, as with most forms of dwarfism, a pregnant achondroplast will usually require Caesarian section for delivery. This will apply whether the infant is affected or normal, as a dwarfed baby is still large in relation to the small dimension of the achondroplastic mother's pelvic outlet. The problems of pregnancy and obstetric management in achondroplasia has been reviewed by Lattanzi & Harger (1982).

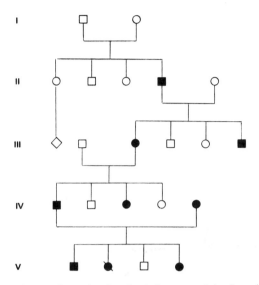

**Fig. 1.8** Achondroplasia: a pedigree showing classical autosomal dominant inheritance. The affected male in generation II was the youngest of his sibship. As his parents were normal, it is likely that he represented a new mutation of the gene. In generation IV the affected son married a female with achondroplasia. Two of their children had achondroplasia, one had lethal homozygous achondroplasia and one was normal. *Key:* □ normal male; ○ normal female; ■ affected male; ● affected female; \ deceased.

## HYPOCHONDROPLASIA

Following the initial description of hypochondroplasia by Léri & Linossier (1924) the syndromic identity of the condition was established by Walker et al (1971) and Kozlowski (1973). Although there have not been a large number of reports hypochondroplasia is relatively common; Specht & Daentl (1975) described 6 cases and reviewed the manifestations in 35 previously reported patients. Hall & Spranger (1979) were able to assemble a series of 39 affected persons and Wynne-Davies et al (1981) identified 24 British hypochondroplasts.

### Clinical and radiographic features

Hypochondroplasts resemble achondroplasts but, in general, the stigmata are much milder. In particular, the head and face are unaffected and the fingers do not have a trident configuration. Birth weight and length are normal and adult height may exceed 150 cm. Skeletal disproportion, mild lumbar lordosis and limitation of full extension of the elbow joints become evident in early childhood. Bowing of the legs may be present in infancy but this abnormality, which is usually of mild degree, often resolves spontaneously. The only important complication is mental deficiency, which

**Fig. 1.9** Hypochondroplasia: The smallest strongman in the world. Short-limbed dwarfism, with marked muscular development and normal craniofacial appearance. (From Heselson et al 1979 Clinical Radiology 30: 79–85.)

occurs in about 20% of hypochondroplasts. Orthopaedic problems in hypochondroplasia have been reviewed by Scott (1976).

The severity of the condition is very variable. At one end of the spectrum it may be difficult to distinguish patients from true achondroplasts, while at the other extreme they may approach normality. Indeed, one hypochondroplast known to the author plays for an international rugby team! Another enjoys a successful career as a circus strongman.

The radiographic features are similar to those of achondroplasia but of a milder degree. In particular, the lumbar vertebrae may show some narrowing of the interpeduncular distances and the femoral necks are short. Distal lengthening of the fibula is a useful diagnostic sign (Frydman et al 1974). The findings of skeletal surveys of 12 hypochondroplasts have been reviewed by Glasgow et al (1978) and Heselson et al (1979).

*Genetics*

There have been several well-documented family studies which indicate that hypochondroplasia is transmitted as an autosomal dominant (Beals 1969).

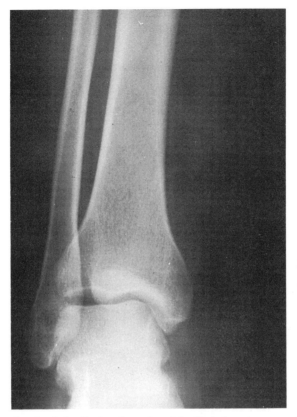

**Fig. 1.10** Hypochondroplasia: distal prolongation of the fibula is an important diagnostic feature. (From Heselson et al 1979 Clinical Radiology 30: 79–85.)

However, there is considerable phenotypic variation and hypochondroplasia could well be heterogeneous. In the author's own experience, only one out of nine unequivocally affected probands had a parent with the condition. The fact that some affected individuals are virtually normal is of practical importance in genetic counselling. On a basis of the clinical stigmata of a dwarfed child derived from an achondroplastic father and a hypochondro-plastic mother, McKusick et al (1973) have suggested that the gene for these two conditions may be allelic (i.e. at the same locus on a particular chromosome). A mother with hypochondroplasia and her husband affected by multiple exostoses, produced a daughter with both conditions (Domin-quez et al 1984). This situation is an example of concurrent inheritance of two separate autosomal dominant traits.

Hypochondroplasia has been diagnosed by ultrasonic measurement of the lengths of fetal limb bones at 22 weeks gestation in the pregnancy of a normal woman married to an affected male. After termination the diagnosis was confirmed by radiographs of the aborted fetus (Stoll et al 1985). In view of the great phenotypic variability it cannot be assumed that antenatal diagnosis would always be possible by this technique.

## ASPHYXIATING THORACIC DYSPLASIA

Asphyxiating thoracic dysplasia (ATD) was first described in a brother and sister by Jeune et al (1955). As maldevelopment of the thorax occurs in several other disorders the eponymous designation 'Jeune thoracic dysplasia' has considerable merit.

### Clinical and radiographic features

Narrowness and immobility of the thoracic cage are the major features of the condition. Polydactyly, which is sometimes present, may lead to confusion with the Ellis-van Creveld syndrome. Radiographically the ribs

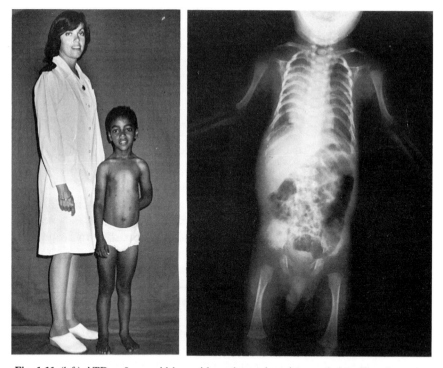

**Fig. 1.11** (left) ATD: a 9-year-old boy with moderate thoracic constriction. The chest deformity was much more obvious during infancy. His height is below the third percentile, but apart from recurrent chest infections, he enjoys good health.

**Fig. 1.12** (right) ATD: radiography of an affected infant showing a 'bell-shaped' thorax, due to shortening of the upper ribs. The ilia are broad, with irregular acetabular margins.

are short and horizontal and sternal ossification is often incomplete in the neonate. The pelvis is broad and the tubular bones may be shortened.

The condition is compatible with life, although respiratory insufficiency and infection are often fatal during infancy. There is considerable variation in the severity of the thoracic constriction and, in some infants, it may be of only minor degree. In later childhood the shape of the thoracic cage tends to revert to normality and the threat to life diminishes. Renal failure, with albuminuria and hypertension, supervenes in the second decade. The only report of an affected adult concerns a 32-year old American Indian who eventually developed hepatic and renal complications (Friedman et al 1975). Progressive renal and hepatic failure was also a feature of 10 cases studied by Oberklaid et al (1977).

*Genetics*

Apart from the original report by Jeune et al (1955), sibs with normal parents have been described by Niemann et al (1963), Maroteaux & Savart (1964) and Hanissian et al (1967). On this basis it is likely that inheritance is autosomal recessive. Kaufman & Kirkpatrick (1974) stated that up to 80% of affected individuals die in infancy, thus promoting speculation as to whether survivors have a genetically distinct form of the disorder. This concept of heterogeneity gained further support when Kozlowski & Masel (1976) described two children in whom the diagnosis was reached by chance at the age of 4, when radiographs were obtained for unrelated reasons. Prenatal diagnosis by ultrasonography has been accomplished (see Ch. 3).

Shokeir et al (1971) and Barnes et al (1971) recognised minor manifestations in parents of affected children and suggested that the gene might occasionally be expressed in the heterozygote. Bankier & Danks (1983) reported an affected mother and son, proposed that inheritance was autosomal dominant and suggested this condition was phenotypically distinct from the disorder previously described by Barnes. This latter family was followed up by Burn et al (1986), the phenotype was expanded, autosomal dominant inheritance was confirmed and a clear distinction was made from the Jeune syndrome. The title 'thoracolaryngopelvic dysplasia' was employed, with the subordinate attachment of the eponym 'Barnes'.

Anomalous patients with thoracic dystrophy and metaphyseal abnormalities have been reported. For instance Kaufman & Kirkpatrick (1974) studied two brothers, born with narrow thoraces, who developed pancreatic insufficiency, cyclical neutropenia and metaphyseal changes. This syndrome complex is similar to that of the Shwachman metaphyseal chondrodysplasia.

From the foregoing it seems likely that there are two uncommon autosomal dominant forms of asphyxiating thoracic dysplasia in addition to the classical potentially lethal autosomal recessive type and rare atypical varieties.

## CHONDROECTODERMAL DYSPLASIA (Ellis-van Creveld)*

Chondroectodermal dysplasia is a well-defined entity in which short-limbed dwarfism and polydactyly are associated with structural cardiac anomalies. Since the original report of Ellis & van Creveld (1940) more than 120 cases have been described. The largest series is that of McKusick et al (1964) who identified the disorder in an inbred religious isolate, the Amish of Pennsylvania, USA.

### Clinical and radiographic features

Chondroectodermal dysplasia presents as disproportionate dwarfism in the newborn. The components of the syndrome are distal limb shortening, postaxial polydactyly of the hands and dysplasia of the nails and teeth. The thorax may be constricted and cardiac abnormalities, particularly of the atria, are present in about 50% of patients. A significant proportion of patients die from the consequences of these cardiac or thoracic malformations. Genu valgum is a common complication in later childhood. Adult height is very variable, stature sometimes being virtually normal.

In infancy the radiographic changes in the pelvis and thorax are very similar to those seen in asphyxiating thoracic dysplasia. Bony fusions in the carpus and hypoplasia of the phalanges serve as distinguishing features.

### Genetics

Metrakos & Fraser (1954) recognised the consanguinity of the parents of a child reported by Ellis & van Creveld (1940), reviewed 10 other cases and suggested that inheritance of chondroectodermal dysplasia was autosomal recessive. This mode of transmission was confirmed by McKusick et al (1964) following their studies in the Amish community. Murdoch & Walker (1969) amplified the Amish investigations and ascertained 61 cases in 33 sibships. These individuals were all related to a common ancestral couple who had emigrated to the USA from Europe in 1744.

Recognition of polydactyly as a diagnostic marker by means of fetoscopy has been accomplished by Mahoney & Hobbins (1977) and by Bui et al (1984). The latter group of investigators monitored two consecutive pregnancies of consanguineous parents who had previously produced an affected child. The diagnosis was confirmed in one instance and excluded in the other. Non-invasive techniques are to be preferred and using ultrasonography, Muller & Cremin (1985) recognised reduced fetal thoracic dimen-

---

* Richard Ellis (1902–1966) spent his early career at Guy's Hospital, London and subsequently became Professor of Paediatrics at the University of Edinburgh. Simon van Creveld (1894–1971) was a paediatrician in Amsterdam and an authority on haemophilia. It is said that they met in a train while travelling to a medical congress and wrote the description of the condition which bears their names during the journey.

sions, shortened tubular bones and postaxial polydactyly of the hands at 32 weeks of gestation. The diagnosis was subsequently confirmed after delivery, by means of conventional radiography and autopsy.

The optimal approach to antenatal diagnosis of chondroectodermal dysplasia has not yet been fully resolved. The cardiac defects, if present, are readily recognisable by ultrasound but the other syndromic components are variable and could remain undetected especially in early pregnancy. Fetoscopy has the drawback of being invasive and technically difficult. In addition, the potential risks that this procedure places upon an unaffected fetus and the fact that the disorder can be comparatively benign enter into the balance when decisions concerning pregnancy monitoring have to be faced.

## MULTIPLE EPIPHYSEAL DYSPLASIA

In multiple epiphyseal dysplasia (MED) changes are maximal in the epiphyses, while involvement of the metaphyses and axial skeleton is minimal or absent. Many cases have been reported and the various types of MED collectively represent one of the most common forms of skeletal dysplasia.

The development of knowledge concerning MED owes a great deal to the work of Sir Thomas Fairbank. In his early descriptions Fairbank (1935) used the term 'epiphyseal dysplasia' and a decade later introduced the designation 'dysplasia ephysealis multiplex' (Fairbank 1946). The clinical and radiological features of 26 patients were subsequently depicted in the 'Atlas of General Affections of the Skeleton' (Fairbank 1951).

It is becoming increasingly evident that MED is very heterogeneous and several overlapping classifications have been proposed. This problem has been discussed in detail by Lie et al (1974). The confusion is compounded by the great variation in phenotypic features which may be present in affected members of the same kindred.

### Clinical and radiographic features

Diagnosis may be difficult, as the height of some patients is within the normal range. However, shortness of stature, which is of mild degree, usually becomes evident in mid-childhood. In some instances the hands are broad, with stubby digits and foreshortened nails. The facies and intelligence are normal and no consistent extraskeletal manifestations have been recorded. Degenerative arthritis with onset in early adult life is a common complication, but general health is otherwise good.

As with many of the genetic skeletal dysplasias, the radiographic features in MED are age related. In the infant the epiphyses may have a 'stippled' appearance. With advancing age irregularity and fragmentation of the epiphyses of the long bones becomes increasingly evident. These changes

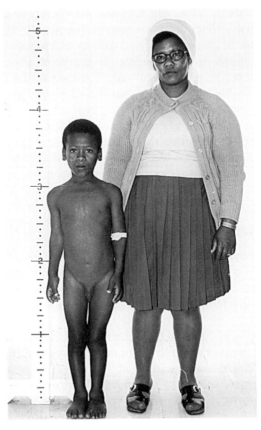

**Fig. 1.13** MED: this 9-year-old boy, depicted with his mother, has few problems other than short stature.

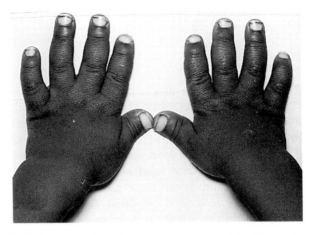

**Fig. 1.14** MED: in some patients the digits are stubby and the finger nails are foreshortened.

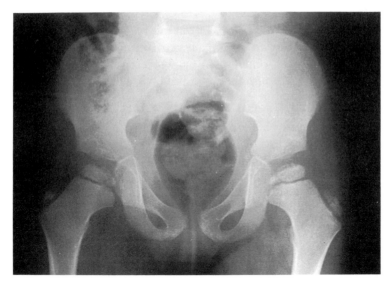

**Fig. 1.15** MED: the hip joints show Perthe-like changes, with flattening and irregularity of the femoral capital epiphyses.

may be particularly marked in the femoral heads. Changes in the digits are very variable but characteristically the tubular bones are shortened, while the carpal and tarsal bones may be distorted. Mild spinal abnormalities are sometimes present, the vertebral bodies having irregular surfaces and slight anterior wedging. The skull and axial skeleton are otherwise normal.

In adolescence the radiological changes in the hip joints resemble those of Perthes disease and a misdiagnosis may be made if the generalised nature of the articular abnormality is overlooked. Bilateral involvement, or radiographic demonstration of widespread epiphyseal changes, are useful diagnostic indicators. MED enters into the differential diagnosis in all instances of 'familial' Perthes disease. Similarly, clinically unrecognised mild forms of MED might be responsible for a proportion of cases of 'idiopathic' osteoarthritis of the hip joint.

*Genetics*

In terms of clinical manifestations MED has been divided into severe 'Fairbank' and mild 'Ribbing' types. However, there is considerable overlap of the stigmata, even within the same kindred, and it is likely that this particular subgrouping has no genetic basis. Nevertheless, there is no doubt that MED is very heterogeneous.

Juberg & Holt (1968) reviewed the literature and published a list of reports of families with dominant transmission of MED and of other kindreds where affected sibs with normal parents indicated possible autosomal recessive inheritance. It must be emphasised that in most cases MED is an autosomal dominant trait. Amir et al (1985) reported an impressive family in which 45 persons in six generations had the disorder.

Delineation has been complicated by problems of incomplete ascertainment, particularly when stature has been virtually normal and by the considerable variability of phenotypic expression within a single kindred. For instance, Diamond (1974) reported a family in which 32 individuals in four generations had a wide variety of changes in the hip and spine. Clinical inconsistency of this nature led Barrie et al (1958) to postulate that the disorder might be mediated by two separate genes. However, this contention has attracted little support.

A family in which six persons had MED was reported by Mena & Pearson (1976). Other family members had joint problems and no clear pattern of inheritance could be discerned. It is possible that there is a form of MED in which articular complications predominate; Patrone & Kredich (1985) encountered four children with classical MED in whom chronic arthritis developed during the first decade.

Linkage studies undertaken in 12 affected individuals in two generations of a kindred were unfruitful (Hoefnagel et al 1967) and molecular studies of the type II collagen gene in 17 affected persons in three generations of a British family were uninformative (Wordsworth et al 1986). So far, the MED gene has not been assigned to any particular chromosome and the basic defect is unknown.

**Other forms of MED**

Autosomal dominant inheritance of epiphyseal dysplasia which is confined to the hip joint has been reported (Monty 1962, Wamoscher & Farhi 1963). This entity is sometimes termed 'familial Perthes disease' but it is ill-defined and has not been accorded syndromic status. It must be streassed that the genetic component of the common form of Perthes disease is very small. Indeed, following a survey of the families of 323 affected individuals in South Wales, Harper et al (1976) calculated that the risk of recurrence in sibs was under 1% while the risk to children of affected parents was about 3%. Meyer arthropathy or dysplasia epiphysealis capitis femoris (Meyer 1964), which is sometimes regarded as a variant of Perthes disease, also presents with hip problems in early childhood. Involvement may be bilateral but the changes are usually sequential rather than concurrent.

Namaqualand hip dysplasia is a unique condition which has been identified in 45 persons in five generations of a kindred of mixed ancestry in South Africa. Discomfort in the hip joints develops in childhood and the course is progressive, leading to handicap in middle age. General health is good, height is not reduced and there is no extra-skeletal involvement. The major changes are in the femoral capital epiphyses which are flattened and fragmented; secondary degenerative arthropathy develops at a later stage.

Platyspondyly of variable but mild degree is present in about 60% of affected persons. Pedigree data indicate autosomal dominant inheritance with a reasonably consistent phenotypic expression (Beighton et al 1984).

Epiphyseal dysplasia is a component of a number of rare syndromes. Lowry & Wood (1975) described two brothers with generalised epiphyseal abnormalities, short stature, congenital nystagmus and microcephaly. The authors suggested that inheritance of the condition was either X-linked or

**Fig. 1.16** Namaqualand hip dysplasia: a group of affected and unaffected family members. (From Beighton et al 1984 American Journal of Medical Genetics 19: 161–169.)

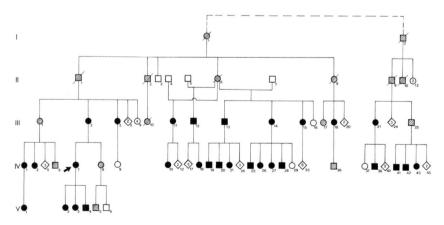

**Fig. 1.17** Namaqualand hip dysplasia: the pedigree of the kindred. 45 persons in 5 generations had the condition. (From Beighton et al 1984 American Journal of Medical Genetics 19: 161–169.)

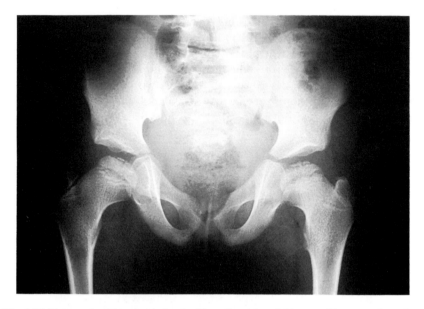

**Fig. 1.18** Namaqualand hip dysplasia: the hips of a girl aged 10 years. Fragmentation of the femoral capital epiphyses. (From Beighton et al 1984 American Journal of Medical Genetics 19: 161–169.)

autosomal recessive. Pfeiffer et al (1973) reported three brothers, two of whom were monozygous twins, with dysplasia of the femoral head, severe myopia and perceptive deafness. The parents were distantly related and it is possible that inheritance was autosomal recessive.

## CHONDRODYSPLASIA PUNCTATA

Autosomal dominant, X-linked dominant and autosomal recessive forms of chondrodysplasia punctata are recognised. The eponym 'Conradi-Hünerman' is conventionally applied to the dominant type, while the designation 'severe rhizomelic' is used for the recessive variety. Other terms which have been used indiscriminately include 'Conradi syndrome', 'stippled epiphyses', 'dysplasia epiphysealis punctata' and 'chondrodystrophia calcificans congenita'.

Stippling of the epiphyses may occur as a harmless, isolated anomaly, or in the early stages of conditions such as multiple epiphyseal dysplasia, spondyloepiphyseal dysplasia, hypothyroidism, cerebrohepatorenal syndrome, trisomy 18 and trisomy 12.

Maternal therapy during early pregnancy with the anticoagulant drug, warfarin, can produce a phenocopy of chondrodysplasia punctata in the fetus. In a review of eight affected infants, Pauli et al (1976) tabulated the clinical and radiographic manifestations, which were virtually indistinguish-

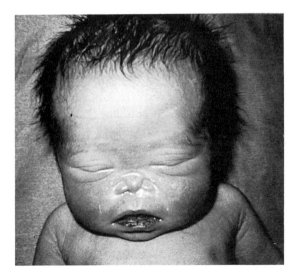

**Fig. 1.19** Chondrodysplasia punctata, AD type: a stillborn infant with a flattened face, a depressed nasal bridge and ichthyotic dermal lesions. These non-specific changes are present in all forms of the condition.

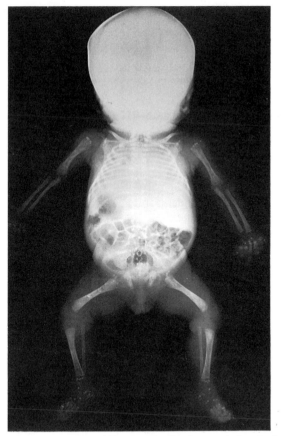

**Fig. 1.20** Chondrodysplasia punctata, AD type: stippling is very obvious in the epiphyses and in the lower spine.

able from those of the genetic forms of the disorder. Maternal alcohol ingestion in the first trimester can produce a similar appearance (Maroteaux et
al 1984).

A distinction must be drawn between the non-specific usage of the term
'stippled epiphyses' as a description of a radiological feature and its precise
application as a name for a distinct disease entity. The situation may be
summarised in the following way:

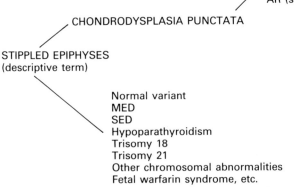

AD (Conradi-Hunerman type)
XLD (Happle type)
Sporadic mild (Sheffield type)
AR (severe rhizomelic type)

CHONDRODYSPLASIA PUNCTATA

STIPPLED EPIPHYSES
(descriptive term)

Normal variant
MED
SED
Hypoparathyroidism
Trisomy 18
Trisomy 21
Other chromosomal abnormalities
Fetal warfarin syndrome, etc.

## Autosomal dominant type (Conradi-Hünerman)

Affected infants have a flat face with a depressed nasal bridge. Cataracts,
alopecia and ichthyosis are inconsistent features. The clinical course is variable but survival to adulthood is usual. Structural abnormalities of the
vertebral bodies may lead to spinal malalignment. Asymmetric shortening
of the limbs may pose problems in later life, when degenerative osteoarthropathy in weight-bearing joints is a common complication.

In the newborn infant stippling is radiographically evident in the
epiphyses of the long bones and in the larynx, wrists, ankles and spine.
Fairbank (1927) likened the radiographic appearances to the effect
produced by 'the flicking of paint from a brush onto a clear surface'. The
cartilagenous stippling is a transient phenomenon which disappears by the
second year. In the older child the epiphyses are flattened and widened.
Structural anomalies of the vertebral bodies may be an additional feature
of the Conradi-Hünerman syndrome.

### Genetics

Generation to generation transmission has been reported by a number of
authors, including Silverman (1969) and Bergstrom et al (1972). This
autosomal dominant form of chondrodysplasia punctata is comparatively

common. Phenotypic expression is variable and it is likely that there is heterogeneity. In an extensive review Spranger et al (1971) classified the Conradi-Hünerman form into three subgroups, although they were unable to settle the question of genetic heterogeneity. The mild type which Sheffield et al (1976) recognised in 23 unrelated individuals in Australia might represent a separate entity.

Many cases are sporadic and the absence of clear-cut phenotypic features can make categorisation a difficult matter. Another problem arises from the fact that the stippling might have disappeared by the time the child is brought to the genetic clinic. In this situation it can be very difficult to reach a firm diagnosis and counselling may have to be speculative. The condition has been recognised radiographically in the fetus during the third trimester of pregnancy (Hyndman et al 1976). However, it is unlikely that this technique would permit antenatal diagnosis at an earlier stage.

## X-linked dominant type

Happle et al (1977) reported an X-linked dominant form of chondrodysplasia punctata, which is lethal in males. The manifestations resemble those of the autosomal dominant type, with the addition of widespread atrophic and pigmented dermal lesions. Following a review of the literature, Happle (1979) identified 35 cases, all of whom were female and further delineated the phenotype. The asymmetry of the lesions was ascribed to the process of Lyonisation (i.e. random X-chromosomal inactivation). The histological and ultrastructural changes in the skin of affected females aged 4 weeks and 14 years were documented in detail by Kolde & Happle (1984) and it was emphasised that these features permitted diagnostic confirmation. The mode of genetic transmission was substantiated by Mueller et al (1985) when they reported a family with six affected females in three generations. Affected males are evidently lost by miscarriage in early pregnancy as the condition has not been recognised in any newborn boy.

## Other forms of chondrodysplasia punctata

An X-linked form of chondrodysplasia punctata in two families was reported by Curry et al (1984). Affected males had the conventional manifestations plus mental retardation, while the females were normal apart from some diminution in stature. A small deletion at the terminal end of the short arm of the X chromosome was identified as the basic defect in this disorder. Another atypical form of chondrodysplasia punctata has been attributed to a duplication of the short arm of chromosome 16 (Hunter et al 1985) and an entity comprising mesomelic dysplasia and punctate epiphyseal calcification has been proposed by Burck (1982).

The autosomal recessive severe rhizomelic form of chondrodysplasia punctata is discussed in detail in Chapter 3.

## METAPHYSEAL CHONDRODYSPLASIA

The metaphyseal chondrodysplasias are a group of conditions in which abnormalities of the metaphyses predominate while the epiphyses, skull and trunk are essentially normal. Immunological incompetence and endocrine dysfunction are important facets of several of these disorders. A number of forms of metaphyseal dysplasia with eponymous or descriptive designations are recognised. The Schmid type is relatively common, the McKusick type is well known but uncommon and the remainder are rare. The following will be discussed:

1. Jansen type
2. Schmid type
3. McKusick type (cartilage-hair hypoplasia)
4. Shwachman (malabsorption and neutropenia)
5. Davis (thymolymphopenia)
6. Other types of metaphyseal chondrodysplasia.

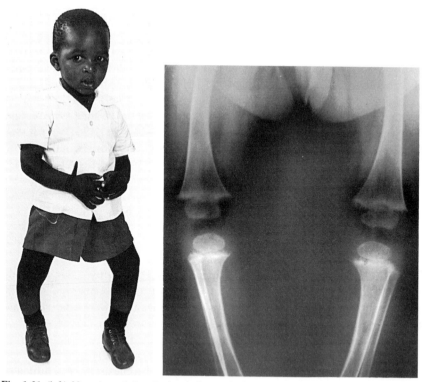

**Fig. 1.21** (left) Metaphyseal chondrodysplasia: marked bowing of the legs and shortness of stature are the most obvious clinical features.

**Fig. 1.22** (right) Metaphyseal chondrodysplasia: the metaphyseal irregularities are reminiscent of dietary and metabolic rickets.

The bowing of the legs and irregularity of the metaphyses bear a close resemblance to the stigmata of dietary and metabolic rickets. In this context the differential diagnosis and manifestations of the metaphyseal chondrodysplasias have been reviewed by Spranger (1976). The skeleton of 'bowed Joseph' which resides in the Museum of the Department of Anatomy of the University of Edinburgh shows changes of this type. Bowed Joseph led the Meal Riots in Edinburgh and died in 1780.

The diagnostic labels 'metaphyseal dysplasia' and 'dysostosis' were employed by Sir Thomas Fairbank (1876–1961) for a set of annotated radiographs of 15 cases which now repose in the Fairbank Collection of the Radiology Museum in the Royal National Orthopaedic Hospital, London. These patients were reviewed more than two decades after the original diagnosis and it transpired that seven of them had been 'battered babies'. This syndrome had not been delineated at the time of Sir Thomas' original apraisal but from his meticulous case notes, it is evident that he had considered the possibility that their metaphyseal changes might have resulted from trauma (Horan & Beighton 1980).

### Jansen* type

The form of metaphyseal chondrodysplasia reported by Jansen (1934) was the first to be clearly delineated and it has received recognition out of all proportion to its prevalence. Less than 10 cases have been described.

Affected individuals are severely dwarfed and disabled, with bowing of the lower limbs and knobbly bone ends. Joint movements are restricted and club feet may be present. Asymptomatic hyperglycaemia is a consistent but unexplained feature. Deafness in adulthood may be the consequence of sclerosis of the base of the skull (Holthausen et al 1975). Mild radiographic changes are present at birth and the disorder is clinically evident by the age of 5 years. Radiographic abnormalities in the neonate include generalised osteoporosis, uneven expansion of the metaphyses and ribbon-like ribs. By mid-childhood the architecture of the metaphyses is grossly disturbed. However, these changes tend to regress in adult life (Haas et al 1969, Kikuchi et al 1976, Nazara et al 1981).

The autosomal dominant mode of inheritance was suggested when Lenz (1969) described an affected mother and daughter and this was confirmed in a further report by Charrow & Poznanski (1984). Other reported cases have all been sporadic, Kikuchi et al (1976) and it seems likely that they represent new mutations.

---

* Murk Jansen (1867–1935) was a Dutch orthopaedic surgeon and a founder of the speciality in his country. He had an international reputation and was the author of several monographs. In his original case description he used the term 'atypical achondroplasia' and ascribed the bony abnormalities to 'maternal exhaustion' consequent upon numerous pregnancies.

## Schmid type

This entity, which was first described by Schmid (1949), is by far the most common of all the metaphyseal chondrodysplasias. Genu varum and moderately short stature are the main stigmata. These abnormalities become apparent in early childhood and a misdiagnosis of dietary or hypophospha-taemic rickets is not unusual. Radiographically the metaphyses are expanded and irregular. These changes are maximal at the hip and knee joints, where varus deformities may be evident. The skeleton is otherwise virtually normal. The radiological changes at various stages of development, from early infancy to adulthood, have been reviewed by Beluffi et al (1982).

Several families with generation to generation transmission have been reported and it is evident that the Schmid form of metaphyseal chondro-dysplasia is inherited as an autosomal dominant (Maroteaux & Lamy 1958, Debray et al 1975). There is considerable variation in the degree of pheno-typic expression in any kindred. Of particular interest is the presence of the condition in more than 40 individuals in four generations of a Mormon family, who were reported by Stephens (1943) under the designation 'achondroplasia'. In keeping with the concept of new dominant mutation, the paternal age effect has been noted in some sporadic cases (Rosenbloom & Smith 1965).

## McKusick type

This form of metaphyseal chondrodysplasia attains maximum prevalence among the Amish of Pennsylvania, USA (McKusick et al 1965). Although originally termed 'cartilage-hair hypoplasia', it was later redesignated 'metaphyseal chondrodysplasia — McKusick type'. The condition is also common in Finland, where 33 cases have been recognised in 28 families (Kaitila & Perheentupa 1980).

Affected individuals have a moderate degree of short stature, with thin, sparse hair on the body and head. The finger joints are lax and the digits are stubby, with wide, foreshortened nails. Distal prolongation of the fibula may produce inversion deformity of the ankle joint. Cellular immunity is deficient (Lux et al 1970, Virolainen et al 1978, Wilson et al 1978) and patients are prone to infection, especially during infancy and early child-hood. They are at particular risk from chickenpox, which may be severe and lethal. Vaccination against smallpox is contraindicated. An additional unusual feature is a liability to megacolon and intestinal malabsorption. These problems tend to regress in later life.

Radiographically, the metaphyses are irregular, with cystic changes across the entire width of the bone. A central depression in the distal metaphysis of the femur produces a 'scalloped' appearance. The height of the lumbar vertebrae may be increased but, apart from some flaring of the ribs, the axial skeleton is otherwise normal (Ray & Dorst 1973).

In their studies of the Amish, McKusick et al (1965) documented 53 sibships in which at least one individual was affected. These investigators calculated a population prevalence of 1–2% and demonstrated that inheritance was autosomal recessive. Subsequently, Lowry et al (1970) suggested that as there was an excess of females in the published series, phenotypic expression of the gene might be influenced by the sex of the patient.

## Metaphyseal chondrodysplasia with exocrine pancreatic insufficiency and cyclic neutropenia (Shwachman-Bodian type)

The Shwachman-Bodian form of metaphyseal chondrodysplasia is a rare disorder in which short stature is associated with pancreatic insufficiency, malabsorption and neutropenia (Shwachman et al 1964, Taybi, et al 1969). There is still doubt as to the precise relationship of this condition with other non-skeletal disorders in which pancreatic and marrow dysfunction coexist (Shwachman & Holsclaw 1972). About 50 cases have been reported.

Patients are dwarfed, with crippling due to hip joint dysplasia. Deficiency of pancreatic enzymes leads to steatorrhoea, which develops during infancy and regresses in later childhood. The neutrophil count may be consistently or intermittently low and anaemia and thrombocytopenia may also occur. Pyogenic infection in various sites is a recurrent problem. Management includes antibiotic therapy for recurrent infections and pancreatic extract for intestinal malabsorption. Femoral osteotomy may be required for the hip deformity. The ultimate prognosis is poor in the inadequately treated patient.

Radiographically, wide-spread metaphyseal irregularity is a prominent feature. The hip joints are severely affected and a valgus deformity is initially present. However, slipping and restabilisation of the epiphyses of the femoral heads may ultimately lead to coxa vara. Leukaemia may be a late complication (Woods et al 1981).

Affected siblings have been reported by Burke et al (1967) and Danks et al (1976) and the pedigree data are consistent with autosomal recessive inheritance.

## Metaphyseal chondrodysplasia with thymolymphopenia (Davis type)

Although the original reports were entitled 'Swiss-type agammaglobulineamia and achondroplasia' (Davis 1966, Fulginiti et al 1967) it is now evident that this rare condition is a distinct entity which can be classified with the metaphyseal chondrodysplasias.

Lymphopenia and agammaglobulinaemia are associated with ectodermal dysplasia. In particular, the hair and eyebrows are often absent, while ichthyosis and erythroderma are present. The tubular bones are shortened with wide but regular metaphyses. The pelvic acetabulae are horizontal and the vertebral end plates are sclerosed (Alexander & Dunbar 1968). The axial skeleton and the extremities are otherwise normal. Overwhelming infection may cause death in infancy.

Gatti et al (1969) reported two sibs with normal parents. Other cases have been sporadic. The genetic basis is uncertain but inheritance is probably autosomal recessive.

## Other types of metaphyseal chondrodysplasia

The Spahr form of metaphyseal chondrodysplasia is clinically and radiographically indistinguishable from the Schmid type. However, in the single kindred which was reported by Spahr & Spahr-Hartmann (1961), four affected sibs had consanguineous parents and inheritance was presumed to be autosomal recessive.

Vaandrager (1960) observed a form of metaphyseal dysplasia in four members of two generations of a kindred. Kozlowski & Sikorska (1970) reported similar cases under the designation 'Vaandrager-Pena type of metaphyseal chondrodysplasia'. However, the affected brother and sister studied by Pena (1965) seem to have had yet another distinct disorder. Similarly, the mild metaphyseal dysplasia described by Kozlowski (1964) may represent a separate entity.

A syndrome of congenital lethal metaphyseal chondrodysplasia was diagnosed by Sedaghatian (1980) in two brothers and a sister from an Iranian family. These siblings had severe changes in their metaphyseas and died from cardiorespiratory problems soon after birth. Inheritance was presumably autosomal recessive.

The Kaitila type of metaphyseal chondrodysplasia was recognised in a Finnish brother and sister (Kaitila et al 1980). These young adults had gross abnormalities in their metaphyses, with limb shortening, stubby digits and mild but progressive kyphoscoliosis. Tracheobronchiomalacia caused respiratory problems and necessitated surgical measures. The parents of these siblings were normal and it is likely that inheritance was autosomal recessive. It is of considerable academic interest that the McKusick type of metaphyseal chondroplasia (vide supra) is also common in Finland and the question arises as to whether there is any fundamental genetic relationship between these two disorders.

Metaphyseal dysplasia associated with retinitis pigmentosa and brachydactyly, possibly inherited as an autosomal recessive trait, was reported in British siblings by Phillips et al (1981).

An autosomal dominant syndrome of metaphyseal dysplasia with maxillary hypoplasia and brachydactyly in a French Canadian family was investigated by Halal et al (1982).

A male infant with mild metaphyseal dysostosis and congenital nystagmus was found to have the fragile X syndrome (Williams et al 1986). The pathogenesis of this disorder is uncertain.

Other rare forms of metaphyseal chondrodysplasia to which eponyms have been applied include those of Maroteaux, Roy, Wiedemann and Spranger, and Rimoin. There is undoubtedly considerable overlap between early reports of ostensibly different conditions. In some instances, the same designation has been applied to disorders which may well be separate entities. The classification and nomenclature of the metaphyseal chondrodysplasias is therefore neither clear-cut nor complete.

## MESOMELIC DYSPLASIA

The term 'mesomelic dysplasia' pertains to short stature with limb shortening which is most pronounced in the forearms and shins. Several distinct forms are known and reports of atypical cases are probably indicative of further heterogeneity. The following types of mesomelic dysplasia are considered in this section:

1. Nievergelt
2. Langer
3. Other types of mesomelic dysplasia

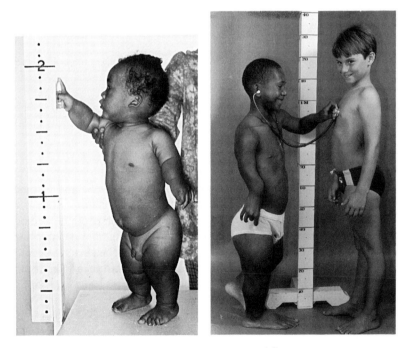

**Fig. 1.23** (left) Mesomelic dysplasia: a 2-year-old boy with the Langer type of mesomelic dysplasia. Shortening is maximal in the middle segments of the limbs.

**Fig. 1.24** (right) Mesomelic dysplasia: the boy depicted in Figure 1.23, now aged 13 years. Shortening and malformation of the forearms and lower legs are evident. His unaffected friend is 8 years of age. (From Goldblatt et al 1987 Clinical Genetics 31: 19–25.)

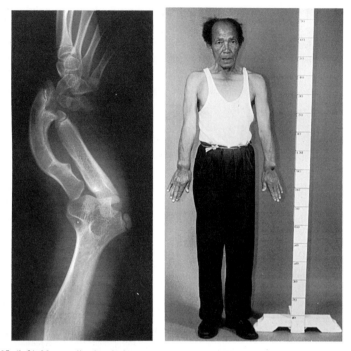

**Fig. 1.25** (left) Mesomelic dysplasia: antero-posterior radiograph of the forearm of the affected boy. The radius and ulna are very short and dysplastic. (From Goldblatt et al 1987 Clinical Genetics 31: 19–25.)

**Fig. 1.26** (right) Mesomelic dysplasia: a heterozygote for Langer mesomelic dysplasia, (the uncle of the boy shown in Figure 1.23) demonstrating a Madelung deformity of the forearms. (From Goldblatt et al 1987 Clinical Genetics 31: 19–25.)

Descriptive anatomical terms which are used in the context of short-limbed dwarfism include micromelia, rhizomelia, mesomelia and acromelia. Micromelia denotes shortening of all segments of a limb, while rhizo-meso- and acro- respectively imply maximal involvement of the proximal, middle and distal limb segments. Compound designations such as acro-mesomelic micromelia are also employed. The mesomelic skeletal dysplasias have been reviewed by Kaitila et al (1976) and Maroteaux & Spranger (1977).

### Nievergelt type

The only reports of this extremely rare condition concern four members of a kindred studied by Nievergelt (1944) and affected boys described by Solonen & Sulamaa (1958); Young & Wood (1975). The pedigree of Niev-ergelt's original family was updated by Hess et al (1978). The major clinical features are gross shortening of the distal portions of the limbs, flexion deformities of the fingers and elbows, genu valgum and club feet. Radiographically, the tibia has a pathognomonic rhomboidal configuration.

Good evidence for autosomal dominant inheritance was provided by a severely dwarfed male who transmitted the disorder to three sons, all by different mothers (Nievergelt 1944). It is apparent that this individual's initiative amply compensated for his disability!

## Langer type

Using the designation 'mesomelic dwarfism of the hypoplastic ulna, fibula, mandible type' Langer (1967) reported two dwarfed individuals with severe shortening of the limbs and underdevelopment of the jaw. These persons were participating in a convention of the Little People of America; over the years this admirable society has played an important role in syndromic delineation! In a review of the disorder, Silverman (1975) mentioned that radiographically, ossification was defective in the distal ulna and the proximal fibula.

Although mandibular hypoplasia has been emphasised in the descriptive title, not all affected persons have this feature. Indeed, there is considerable doubt as to whether a small lower jaw is a genuine syndromic component.

Böök (1950) gave a detailed account of a family in Northern Sweden in which several persons had mild mesomelia, which he designated 'achondrohypoplasia'. Two affected cousins in this kindred produced a son with a severe dwarfing disorder which Böök regarded as being similar to the entity reported by Brailsford (1953). With hindsight, the boy's condition is recognisable as Langer mesomelic dysplasia, while the parents probably had dyschondrosteosis. On this basis, Böök is to be credited with recognising that the former condition is the homozygous state of the latter.

The concept that Langer mesomelic dysplasia was homozygous dyschondrosteosis received further support when Espiritu et al (1975) described two affected siblings whose parents had dyschondrosteosis, while Fryns & van de Berghe (1979), Kanze & Klemm (1980) and Goldblatt et al (1987) made similar observations. The situation is not entirely clearcut as there is blurring of the syndromic boundaries between isolated Madelung deformity of the forearms which is inherited as an autosomal dominant trait and mild dyschondrosteosis, in which the Madelung deformity is the major manifestation. Indeed, it is entirely possible that these latter conditions are examples of variable expression of the same abnormal gene.

## Other types of mesomelic dysplasia

Several varieties of mesomelic dwarfism have been reported in individuals or small kindreds and each has characteristic features which permit recognition.

1. *Werner type* of mesomelic dysplasia, which is inherited as an autosomal

dominant trait, comprises gross tibial hypoplasia associated with polydactyly and absence of the thumbs (Werner 1915, Pashayan et al 1971).

2. *Robinow or 'fetal face' syndrome* consists of limb shortening, hypoplasia
of the genitals, an abnormal facies and anomalies of the vertebrae and ribs.
A firm radiographic diagnosis can be made by recognition of the combination of mesomelia, hemivertebrae, rib fusion and bifid terminal phalanges. In their original paper Robinow (1969) reported involvement in nine
persons in six generations of a family in North America. Inheritance was
apparently autosomal dominant but there was no male to male transmission
and X-linked dominant transmission cannot be ruled out. A description of
a brother and sister with the Robinow syndrome, whose parents were
normal, reopened the question of the mode of genetic transmission
(Wadlington & Tucker 1973). Robinow (1973) provided further data and
subsequent reports were mainly concerned with phenotypic delineation of
sporadic cases (Schinzel et al 1974, Kelly et al 1975). A family in which
a father and his two children had the facial features of the Robinow
syndrome, in the absence of limb involvement, was described by Bain et
al (1986). These authors proposed the existence of mild autosomal dominant
and severe autosomal recessive forms of the disorder.

Clinical diagnosis reached new levels of sophistication when Friedman
(1985) in an article entitled 'Umbilical dysmorphology: The importance of
contemplating the belly button' drew attention to abnormalities of position
and epithelialisation of the umbilicus in the Robinow syndrome.

3. *Reinhardt-Pfeiffer type* of mesomelia is characterised by bowing and
shortening of the bones of the forearms and shins, with variable synostosis
in the carpus and tarsus (Reinhardt & Pfeiffer 1967). In the single kindred
which has been reported, 14 members of four generations were affected.
The pattern of transmission was consistent with autosomal dominant
inheritance (Reinhardt 1976).

4. Leroy et al (1975) described a form of mesomelic dysplasia in a father
and his two sons. These individuals had severe shortening and bowing of
the tibia with milder changes in the radius, ulna and fibula. The skeleton
was otherwise normal. It is probable that inheritance is autosomal
dominant.

5. Mesomelic dysplasia with predominant shortening of the ulna was
reported by Burck et al (1980). The girl in question, aged 8 years, also had
mild micrognathia, shortening of the metacarpals and articular contractures.
A second sporadic girl with similar features was subsequently described by
Burck (1982) and attention was drawn to the presence of punctate epiphyseal calcifications in both. The mode of inheritance is unknown.

## ACROMESOMELIC DYSPLASIA

Maroteaux et al (1971) described an autosomal recessive dwarfism syndrome
in which shortening and hypoplasia was maximal in the fibula, proximal

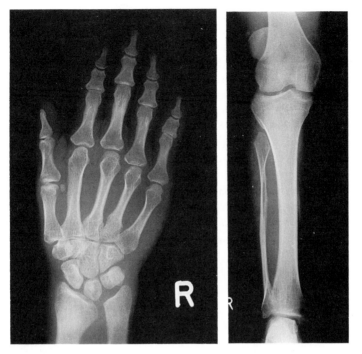

**Fig. 1.27** (left) Acromesomelic dysplasia: the tubular bones of the hands are short and the wrist joint is angulated. (From Beighton P 1974 Clinical Genetics 5: 363–367.)

**Fig. 1.28** (right) Acromesomelic dysplasia: the tibia is short and the fibula is gracile. (From Beighton P 1974 Clinical Genetics 5: 363–367.)

radius and distal ulna and in the tubular bones of the digits. The term 'acromesomelia' is applicable to this distribution of abnormalities and the eponym 'Maroteaux' is applied to this disorder. In a further case report, Campailla & Martinelli (1971) documented an Italian brother and sister with acromesomelia. It is uncertain whether or not these conditions are the same entity as the severity of the changes differed between the affected families. A further report of acromesomelia concerned five sisters in Cape Town (Beighton 1974). The consanguinity of their unaffected parents is in keeping with the autosomal recessive mode of inheritance.

Acromesomelic dysplasia was reviewed in detail by Langer et al (1977) in an 11 author, multi-centre report of 8 affected children and 2 adults with the disorder. The clinical and radiographic manifestations were described in detail and evidence for autosomal recessive inheritance was presented. Additional cases in the literature were identified, including those reported by Lannois (1902) and Goodman et al (1975), while the case reported by Hunter & Thompson (1972) as acromesomelic dwarfism was consigned to a different category. Borelli et al (1983) described yet another affected kindred.

A kindred in which at least 10 persons have acromesomelia has been identified on the island of St Helena in the Atlantic ocean. This autosomal recessive disorder was originally reported as 'brachydactylous dwarfism' (Shine 1970). Reappraisal of the original radiographs has permitted revision of the original diagnosis. The digital and forearm changes are intermediate between those of the Maroteaux and Campailla-Martinelli forms of acro-mesomelia and the issue of homogeneity versus heterogeneity thus arises.

Acro-coxo-mesomelic dwarfism comprises shortening of the bones of the forearms and digits, with articular rigidity and dislocations of the hip joint (Plauchu et al 1984). In the only reported family, inheritance is autosomal recessive.

Acromesomelia was associated with fusions in the carpus and tarsus, bipartate calcanei and gross hypoplasia of the middle phalanges in a family reported by Osebold et al (1985). Seven persons in the kindred were affected and inheritance was autosomal dominant.

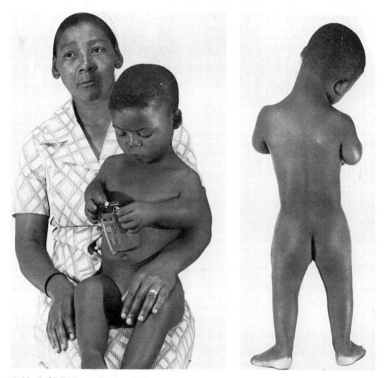

**Fig. 1.29** (left) Rhizomelic dysplasia: a 5-year-old boy with predominant shortening in the proximal segments of the limbs.

**Fig. 1.30** (right) Rhizomelic dysplasia: the rhizomelia is particularly noticeable when the child is viewed from behind.

## RHIZOMELIC DYSPLASIA

Proximal limb shortening or rhizomelia is a feature of several well-known forms of dwarfism. In addition, rhizomelia predominates in a number of rare malformation syndromes.

Under the designation 'humerospinal dysostosis with congenital heart disease' Kozlowski et al (1974) described two infants with rhizomelia, distal bifurcation of the humeri, subluxation of the elbow joints, coronal cleft vertebrae, talipes equinovarus and congenital heart disease. This brother and sister had the same mother but different fathers. As the parents were unaffected and non-consanguineous, it is likely that inheritance was autosomal dominant and that the gene was non-penetrant in the mother.

Patterson & Lowry (1975) reported a new dwarfing syndrome in which gross shortening of the humeri and coxa vara were the major features. As the patient in question was 86 years of age, it is apparent that the disorder is relatively innocuous!

Other reports concern an adolescent Taiwanese boy with severe humeral shortening (Yang & Lenz 1976) and a consanguineous Arab family in which

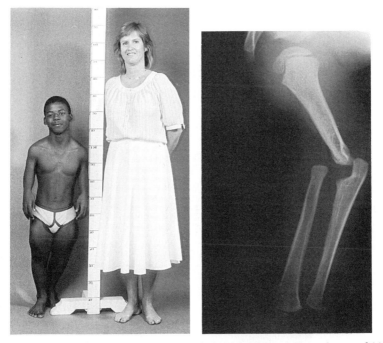

**Fig. 1.31** (left) Rhizomelic dysplasia: the boy depicted in Figure 1.29, at the age of 16 years. Short stature and rhizomelia are evident. (From Viljoen et al 1987 American Journal of Medical Genetics 4: 941–947.)

**Fig. 1.32** (right) Rhizomelic dysplasia: the distal portion of the humerus is very underdeveloped. (From Viljoen et al 1987 American Journal of Medical Genetics 4: 941–947.)

two siblings had a lethal syndrome of humeral rhizomelia together with articular contractures, craniofacial abnormalities and cardiac defects (Urbach et al 1986). At present it is uncertain whether these conditions are separate entities or examples of wide phenotypic variation of the same disorder.

In a remarkable Cape Town family a boy with severe humeral rhizomelia and normal legs is closely related to three siblings with gross reduction defects of the lower limbs (Viljoen et al 1987). In view of the anatomical parallels and the rarity of rhizomelia, it is probable that the conditions reflect the same underlying mutation but the great phenotypic disparity defies explanation.

## GREBE CHONDRODYSPLASIA

Grebe chondrodysplasia is an unusual form of acromesomelia in which gross limb and digital shortening predominate. The term 'achondrogenesis' was employed in the original case description of two affected sisters in Germany (Grebe 1952) and there was subsequently terminological confusion with the forms of lethal short-limbed dwarfism which also bear that title (see Ch. 3). This problem has now been resolved by the adoption of the eponymous designation 'Grebe chondrodysplasia' for the condition discussed in this section.

**Fig. 1.33** Grebe chondrodysplasia: a consanguineous Hindu family from Andhra Pradesh, India, in which several relatives have gross acromesomelia. (From Khan & Khan 1982 Progress in Clinical and Biological Research, Vol. 104, Skeletal dysplasias. Liss, New York. By courtesy of the authors.)

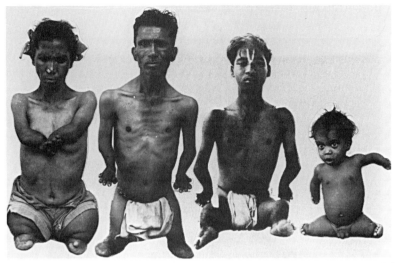

**Fig. 1.34** Grebe chondrodysplasia: severe shortening of the forearms, lower legs and digits. (From Khan & Khan 1982 Progress in Clinical and Biological Research, Vol 104, Skeletal dysplasias. Liss, New York. By courtesy of the authors.)

**Fig. 1.35** Grebe chondrodysplasia: an old family photograph of an affected male with a supernumerary digit on the right hand. Despite his disability, this gentleman had a successful career as a publican. (From Kumar et al 1984 Clinical Genetics 25: 68–72. By courtesy of the authors.)

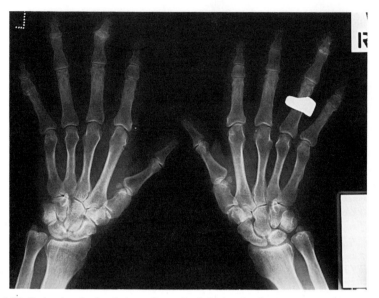

**Fig. 1.36** Grebe chondrodysplasia: radiograph of the hands of a descendant of the affected person depicted in Figure 1.35. Dysplastic changes, notably shortening of the middle phalanges of the 2nd and 5th fingers are manifestations of her status as an obligate heterozygote. (From Kumar et al 1984 Clinical Genetics 25: 68–72. By courtesy of the authors.)

Persons with Grebe chondrodysplasia have profound shortening of all tubular bones of the limbs. The changes are most severe in the extremities and most marked in the lower limbs. Brachydactyly is a significant feature and polydactyly is present in about 60% of affected individuals. The supernumerary digits may be on the hands or feet and involvement is not necessarily symmetrical. The head, face, trunk and spine are normal, intellect is unimpaired and there are no systemic ramifications. Despite the severe dwarfism some affected persons adapt to their disability. This point was clearly brought out by Kumar et al (1984) in their report of an affected publican who was able to subdue his unruly customers and keep an orderly house despite his small size!

About 70 affected persons have been reported, the majority in a consanguineous Brazilian community of Portuguese-Negro ancestry (Quelco-Salgado 1964, 1968). The gene evidently has a wide ethinic and geographic distribution as it has been encountered in the Tamil population of South India (Korula & Gundappa 1963), an endogamous Hindu community in Andhra Pradesh, India (Khan & Khan 1982), in England (Kumar et al 1984) and in an inbred Miao family in Yunnan, China (Feng Bo et al 1985).

Autosomal recessive inheritance has been established and minor abnormalities in the hands and feet of heterozygotes have been well documented (Curtis 1986). These variable changes include short metacarpals, brachydactyly and polydactyly. In view of the serious nature of the condition in the homozygote these digital anomalies are of considerable potential importance in affected families as indicators of heterozygosity for the Grebe chondrodysplasia gene.

The severity of the digital and limb shortening in Grebe chondrodysplasia would permit diagnosis by ultrasonography in a fetus in early pregnancy. As yet, however, there have been no reports of antenatal recognition of the disorder.

REFERENCES

**Preamble**
Dorst J P, Scott C I, Hall J G 1972 The radiologic assessment of short stature–dwarfism. Radiologic Clinics of North America 10/2: 393
Horan F, Beighton P 1982 Orthopaedic problems in inherited skeletal disorders. Springer-Verlag, Berlin
Kaufman H J C 1976 Classification of the skeletal dysplasias and the radiologic approach to their differentiation. Clinical Orthopaedics and Related Research 114: 12
Kopits S E 1976 Orthopaedic complications of dwarfism. Clinical Orthopaedics and Related Research 114: 153
Kozlowski K 1976 Bone dysplasias in radiographic diagnosis. Pediatric Radiology 4/2: 66
Lachman R S, Rimoin D L, Hollister D W 1974 Hip arthrography in the epiphyseal dysplasias. Birth Defects: Original Article Series 10/12: 186
Nelson M A 1970 Orthopaedic aspects of the chondrodystrophies. The dwarf and his orthopaedic problems. Annals of the Royal College of Surgeons of England 47: 185
Rimoin D L, Silberberg R, Hollister D W 1976 Chondro-osseous pathology in the chondrodystrophies. Clinical Orthopaedics and Related Research 114: 137
Stanescu V, Stanescu R, Maroteaux P 1984 Pathogenic mechanisms in osteochondrodysplasias. Journal of Bone and Joint Surgery 66: 817–836

**Achondroplasia**
Beighton P, Bathfield C A 1981 Gibal achondroplasia. Journal of Bone and Joint Surgery 63B: 328–329
Blondeau M, Brunet D, Blanche J M, Debauchez C, Etienne M 1984 Compression of the cervical spinal cord in achondroplasia. Semaine des Hôpitaux de Paris 8/60(II): 771–775
Bowen P 1974 Achondroplasia in two sisters with normal parents. Birth Defects: Original Article Series 10/12: 31
Collipp P J, Gupta K K, Beller E 1976 Achondroplasia: Parental age. New York State Journal of Medicine 76: 1810–1811
Elejalde B R, Elejalde M M, Hamilton P R, Lombardi J M 1983 Prenatal diagnosis in two pregnancies of an achondroplastic woman. American Journal of Medical Genetics 15: 437–439
Eng C E L, Pauli R M, Strom C M 1985 Nonrandom association of a type II procollagen genotype with achondroplasia. Proceedings of National Academy of Science USA 82: 5465–5469
Filly R A, Globus M S, Carey J C, Hall J G 1981 Short limbed dwarfism: Ultrasonic diagnosis by measurement of fetal femoral length. Radiology 138: 653–656
Fitzsimmons J S 1985 Brief clinical report: Familial recurrence of achondroplasia. American Journal of Medical Genetics 22: 609–613
Fryns J P, Kleczkowska A, Verresen H, van den Berghe H 1983 Germinal mosaicism in achondroplasia: a family with 3 affected siblings of normal parents. Clinical Genetics 24: 156–158
Gardner R J M 1977 A new estimate of the achondroplasia mutation rate. Clinical Genetics 11: 31
Hall J G, Globus M S, Graham C B, Pagon R A, Luthy D A, Filly R A 1979 Failure of early prenatal diagnosis in classic achondroplasia. American Journal of Medical Genetics 3: 271–275

Hall J G, Dorst J P, Taybi H, Langer L O, Scott C I, McKusick V A 1969 Two probable cases of homozygosity for the achondroplasia gene. Birth Defects: Original Article Series 5(4): 24

Hecht J T, Butler J J, Scott C I 1984 Long-term neurological sequelae in achondroplasia. European Journal of Pediatrics 143: 58–60

Hecht J T, Nelson F W, Butler I J, Horton W A, Scott C I, Wassman E R, Mehringer C M, Rimoin D L, Pauli R M 1985 Computerized tomography of the foramen magnum: Achondroplastic values compared to normal standards. American Journal of Medical Genetics 20: 355–360

Horton W A, Rotter J I, Rimoin D L, Scott C I, Hall J G 1978 Standard growth curves for achondroplasia. Journal of Pediatrics 93: 435–438

Kahanovitz N, Rimoin D L, Sillence D O 1982 The clinical spectrum of lumbar spine disease in achondroplasia. Spine 7: 137–140

Langer L O Jr, Baumann P A, Gorlin R J 1967 Achondroplasia. American Journal of Roentgenology, Radium Therapy and Nuclear Medicine 100: 12

Langer L O, Spranger J W, Greinacher I, Herdman R C 1969 Thanatotrophic dwarfism. Radiology 92: 285

Lattanzi D R, Harger J H 1982 Achondroplasia and pregnancy. Journal of Reproductive Medicine 27: 363–366

Mørch E T 1941 Chondrodystrophic dwarfs in Denmark, Opera ex Demo. Biologiae Hereditariae Humanae Vol 3. Enjar Munksgaard, Copenhagen

Murdoch J L, Walker B A, Hall J G, Abbey H, Smith K K, McKusick V A 1970 Achondroplasia — a genetic and statistical survey. Annals of Human Genetics 33: 227

Opitz J M 1969 Delayed mutation in achondroplasia? Birth Defects: Original Article Series 5/4: 20

Opitz J M 1985 Editorial comment: 'Unstable premutation' in achondroplasia: Penetrance vs phenotrance. American Journal of Medical Genetics 19: 251–254

Potter E L, Coverstone V A 1948 Chondrodystrophy fetalis. American Journal of Obstetrics and Gynecology 56: 790

Reiser C A, Pauli R M, Hall J G 1985 Achondroplasia: Unexpected familial recurrence. American Journal of Medical Genetics 19: 245–250

Scott C I 1976 Achondroplastic and hypochondroplastic dwarfism. Clinical Orthopaedics and Related Research 114: 18

Stevenson A C 1957 Achondroplasia: an account of the condition in Northern Ireland. American Journal of Human Genetics 9: 81

Stoll C, Roth M P, Bigel P 1982 A reexamination of parental age effect on the occurrence of new mutations for achondroplasia. In: Papadatos J, Bartsocas C S (eds), Skeletal Dysplasias. Liss, New York, p 419–426

Strom C M 1984 Achondroplasia due to DNA insertion into the type II collagen gene. Pediatric Research 18: 226

Todororov A B, Bolling D R 1974 A computerised file for studying growth development in achondroplasia. Birth Defects: Original Article Series 10/9: 241

Warkany J 1971 In: Congenital malformations: notes and comments. Year Book Medical Publishers: Chicago

Wordsworth B P, Ogilvie D J, Sykes B C, Thompson E 1986 Genetic linkage studies of the type II collagen gene and achondroplasia. British Journal of Rheumatology 25: 105

Wynne-Davies R, Walsh W K, Gormley J 1981 Achondroplasia and hypochondroplasia. Clinical variation and spinal stenosis. Journal of Bone and Joint Surgery 63: 508–515

Wynne-Davies R, Gormley J 1985 The prevalance of skeletal dysplasias. Journal of Bone and Joint Surgery 67: 133–137

**Hypochondroplasia**

Beals R K 1969 Hypochondroplasia: a report of five kindreds. Journal of Bone and Joint Surgery 51A: 728

Dominquez R, Young L W, Steele M W, Giordany B R 1984 Multiple exostotic hyponchondroplasia: Syndrome of combined hypochondroplasia and multiple exostoses. Pediatric Radiology 14: 356–359

Frydman M, Hertz M, Goodman R M 1974 The genetic entity of hypochondroplasia, Clinical Genetics 5: 223

Glasgow J F, Nevin N C, Thomas P S 1978 Hypochondroplasia. Archives of Disease in Childhood 53: 868–872

Hall B D, Spranger J 1979 Hypochondroplasia: Clinical and radiological aspects in 39 cases.
Radiology 133: 95–100
Heselson N G, Beighton P, Cremin B J 1979 The radiographic manifestations of
hypochondroplasia. Clinical Radiology 30: 79–85
Kozlowski K 1973 Hypochondroplasia. Progress in Pediatric Radiology 4: 238–249
Léri A, Linossier Mle 1924 Hypochondroplasia héréditaire. Bulletin de la Société Médicale
des Hôpitaux de Paris 48: 1780–1787
McKusick V A, Kelley T E, Dorst J P 1973 Observations suggesting allelism of the
achondroplasia and hypochondroplasia genes. Journal of Medical Genetics 10: 11
Scott C I 1976 Achondroplastic and hypochondroplastic dwarfism. Clinical Orthopaedics
and Related Research 114: 18
Specht E E, Daentl D L 1975 Hypochondroplasia. Clinical Orthopaedics and Related
Research 110: 249
Stoll C, Manini P, Bloch J, Roth M-P 1985 Prenatal diagnosis of hypochondroplasia.
Prenatal Diagnosis 5: 423–426
Walker B A, Murdoch J L, McKusick V A, Langer L O, Beals R K 1971
Hypochondroplasia. American Journal of Diseases of Children 122: 95–104
Wynne-Davies R, Walsh W K, Gormley J 1981 Achondroplasia and hypochondroplasia.
Journal of Bone and Joint Surgery 63: 508–515

**Asphyxiating thoracic dysplasia**
Bankier A, Danks, D M 1983 Thoracic-pelvic dysostosis: a 'new' autosomal dominant form.
Journal of Medical Genetics 20: 276–279
Barnes N D, Hull D, Milner A D, Waterston D J 1971 Chest reconstruction in thoracic
dystrophy. Archives of Disease in Childhood 46: 833
Burn J, Hall C, Marsden D, Matthew D J 1986 Autosomal dominant thoracolaryngopelvic
dysplasia: Barnes syndrome. Journal of Medical Genetics 23: 345–349
Friedman J M, Kaplan H G, Hall J G 1975 The Jeune syndrome (asphyxiating thoracic
dystrophy) in an adult. American Journal of Medicine 59/6: 857
Hanissian A S, Riggs W W, Thomas D A 1967 Infantile thoracic dystrophy. A variant of the
Ellis-van Creveld syndrome. Journal of Pediatrics 71: 855
Jeune M, Beraud C, Carron R 1955 Dystrophie thoracique asphyxiante de caratère familial.
Archives Françaises de Pédiatrie 12: 886
Kaufman H J, Kirkpatrick J A Jr 1974 Jeune thoracic dysplasia — a spectrum of
disorders? Birth Defects: Original Article Series 10/9: 101
Kozlowski K, Masel J 1976 Asphyxiating thoracic dystrophy without respiratory distress.
Pediatric Radiology 5: 30
Maroteaux P, Savart P 1964 La dystrophie thoracique asphyxiante. Etude radiologique et
rapports avec le syndrôme d'Ellis van Creveld. Annals of Radiology 7: 332
Naumoff P, Young L W, Mazer J, Amortegui A J 1977 Short rib-polydactyly syndrome
type 3. Radiology 122/2: 443
Niemann N, Mangiaux M, Rauber G, Pernot C, Bretagne-de Kersauson M C 1963
Dystrophie thoracique asphyxiante du nouveau-ne. Pédiatrie 18: 387
Oberklaid F, Danks D M, Mayne V, Campbell P 1977 Asphyxiating thoracic dysplasia.
Clinical, radiological and pathological information on 10 patients. Archives of Disease in
Childhood 52: 758–765
Shokeir M H K, Houston C S, Awen C F 1971 Asphyxiating thoracic chondrodystrophy.
Association with renal disease and evidence for possible heterozygous expression. Journal
of Medical Genetics 8: 107

**Chondroectodermal dysplasia (Ellis-van Creveld)**
Bui T H, Marsk L, Eklof O, Theorell K 1984 Prenatal diagnosis of chondroectodermal
dysplasia with fetoscopy. Prenatal Diagnosis 4: 155–159
Ellis R W B, Van Creveld S 1940 A syndrome characterised by ectodermal dysplasia,
polydactyly, chondroplasia and congenital morbus cordis. Archives of Disease in
Childhood 15: 65
McKusick V A, Egeland J A, Eldridge R, Krusen D E 1964 Dwarfism in the Amish. The
Ellis-van Creveld syndrome. Bulletin of the Johns Hopkins Hospital 115: 306
Mahoney M J, Hobbins J C 1977 Prenatal diagnosis of chondroectodermal dysplasia
(Ellis-van Creveld syndrome) using fetoscopy and ultrasound. New England Journal of
Medicine 297: 258–260

Metrakos J D, Fraser F C 1954 Evidence for a heredity factor in chondroectodermal dysplasia (Ellis-van Creveld syndrome). American Journal of Human Genetics 6: 260

Muller L M, Cremin B J 1985 Ultrasonic demonstration of fetal skeletal dysplasia. South African Medical Journal 67: 222–226

Murdoch J L, Walker B A 1969 Ellis-van Creveld syndrome. Birth Defects: Original Article Series 5/4: 279

## Multiple epiphyseal dysplasia

Amir D, Mogle P, Weinberg H 1985 Multiple epiphysial dysplasia in one family. A further review of seven generations. Journal of Bone and Joint Surgery 67: 809–813

Barrie H, Carter C, Sutcliffe J 1958 Multiple epiphyseal dysplasia. British Medical Journal ii: 133

Beighton P, Christy G, Learmonth I D 1984 Namaqualand hip dysplasia: An autosomal dominant entity. American Journal of Medical Genetics 19: 161–169

Diamond L S 1974 Pleomorphism of hip disease in a large kinship with spondylo- and multiple epiphyseal dysplasia. Birth Defects: Original Article Series 10/12: 406

Fairbank H A T 1935 Generalised diseases of the skeleton. Proceedings of the Royal Society of Medicine 28: 1611

Fairbank H A T 1946 Dysplasia epiphysealis multiplex. Proceedings of Royal Society of Medicine 39: 315–317

Fairbank H A T 1951 An Atlas of General Affections of the Skeleton. Churchill Livingstone, Edinburgh

Harper P S, Brotherton B J, Cochlin D 1976 Genetic risks in Perthes disease. Clinical Genetics 10: 178

Hoefnagel D, Sycamore L K, Russell S W, Bucknall W E 1967 Hereditary multiple epiphyseal dysplasia. Annals of Human Genetics 30: 201

Juberg R C, Holt J F 1968 Inheritance of multiple epiphyseal dysplasia tarda. American Journal of Human Genetics 20: 549

Lie S O, Siggers D C, Dorst J P, Kopits D E 1974 Unusual multiple epiphyseal dysplasias. Birth Defects: Original Article Series 12: 165

Lowry R B, Wood B J 1975 Syndrome of epiphyseal dysplasia, short stature, microcephaly and nystagmus. Clinical Genetics 8/4: 269

Mena H R, Pearson E O 1976 Multiple epiphyseal dysplasia. A family case report. Journal of the American Medical Association 6: 2629–2633

Meyer J 1964 Dysplasia epiphyseal capitis femoris. Acta Orthopaedica Scandinavica 34: 183–197

Monty C P 1962 Familial Perthes disease resembling multiple epiphyseal dysplasia. Journal of Bone and Joint Surgery 44B: 565

Patrone N A, Kredich D W 1985 Arthritis in children with multiple epiphyseal dysplasia. Journal of Rheumatology 12: 145–149

Pfeiffer R A, Jünemann G, Polster J, Bauer H 1973 Epiphyseal dysplasia of the femoral head, severe myopia and perceptive hearing loss in three brothers. Clinical Genetics 4: 141

Wamoscher Z, Farhi A 1963 Hereditary Legg-Calvé-Perthes disease. American Journal of Diseases of Children 106: 131

Wordsworth B P, Ogilvie D J, Smith M, Sykes B C 1986 Multiple epiphyseal dysplasia and the cartilage collagen gene. British Society for Rheumatology: Abstracts 25: 119

## Chondrodysplasia punctata

Maroteaux P, Lavollay B, Bomsell F, Gautry P, Vigneron J, Walbaum R 1984 Chondrodysplasia punctata and maternal alcohol intoxication. Archives Françaises de Pédiatrie 41: 547–550

Pauli M P, Madden J D, Kranzler K J, Culpepper W, Port R 1976 Warfarin therapy initiated during pregnancy and phenotypic chondrodysplasia punctata. Journal of Paediatrics 88: 506

### Autosomal dominant type (Conradi-Hünerman)

Bergstrom K, Gustavson K H, Jorulf H 1972 Chondrodystrophia calcificans congenita in a mother and her child. Clinical Genetics 3: 158

Fairbank H A T 1927 Some general diseases of the skeleton. British Journal of Surgery 15: 120

Happle R, Matthiass H H, Macher E 1977 Sex-linked chondrodysplasia punctata? Clinical Genetics 11: 73

Hyndman W B, Alexander D S, Mackie K W 1976 Chondrodystrophia calcificans congenita (Conradi-Hünermann syndrome). Report of a case recognised antenatally. Clinical Pediatrics 15/4

Sheffield L J, Danks D M, Mayne V, Hutchinson A L 1976 Chondrodysplasia punctata — 23 cases of a mild and relatively common variety. Journal of Pediatrics 89/6: 916

Silverman F N 1969 Discussion on the relation between stippled epiphyses and the multiplex form of epiphyseal dysplasia. Birth Defects: Original Article Series 5/4: 68

Spranger J, Opitz J M, Bidder U 1971 Heterogeneity of chondrodysplasia punctata. Humangenetik 11: 190

*X-linked dominant type*

Happle R 1979 X-linked dominant chondrodysplasia punctata. Review of literature and report of a case. Human Genetics 53: 65–73

Kolde G, Happle R 1984 Histologic and ultrastructural features of the ichthyotic skin in X-linked dominant chondrodysplasia punctata. Acta Dermato-Venereologica (Stockholm) 64: 389–394

Mueller R F, Crowle P M, Jones R A, Davison B C 1985 X-linked dominant chondrodysplasia punctata: A case report and family studies. American Journal of Medical Genetics 20: 137–144

*Other forms of chondrodysplasia punctata*

Burck U 1982 Mesomelic dysplasia with punctate epiphyseal calcifications: A new entity of chondrodysplasia punctata? European Journal of Pediatrics 138: 67–72

Curry C J, Magenis R E, Brown M, Lanman J T, Tsai J et al 1984 Inherited chondrodysplasia punctata due to a deletion of the terminal short arm of an X-chromosome. New England Journal of Medicine 18: 1010–1015

Hunter A G, Rimoin D L, Koch U M, MacDonald G J, Cox D M, Lachman R S, Adomian G 1985 Chondrodysplasia punctata in an infant with duplication 16p due to a 7;16 translocation. American Journal of Medical Genetics 21: 581–589

**Metaphyseal chondrodysplasia**
*Preamble*

Horan F T, Beighton P H 1980 Infantile metaphyseal dysplasia or 'battered babies'? Journal of Bone and Joint Surgery 62: 243–247

Spranger J W 1976 Metaphyseal chondrodysplasias. Birth Defects: Original Article Series 12/6: 33

*Jansen type*

Charrow J, Poznanski A K 1984 The Jansen type of metaphyseal chondrodysplasia. Confirmation of dominant inheritance and review of radiographic manifestations in the newborn and adult. American Journal of Medical Genetics 18: 321–327

Haas W H D, Boer W de Griffioen F 1969 Metaphyseal dysostosis; a late follow-up of the first reported case. Journal of Bone and Joint Surgery 51B: 2–0

Holthusen W, Holt J F, Stoeckenius M 1975 The skull in metaphyseal chondrodysplasia type Jansen. Paediatric Radiology 3/3: 137

Jansen M 1934 Uber atypische chondrodystrophie (achondroplasie) und über eine noch nicht beschriebene angeborene Wachstumsstörung des Knochensystems: metaphysäre Dysostosis. Zeitschrift für Orthopädische Chirurgie 61: 253

Kikuchi S, Hasue M, Watanabe M, Hasebe K 1976 Metaphyseal dysostosis (Jansen type). Journal of Bone and Joint Surgery 58B: 102–106

Lenz W D 1969 Discussion in first conference on the clinical delineation of birth defects. Birth Defects: Original Article Series 5/4: 71

Nazara Z, Hernandez A, Corona-Rivera E, Vaca G, Panduro A, Martinez-Basalo C, Cantu J H 1981 Further clinical and radiological features in metaphyseal chondrodysplasia Jansen type I. Radiology 140: 697–700

*Schmid type*

Beluffi G, Fiori P, Schifino A, Notarangelo L D, Giardini D, Bozzola M, Montarari C, Martini A 1982 Metaphyseal dysplasia, type Schmid. In: Papadatos J, Bartsocas C S (eds) Skeletal Dysplasias. Liss, New York, p 103–110

Debray H, Poissonnier M, Brault J, D'Angely S 1975 Metaphyseal chondrodysplasia. Annals of Pediatrics 51/3: 253

Maroteaux P, Lamy M 1958 La dysostose métaphysaire. Semaine des Hôpitaux de Paris 34: 1729

Schmid F 1949 Beitrag zur Dysostosis enchondralis metaphysaria. Monatsschrift für Kinderheilkunde 97: 393

Stephens F E 1943 An achondroplastic mutation and the nature of its inheritance. Journal of Heredity 34: 229

*McKusick type*

Kaitila I, Perheentupa J 1980 Cartilage-hair hypoplasia (CHH). In: Eriksson A W, Forsius H R, Nevanlinna H R, Workman P L, Norio R K (eds) Population Structure and Genetic disorders. Academic, New York, p 588–591

Lowry R B, Wood B J, Birkbeck J A, Padwick P H 1970 Cartilage-hair hypoplasia. A rare and recessive cause of dwarfism. Clinical Pediatrics 9: 44

Lux S E, Johnston R B Jr, August C S, Say B, Penchaszadeh V B, Rosen F S, McKusick V A 1970 Neutropenia and abnormal cellular immunity in cartilage-hair hypoplasia. New England Journal of Medicine 282: 234

Maroteaux P, Savart P, Lefebvre J, Royer P 1963 Les formes partielles de la dysostose metaphysaire. Presse Médicale 71: 1523

McKusick V A, Eldridge R, Hostetler J A, Ruangwit J A, Egeland J A 1965 Dwarfism in the Amish. II. Cartilage-hair hypoplasia. Bulletin of the Johns Hopkins Hospital 116: 285

Ray H C, Dorst J P 1973 Cartilage-hair hypoplasia. Progress in Pediatric Radiology 4: 297–298

Virolainen M, Savilahti E, Kaitila I, Perheentupa J 1978 Cellular and humoral immunity in cartilage-hair hypoplasia. Pediatric Research 12: 961–966

Wilson W G, Aylsworth A, Folds J D, Whisnant J K 1978 Cartilage-hair hypoplasia (metaphyseal chondrodysplasia, type McKusick) with combined immune deficiency: variable expression and development of immunologic functions in sibs. Birth Defects Original Article Series XIV(6A): 117–129

*Shwachman-Bodian type*

Bodian M, Sheldon W, Lightwood R 1964 Congenital hypoplasia of the exocrine pancreas. Acta Paediatrica 53: 282

Burke V, Colebatch J H, Anderson C M, Simions M J 1967 Association of pancreatic insufficiency and chronic neutropenia in childhood. Archives of Disease in Childhood 42: 147

Danks D M, Haslam R H, Mayne V, Kaufmann H J, Holtzapple P G 1976 Metaphyseal chondrodysplasia, neutropenia and pancreatic insufficiency presenting with respiratory distress in the neonatal period. Archives of Disease in Childhood 51: 697–701

Shwachman H, Holsclaw D 1972 Some clinical observations on the Shwachman syndrome (pancreatic insufficiency and bone marrow hypoplasia). Birth Defects: Original Article Series 8/3: 46

Shwachman H, Diamond L K, Oski F A, Khaw K T 1964 The syndrome of pancreatic insufficiency and bone marrow dysfunction. Journal of Pediatrics 65: 645–663

Stanley P, Sutcliffe J 1973 Metaphyseal chondrodysplasia with dwarfism, pancreatic insufficiency and neutropenia. Pediatric Radiology 1: 119

Taybi H, Mitchell A D, Friedman G D 1969 Metaphyseal dysostosis and the associated syndrome of pancreatic insufficiency and blood disorders. Radiology 93: 563

Woods W G, Roloff J S, Lukens J N, Krivit W 1981 The occurrence of leukemia in patients with the Shwachman syndrome. Journal of Pediatrics 99: 425–428

*Davis type*

Alexander W J, Dunbar J S 1968 Unusual bone changes in thymic alymphoplasia. Annals of Radiology 2: 289

Davis J A 1966 A case of Swiss-type agammaglobulinaemia and achondroplasia. British Medical Journal ii: 1371

Fulginiti V A, Hathaway W E, Pearlman D S, Kempe C H 1967 Agammaglobulinaemia and achondroplasia (Letter) British Medical Journal ii: 242

Gatti R A, Platt N, Pomerance H H, Hong R, Langer L O, Kay H E M, Good R A 1969 Hereditary lymphopenic agammaglobulinaemia associated with a distinctive form of short-limbed dwarfism and ectodermal dysplasia. Journal of Pediatrics 75: 675

*Other types*

Halal F, Picard J I, Raymond-Tremblay D, de Bosset P 1982 Metaphyseal dysplasia with maxillary hypoplasia and brachydactyly. American Journal of Medical Genetics 13: 71–79

Kaitila I I, Halttunen P, Snellman O, Takkunen O 1982 A new form of metaphyseal chondrodysplasia in two sibs: Surgical treatment of tracheobronchial malacia and scoliosis. American Journal of Medical Genetics 11: 415–424

Kozlowski K 1964 Metaphyseal dysostosis. Report of five familial and two sporadic cases of mild type. American Journal of Roentgenology, Radium Therapy and Nuclear Medicine 91: 602

Kozlowski K, Sikorska B 1970 Dysplasia metaphysaria type Vaandrager-Pena. Zeitschrift für Kinderheilkunde 108: 165

Pena J 1965 Disostosis metafisaria. Una revision. Con aportacion do una observacion familar. Una formá mieva de la enfermedael. Radiologia 47: 3

Phillips C I, Wynne-Davies R, Stokoe N L, Newton M 1981 Retinitis pigmentosa, metaphyseal chondrodysplasia and brachydactyly: An affected brother and sister. Journal of Medical Genetics 18: 46–49

Sedaghatian M R 1980 Congenital lethal metaphyseal chondrodysplasia: A newly recognized complex autosomal recessive disorder. American Journal of Medical Genetics 6: 269–274

Spahr A, Spahr-Hartmann I 1961 Dysostose metaphysaire familiale Etude de 4 cas dans une fratrie. Helvetica Paediatrica Acta 16: 836

Vaandrager G J 1960 Metafysaire dysostosis. Nederlands Tydschrift voor Geneeskunde 104: 547

Williams A W, Cantu E S, Frias J L 1986 Brief clinical report: Metaphyseal dysostosis and congenital nystagmus in a male infant with the fragile X syndrome. American Journal of Medical Genetics 23: 207–211

**Mesomelic dysplasia**
*Preamble*

Kaitila I I, Leisti J T, Rimoin D L 1976 Mesomelic skeletal dysplasia. Clinical Orthopaedics and Related Research 114: 94

Maroteaux P, Spranger J 1977 Essai de classification des chondrodysplasies a predominance mesomelique. Archives Françaises de Pédiatrie 34: 945–958

*Nievergelt type*

Hess O M, Goebel N H, Streuli R 1978 Familiaerer mesomeler Kleinwuchs (Nievergelt-Syndrome). Schweizerische Medizinische Wochenschrift 108: 1202–1206

Nievergelt K 1944 Positiver Vaterschaftsnachweis auf Grund erblicher Missbildungen der Extremitaten. Archiv der Julius Klaus-Stiftung für Vererbungsforschung Sozialanthropologie und Rassenhygiene 19: 157

Solonen K A, Sulamaa M 1958 Nievergelt syndrome and its treatment. Annales Chirugiae et Gynaecologiae Fenniae 47: 142

Young L W, Wood B P 1975 Nievergelt syndrome (mesomelic dwarfism type Nievergelt). Birth Defects: Original Article Series 11/5: 81

*Langer type*

Böok J A 1950 A clinical genetical study of disturbed skeletal growth (chondrohypoplasia). Hereditas 36: 161

Brailsford J F 1953 Dystrophies of the skeleton. British Journal of Radiology 8: 533

Espiritu C, Chen H, Wooley P V 1975 Mesomelic dwarfism as the homozygous expression of dyschondrosteosis. American Journal of Diseases of Children 129: 375

Fryns J P, van de Berghe H 1979 Langer type of mesomelic dwarfism as the possible homozygous expression of dyschondrosteosis. Human Genetics 46: 21–27

Goldblatt J, Wallis C, Viljoen D, Beighton P 1987 Heterozygous manifestations of Langer mesomelic dysplasia. Clinical Genetics 31: 19–25

Kanze J, Klemm T 1980 Mesomelic dysplasia, type Langer — A homozygous state for dyschondrosteosis. European Journal of Pediatrics 134: 269–272

Langer L O 1967 Mesomelic dwarfism of the hypoplastic ulna, fibula and mandibular type. Radiology 89: 654

Silverman F N 1975 Intrinsic diseases of bones. In: Kaufman H J (ed) Progress in Paediatric Radiology Vol. 4. Karger, Basel, p 546

*Other types*

Bain M D, Winter R M, Burns J 1986 Robinow syndrome without mesomelic 'brachymelia': A report of 5 cases. Journal of Medical Genetics 23: 350–354

Burck U 1982 Mesomelic dysplasia with punctate epiphyseal calcifications — A new entity of chondrodysplasia punctata? European Journal of Pediatrics 138: 67–72

Burck U, Schaefer E, Held K R 1980 Mesomelic dysplasia with short ulna, long fibula, brachymetacarpy and micrognathia. Pediatric Radiology 9: 161–165

Eaton G O, McKusick V A 1969 A seemingly unique polydactyly syndactyly syndrome in four persons in three generations. Birth Defects: Original Article Series 5/3: 221

Friedman J M 1985 Umbilical dysmorphology. Clinical Genetics 28: 343–347

Giedion A, Mattaglia G F, Bellini F, Fancone G 1975 The radiological diagnosis of the fetal face (Robinow) syndrome (mesomelic dwarfism and small genitalia). Report of three cases. Helvetica Paediatrica Acta 30/4–5: 409

Kelly T E, Benson R, Temtamy S A, Plotnick L, Levin S 1975 The Robinow syndrome: An isolated case with a detailed study of the phenotype. American Journal of Diseases in Children 129: 383–386

Leroy J G, De Vos J, Timmermans J 1975 Dominant mesomelic dwarfism of the hypoplastic tibia, radius type. Clinical Genetics 7/4: 280

Pashayan H, Fraser F C, McIntyre J M, Dunbar J S 1971 Bilateral aplasia of the tibia, polydactyly and absent thumb in father and daughter. Journal of Bone and Joint Surgery 53B: 495

Reinhardt K 1976 A dominant-autosomal transmitted micromesomelia with dysplasia of radius and ulna (Reinhardt-Pfeiffer syndrome). V International Congress of Human Genetics, Mexico, D. F.

Reinhardt K, Pfeiffer R A 1967 Ulno-fibulare Dysplasie. Eine autosomal-dominant vererbte Mikromesomelie ähnlich dem Nievergeltsyndrome. Fortschritte auf dem Gebiete der Röntgenstrahlen und der Nuklearmedizin 107: 379

Robinow M 1973 Syndrome's progress. American Journal of Diseases of Children 126: 150

Robinow M, Silverman F N, Smith H D 1969 A newly recognised dwarfing syndrome. American Journal of Diseases of Children 117: 645

Schinzel A, Zellweger H, Grella A, Prada A 1974 Fetal face syndrome with acral dysostosis. Helvetica Paediatrica Acta 29: 55–60

Wadlington W B, Tucker V L, Schminke R N 1973 Mesomelic dwarfism with hemivertebrae and small genitalia (the Robinow syndrome). American Journal of Diseases of Children 126: 202

Werner P 1915 Ueber einen seltenen Fall von Zwergwuchs. Archiv für Gynaekologie 104: 278

**Acromesomelic dysplasia**

Beighton P 1974 Autosomal recessive inheritance in the mesomelic dwarfism of Campailla and Martinelli. Clinical Genetics 5: 363

Borelli P, Fasanelli S, Marini R 1983 Acromesomelic dwarfism in a child with an interesting family history. Pediatric Radiology 13: 165–168

Campailla E, Martinelli B 1971 Deficit staturate con micromesomelia. Presentazione di due case familiari. Minerva Orthopédica 22: 180

Goodman R M, Weinberg U, Hertz M, Rosenthal T, Hertz R 1975 Peripheral dysostosis: An autosomal recessive form. Birth Defects Original Article Series X(12): 137–146

Hunter A G, Thompson M W 1972 Acromesomelic dwarfism: Description of a patient and comparison with previously reported cases. Human Genetics 34: 197–113

Langer L O, Beals R K, Solomon I L, Bard P A, Bard L A, Rissman E M, Rogers J G, Dorst J P, Hall J G, Sparkes E O R S, Franken E D 1977 Acromesomelic dwarfism: Manifestations in childhood. American Journal of Medical Genetics 1: 87–100

Lannois M 1902 Deux cas de nanisme achondroplasique chez le frère et la soeur. Lyon Médicale 98: 1839–1842

Maroteaux P, Martinelli B, Campailla E 1971 Le nanisme acromesomelique. Presse Médicale 79: 1839

Osebold W R, Remondini D J, Lester E L, Spranger J W, Opitz J M 1985 An autosomal dominant syndrome of short stature with mesomelic shortness of limbs, abnormal carpal and tarsal bones, hypoplastic middle phalanges and bipartite calcanei. American Journal of Medical Genetics 22: 791–809

Plauchu H, Maisonneuve D, Floret D 1984 Le nanisme acro-coxo-mésomelique: Variété nouvelle de nanisme récessif autosomique. Annales de Génétique (Paris) 27: 83–87

Shine I 1970 Serendipity in St Helena: A genetical and medical study of an isolated community. Pergamon, Oxford

**Rhizomelic dysplasia**

Kozlowski K S, Celermajer J M, Tink A R 1974 Humero-spinal dysostosis with congenital heart disease. American Journal of Diseases of Children 127: 407

Patterson C, Lowry R B 1975 A new dwarfing syndrome with extreme shortening of humeri and severe coxa vara. Radiology 114/2: 341

Urbach D, Hertz M, Shine M, Goodman R M 1986 A new skeletal dysplasia syndrome with rhizomelia of the humeri and other malformations. Clinical Genetics 29: 83–87

Viljoen D, Goldblatt J, Wallis C, Beighton P 1987 Familial rhizomelic dysplasia. American Journal of Medical Genetics 4: 941–947

Yang T S, Lenz W 1976 Uber symmetrische Verkurzung der Humeri. Padiatrie und Padologie 11: 12–16

**Grebe achondrodysplasia**

Curtis D 1986 Heterozygote expression in Grebe chondrodysplasia (Letter to the Editor). Clinical Genetics 29: 455–456

Feng Bo, Chen Renbiao, Luo Jianguo, Chen Ruigan, Zheng Yingming 1985 A kindred of Miao Nationality affected with Grebe-Quelco-Salgado achondrogenesis. Acta Genetica Sinica 12: 378–386

Grebe H 1952 Die achondrogenesis: Ein einfach rezessives Erbmerkmal Folia Hereditera Pathologica 2: 23–29

Khan P M, Khan A 1982 Grebe chondrodysplasia in three generations of an Andhra family in India. Progress in Clinical and Biological Research, Vol 104. Skeletal Dysplasias. Liss, New York, p 69–80

Korula J, Gundappa M P 1963 Congenital deformities of the limbs in different members of a family. Journal of the Indian Medical Association 41: 559

Kumar D, Curtis D, Blank C E 1984 Grebe chondrodysplasia and brachydactyly in a family. Clinical Genetics 25: 68–72

Quelco-Salgado A 1964 A new type of dwarfism with various bone aplasias and hypoplasias of the extremities. Acta Genetica (Basel) 14: 63–66

Quelco-Salgado A 1968 A rare genetic syndrome. Lancet 1: 1430

# 2

# Dwarfing skeletal dysplasias with significant spinal involvement

The conditions considered in this section are all characterised by dwarfism and spinal abnormalities. Many of them are heterogeneous and there is little doubt that other entities in this general category await delineation.

1. Pseudoachondroplasia
2. Spondyloepiphyseal dysplasia
   a. Congenita
   b. Tarda
   c. Tarda with progressive arthropathy (SEDTPA)
   d. Mselini joint disease
3. Spondylometaphyseal dysplasia
   a. Brachyolmia
4. Spondyloepimetaphyseal dysplasia
   a. Spondyloepiphysometaphyseal dysplasia, Irapa type (SEMDIT)
   b. Spondylometaepiphyseal dysplasia congenita, type Strudwick
   c. Spondyloepimetaphyseal dysplasia with joint laxity (SEMDJL)
5. Schwartz syndrome
6. Metatropic dysplasia
7. Kniest syndrome
8. Diastrophic dysplasia
9. Dyggve-Melchior-Clausen dysplasia
10. Parastremmatic dysplasia
11. Opsismodysplasia
12. Osteoglophonic dysplasia

It is tempting to speculate that Richard Crookback, Duke of Gloucester, who became King Richard III, might have been afflicted with an osteochondrodysplasia of this type. His prematurity and the presence of clinically obvious malformations at the time of birth are clues which might lead to a more precise diagnosis.

> 'deformed, unfinished, sent before my time in this breathing world scarce half made-up'.
>
> Richard III. Act I, Scene 1, Shakespeare.

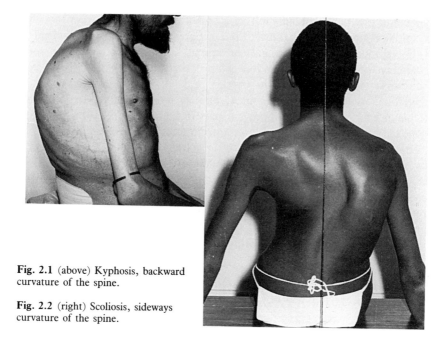

**Fig. 2.1** (above) Kyphosis, backward curvature of the spine.

**Fig. 2.2** (right) Scoliosis, sideways curvature of the spine.

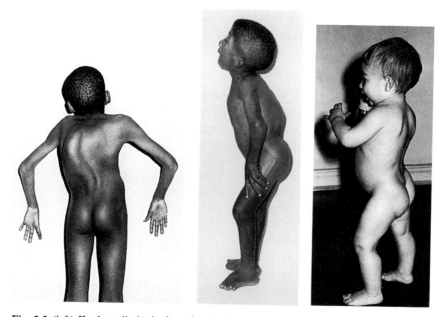

**Fig. 2.3** (left) Kyphoscoliosis, backward and sideways curvature of the spine

**Fig. 2.4** (centre) Gibbus, localised backward angulation of the spine

**Fig. 2.5** (right) Lordosis, forwards curvature of the spine. This is a normal feature in infancy.

The major practical problems in this group of disorders are the consequence of spinal abnormality. Extensive and progressive deformity may lead to cardiorespiratory embarrassment or spinal cord compression. In some instances, hypoplasia of the odontoid process predisposes to atlanto-axial dislocation and hyperextension of the neck during anaesthesia may be dangerous. As orthopaedic measures, such as prosthetic joint replacement, are playing an increasingly important part in the management of the osteochondrodysplasias, this hazard is of considerable practical significance.

The term 'Morquio syndrome' is often loosely and incorrectly applied to any short-limbed dwarf with spinal abnormalities. In this way, conditions such as spondyloepiphyseal dysplasia, pseudoachondroplasia, spondylometaphyseal dysplasia and the mucopolysaccharidoses have been erroneously grouped together. In the strict sense, the eponym 'Morquio syndrome' is applicable only to a distinct entity, mucopolysaccharidosis (MPS) type IV. Nevertheless, this nosological problem is still the cause of considerable confusion.

## PSEUDOACHONDROPLASIA

Maroteaux & Lamy (1959) recognised that this disorder was distinct from true achondroplasia and the various types of spondyloepiphyseal dysplasia. Pseudoachondroplasia is a relatively common osteochondrodystrophy and numerous cases have been reported. The condition was reviewed in detail by Hall (1975).

### Clinical and radiographic features

Growth retardation is apparent in early childhood. Body proportions resemble those of achondroplasia, with disproportionate shortening and deformity of the limbs and foreshortening of the digits. Lumbar lordosis and scoliosis are sometimes present and the joints may be hypermobile. There is no craniofacial involvement. Secondary osteoarthritis supervenes in early adulthood, particularly in the weight bearing joints. Management centres upon orthopaedic correction of hip and knee problems (Kopits et al 1974, Kopits 1976). Growth curves which have been charted for 61 persons with pseudoachondroplasia (Horton et al 1982) are of value for prediction of height in adulthood.

Radiographic abnormalities appear during late infancy and evolve throughout childhood. The epiphyses of the long bones are irregular, while the adjacent metaphyses are cup-shaped and widened. The diaphyses are relatively broad and may be misshapen. The tubular bones of the hands are shortened and the phalanges have broad bases. During the developmental phase, the vertebrae are irregular and biconvex, with central projections. However, these changes are much less marked in the adult. In the pelvis the acetabulae are flattened and the pubis and ischium are hypoplastic.

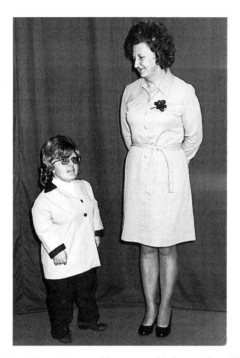

**Fig. 2.6** Pseudoachondroplasia: a 22-year-old woman with her mother. The kindred are normal and it is likely that this patient represents a new mutation for the severe autosomal dominant form of the condition (Heselson N G, Cremin B J, Beighton P 1977 British Journal of Radiology 50: 473.)

*Genetics*

Hall & Dorst (1969) reviewed the manifestations in 32 patients from 12 kindreds and delineated four forms of pseudoachondroplasia. According to this classification, types I and II were the autosomal dominant and recessive 'Kozlowski' types, while types III and IV were the autosomal dominant and recessive 'Maroteaux-Lamy' forms. However, the four subdivisions are not universally accepted and delineation cannot be regarded as complete.

McKusick (1975) suggested that the two dominant forms might be allelic. Generation to generation transmission in a manner compatible with autosomal dominant inheritance has been recorded in two South African families by Heselson et al (1977) and in a French kindred by Fontaine et al (1979).

Dennis & Renton (1975) described a kindred in which four out of seven sibs had the severe autosomal recessive form of pseudoachondroplasia. The authors note that the parents were of short stature without any additional signs of pseudoachondroplasia and they suggested that there might be partial manifestation of the abnormal gene in the heterozygote. Other reports of autosomal recessive pseudoachondroplasia concern the two

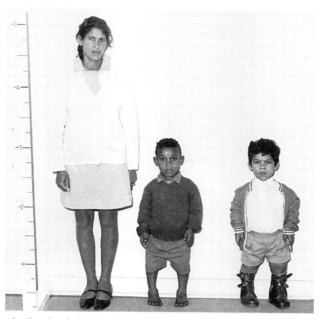

**Fig. 2.7** Pseudoachondroplasia: two affected brothers, aged 14 and 19 with their mother. The parents and six other sibs were normal and it is probable that these young men had the severe autosomal recessive form of the disorder.

**Fig. 2.8** Pseudoachondroplasia: an 11-year-old boy with his unaffected younger brother. The limbs are short, but in distinction to achondroplasia, the head and face are normal.

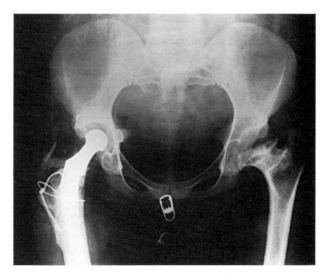

**Fig. 2.9** Pseudoachondroplasia: prosthetic hip joint replacement is proving to be of value. (Courtesy of Mr F. J. Heddon, FRCS, Durban.)

brothers from the mixed ancestry population of Cape Town (Heselson et al 1977) and an Indian boy with consanguineous parents in Britain (Young & Moore 1984).

The majority of persons with pseudoachondroplasia are sporadic and the question of new dominant mutation versus autosomal recessive inheritance sometimes arises. Lachman et al (1975) emphasised that in clinical practice, it is not always possible to assign a sporadic individual to a specific category. As the autosomal dominant and autosomal recessive forms cannot be recognised on clinical or radiological grounds alone, the genetic counsellor may be faced with a difficult situation when discussing recurrence risks for pseudoachondroplasia.

Maynard et al (1972) observed inclusion material in the endoplasmic reticulum of the chondrocytes from affected individuals. In further studies of five patients with various types of pseudoachondroplasia Cranley et al (1975) found consistent histological abnormalities in the cartilage and confirmed the previous reports of intracellular inclusion bodies. The possible nature of the biochemical defect was discussed by Stanescu et al (1982). Following sophisticated studies of cartilage obtained at iliac crest biopsy from three patients, Pedrini-Mille et al (1984) postulated that pseudoachondroplasia was a generalised disorder of cartilage in which the proteoglycans are abnormal. These histological and biochemical changes are characteristic of the pseudoachondroplasias as a whole and they do not permit differentiation of the various types of the condition.

## SPONDYLOEPIPHYSEAL DYSPLASIA

The predominant features of the spondyloepiphyseal dysplasias (SED) are dwarfism and spinal deformity. The changes are maximal in the vertebrae and in the epiphyses of the long bones, while involvement of the metaphyses is of lesser degree.

**Fig. 2.10** (left) SED: a boy aged 15 years with stunted stature and truncal shortening.

**Fig. 2.11** (below) SED: severe changes in the hip joints of an affected adult.

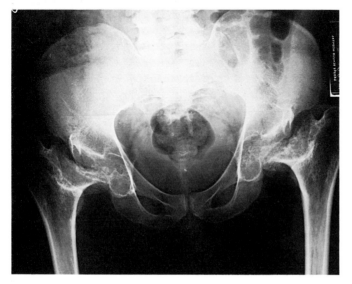

Rubin (1964) suggested that SED could be subdivided into 'congenita' and 'tarda' forms. In the 'congenita' type stigmata are present at birth, while in the 'tarda' type the disorder becomes evident in later childhood. These categories have met with general acceptance. In an alternative classification Maroteaux (1969) described three forms of SED on the basis of the anatomical distribution of the skeletal changes. Pseudoachondroplasia and spondylometaphyseal dysplasia have been split off from SED and recognised as entities in their own right but there is no doubt that considerable heterogeneity still exists within the SED group of conditions. Indeed, in the author's own experience 'atypical' or 'unclassifiable' forms of SED are encountered more frequently than the traditional types.

### Spondyloepiphyseal dysplasia congenita

In spondyloepiphyseal dysplasia congenita (SEDC) spinal and epiphyseal changes are present in the neonate. Spranger & Weidemann (1966) recognised the distinction between this condition and the Morquio syndrome and subsequently Spranger & Langer (1970) described a series of 29 cases.

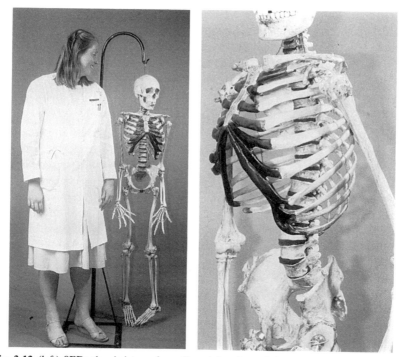

**Fig. 2.12** (left) SED: the skeleton of an affected female aged 60 years. In contrast to the changes in the spine and proximal epiphyses, the extremities are normal.

**Fig. 2.13** (right) SED: the vertebral bodies are flat. Severe involvement of the hip joints necessitated replacement with prostheses.

*Clinical and radiographic features*

Dwarfism, hypertelorism, cleft palate and talipes equinovarus are evident at birth. Extension of the elbow joints is limited and the hips are sometimes dislocated. Severe kyphoscoliosis and thoracic deformity develop in childhood. Leg length may be disproportionate, with genu valgum, genu recurvatum, lateral displacement of the patellae and metatarsus adductus. Ophthalmological problems occur in more than 50% of the affected children and retinal detachment in association with myopia represents a threat to vision (Murray et al 1985). The adult with SED congenita (SEDC) is usually less than 140 cm in height and has a flat face, short neck and barrel chest. Secondary osteoarthritis develops in the weight bearing joints and backache is a frequent problem. Spinal and thoracic deformity lead to cardiorespiratory embarrassment and cor pulmonale may supervene in middle age. The manifestations of SEDC have been reviewed by Williams & Cranley (1974) and Spranger (1975).

Radiographic changes in the newborn are particularly obvious in the spine and pelvis. Shortening of the tubular bones is not prominent but there is a generalised delay in development of ossification centres. In childhood the vertebrae are flattened and irregular and progressive dorsal kyphoscoliosis develops. The odontoid process may be hypoplastic but the skull is usually normal. The epiphyses of the large joints are irregular while the metaphyses are involved to a lesser extent. The hip joints become increasingly dysplastic and severe degenerative changes may be present by adulthood.

*Genetics*

SED congenita is usually inherited as an autosomal dominant and kindreds with generation to generation transmission have been reported by Spranger & Wiedemann (1966) and Spranger & Langer (1970). The inter-familial inconsistency of clinical stigmata may be indicative of heterogeneity. Furthermore, there is good evidence for an autosomal recessive form of SEDC as two pairs of siblings with normal parents have been encountered (Harrod et al 1984). In both instances, counselling was on a basis of probable new dominant mutation and the birth of a second affected child came as a surprise to the parents and the medical attendants. The clinical and radiographic changes are indistinguishable from classical autosomal dominant SEDC.

The question of syndromic identity of SEDC was reviewed by Kozlowski et al (1977). Minor changes which are sometimes present in the metaphyses confuse the nosological issue and there is debate concerning exact syndromic boundaries (Spranger & Maroteaux 1982, Anderson et al 1982).

**Spondyloepiphyseal dysplasia tarda**

The manifestations of spondyloepiphyseal dysplasia tarda (SEDT) appear in mid-childhood. Involvement is predominantly spinal, with shortening of the trunk relative to the limbs. The condition was delineated by Maroteaux et al (1957). However, a number of earlier reports under a variety of designations can be recognised in the literature.

## Clinical and radiographic features

Clinical manifestations are variable and at the mild end of the spectrum affected individuals may be recognised only by demonstration of radiographic changes in the spine. Dorsal kyphoscoliosis usually develops in mid-childhood and the trunk becomes progressively shortened, although the limbs remain relatively uninvolved. The hamstrings tighten and pain in the legs and back is a common problem. Progressive degenerative osteoarthropathy of the spine and hip joints may cause severe disability in middle age. Two instances of the occurrence of malignant neoplasms of the bone in SEDT were recorded in an international survey (Tsuruta & Ogihara 1984).

Radiographically the skeleton is virtually normal until the age of 5. Later, generalised platyspondyly develops, with kyphoscoliosis and thoracic cage deformity. Heaping up of the posterior part of the upper surfaces of the bodies of the lumbar vertebrae is a pathognomonic feature (Langer 1964). The articular surfaces of the large joints become flattened and dysplastic, the changes being maximal in the hip joints.

## Genetics

Bannerman et al (1971) updated an extensive pedigree published by Jacobsen (1939) in which SEDT was inherited as an X-linked recessive trait. These investigators also demonstrated that the genes for SEDT and the Xg blood group were not closely linked. (The Xg blood group locus is situated on the X chromosome and serves as a useful marker in linkage studies.) In view of the increasing availability of polymorphic DNA probes for the X chromosome, it is likely that gene localisation will be achieved in the near future by means of molecular techniques. Although SEDT is well known there is a paucity of reports of large affected kindreds and it is questionable as to whether the disorder is as common as generally supposed.

There is good evidence for autosomal dominant and autosomal recessive inheritance of late onset SED in some kindreds. Affected sibs with normal consanguineous parents were described by Klenerman (1961) and Martin et al (1970). This form of SEDT seems to be inherited as an autosomal recessive. O'Brien et al (1976) described a girl with SEDT in whom a deficiency of beta-galactosidase activity was demonstrated in cultured fibroblasts. The activity of this enzyme in fibroblasts from both unaffected parents was approximately 50% of normal. On this evidence, it is likely that these parents were heterozygous for the abnormal gene. Generation to generation transmission of SEDT consistent with autosomal dominant inheritance was reported by Moldauer et al (1962). A boy with SEDT, marked dwarfism, craniofacial abnormalities and eye involvement, born to consanguineous parents, seems to represent yet another distinct autosomal recessive entity (Jones et al 1986).

The clinical and radiographic features of these uncommon forms of SEDT are by no means clear-cut and diagnostic precision rests largely upon recognition of the pattern of transmission within a kindred. For this reason genetic counselling in the case of a sporadic individual should be undertaken with caution.

## Spondyloepiphyseal dysplasia tarda with progressive arthropathy (SEDTPA)

SEDTPA presents clinically in mid-childhood as atypical rheumatoid arthritis. The course is progressive and the development of contractures leads to severe handicap. Stature is not significantly reduced and there is no involvement of extra-articular tissue. Despite the inflammatory manifestations, haematological and serological parameters remain within normal limits. The disorder can be differentiated from rheumatoid arthritis by the radiological demonstration of vertebral changes which are indistinguishable from those of classical SEDT.

The condition was recognised by Wynne-Davies et al (1982) in 15 patients from nine families and pedigree data were suggestive of autosomal recessive inheritance. The disorder had been previously reported by Spranger et al (1980) and it was further delineated by Spranger et al (1983). The autosomal recessive mode of transmission was confirmed when Al-Awadi et al (1984) detected eight affected persons in a large consanguineous Arab kindred in Kuwait.

## Mselini joint disease

Mselini joint disease is a progressive, degenerative arthropathy which affects several hundred persons in northern Zululand (Wittman & Fellingham 1970, Fellingham et al 1973, Yach & Botha 1985). The term 'Mselini' is derived from the name of the region inhabited by this tribal community. The condition presents in childhood with discomfort in the large joints and progresses to cause crippling by adulthood. Apart from supportive measures the only effective form of management is prosthetic joint replacement (Du Toit 1979). The early radiological changes are in the epiphyses and spine and for this reason Mselini joint disease can be regarded as a form of SED (Lockitch et al 1973). Furthermore, a minority of persons with the condition are dwarfed and in the author's experience their clinical appearance is similar to the conventional form of SED.

Mselini joint disease clusters in families but no clear pattern of Mendelian inheritance can be discerned and it is uncertain whether the disorder has a genetic basis. There is similarity with the Kashin-Beck disease, which is endemic in Siberia and China; this condition is the result of fungal contamination of food (Nesterov 1964). However, intensive investigations over a 15 year period have failed to reveal any environmental determinant and the aetiology of Mselini joint disease remains enigmatic.

## SPONDYLOMETAPHYSEAL DYSPLASIA

Spondylometaphyseal dysplasia (SMD) was delineated by Kozlowski et al (1967) and reviewed in detail by Kozlowski (1976). SMD has been a source of great diagnostic and terminologic confusion and it is likely that there is considerable heterogeneity.

### Clinical and radiographic features

In SMD metaphyseal abnormalities are associated with spinal changes. The predominant features are short-trunked dwarfism, kyphoscoliosis, pectus carinatum, limited movements of the hips and elbows and knee deformity.

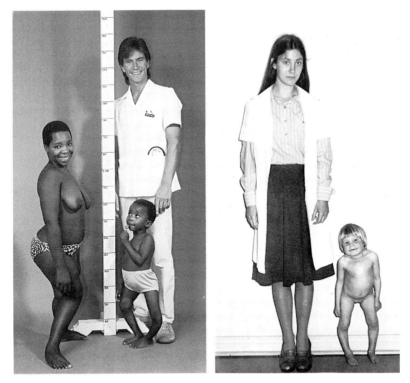

**Fig. 2.14** (left) SMD: a mother and child with short stature and leg malalignment. The mother has gross knock knees whereas her daughter has severe bow legs.

**Fig. 2.15** (right) SMD: a 5-year-old girl with dwarfism, a short neck, thoracic deformity and marked genu varum.

The skeleton is radiographically normal at birth. Platyspondyly, which is maximal in the thoracic vertebrae, develops in early childhood. The metaphyses of the long bones become irregular at this time and in some instances show multiple radiolucent areas. Coxa vara may be associated with a short femoral neck. The epiphyses are uninvolved. The radiographic changes in SMD have been reviewed by Riggs & Summitt (1971) and Thomas & Nevin (1977).

*Genetics*

The autosomal dominant 'common' or 'Kozlowski' form of SMD is well recognised. More than 40 cases have been documented and syndromic identity is firmly established (Kozlowski et al 1980). In addition, there have been several instances of atypical SMD in which pedigree data are suggestive of autosomal recessive inheritance (Kozlowski et al 1976, Gustavson et al 1978, Kozlowski et al 1979). There is considerable phenotypic disparity in these patients and it seems likely that there are several distinct forms of autosomal recessive SMD.

## Brachyolmia

Brachyolmia is a rare disorder in which changes in the vertebrae predominate (Fontaine et al 1975). Although autosomal dominant and autosomal recessive forms are listed in the Paris Nomenclature, Kozlowski et al (1979) have questioned the autonomy of this disorder. Based upon their extensive experience, these authors pointed out that scattered metaphyseal changes are always present in affected persons and that it would be correct to regard brachyolmia as a form of SMD.

## SPONDYLOEPIMETAPHYSEAL DYSPLASIA

Several rare syndromes which are characterised by involvement of the epiphyses, metaphyses and spine derive their names from the predominant anatomical localisation of the changes. The situation is confused as there is phenotypic overlap in many of these conditions but the disorders mentioned below are reasonably well established.

### Spondyloepiphysometaphyseal dysplasia, Irapa type (SEMDIT)

SEMDIT was recognised by Arias et al (1976) in an Indian community in Venezuela and further cases were reported in a Mexican family of mixed ancestry (Hernandez et al 1980). Inheritance is autosomal recessive.

### Spondylometaepiphyseal dysplasia congenita, type Strudwick

This condition derives its eponym from a boy reported by Murdoch &

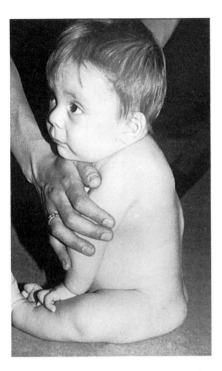

**Fig. 2.16** (left) SEMDJL: a male infant with a kypho-scoliosis and a short neck. He has the characteristic oval face, with a stubby nose and longer upper lip. (From Beighton et al 1983 South African Medical Journal 64: 772–775.)

**Fig. 2.17** (below) SEMDJL: the hands of an affected boy, showing gross joint laxity. (From Beighton et al 1984 Clinical Genetics 26: 308–317.)

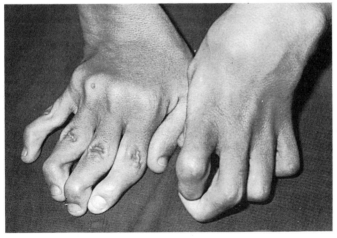

Walker (1969) as having a 'new form of spondylometaphyseal dysplasia'. Further patients were recorded by Anderson et al (1982) and the radiological appearance of 'dappled metaphyses' as a diagnostic feature was emphasised. The independent status of this autosomal recessive disorder was questioned by Spranger & Maroteaux (1983) but supported by Kousseff & Nichols (1984).

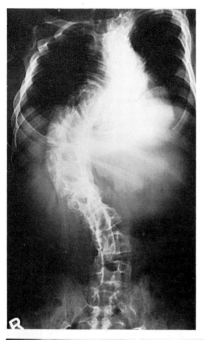

Fig. 2.18 (left) SEMDJL: irregular vertebral bodies and severe spinal malalignment. (From Beighton et al 1984 Clinical Genetics 26: 308–317.)

Fig. 2.19 (below) SEMDJL: the brother of the boy depicted in Figure 2.17. The acetabulae are flat and the iliac wings are wide. The femoral capital epiphyses are hypoplastic and coxa valga is present. (From Kozlowski et al 1984 Röfo 142: 337–341.)

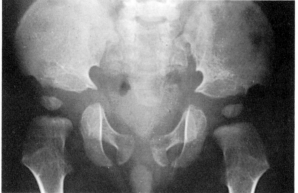

## Spondyloepimetaphyseal dysplasia with joint laxity (SEMDJL)

Beighton & Kozlowski (1980) delineated SEMDJL following appraisal of seven South African children in whom skeletal abnormalities were associated with dwarfism and gross joint laxity. Dislocation of the hips and club feet are often present at birth and progressive spinal malalignment leads to spinal cord compression and cardiorespiratory embarrassment. A high proportion of affected children die during the first decade and only two are known to have survived to adulthood. The ligamentous laxity makes surgical management difficult and the results of operations for spinal stabilisation are usually indifferent.

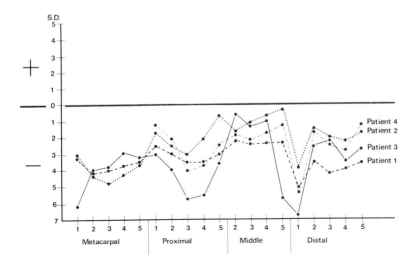

**Fig. 2.20** SEMDJL: pattern profile analysis showing that 4 affected persons all have similar shortening of the tubular bones of their hands. (From Beighton & Kozlowski 1980 Skeletal Radiology 5: 205–212.)

The phenotypic features of this autosomal recessive disorder were reviewed in detail by Beighton et al (1983, 1984) and the radiographic appearances were depicted by Kozlowski & Beighton (1984). More than 20 cases have now been recognised in the Afrikaans speaking community of South Africa. Many of the affected families have German progenitors and it is likely that the gene originated from this source.

## SCHWARTZ SYNDROME

The Schwartz syndrome is a unique disorder in which myotonia coexists with dysplasia of the skeleton. The short eponym is preferable to the long descriptive designation 'spondyloepimetaphyseal dysplasia with myotonia', although the term 'myotonic chondrodysplasia' represents a reasonable compromise. The first recognisable cases were a pair of sisters investigated by Catel (1951) and at the present time about 30 affected persons, mainly children, have been reported.

### Clinical and radiographic features

The main features are short stature, stiff joints, spinal malalignment and pectus carinatum. Blepharophimosis is present and the face is immobile and 'mask-like'. Myotonia progresses until a plateau is reached in mid childhood. Operative treatment for dislocation of the hips and talipes equinovarus has been successful. However, the anaesthetist may encounter problems on endotracheal intubation from the small size of the mouth,

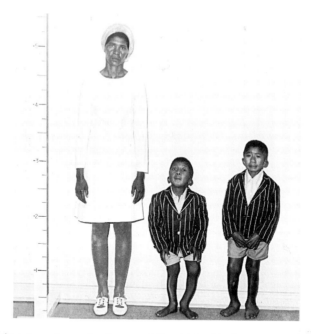

**Fig. 2.21** Schwartz syndrome: brothers, aged 7 and 10, with short stature and limb deformity. (From Beighton & Kozlowski 1980 Skeletal Radiology 5: 205–212.)

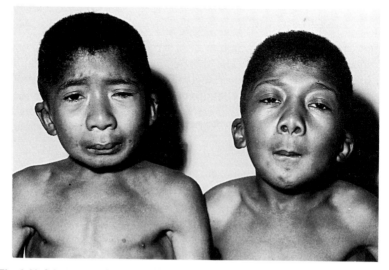

**Fig. 2.22** Schwartz syndrome: the face is immobile and mask-like. (From Horan F T, Beighton P 1975 Journal of Bone and Joint Surgery 57: 544.)

**SCHWARTZ SYNDROME**

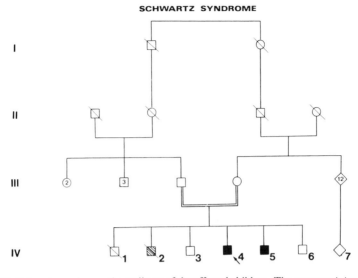

**Fig. 2.23** Schwartz syndrome: the pedigree of the affected children. The consanguinity of their parents is evidence in favour of autosomal recessive inheritance. Key: □ normal male; ○ normal female; ■ affected male; ● affected female; \ deceased. (From Beighton P 1973 Clinical Genetics 4: 548.)

rigidity of the temporomandibular joint and shortness of the neck (Horan & Beighton 1975).

Radiographically the skeleton is undermineralised. The vertebrae are flattened and anterior wedging may develop. The epiphyses and metaphyses, particularly of the large joints, are dysplastic. The hips may be dislocated, with varus or valgus deformity of the femoral necks.

*Genetics*

Schwartz & Jampel (1962) described an affected brother and sister and other cases have been reported by Huttenlocher et al (1969), Mereu et al (1969), Kozlowski & Wise (1974), Greze et al (1975), Cadilhac et al (1975) and Scaff et al (1979). All the parents were normal but consanguinity was present in the kindreds reported by Saadat et al (1972) from the Middle East and by Beighton (1973) from South Africa. There is little doubt that the Schwartz syndrome is inherited as an autosomal recessive.

Simpson & Degnan (1975) described a boy with features resembling the Schwartz syndrome, but lacking myotonia and with blepharophimosis of mild degree. The authors suggested that this child probably had the same condition as the infants reported by Marden & Walker (1966) and Fitch et al (1971) and that this disorder was distinct from the true Schwartz syndrome. Temtamy et al (1975), using the designation 'Marden-Walker syndrome' reported two affected cousins, both of whom were the offspring of consanguineous marriages and suggested that inheritance was autosomal recessive.

A further atypical form of the Schwartz syndrome is represented by four Brazilian siblings studied by Richieri-Costa et al (1984). The patients had mental retardation but lacked involvement of the face and joints. Inheritance is clearly autosomal recessive but the syndromic status is uncertain.

## METATROPIC DYSPLASIA

Metatropic dysplasia was described by Maroteaux et al (1966). The term 'metatropic' pertains to the reversal of bodily proportions which occurs during early childhood and it is derived from the Greek 'metatropos' meaning 'changing pattern'. The manifestations have been reviewed by Larose & Gay (1969), Jenkins et al (1970) Bailey (1971), Gefferth (1973) and Rimoin et al (1976).

### Clinical and radiographic features

Affected neonates have a relatively long trunk, a narrow cylindrical thorax and short limbs. The face is normal but the palate may be cleft while the joints are knobbly and stiff. A coccygeal cutaneous fold, or 'tail', which is sometimes present, represents a valuable diagnostic feature. In early childhood growth of the spine is retarded and kyphoscoliosis becomes pronounced when the child begins to walk. In this way the trunk becomes short in comparison with the limbs. Spinal deformity may be very marked and patients are severely dwarfed and crippled. Atlanto-axial instability

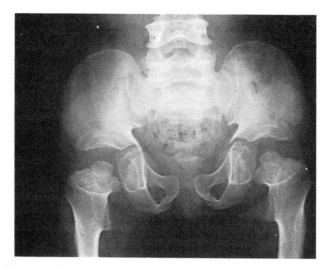

**Fig. 2.24** Metatropic dysplasia: the configuration of the proximal region of the femur resembles a mediaeval battle-axe.

poses a severe threat to the spinal cord in the cervical region. Death in infancy from respiratory distress has been recorded (Perri 1978, Belik et al 1985).

Gross widening of the intervertebral disc spaces is apparent at birth, in conjunction with marked platyspondyly. The metaphyses are very broad. The iliac crests are crescentic and the acetabular roofs are horizontal and irregular. In later childhood the vertebrae become wedged and the proximal end of the femur takes on a configuration which has been likened to a 'halberd' or 'battle axe'.

### Genetics

The majority of case reports have concerned sporadic individuals of either sex, although affected brothers with normal parents have been encountered by Michail et al (1956) and Crowle et al (1970). Kozlowski et al (1976) evaluated published reports, added six of their own cases and concluded that there were 'variants' of metatropic dysplasia which were probably indicative of heterogeneity. Beck et al (1983) postulated that there were three forms of metatropic dysplasia, autosomal recessive non-lethal, autosomal dominant non-lethal and an autosomal recessive variety which is lethal in the perinatal period.

## KNIEST DYSPLASIA

This rare disorder was delineated by Kniest (1952) and further cases were reported by Maroteaux & Spranger (1973) and Siggers et al (1974). McKusick (1975) proposed the designation 'metatropic dwarfism type II' but the title 'Kniest dysplasia' has been preferred in the Paris Nomenclature.

### Clinical and radiographic features

The face is round and flattened and the eyes are prominent. Inguinal hernia, cleft palate and club feet are inconsistent components of the syndrome. In early childhood short stature, kyphoscoliosis and stiff joints are evident. The thorax is broad and the trunk is short. Deafness, myopia and retinal detachment are important complications.

The skeleton is generally osteoporotic. In infancy the vertebral bodies are flattened with coronal clefts and anterior wedging. The epiphyses and metaphyses of the tubular bones are bulky and flared. The femoral necks are broad and short and the femoral capital epiphyses remain unossified throughout childhood (Lachman et al 1975).

Specific histological and ultrastructural changes have been identified in the collagen of the cartilage (Rimoin et al 1973). Inclusions in the endoplasmic reticulum of the chondrocytes are a notable feature and the proteoglycans of cartilage are abnormal (Rimoin et al 1976). Excess urinary excretion of keratan sulphate has been demonstrated in an affected mother and daughter (Brill et al 1975) but the pathogenetic significance of this observation is uncertain.

*Genetics*

In a review of the Kniest syndrome Siggers et al (1974) pointed out that apart from a pair of male identical twins, all reported patients have been sporadic. Subsequently Kim et al (1975) and Gnamey et al (1976) described affected mothers and daughters. So far there have been no reports of male to male transmission and the autosomal status of the abnormal gene has not yet been confirmed.

Three sporadic infants with a lethal condition which resembles Kniest dysplasia were described by Langer et al (1976). These authors identified an earlier report of affected siblings and suggested that inheritance was autosomal recessive.

## DIASTROPHIC DYSPLASIA

Diastrophic dysplasia was delineated by Lamy & Maroteaux (1960). The term 'diastrophic' adapted from the Greek word meaning 'twisted' aptly fits

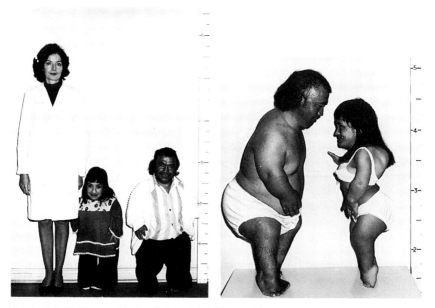

**Fig. 2.25** (left) Diastrophic dysplasia: adult siblings with severe limb shortening.

**Fig. 2.26** (right) Diastrophic dysplasia: the sister has a marked dorsal kyphoscoliosis although her brother's spine is virtually normal. Both have gross rigid talipes equinovarus.

the disorder. Diastrophic dysplasia is not uncommon and over 200 cases have been reported from the USA including a series of 51 described by Walker et al (1972) and more than 80 from Finland (Kaitila 1980).

*Clinical and radiographic features*

Shortness of stature and micromelia are obvious at birth. Severe talipes equinovarus is a universal finding, while cleft palate is present in about 50% of patients. The hands are broad with rigidity of the interphalangeal joints. The first metacarpal is short and the thumb is subluxed into the 'hitch-hiker' position. Curious episodes of spontaneous swelling and inflammation of the pinna of the ear during early childhood lead to a 'cauliflower' appearance. Although the face is normal, affected individuals bear a close resemblance to each other. Life expectancy is reasonably good but dwarfing is extreme and deformity is severe. Kyphoscoliosis may predispose to cardiopulmonary complications and spinal cord compression. The orthopaedic management of diastrophic dysplasia has been discussed by Hollister & Lachman (1976) and Horton et al (1978).

The spine is relatively normal at birth but lumbar lordosis and kyphoscoliosis develop during childhood. The femoral heads are flattened and the hip joints are usually dislocated. The long bones are reduced in length, the epiphyses are irregular and the metaphyses are flared. Mineralisation of the skeleton may be delayed. The radiographic criteria for diagnosis in the newborn have been reviewed by Saule (1975).

*Genetics*

A considerable body of evidence indicates that diastrophic dysplasia is inherited as an autosomal recessive. There have been several descriptions of multiple affected sibs, all with normal parents (Jackson 1951, Lamy & Maroteaux 1960, Paul et al 1965). Parental consanguinity was mentioned by Taybi (1963). The only reports of affected females who have reproduced concern two women who gave birth to normal children (Walker et al 1972).

Rimoin (1975) reviewed the features of a 'diastrophic variant' or 'pseudo-diastrophic' dysplasia, an uncommon entity in which clinical and radiographic stigmata resemble those of mild diastrophic dysplasia. From their study of 20 patients, including three pairs of sibs, Horton et al (1976) concluded that inheritance was probably autosomal recessive. As the histological appearances of cartilage are identical to those of diastrophic dysplasia, these authors suggested that the genes which determine these conditions might be allelic. Controversy concerning separate existence of two forms of diastrophic dysplasia was resolved following the recognition of mild and severely affected persons in the same kindred and this 'variant' was formally relegated to oblivion by a group of interested experts (Lachman et al 1981)

The issue of heterogeneity arose again when Gustavson et al (1985) reported six instances of early death from respiratory and cardiac insufficiency. Four of these infants had cardiac defects and three were mentally retarded. Their radiographic features differed from classical diastrophic dysplasia and the authors suggested that this lethal condition might be a distinct entity. Two siblings in this series of six cases had trisomy 18 mosaicism and the authors pointed out that this observation provided evidence for the localisation of the faulty gene to that chromosome.

In view of the potential severity of manifestations, the question of antenatal diagnosis arises. This has been accomplished in early pregnancy by ultrasonic techniques (O'Brien et al 1980). Kaitila et al (1980) successfully monitored four 'at risk' pregnancies, in one of which a positive diagnosis was confirmed at 19 weeks gestation.

## DYGGVE-MELCHIOR-CLAUSEN DYSPLASIA (DMC)

Dyggve et al (1962) reported three mentally retarded dwarfed sibs from Greenland. The clinical features resembled those of the Morquio syndrome and it was suggested that the condition might be mucopolysaccharidosis. However, initial reports of mucopolysacchariduria could not be confirmed on repeated urinary testing. Later, radioactive sulphate uptake by fibroblasts was shown to be normal and Spranger et al (1975) concluded that the syndrome was not a mucopolysaccharidosis. Abnormal serum alpha-2 macroglobulins were identified by Rostagi et al (1980) but the significance of this observation is uncertain.

### Clinical and radiographic features

Major characteristics are short-limbed dwarfism with a short neck, barrel chest, lumbar lordosis, genu valgum and a crouching stance. The majority of patients have been mentally defective. Important radiographic features are platyspondyly, in association with metaphyseal and epiphyseal changes in the long bones. The acetabulae are dysplastic and the hips may be dislocated. During childhood the margin of the iliac crests have a pathognomonic lace-like configuration. This appearance does not persist into adult life.

### Genetics

The patients mentioned in the initial report were the product of an uncle-niece relationship (Dyggve et al 1962) and affected sibs with consanguineous parents have also been observed by Spranger et al (1975).

In South African institutions for the mentally defective, the author has encountered two brothers and a sister with the condition. Their parents, who were of Lebanese extraction, were first cousins. At least three other

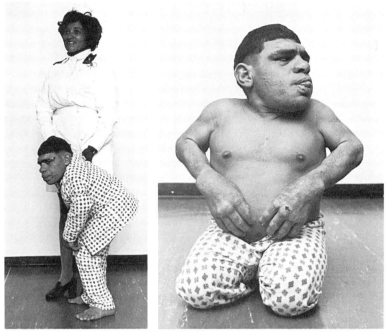

**Fig. 2.27** (left) DMC syndrome: an adult male with short stature and mental deficiency. His brother and sister also had the condition. The unaffected parents were consanguineous.

**Fig. 2.28** (right) DMC syndrome: the neck is short, and the chest is barrel-shaped.

descriptions have concerned Lebanese kindreds and the gene evidently reaches a relatively high frequency in this population (Naffah 1976, Bonafede & Beighton 1978). At the present time about 30 case reports can be found in the literature. Sex distribution is approximately equal and there has been no instance of generation to generation transmission. There is little doubt that inheritance is autosomal recessive.

Three affected sibs of Japanese stock, including a pair of non-identical twins, were reported by Smith & McCort (1959). These children were mentally normal, but otherwise they had the typical features of the syndrome. Further instances of normal intelligence were reported by Spranger et al (1976) and they concluded that this condition was a distinct autosomal recessive entity, for which they proposed the designation 'Smith-McCort' dwarfism.

## PARASTREMMATIC DYSPLASIA

The designation of this disorder is derived from the Greek 'parastremma' meaning 'distorted limb' (Langer et al 1970). In this rare entity, dwarfism and kyphoscoliosis are associated with severe malformations of the extremities.

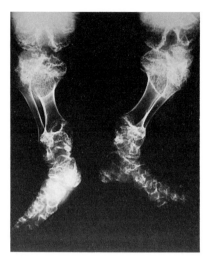

**Fig. 2.29** (left) Parastremmatic dysplasia a 5-year-old girl with dwarfism and severe limb deformity.

**Fig. 2.30** (above) Parastremmatic dysplasia: radiographically, the skeleton has a pathognomonic 'flocky' appearance. (From Horan F T, Beighton P 1976 Journal of Bone and Joint Surgery 58: 343.)

### Clinical and radiographic features

The forehead is high, with brachycephaly and a temporal bulge. Scoliosis appears in early infancy and becomes increasingly severe. The extremities are short with bilateral genu valgum, bowing of the shins, osseous enlargement of the knees and contractures of the hip joints.

The skeleton is grossly undermineralised. The endochondral bones are lucent, widened and coarsely trabeculated and contain areas of irregular stippling. The bone has a pathognomonic 'flocky' appearance. The vertebral bodies are flattened and irregular and the pelvic bones are very dysplastic. The metaphyses and epiphyses of the tubular bones are grossly deformed. The femoral necks are short and the femoral heads are distorted.

### Genetics

Seven cases have been reported, including three unrelated females (Langer et al 1970), a father and daughter (Rask 1963) and a girl of mixed ancestry in Cape Town (Horan & Beighton 1976). Neither affected sibs nor parental consanguinity have been recorded. It is generally assumed that inheritance is autosomal dominant and that sporadic individuals represent new mutations.

## OPSISMODYSPLASIA

Opsismodysplasia was delineated by Maroteaux et al (1984) when they reported three children with rhizomelia, short stature, brachydactyly and depression of the nasal bridge. Radiographically, skeletal maturation was very retarded and the bones of the hands and feet were extremely short, with concave metaphyses. The vertebral bodies were flat, although the pedicles were normal. The authors proposed the name 'opsismodysplasia' from the Greek word for 'delay in maturity'.

In older children, dwarfism is severe but mentality is normal. Pulmonary problems are sometimes a feature; Maroteaux et al (1984) mentioned an affected sibling of one of their patients who died in infancy from respiratory distress. These authors also alluded to a previous report of a recognisable case (Zonana et al 1977), noted parental consanguinity in one of their own families and concluded that inheritance was probably autosomal recessive.

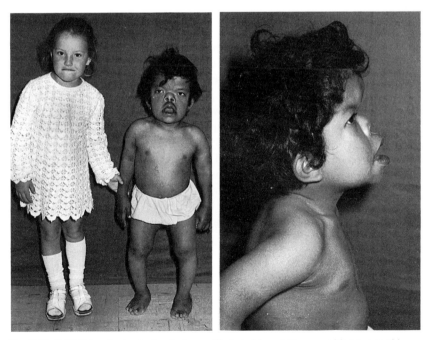

**Fig. 2.31** (left) Osteoglophonic dysplasia: an affected girl aged 12 years with a 4-year-old child. (From Beighton et al 1980 Paediatric Radiology 10: 46–50.)

**Fig. 2.32** (right) Osteoglophonic dysplasia: prominence of the forehead and jaws together with mid-facial hypoplasia produces a characteristic facies.

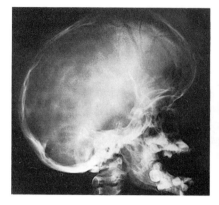

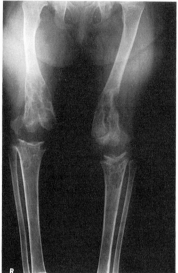

**Fig. 2.33** (above) Osteoglophonic dysplasia. Skull radiograph shows marked acrocephaly.

**Fig. 2.34** (right) Osteoglophonic dysplasia: areas of lucency in the metaphyses.

## OSTEOGLOPHONIC DYSPLASIA

Osteoglophonic dysplasia is a rare disorder in which dwarfism and severe abnormalities of the craniofacial structures are associated with a bizarre skeletal dysplasia. Intelligence is normal and there are no systemic ramifications. The main radiographic features are craniostenosis, metaphyseal lucencies and platyspondyly. On the basis of the changes in the metaphyses, Professor Jurgen Spranger proposed the designation 'osteoglophonic' to denote the unusual 'hollowed out' appearances.

Sporadic cases were reported by Sir Thomas Fairbank (1959) and Keats et al (1975) and a description of the clinical and radiographic features of an affected South African girl aged 12 years was given by Beighton et al (1980). In addition, an infant with the classical stigmata of osteoglophonic dysplasia has recently been investigated in Coimbra, Portugal. An affected father and son were documented by Kelley et al (1983) and on this basis it is possible that inheritance is autosomal dominant.

REFERENCES

**Pseudoachondroplasia**
Cranley R E, Williams B R, Kopits S E, Dorst J P 1975 Pseudoachondroplastic dysplasia: five cases representing clinical, roentgenographic and histologic heterogeneity. Birth Defects: Original Article Series 11/6: 205
Dennis N R, Renton P 1975 The severe recessive form of pseudoachondroplastic dysplasia. Pediatric Radiology 3/3: 169
Fontaine G, Gourguechon A, Smith M 1979 A dominant form of pseudoachondroplastic dysplasia: A familial case. Presse Médicale 8: 3962–3963
Hall J G 1975 Pseudoachondroplasia. Birth Defects 11(6): 187–202

Hall J G, Dorst J P 1969 Four types of pseudoachondroplastic spondyloepiphyseal dysplasia (SED). Birth Defects: Original Article Series 5/4: 242

Heselson N G, Cremin B J, Beighton P 1977 Pseudoachondroplasia: A report of 13 cases. British Journal of Radiology 50: 473–482

Horton W A, Hall J G, Scott C I, Pyeritz R E, Rimoin D L 1982 Growth curves for height for diastrophic dysplasia, spondyloepiphyseal dysplasia and pseudoachondroplasia. American Journal of Diseases of Children 136: 316–319

Kopits S E 1976 Orthopedic complications of dwarfism. Clinical Orthopaedics 114: 153–179

Kopits S E, Lindstrom J A, McKusick V A 1974 Pseudoachondroplastic dysplasia: pathodynamics and management. Birth Defects: Original Article Series 10/12: 341

Lachman R S, Rimoin D L, Hall J G 1975 Difficulties in the classification of the epiphyseal dysplasias. Birth Defects: Original Article Series 11/6: 231

McKusick V A 1975 Pseudoachondroplastic dysplasia I (formerly pseudoachondroplastic spondyloepiphyseal dysplasia). Mendelian inheritance in man, 5th edn. Johns Hopkins University Press, Baltimore, p 279

Maroteaux P, Lamy M 1959 Les formes pseudoachondroplastiques des dysplasies spondyloepiphysaires. Presse Médicale 67: 383

Maynard J A, Cooper R R, Ponseti I V 1972 A unique rough surfaced endoplasmic reticulum inclusion in pseudoachondroplasia. Laboratory Investigations 26: 40

Pedrini-Mille A, Maynard J A, Pedrini V A 1984 Pseudoachondroplasia: Biochemical and histochemical studies of cartilage-2. Journal of Bone and Joint Surgery 66: 1408–1414

Stanescu V, Maroteaux P, Stanescu R 1982 The biochemical defect of pseudoachondroplasia. European Journal of Pediatrics 138: 221–225

Young I D, Moore J R 1984 Severe pseudoachondroplasia with parental consanguinity. Journal of Medical Genetics 22: 150–153

**Spondyloepiphyseal dysplasia**

*Preamble*

Maroteaux P 1969 Spondyloepiphyseal dysplasias and metatropic dwarfism. Birth Defects: Original Article Series 5/4: 35

Rubin P 1964 In: Dynamic classification of Bone Dysplasias. Year Book Publishers, Chicago

*SEDC*

Anderson C E, Sillence D O, Lachman R S, Toomey K, Bull M, Dorst J, Rimoin D L 1982 Spondylo-metepiphyseal dysplasia, Strudwick type. American Journal of Medical Genetics 13: 243–256

Harrod M J, Friedman J M, Currarino G, Pauli R M, Langer L O 1984 Genetic heterogeneity in spondyloepiphyseal dysplasia congenita. American Journal of Medical Genetics 18: 311–320

Kozlowski K, Masel J, Nolte K 1977 Dysplasia spondylo-epiphysealis congenita Spranger-Wiedemann. A critical analysis. Australasian Radiology 21: 260–280

Murray T G, Green W R, Maumenee I H, Kopits S E 1985 Spondyloepiphyseal dysplasia congenita: Light and electron microscopic studies of the eye. Archives of Ophthalmology 103: 407–411

Spranger J 1975 Spondyloepiphyseal dysplasias. Birth Defects 11(6): 177–182

Spranger J W, Maroteaux P 1982 Editorial comment: Genetic heterogeneity of spondyloepiphyseal dysplasia congenita? American Journal of Medical Genetics 13: 241–242

Spranger J, Wiedemann H R 1966 Dysplasia spondyloepiphysaria congenita. Helvetica Paediatrica Acta 21: 598

Spranger J, Langer L O 1970 Spondyloepiphyseal dysplasia congenita. Radiology 94: 313

Williams B R, Cranley R E 1974 Morphologic observations on four cases of SED congenita. Birth Defects 10(4): 75–87

*SEDT*

Bannerman R M, Ingali A B, Mohn J F 1971 X-linked spondyloepiphyseal dysplasia tarda: clinical and linkage data. Journal of Medical Genetics 8: 291

Jacobsen A W 1939 Hereditary osteochondro-dystrophia deformans. A family with 20 members affected in 5 generations. Journal of the American Medical Association 113: 121–124

Jones K L, Jones K L, Miller K 1986 A new skeletal dysplasia syndrome with dwarfism, craniofacial anomalies and unique radiographic findings. American Journal of Medical Genetics 23: 751–757

Klenerman L 1961 An adult case of chondro-osteodystrophy. Proceedings of the Royal Society of Medicine 54: 71

Langer L O Jr 1964 Spondyloepiphyseal dysplasia tarda. Hereditary chondrodysplasia with characteristic vertebral configuration in the adult. Radiology 82: 833–839

Maroteaux P, Lamy M, Bernard J 1957 La dysplasie spondylo-epiphysaire tardive. Presse Médicale 65: 1205

Martin J R, Macewan D W, Blais J A, Metrakos J, Gold P, Langer F, Hill R O 1970 Platyspondyly, polyarticular osteoarthritis, and absent beta-2-globulin in two brothers. Arthritis and Rheumatism 13: 53

Moldauer M, Hanelin J, Bauer W 1962 Familial precocious degenerative arthritis and the natural history of osteochondrodystrophy. In: Blumenthal H T (ed) Medical and clinical aspects of aging. Columbia University Press, New York, p 226

O'Brien J S, Gugler E, Giedion A, Weissman U, Herschkowtiz N, Meier C, Leroy J 1976 Spondyloepiphyseal dysplasia, corneal clouding, normal intelligence and acid beta-galactosidase deficiency. Clinical Genetics 9: 495

Tsuruta T, Ogihara Y 1984 Malignant tumours arising in patients with congenital bone diseases. Investigations by a questionnaire. Japanese Journal of Human Genetics 29: 31–37

*SEDTPA*

Al-Awadi S A, Farag T I, Naguib K, El-Khalifa M Y, Cuschieri A, Hosny G; Zahran M, Al-Ansari A G 1984 Spondyloepiphyseal dysplasia tarda with progressive arthropathy. Journal of Medical Genetics 21: 193–196

Spranger J, Albert C, Schilling F 1980 A progressive connective tissue disease with features of juvenile rheumatoid arthritis and osteochondrodysplasia. European Journal of Pediatrics 133: 187

Spranger J, Albert C, Schilling F, Bartsocas C, Stoss H 1983 Progressive pseudorheumatoid arthritis of childhood (PPAC). European Journal of Pediatrics 140: 34–40

Wynne-Davies R, Hall C, Ansell B M 1982 Spondylo-epiphyseal dysplasia tarda with progressive arthropathy. A new disorder of autosomal recessive inheritance. Journal of Bone and Joint Surgery (British) 64: 442–445

*Mseleni joint disease*

Du Toit G T 1979 Hip disease of Mseleni. Clinical Orthopaedics 141: 223–236

Fellingham S A, Elphinstone C D, Wittmann W 1973 Mseleni joint disease: Background and prevalence. South African Medical Journal 47: 2173–2180

Lockitch G, Fellingham S A, Elphinstone C D 1973 Mseleni joint disease: A radiological study of two affected families. South African Medical Journal 47: 2366–2376

Nesterov A I 1964 The clinical course of Kashin-Beck disease. Arthritis and Rheumatism 7: 29–40

Wittman W, Fellingham S F 1970 Unusual hip disease in remote part of Zululand. Lancet i: 842–843

Yach D, Botha J L 1985 Mseleni joint disease in 1981. International Journal of Epidemiology 14: 276–284

**Spondylometaphyseal dysplasia**

Fontaine G, Maroteaux P, Farriaux J P, Bosquet M 1975 La dysplasie spondylaire pure ou brachyolmie. Archives Francaises de Pédiatrie 32: 695–708

Gustavson K H et al 1978 Spondylometaphyseal dysplasia in two sibs of normal parents. Pediatric Radiology 7: 90–96

Kozlowski K 1973 Spondylometaphyseal dysplasia. Progress in Pediatric Radiology 4: 299

Kozlowski K 1976 Metaphyseal and spondylometaphyseal chondrodysplasias. Clinical Orthopaedics and Related Research 114: 83

Kozlowski K, Cremin B, Beighton P 1980 Variability of spondylo-metaphyseal dysplasia, common type. Radiological Diagnosis 21: 682–686

Kozlowski K, Maroteaux P, Spranger J 1967 La dysostose spondylometaphysaire. Presse Médicale 75: 2769

Kozlowski K, Prokop B E, Scougall J S, Silink M, Vines R H 1979 Spondylometaphyseal dysplasia: Report of a case of common type and three cases of 'new varieties'. Fortschritte auf dem Gebiete der Roentgenstrahlen 130: 220–230

Kozlowski K, Barylak A, Middleton R W, Rybakowa M, Thomas P, Walecki J 1976 Spondylo-metaphyseal dysplasias: Report of a case of common type and three pairs of siblings of new varieties. Australasian Radiology 20: 154–166

Riggs W, Summitt R L 1971 Spondylometaphyseal dysplasia (Kozlowski). Radiology 101: 375–281

Thomas P S, Nevin N C 1977 Spondylometaphyseal dysplasia. American Journal of Roentgenology 128: 89

**Spondyloepimetaphyseal dysplasia**

*Spondyloepiphysometaphyseal dysplasia, Irapa type (SEMDIT)*

Arias S, Mota M, Pinto-Cisternas J 1976 L'ostéochondrodysplasie spondylo-metaphysaire type Irapa: Nouveau nanisme avec rachis et metatarsiens courts. Nouvelle Presse Médicale 5: 319–323

Hernandez A, Ramirez M L, Nazara Z, Ocampo R, Ibarra B, Cantu J M 1980 Autosomal recessive spondylo-epi-metaphyseal dysplasia (Irapa type) in a Mexican family: Delineation of the syndrome. American Journal of Medical Genetics 5: 179–188


*Spondylometaepiphyseal dysplasia congenita, type Strudwick*

Anderson C E, Sillence D O, Lachman R S, Toomey K, Bull M, Dorst J, Rimoin D 1982 Spondylometepiphyseal dysplasia, Strudwick type. American Journal of Medical Genetics 13: 243–256

Kouseff B G, Nichols P 1984 Letter to editor. American Journal of Medical Genetics 17: 547–550

Murdoch J L, Walker B A 1969 A 'new' form of spondylometaphyseal dysplasia. Birth Defects Original Article Series V(4): 368–370

Spranger J W, Maroteaux P 1983 Editorial comment: Genetic heterogeneity of spondylometepiphyseal dysplasia congenita? American Journal of Medical Genetics 14: 601–602


*Spondyloepimetaphyseal dysplasia with joint laxity (SEMDJL)*

Beighton P, Kozlowski K 1980 Spondylo-epi-metaphyseal dysplasia with joint laxity and severe, progressive kyphoscoliosis. Skeletal Radiology 5: 205–212

Beighton P, Gericke G, Kozlowski K, Grobler L 1984 The manifestations and natural history of spondylo-epi-metaphyseal dysplasia with joint laxity. Clinical Genetics 26: 308–317

Beighton P, Kozlowski K, Gericke G, Wallis G, Grobler L 1983 Spondylo-epi-metaphyseal dysplasia with joint laxity and severe, progressive kyphoscoliosis. South African Medical Journal 64: 772–776

Kozlowski K, Beighton P 1984 Radiographic features of spondylo-epi-metaphyseal dysplasia with joint laxity and progressive kyphoscoliosis. Review of 19 cases. ROFO 141: 337–341

**Schwartz syndrome**

Beighton P 1973 The Schwartz syndrome in Southern Africa. Clinical Genetics 4: 548

Cadilhac J, Baldet P, Greze J, Duday H 1975 E.M.F. studies of two familial cases of the Schwartz and Jampel syndrome (osteo-chondro-muscular dystrophy with myotonia). Electromyography and Clinical Neurophysiology 15/1: 5

Catel W 1951 Diffentialdiagnosticsche syptomatologie von krankheiten des kindesalters. Klinische Vorlesungen. Thieme, Stuttgart, p 48

Fitch N, Karpati G, Pinsky L 1971 Congenital blepharophimosis, joint contractures and muscular hypotonia. Neurology 21: 1214

Greze J, Baldet P, Dumas R 1975 Schwartz Jampel's osteo-chondro-muscular dystrophy. Two familial cases. Archives Francaises de Pédiatrie 32/1: 59

Horan F, Beighton P 1975 Orthopaedic aspects of the Schwartz syndrome. Journal of Bone and Joint Surgery 57A/4: 542

Huttenlocher P R, Landwrith J, Hanson V, Gallagher B B, Bench K 1969 Osteo-chondro-muscular dystrophy. A disorder manifested by multiple skeletal deformities, and dystrophic changes in muscle. Pediatrics 44: 945

Kozlowski K, Wise G 1974 Spondylo-epi-metaphyseal dysplasia with myotonia. A radiographic study. (Catel-Jampel syndrome, Schwartz-Jampel syndrome, Aberfeld syndrome, chondrodystrophic myotonia). Radiologia Diagnostica 6: 817

Kuriyama M, Shinmyozu K, Osame M, Kawahira M, Igata A 1985 Schwartz-Jampel syndrome associated with von Willebrand's disease. Journal of Neurology, Neurosurgery and Psychiatry 232: 49–51

Marden P M, Walker W A 1966 A new generalised connective tissue syndrome. American Journal of Diseases of Children 112: 225

Mereu T R, Porter I H, Hug G 1969 Myotonia, shortness of stature and hips dysplasia: Schwartz-Jampel syndrome. American Journal of Diseases of Children 117: 470

Richieri-Costa A, da Silva S M, Frota-Pessoa O 1984 Late infantile autosomal recessive myotonia, mental retardation and skeletal abnormalities: A new autosomal recessive syndrome. Journal of Medical Genetics 21: 103–107

Saadat M, Mokfi H, Vakil H, Ziai M 1972 Schwartz syndrome: myotonia with blepharophimosis and limitation of joints. Journal of Pediatrics 81: 348

Scaff M, Mendonca L I, Levy J A, Canelas H M 1979 Chondrodystropic myotonia: Electromyographic and cardiac features of a case. Acta Neurologica Scandinavica 60: 243–249

Schwartz O, Jampel R S 1962 Congenital blepharophimosis associated with a unique generalised myopathy. Archives of Ophthalmology 68: 52

Simpson J L, Degnan M 1975 A child with facial and skeletal dysmorphism reminiscent of Schwartz syndrome. Birth Defects: Original Article Series 11/2: 456

Temtamy S A, Shoukry A S, Raafat M, Mihareb S 1975 Probable Marden-Walker syndrome: evidence for autosomal recessive inheritance. Birth Defects: Original Article Series 1112: 104

**Metatropic dysplasia**

Bailey J A 1971 Forms of dwarfism recognisable at birth. Clinical Orthopaedics and Related Research, 76: 150

Beck M, Roubicek M, Rogers J G et al 1983 Heterogeneity of metatropic dysplasia. European Journal of Pediatrics 140: 231–237

Belik J, Anday E K, Kaplan F, Zackai E 1985 Respiratory complications of metatropic dwarfism. Dysmorphology 24: 504–511

Crowle P, Astley R, Insley J 1976 A form of metatropic dwarfism in two brothers. Pediatric Radiology 4/3: 172

Gefferth K 1973 Metatropic dwarfism. In: Kaufman H J Progress in Pediatric Radiology Vol. 4. Karger, Basel, p 137

Jenkins P, Smith M B, McKinnel J S 1970 Metatropic dwarfism. British Journal of Radiology 43: 561

Kozlowski K, Morris L, Reinwein H, Sprague P, Tamaela L A 1976 Metatropic dwarfism and its variants: Report of six cases. Australasian Radiology 20: 367–385

Larose J H, Gay B B 1969 Metatropic dwarfism. American Journal of Roentgenology 106: 156–161

Maroteaux P, Spranger J, Wiedemann H 1966 Metatropischer Zwergwuchs. Archiv fur Kinderheilkunde 173: 211

Michail J, Matsovkas J, Theodorou S, Houliaras K 1956 Maladie de Morquio (osteochondrodystrophie polyepiphysaire deformante) chez deux frères. Helvetica Paediatrica Acta 2: 403

Perri G 1978 A severe form of metatropic dwarfism. Pediatric Radiology 7: 183–185

Rimoin D L, Siggers D C, Lachman R S, Silberberg R 1976 Metatropic dwarfism, the Kniest syndrome and the pseudoachondroplastic dysplasias. Clinical Orthopaedics and Related Research 114: 70

**Kniest dysplasia**

Brill P W, Kim H J, Beratis N G, Hirschhorn K 1975 Skeletal abnormalities in the Kniest syndrome with mucopolysacchariduria. American Journal of Roentgenology, Radium Therapy and Nuclear Medicine 125/3: 731

Gnamey D, Farriaux J P, Fontaine G 1976 Kniest's disease. A familial case report. Archives Francaises de Pédiatrie 33/2: 143

Kim H J, Beratis N G, Brill P 1975 Kniest syndrome with dominant inheritance and mucopolysacchariduria. American Journal of Human Genetics 27/6: 755

Kniest W 1952 Zur Abgrenzung der Dysostosis enchondralis von der Chondrodystrophie. Zeitschrift fur Kinderheilkunde 70: 633

Lachman R S, Rimoin D L, Hollister D W, Siggers D C, McAlister W, Kaufman R L,
Langer L O 1975 The Kniest syndrome. American Journal of Roentgenology 123: 805
Langer L O, Gonzalez-Ramos M, Chen H, Espiritu C E, Courtney N W, Opitz J M 1976
A severe infantile micromelic chondrodysplasia which resembles Kniest disease. European
Journal of Pediatrics 123: 29–38
McKusick V A 1975 Metatropic dwarfism. Mendelian inheritance in man, 4th edn. Johns
Hopkins, Baltimore, p 496
Maroteaux P, Spranger J 1973 La maladie de Kniest. Archives Francaises de Pédiatrie
30: 735–750
Rimoin D L, Hollister D W, Silberberg R, Lachman R S, McAlister W, Kaufman R 1973
The Kniest (Swiss cheese cartilage) syndrome: Clinical, radiographic, histologic and
ultrastructural studies. Clinical Orthopaedics and Related Research 21: 296
Rimoin D L, Siggers D C, Lachman R S, Silberberg R 1976 Metatropic dwarfism, the
Kniest syndrome and the pseudoachondroplastic dysplasias. Clinical Orthopaedics and
Related Research 114: 70–82
Siggers D, Rimoin D, Dorst J, Doty S, Williams B, Hollister D, Silberberg R, Cranley R,
Kaufman R, McKusick V 1974 The Kniest syndrome. Birth Defects: Original Article
Series 10/9: 193

**Diastrophic dysplasia**

Gustavson K, Holmgren G, Jagell S, Jorulf H 1985 Lethal and non-lethal diastrophic
dysplasia. Clinical Genetics 28: 321–334
Hollister D W, Lachman R S 1976 Diastrophic dwarfism. Clinical Orthopaedics and
Related Research 114: 61
Horton W A, Rimoin D L, Lachman R S, Hollister D W, Dorst J P, Skovby F, Scott C I,
Hall J G 1976 The diastrophic variant. Fifth International Congress of Human Genetics,
Mexico, D.F.
Jackson W P U 1951 Irregular familial chondro-osseous defect. Journal of Bone and Joint
Surgery 33B: 420
Kaitila I 1980 Diastrophic dysplasia. In: Eriksson A W, Forsius H, Nevanlinna H R,
Workman P L, Norio R K (Eds) Population Structure and Genetic Disorders. Academic
Press, New York p 610–613
Kaitila I, Ammala P, Karjalainen O, Liukkonen S, Rapola J 1983 Early prenatal detection
of diastrophic dysplasia. Prenatal Diagnosis 3: 237–244
Lachman R, Sillence D, Rimoin D, Horton W, Hall J, Scott J, Spranger J, Langer R 1981
Diastrophic dysplasia: The death of a variant. Radiology 140: 79–86
Lamy M, Maroteaux P 1960 Le nanisme diastrophique. Presse Médicale 68: 1977–1980
O'Brien G D, Rodeck C, Queenan J R 1980 Early prenatal diagnosis of diastrophic
dwarfism by ultrasound. British Medical Journal 280: 1300
Paul S S, Rao P L, Mullick P, Saigal S 1965 Diastrophic dwarfism. A little known disease
entity. Clinical Pediatrics 4: 95
Rimoin D L 1975 The chondrodystrophies. Advances in Genetics 5: 1
Saule H 1975 Diastrophic dwarfism. Radiologe 15/2: 50
Taybi H 1963 Diastrophic dwarfism. Radiology 80: 1
Walker B A, Scott C I, Hall J G, Murdoch J L, McKusick V 1972 Diastrophic dwarfism.
Medicine 51: 41

**Dyggve-Melchior-Clausen dysplasia (DMC)**

Bonafede R P, Beighton P 1978 The Dyggve-Melchior-Clausen syndrome in adult siblings.
Clinical Genetics 14: 24–30
Dyggve H V, Melchior J C, Clausen J 1962 Morquio-Ulrich's disease. An inborn error of
metabolism? Archives of Disease in Childhood 37: 525
Naffah J 1976 The Dyggve-Melchior-Clausen syndrome. American Journal of Human
Genetics 28/6: 607
Rostagi S C, Clausen J, Melchior J C, Dyggve H V 1980 Abnormal serum alpha-2-
macroglobulin in Dyggve-Melchior-Clausen syndrome. Journal of Clinical Biochemistry
18: 67–68
Smith R, McCort J J 1959 Osteochondrodystrophy (Morquio-Brailsford type); occurrence in
three siblings. California Medicine 88: 53

Spranger J W, Bierbaum B, Herrmann J 1976 Heterogeneity of Dyggve-Melchior-Clausen dwarfism. Human Genetics 33: 279–287
Spranger J, Maroteaux P, Der Kaloustian V M 1975 The Dyggve-Melchior-Clausen syndrome. Radiology 114/2: 415

**Parastremmatic dysplasia**
Horan F, Beighton P 1976 Parastremmatic dwarfism. Journal of Bone and Joint Surgery 58B: 343
Langer L O, Petersen D, Spranger J 1970 An unusual bone dysplasia: parastremmatic dwarfism. American Journal of Roentgenology, Radium Therapy and Nuclear Medicine 110: 550
Rask M R 1963 Morquio-Brailsford osteochondrodystrophy and osteogenesis imperfecta: report of a patient with both conditions. Journal of Bone and Joint Surgery 45A: 561

**Opsismodysplasia**
Maroteaux P, Stanescu V, Stanescu R, Le Marec B, Moraine C, Lejarraga H 1984 Opsismodysplasia: A new type of chondrodysplasia with predominant involvement of the bones of the hand and the vertebrae. American Journal of Medical Genetics 19: 171–182
Zonana J, Rimoin D L, Lachman R S, Cohen A H 1977 A unique chondrodysplasia secondary to a defect in chondroosseous transformation. Liss, New York for the National Foundation March of Dimes AS XIII(30): 155–163

**Osteoglophonic dysplasia**
Beighton P, Cremin B J, Kozlowski K 1980 Osteoglophonic dwarfism. Pediatric Radiology 10: 46–50
Fairbank T 1959 An Atlas of General Affections of the Skeleton. Livingstone, Edinburgh, p 181
Keats T E, Smith T H, Sweet D E 1975 Craniofacial dysostosis with fibrous metaphyseal defects. American Journal of Roentgenology 124: 271
Kelley R I, Borns P F, Nichols D, Zackai E H 1983 Osteoglophonic dwarfism in two generations. Journal of Medical Genetics 20: 436–44

# 3

# Lethal short-limbed dwarfism in the newborn

The neonatal skeletal dysplasias can be broadly categorised according to lethality or compatibility with survival. In an alternative classification the disorders are divided into those which are recognisable at birth and those which only become evident later in childhood. The skeletal dysplasias which present in the neonate as potentially lethal short-limbed dwarfism form the subject of this chapter.

1. Achondrogenesis
2. Thanatophoric dysplasia
3. Homozygous achondroplasia
4. Asphyxiating thoracic dysplasia (Jeune syndrome)
5. Short-rib syndrome (with or without polydactyly)
   a. Type I (Saldino-Noonan type)
   b. Type II (Majewski type)
   c. Type III (Naumoff, lethal thoracic dysplasia)
6. Chondrodysplasia punctata, rhizomelic form
7. Campomelic dysplasia
8. Osteogenesis imperfecta type II
9. Hypophosphatasia lethalis
10. Fibrochondrogenesis
11. Hypochondrogenesis
12. Atelosteogenesis
13. Dyssegmental dysplasia
14. Boomerang dysplasia
15. De la Chapelle dysplasia
16. Schneckenbecken dysplasia
17. Miscellaneous, new, potentially lethal, neonatal dwarfing dysplasias

The newborn child with a skeletal dysplasia poses special problems to the paediatrician and the medical geneticist; numerous rare conditions present in this way and many are lethal. Management is greatly influenced by the prognosis, which in turn demands diagnostic accuracy. Equally, the determination of recurrence risks for further pregnancies is dependent upon the establishment of a firm diagnosis. This can almost always be achieved by

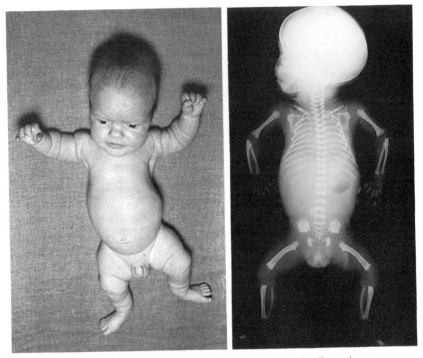

**Fig. 3.1** (left) A neonate with short-limbed dwarfism. Although the diagnosis can sometimes be made on a basis of the clinical features, radiographic studies are required for precision and confirmation. In this instance, the child has achondroplasia, which is usually non-lethal.

**Fig. 3.2** (right) Radiograph of a newborn short-limbed dwarf. The configuration of the pelvis and limb bones confirms the diagnosis of achondroplasia.

radiographic studies and a single whole body 'babygram' is usually sufficient. Excellent radiographs can be obtained of dead infants and it is mandatory that any stillborn child in whom a skeletal dysplasia is suspected should be investigated in this way.

Several of these disorders present as short-limbed dwarfism and in the past a spurious diagnosis of 'achondroplasia' was often made, thus making genetic management a difficult matter. The diversity of these conditions is now generally appreciated; this problem has been reviewed by Maroteaux, et al (1976), Sillence et al (1978). The radiographic manifestations of these disorders have been depicted in the monograph 'Bone Dysplasias of Infancy' (Cremin & Beighton 1978) and a systematic approach to radiodiagnosis is provided in the Gamut Index of Skeletal Dysplasias (Kozlowski & Beighton 1984).

Prenatal diagnosis has now been achieved in the majority of the dwarfing dysplasias. The application of radiographic techniques during pregnancy was discussed by Lachman & Hall (1979) and the role of ultrasound in antenatal diagnosis has been reviewed by Hobbins & Mahoney (1980), Filley & Golbus (1981, 1982) Hobbins et al (1982), Kurtz & Wapner (1983) and Muller & Cremin (1985). Normal values for limb length and other skeletal parameters at various stages of pregnancy, as determined by ultrasound, have been published as centiles by Elejalde & Elejalde (1986). These charts are of great value in the antenatal assessment of the fetal skeleton.

Histopathological investigations are an important adjunct to the diagnostic process but all too frequently these are omitted due to lack of local expertise or problems in obtaining appropriate specimens. The pathological features in the chondrodysplasias have been documented by several research teams, including Rimoin et al (1976), Stanescu et al (1977) and Sillence et al (1979).

A perspective of the relative frequency of the lethal neonatal chondrodysplasias was provided by Connor et al (1985) in a review of 43 cases presenting in Western Scotland in the period 1970 to 1983. A minimum incidence of 1 in 8900 newborns was calculated and 11 different conditions were recognised. In South America a minimum prevalence rate of 2.3 in 10 000 for liveborn and stillborn infants with dwarfing skeletal dysplasias was documented by Orioli et al (1986). Comparable figures had previously been obtained by Camera & Mastroiacovo (1982) from the Italian multicentric birth defects monitoring system.

## ACHONDROGENESIS

The stigmata of achondrogenesis have been well documented and recognition is not difficult. More than 100 cases have now been reported.

With the accumulation of experience, it has become evident that achondrogenesis is heterogeneous. Various subclassifications have been proposed, the division into types I and II being generally accepted. The eponym 'Parenti-Fraccaro' has been applied to the former variety and that of 'Langer-Saldino' to the latter. The main difference between these disorders is the greater severity of under-ossification of the axial skeleton in the Langer-Saldino type (Yang et al 1974). The distinctive clinical and radiographic manifestations of achondrogenesis type I and II have been discussed in detail by Chen et al (1981).

There have been problems with classification and the designation 'achondrogenesis type II' has been erroneously applied to Grebe chondrodysplasia, a condition which is quite distinct from the lethal forms of achondrogenesis (see Ch. 2). Similarly, the term 'achondrogenesis type III' was used for a lethal form of short-limbed dwarfism recognised by Verma et al (1975) in a consanguineous Sikh kindred in India, in which six of nine sibs had been stillborn. In addition to severe thoracic constriction, these infants had micromelia, postaxial polydactyly and genital anomalies. This condition is now categorised as the short-rib syndrome type III, or Naumoff lethal thoracic dysplasia.

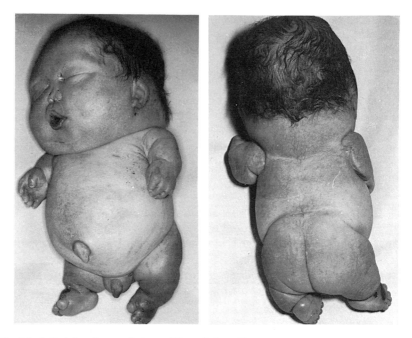

**Fig. 3.3** (left) Achondrogenesis in a stillborn infant. The limbs are very short.

**Fig. 3.4** (right) Achondrogenesis: the hydropic head is disproportionately large.

It is possible that there is further heterogeneity; following review of the radiographic features of 79 cases, Whitley & Gorlin (1983) suggested that there might be at least four distinct forms of achondrogenesis. This issue remains unresolved.

*Clinical, radiographic and histopathological features*

Infants with achondrogenesis are usually stillborn, although a few have survived for a limited period of time. The limbs are very short and the hydropic head and bulging abdomen are disproportionately large. The diagnosis can usually be suspected clinically and confirmed by recognition of the characteristic radiographic changes.

The bones of the skull are relatively normal, in marked contrast to the swollen peri-cranial soft tissues. The spine and pelvis are grossly underossified and the tubular bones are short, with expansion and spiky irregularity of their metaphyses.

If facilities are available categorisation is possible by demonstration of specific histological abnormalities. The ultrastructural features of achondrogenesis type I have been reviewed by several researchers, including Houston et al (1972), Yang et al (1976) and Ornoy et al (1976). The changes

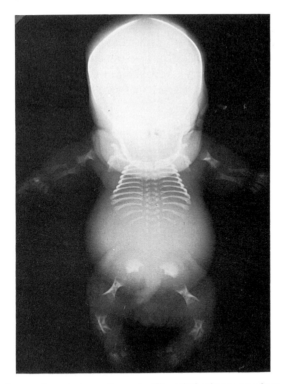

Fig. 3.5 Achondrogenesis: anterior radiograph. The tubular bones are short, with irregular metaphyses. The lumbar vertebrae are undermineralised. (From Cremin B J, Beighton P 1974 British Journal of Radiology 47: 77.)

in type II, which are unique, have been documented by Yang et al (1974) and Yang et al (1976).

### Genetics

Although precise categorisation of cases mentioned in earlier reports is not always possible, there is little doubt that both types of achondrogenesis are inherited as autosomal recessive traits. This mode of transmission was established when families with multiple affected sibs and normal parents were reported by Scott (1972), Houston et al (1972) and Wiedemann et al (1972).

Achondrogenesis has been recognised in utero at a late stage of pregnancy by means of standard radiographic techniques (Maroteaux et al 1976) and by amniography (Globus et al 1977). More recently, the disorder has been detected antenatally with ultrasound by Smith et al (1981), Graham et al (1983), Benacerraf et al (1984), Glenn & Teng (1985), Muller & Cremin (1985) and Connor et al (1985).

## THANATOPHORIC DYSPLASIA

Thanatophoric dysplasia is the most common of the potentially lethal conditions which present with disproportionate shortening of the limbs in the newborn. The designation, derived from the Greek 'thanatophoras', or 'death bearing', was used in the original case description by Maroteaux et al (1976).

### Clinical and radiographic features

A firm diagnosis cannot be made without radiological studies. Indeed, the short limbs, prominent forehead and depressed nasal bridge can easily be mistaken for the stigmata of achondroplasia, while the narrow thorax is reminiscent of asphyxiating thoracic dysplasia. Affected individuals are usually stillborn or die of respiratory insufficiency in the neonatal period. However, survival for 10 weeks, with minimal supportive care, has been reported (Moir & Kozlowski 1976).

The vertebral bodies have an H-shaped appearance in frontal radiographs due to flattening of their mid-portions and relative prominence of their pedicles. The broad pelvis, with horizontal acetabulae, closely resembles that of achondroplasia. The tubular bones are short and broad and the femora are bowed with a 'telephone receiver' configuration.

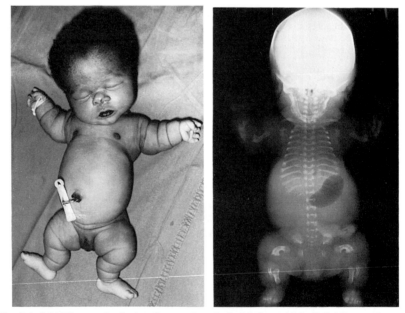

**Fig. 3.6** (left) Thanatophoric dwarfism: a short-limbed infant which died 6 hours after delivery. (From Cremin B J, Beighton P 1974 British Journal of Radiology (47: 77.)

**Fig. 3.7** (right) Thanatophoric dwarfism: the 'telephone receiver' configuration of the femur is pathognomonic. (From Cremin B J, Beighton P 1974 British Journal of Radiology 47: 77.)

The similarity of the pelvic and limb abnormalities in thanatophoric dysplasia and achondroplasia has prompted speculation that there might be some fundamental relationship between these conditions (Langer et al 1969). However, there are marked histological differences (Maroteaux et al 1976, Rimoin et al 1976). Ornoy et al (1985) have suggested that the basic pathogenetic mechanism is focal replacement of the growth plate and periosteum by persisting abnormal mesenchymal-like tissue from which the abnormal bone originates.

*Genetics*

The majority of reports have concerned sporadic cases. In an early review of the literature, Maroteaux et al (1967) postulated that new dominant mutation was the probable genetic basis of the disorder. Following appraisal of accumulated case reports, Pena & Goodman (1973) suggested that inheritance was polygenic, with a 2% recurrence risk. Bouvet et al (1974) reviewed the genetic background of thanatophoric dysplasia; no firm conclusions were reached but these authors suggested that the well-documented but anomalous observation of increased birth-rank without increased parental age might be accounted for by loss of affected fetuses through early spontaneous abortion. The 2:1 male to female ratio remains unexplained.

It is probable that the few familial cases which have been recorded had conditions other than thanatophoric dysplasia and there are no confirmed reports of affected siblings other than sets of monozygous twins (Soto et al 1981, Horton et al 1983, Servilea et al 1984). The most significant factors in genetic counselling are the small likelihood of recurrence and the lethal nature of the disorder.

Thanatophoric dysplasia has been diagnosed in late pregnancy by recognition of the characteristic radiographic features in the fetus (Thompson & Parmely 1971, Bergstrom et al 1971). Ultrasonographic diagnosis has been achieved by O'Malley et al (1972), Cremin & Shaff (1977), Camera et al (1984), Beetham & Reeves (1984) and Elejalde & Elejalde (1985).

The term 'cloverleaf skull' or 'Kleeblattschadel' syndrome has been used synonymously with thanatophoric dysplasia (Young et al 1973). There is no doubt, however, that these abnormalities can occur together or separately and the issue is that of phenotypic variation versus heterogeneity. The cloverleaf skull can certainly exist in isolation and as a component of a variety of other disorders (Feingold et al 1973, Pilz & Swoboda 1975, Temtamy et al 1975, Kozlowski et al 1985). The association with thanatophoric dysplasia is evidently fairly frequent as a review of 51 known cases of the Kleeblattschadel anomaly revealed that about 40% had both anomalies (Hodach et al 1975). In one report maternal rubella has been invoked as a pathogenic factor (Widdig et al 1974). The ultrasonic prenatal diagnosis of the isolated cloverleaf skull abnormality has been reviewed by Salvo

(1981). An infant with thanotophoric dysplasia and the cloverleaf skull (vide infra) was detected prenatally by Mahony et al (1985).

Partington et al (1971) in a report of four cases with the combined abnormalities, two of whom were sibs, postulated autosomal recessive inheritance. Apart from these siblings, there have been no other known instances of familial recurrence, although 26 cases had been reported when the combination was reviewed by Kremens et al (1982). The presence of cloverleaf skull in only one of a pair of monozygotic twins with thanatophoric dysplasia is highly suggestive that this cranial abnormality represents a variable component of a homogeneous syndrome, which carries a low recurrence risk (Horton et al 1983).

## HOMOZYGOUS ACHONDROPLASIA

Achondroplasia is by far the most common of the dwarfing skeletal dysplasias (see Ch. 1). There have been numerous marriages between achondroplasts and in these circumstances there is a 1 in 4 chance that any child of the union will inherit both faulty genes and thus be homozygous for the condition. Affected neonates resemble classical achondroplasts but the radiographic changes are more severe and death from respiratory failure usually supervenes in the early weeks after delivery (Hall et al 1969, Murdoch et al 1970, Rogovits et al 1972, Kozlowski et al 1977, 1978). The condition is not always lethal in the neonatal period and survival of homozygous achondroplasts to 29 weeks, 33 months and 37 months has been documented by Pauli et al (1983).

In view of the severity of the limb shortening, it is likely that homozygous achondroplasia could be diagnosed prenatally by ultrasonography. To date, however, to the best of the author's knowledge, this has not been reported.

## ASPHYXIATING THORACIC DYSPLASIA (JEUNE)

Asphyxiating thoracic dysplasia is sometimes compatible with long-term survival and for this reason the manifestations, natural history and genetics are reviewed in detail in Chapter 1. More than 100 cases have now been reported and syndromic status is well established. The necropsy findings in seven affected neonates were recorded by Turkel et al (1985). In addition to irregular endochondral ossification, these infants all had visceral involvement, notably pulmonary hypoplasia and variable pancreatic and hepatic fibrosis. Prenatal sonographic diagnosis has been accomplished by Schinzel et al (1985) and Elejalde et al (1985). In both instances the pregnancies were terminated and the diagnosis confirmed at autopsy.

## SHORT-RIB SYNDROMES

The short-rib syndromes, formerly short-rib polydactyly syndromes, are a group of potentially lethal disorders in which thoracic constriction is associated with variable preaxial polydactyly, micromelia and abnormalities in other systems. These conditions were originally differentiated from Jeune thoracic dysplasia (Spranger et al 1974). Saldino-Noonan, Majewski and Naumoff forms are recognised and respectively designated type I, II and III. Inheritance is autosomal recessive in each form.

## Type I (Saldino-Noonan)

This entity is characterised by micromelia, postaxial polydactyly, brachydactyly, thoracic narrowing and abnormalities of the cardiovascular system and genitalia. Significant radiographic changes include short horizontal ribs, small iliac bones, metaphyseal spurs on the tubular bones and deficient ossification of the extremities. The disorder was first recognised in two stillborn sibs (Saldino & Noonan 1972).

## Type II (Majewski)

In this condition thoracic, digital, limb and visceral abnormalities resemble those of the Saldino-Noonan form. However, facial clefting and nose and ear deformity are additional features (Majewski et al 1971, Motegi et al

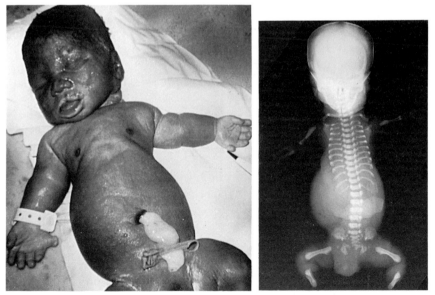

**Fig. 3.8** (left) Saldino-Noonan syndrome: limb shortening, brachydactyly, postaxial polydactyly and thoracic constriction are suggestive of the diagnosis in this stillborn infant.

**Fig. 3.9** (right) Saldino-Noonan syndrome: the diagnosis is confirmed by the radiographic appearance of short horizontal ribs, small iliac bones and metaphyseal spurs on the tubular bones.

1971). It may be distinguished radiographically from type I by the normal appearance of the pelvis and by marked shortening and an ovoid configuration of the tibia (Spranger et al 1974).

## Type III (Naumoff; lethal thoracic dysplasia)

The clinical and radiographic features resemble those of types I and II but, in general, the visceral manifestations are less severe. Thoracic constriction is severe and the genitals may be abnormal (Naumoff et al 1977).

There has been terminological confusion in the short-rib syndromes, as Cherstvoy et al (1980) and Gembruch et al (1985) reversed the numerical designations and eponyms of types I and II. For the sake of conformity the style used in the Paris Nomenclature is to be preferred. Additional varieties have been proposed (Piepkorn et al 1977, Bidot-Lopex et al 1978, Beemer et al 1983) but their autonomous syndromic status is uncertain.

Sillence (1980) drew attention to the phenotypic overlap between these conditions and argued for homogeneity. Walley et al (1983) reviewed the clinical, radiographic and ultrastructural features of the three conventional types and suggested that there might be even further heterogeneity. This issue remains unresolved.

Yang et al (1980) recorded a female preponderance in accumulated cases. The extent of abnormal sexual development is noteworthy; Bernstein et al (1985) found an XY male karyotype in two phenotypic females, while two others had ambiguous genitalia.

Antenatal diagnosis of type I has been achieved by radiography at 29 weeks gestation (Richardson et al 1977) and by serial ultrasonography at 17–20 weeks gestation (Beemer et al 1983). Type II, the Majewski form, was diagnosed by ultrasonography in the 3rd trimester by Mavel (1982) and at 10 weeks by Gembruch et al (1985). Tollager-Carsen & Benzie (1984) suggested that fetoscopy was the method of choice and monitored four at risk pregnancies using this technique. They made a positive diagnosis of the Majewski syndrome in one instance, while two pregnancies were correctly deemed to be normal. In another, the fetus could not be visualised due to technical problems. In view of the lethality of these conditions and the risks inherent in fetoscopy, it would seem that the optimal approach should be non-invasive.

## CHONDRODYSPLASIA PUNCTATA, RHIZOMELIC FORM

The terminology and classification of the chondrodysplasia punctata group of disorders is discussed in Chapter 1, where the common benign autosomal dominant Conradi-Hünnermann type is reviewed in detail. The rare autosomal recessive rhizomelic form, which is discussed below, is a clearly defined entity in which stillbirth or early demise is usual.

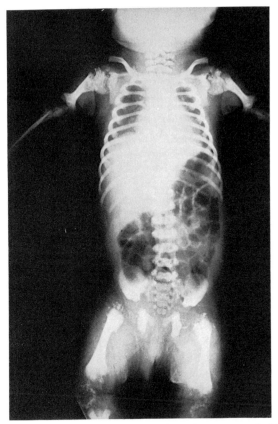

**Fig. 3.10** Chondrodysplasia punctata, AR type; the humeri are symmetrical but very short. Stippling extends beyond the margins of the epiphyses, although the spine is relatively spared. The metaphyses are irregular.

## Clinical and radiographic features

Marked symmetrical rhizomelia (proximal limb shortening) is a significant feature, the changes being maximal in the humeri. Lenticular cataracts and ichthyotic dermal abnormalities are often present. The digits are stubby and joint mobility is limited. Death from respiratory complications usually takes place during the first year. Tracheal stenosis may be a precipitating factor (Kaufman et al 1976).

Radiographically, the epiphyses are stippled but in distinction to the dominant type of the condition, involvement of the spine is of mild degree. Other distinguishng features are the presence of calcification outside the margins of the epiphyses, widening and fraying of the metaphyses and coronal clefts in the vertebral bodies. The pathological features of the rhizomelic form of chondrodysplasia punctata have been reviewed by Gilbert et al (1976) and the radiological findings were depicted, tabulated and discussed by Heselson et al (1978).

*Genetics*

Affected sibs with normal parents have been reported by several authors, including Fraser & Scriver (1954) and Mason & Kozlowski (1973). Sporadic patients from consanguineous matings were described by Mosekilde (1958) and Melnick (1965). In the latter instance, father–daughter incest had taken place. These observations are all indicative of autosomal recessive inheritance. On a basis of four affected infants, including a pair of siblings, born in the west of Scotland during the period 1975–1983, Connor et al (1985) calculated an incidence of 1 in 84 442 and a minimum carrier frequency of 1 in 145. These authors also successfully diagnosed the condition by ultrasound in the second trimester of pregnancy.

Maternal therapy during early pregnancy with the anticoagulant drug, warfarin, can produce a phenocopy of chondrodysplasia punctata in the fetus. In a review of eight affected infants (Pauli et al 1976) tabulated the clinical and radiographic manifestations of warfarin embryopathy; these were virtually indistinguishable from those of the genetic forms of the disorder.

It has recently emerged that rhizomelic chondrodysplasia punctata is the consequence of multiple peroxisomal dysfunction which leads to defective oxidation of phytanic acids. The peroxisomal disorders are a group of inborn errors of metabolism in which functional pathways in cytoplasmic organelles, are abnormal. Other conditions in this category include the Zellweger and Refsum syndromes.

## CAMPOMELIC DYSPLASIA

More than 100 patients with campomelic dysplasia have been described and reports continue to accumulate. The name of the condition, which is derived from the Greek, has the connotation 'bent limb'. The condition was delineated by Spranger et al (1970) and by Maroteaux et al (1971). The manifestations have been reviewed by Becker et al (1975), Eliachor et al (1975), Weiner et al (1976) and Houston et al (1983).

*Clinical and radiographic features*

Campomelic dysplasia is potentially lethal and the majority of reports concern babies who have succumbed to respiratory obstruction and recurrent aspiration of food during the neonatal period. Bowing and shortening of the tubular bones is maximal in the legs. Hypertelorism, micrognathia and a cleft palate are additional features. The feet may have a calcaneovalgus or equinovarus deformity and a subcutaneous dimple is often present over the anterolateral aspect of the tibia. Radiographically, the scapula and fibula may be hypoplastic, while ossification of the vertebral bodies is sometimes defective.

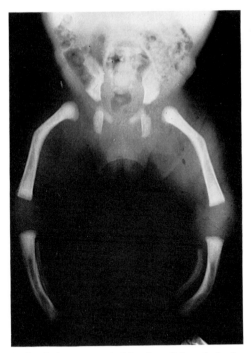

**Fig. 3.11** Campomelic dysplasia: symmetrical bowing of the tubular bones of the legs.

Khajavi et al (1976) examined the radiographs of nine patients, reviewed the literature and concluded that the campomelic syndrome was heterogeneous. These authors suggested that there might be a 'classic' or 'long bone' form and two 'short bone' types, one of which was associated with craniostenosis while the other was normocephalic.

*Genetics*

The majority of reported cases of campomelic dysplasia have been sporadic. Spranger et al (1970) mentioned that they had seen 8 affected infants in a 6 month period and suggested that the condition might be increasing in frequency. There has been considerable speculation concerning the pathogenesis of the disorder and a variety of exogenous factors have been incriminated. Gardner et al (1971) suggested that oral contraceptives might be involved. However, this contention has not been substantiated.

Some investigators have favoured a genetic aetiology, although there has been no general agreement. Bianchine et al (1971) considered that new dominant mutation was a likely cause. Reports of a pair of sisters with the disorder (Stuve & Wiedemann 1971), mixed sex affected siblings (Shafai & Schwartz 1976) and affected children with consanguineous parents (Cremin et al 1973, Winter et al 1985) constitutes evidence for autosomal recessive inheritance. Thurmon et al (1973) encountered campomelic dysplasia in two sibs and their half-sister. The mother of all of these infants had slight tibial bowing. There was a suspicion of consanguinity in the kindred and the authors commented that the pedigree data were consistent with autosomal recessive inheritance, with minor manifestations in the heterozygous mother, or with autosomal dominant inheritance with very variable expression. Hall & Spranger (1980) analysed published data and concluded that autosomal recessive inheritance was probable. It seems reasonable to offer genetic counselling on this basis, provided that the diagnosis is secure.

Discordance between phenotypic and chromosomal sex is a curious feature of campomelic dysplasia and there have been several well documented instances of apparent females with a 46XY chromosomal constitution (Hoefnugel et al 1978, Rolland et al 1975, Hovmoller et al 1977, Slater et al 1985). There have, however, been no examples of phenotypic males with a female 46XX karyotype. The excess of ostensible females in published series can probably be accounted for by this sex reversal situation.

Campomelic dysplasia was diagnosed by ultrasonography at 22 weeks of gestation by Balcar & Bieber (1983) and at 18 weeks in an 'at risk' pregnancy by Winter et al (1985). Nevertheless, ultrasonography does not always provide definitive diagnosis. Slater et al (1985) discussed this problem and concluded that with increasing technical sophistication, earlier detection would become feasible.

## OSTEOGENESIS IMPERFECTA

Osteogenesis imperfecta (OI) is a common and important disorder which is currently the focus of much medical interest. A full account of the manifestations, genetic implications and nosology is given in Chapter 5.

The severe, potentially lethal form previously known as OI congenita, is now conventionally categorised as 'OI type II'. This condition presents in the neonate with angulated limbs, a soft skull (caput membranaceum) and thoracic asymmetry due to multiple rib fractures. The sclerae may be white or any shade of blue. The diagnosis is confirmed by the radiographic demonstration of fractures in the long bones and ribs in all stages of healing, together with Wormian bones in the skull.

The mode of inheritance of OI type II has long been a matter of controversy. Affected siblings with normal parents have been reported (Goldfarb & Ford 1954, Chawla 1964, Wilson 1974, Braga & Passage 1981) and there have been instances of parental consanguinity. Heterogeneity is highly probable but clear-cut phenotypic markers for sub-classification are lacking; Sillence et al (1984), following a study of 48 cases, proposed three separate forms (designated OI II-A, B and C) on a basis of.radiological changes. The

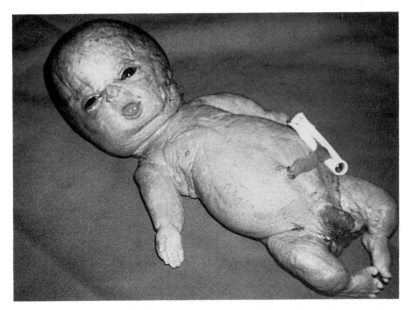

**Fig. 3.12** Osteogenesis imperfecta: a stillborn premature infant. Blue sclerae, a poorly ossified cranial vault (caput membranaceum) and limb deformities were indicative of the diagnosis, which was confirmed radiographically.

majority of cases are the result of new dominant mutation (Maroteaux & Cohen-Solal 1984).

The problem of heterogeneity of OI II is compounded by the fact that infants with the common benign autosomal dominant OI I occasionally present with multiple fractures at birth and may be confused with OI II. Furthermore, infants with the autosomal recessive OI type III often have severe manifestations at birth and may surprise their medical attendants by their capacity for suvival, although subsequent deformity is often gross. Although it is sometimes impossible to accurately categorise a severely affected, sporadic neonate, the prognosis for life can be estimated in terms of an index derived from the numbers of rib fractures which are present (Spranger et al 1982). This index has proved to be of value in the clinical situation.

Molecular techniques are beginning to have an impact in the elucidation of OI and it is probable that there is even greater heterogeneity than was previously supposed. Indeed, in the cases in which the structure of the faulty gene has been investigated in detail, each has proved to have a different mutation!

The laboratory determination of the exact molecular defect in an affected neonate is time consuming and inappropriate for routine use (Bonadio & Byers 1985). An alternative approach involves electrophoretic studies of

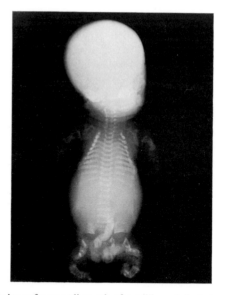

Fig. 3.13 Osteogenesis imperfecta: radiograph of a stillborn infant showing broad crumpled long bones and beaded ribs. These appearances correspond to OI type IIA, in which the majority of affected neonates represent new mutations.

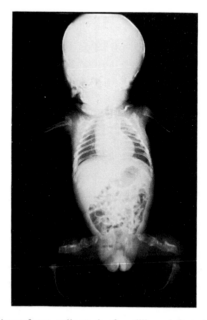

Fig. 3.14 Osteogenesis imperfecta: radiograph of a stillborn infant showing broad crumpled long bones and thin ribs which are virtually normal. This is the uncommon type IIB, which is possibly autosomal recessive.

collagen secreted by cultured fibroblasts of the parents and affected child. In this way if positive results are obtained, the status of the parents regarding heterozygosity for the faulty gene can be elucidated. On this basis the question of autosomal recessive inheritance versus new dominant mutation can be settled and recurrence risks predicted with accuracy.

As the majority of infants with OI II have been sporadic, there have been comparatively few reports of early antenatal diagnosis following planned monitoring of pregnancy. However, OI II has been diagnosed in the third trimester by conventional radiography (Heller et al 1975) and by fetography (Ojitu et al 1976). Ultrasound has proved to be effective for demonstration of limb malalignment consequent upon intra-uterine fracturing and the condition has been recognised antenatally by Milsom et al 1982, Shapiro et al 1982, Stephens et al 1983, Elejalde & Elejalde 1983, Patel et al 1983, Aylsworth et al 1984 and Brown 1984. This technique is now established as the method of choice for the early in utero diagnosis of a severely affected fetus.

## HYPOPHOSPHATASIA LETHALIS

Hypophosphatasia lethalis is an uncommon heterogeneous autosomal recessive disorder which bears many similarities to osteogenesis imperfecta. The phenotypic features and genetic background are reviewed in Chapter 5.

The 'congenita' or 'lethalis' neonatal form has been diagnosed antenatally in at-risk fetuses by ultrasonography (Rudd et al 1976), by recognition of a low alkaline phosphatase level in cultured amniotic fluid cells (Clark et al 1976) and by combining both techniques (Benzie et al 1976). Estimations of alkaline phosphatase activity in amniotic fluid have also facilitated successful antenatal diagnosis (Shin-Buehring et al 1981). Using ultrasonography and radiography, Burck et al (1982) diagnosed an affected fetus at 26 weeks gestation and drew attention to the markedly delayed ossification of the fetal skeleton. Ultrasonic diagnosis at 16 weeks of gestation has been achieved by Wladimiroff et al (1985). It is apparent that pregnancy monitoring by a combination of these techniques is appropriate in families with the severe form of hypophosphatasia.

## FIBROCHONDROGENESIS

Fibrochondrogenesis is a recently delineated skeletal dysplasia. The designation 'fibrochondrogenesis' pertains to the unusual histological appearance of the cartilage matrix, which contains many interwoven fibrous septa. Affected infants have short limbs but there are no other distinguishing features at the clinical level. The diagnosis is confirmed by the radiographic recognition of the characteristic dumbell-shaped configuration of the tubular bones, which are short and broad, with marked metaphyseal flaring. In addition there is platyspondyly with vertebral clefting and short, cupped ribs.

The autosomal recessive mode of inheritance was indicated by the first case, which was the product of an uncle–niece relationship (Lazzaroni-Fossati et al 1978, Lazzaroni-Fossati 1979). Subsequently, two unrelated stillborn infants with the disorder were reported by Eteson et al (1984) and two others by Whitley et al (1984).

## HYPOCHONDROGENESIS

Hypochondrogenesis was delineated by Maroteaux et al (1982) following analysis of the radiographic features of 21 cases and further details were presented by the same authors in the following year (Maroteaux et al 1983). An additional case was documented by Hendrick et al (1983). So far all affected infants have been sporadic and the mode of inheritance is unknown.

The condition presents neonatally as short-limbed dwarfism, the neck is very short and the head appears large in relation to the trunk. Affected infants are stillborn or die from respiratory problems in the first weeks after delivery. The radiological changes, notably delayed ossification of the vertebrae and pelvis, resemble those of spondyloepiphyseal dysplasia congenita but, when severe, they are reminiscent of achondrogenesis. The histological features of hypochondrogenesis also resemble those of achondrogenesis type II and it has been suggested that these conditions represent a spectrum with marked phenotypic variability rather than separate entities (Borochowitz et al 1986). Chondrocytes permit differentiation. All cases have been sporadic and the mode of inheritance is unknown.

## ATELOSTEOGENESIS

Atelosteogenesis is a lethal chondrodysplasia which presents in the neonate with micromelic dwarfism, incurved legs, club feet, dislocated elbows and variable cleft palate. The name of the condition is derived from the Greek word for 'incomplete' and reflects the skeletal underdevelopment. Radiographically, the vertebral bodies are poorly ossified and the humerus and femur are club-shaped, with distal hypoplasia. Specific histological changes include fibrous encapsulation of chondrocytes, areas of cartilagenous degeneration and accumulations of metachromatic material in the epiphyses and growth plate.

The disorder attracted attention when a series of six affected neonates were reported by Maroteaux et al (1982). With hindsight the condition was recognisable in a previous report by Kozlowski et al (1981) which was subsequently republished by Kozlowski & Bateson (1984). Rimoin et al (1980) had also mentioned two cases in an abstract and these reports were expanded by Sillence et al (1982). These latter authors proposed the title 'spondylohumerofemoral dysplasia', with 'giant cell chondrodysplasia' as an alternative. Yang et al (1983) studied a further case and commented that chondrocytic giant cells were a causally non-specific phenomenon. In order to achieve terminological unanimity Opitz (1983) canvassed 10 clinical geneticists for their views on an appropriate name. Six voted for 'atelosteogenesis', three for 'spondylohumerofemoral dysplasia' and one abstained. The term 'atelosteogenesis' is currently enshrined in the Paris nomenclature.

The mechanism of genetic transmission has not yet been established.

## DYSSEGMENTAL DYSPLASIA

Dyssegmental dysplasia presents in the neonate with short-limbed dwarfism, articular rigidity, club feet and leg bowing. The face is round, the nose is flattened and additional variable abnormalities include hydrocephalus, occipital encephalocele, cleft palate and urinary tract malformations. Radiographically, the vertebral bodies show a unique malsegmentation, with ossification centres of different size and shape which has prompted the designation 'dyssegmental dysplasia'. The ilia are small and rounded and the tubular bones are short and broad, with expanded metaphyses. Abnormal collagen peptides have been demonstrated by electrophoresis in one instance (Svejcar 1983). About 20 cases have been recorded, including two sets of affected siblings. Sex distribution is equal and autosomal recessive inheritance is likely.

Fasanelli et al (1985) reviewed the literature and concluded that there were two forms of the disorder. The Silverman type is said to be invariably lethal in the perinatal period, while the Rolland-Desbuquois type was compatible with survival for weeks, or even months. According to these authors, the former type had been reported previously (Silverman 1969, Goodlin & Lowe 1974, Handmaker et al 1979, Gruhn et al 1978, Miething et al 1981) while the latter had featured in reports by Rolland et al (1972) Dinno et al (1976) Langer et al (1976) and Bueno et al (1984).

Although there are clinical and radiographic differences between these two putative entities, these are not absolute and the question of syndromic identity awaits resolution.

## BOOMERANG DYSPLASIA

Boomerang dysplasia has been reported in a total of three cases by Kozlowski et al (1981), Tenconi et al (1983) and Kozlowski et al (1985). The name of the condition relates to the unusual shape of the long bones of the legs. For the enlightenment of those with limited knowledge of antipodean culture, the latter authors define a boomerang as 'an implement of the indigenous aboriginal peoples of Australia, thrown for hunting and pleasure'.

The three reported cases all died in the neonatal period. They had dwarfism with short, bowed, rigid limbs and a characteristic facies. In particular, the nose had a broad root and severe hypoplasia of the nares and septum. Radiographically, the radii and fibulae were absent, while the remaining long bones had the 'boomerang' configuration. The iliac bodies were small and ossification in the lower spine and digits was retarded.

The three affected infants were all sporadic males, derived from Japan, Italy and Australia and, at present, the genetic basis of the disorder is unknown.

## DE LA CHAPELLE DYSPLASIA

A brother and sister with a unique form of lethal neonatal dwarfism were described by de la Chapelle et al (1972). Another affected sister was subsequently born into the same consanguineous Finnish family and the case details, together with those of an unrelated sporadic Belgian boy, were reported by Whitley et al (1986).

The main features of de la Chapelle dysplasia are severe micromelia, limb bowing, equinovarus deformity of the feet, cleft palate and a narrow thorax. Death occurs in the neonatal period from respiratory failure due to laryngeal stenosis, tracheobronchomalacia and pulmonary hypoplasia. Radiographically, the most important diagnostic features are in the ulnae and fibulae, which are very small and have a distinctive triangular configuration. The other tubular bones are short and bowed, the spine is malaligned and the vertebrae are flattened, with anterior projections. Histologically, the cartilage contains lacunar haloes, which are a helpful but not pathognomonic diagnostic indicator.

The occurrence of de la Chapelle dysplasia in three siblings born to unaffected consanguineous parents is highly suggestive of autosomal recessive inheritance.

## SCHNECKENBECKEN DYSPLASIA

The term 'Schneckenbecken dysplasia' is derived from the German for 'snail' pelvis and reflects an imaginative interpretation of the unusual radiographic appearances in this lethal neonatal dysplasia (Borochowitz et al 1986).

Hydramnios is a frequent occurrence in the third trimester of an affected pregnancy and infants with the condition are stillborn. The limbs are short and the head is comparatively large, but these features do not distinguish Schneckenbecken dysplasia from other lethal dwarfing disorders. The radiographic changes in the pelvis are specific; notably a projection from the medial aspect of the ilium which has been likened to the head of a snail, the bone itself being the body. In addition, the vertebral bodies are flattened, the pubic and ischial bones are hypoplastic and precocious ossification is present in the tarsus.

Histological examination of cartilage reveals a characteristic hypercellularity, with chondrocytes which contain large nuclei. These changes permit differentiation from the other forms of neonatal dwarfism.

When delineating the disorder Borochowitz et al (1986) described two families, one of which was consanguineous, each with two affected children. These authors also reached a retrospective diagnosis of Schneckenbecken dysplasia in siblings previously reported by Laxova et al (1973) under the title 'achondrogenesis' and others described by Chemke et al (1971) and Graff et al (1972) as 'familial thanotophoric dysplasia'. Concurrently, Knowles et al (1986) reappraised the family published by Laxova et al (1973) in which a consanguineous Asian couple in England had produced 5 stillborn dwarves out of a total of 13 pregnancies and documented the characteristic radiographic and histological features of Schneckenbecken dysplasia.

The occurrence of affected siblings of both sexes in several consanguineous unions is indicative of autosomal recessive inheritance.


## MISCELLANEOUS, NEW, POTENTIALLY LETHAL, NEONATAL DWARFING DYSPLASIAS

The potentially lethal skeletal dysplasias which present with dwarfism in the newborn constitute a fertile field for syndromic delineation and new entities continue to be described. As reports accumulate, consistent manifestations are recognised and syndromic identity is established. This is based upon clinical and radiographic features and, whenever possible, by the demonstration of specific histological changes.

The following new entities (Table 3.1) are good candidates for elevation to syndromic status in the future.


**Table 3.1**

| | |
|---|---|
| 1. Neonatal death dwarfism, resembling metatropic dysplasia | Colavita & Kozlowski (1984) |
| 2. Short-rib dwarfism, abnormal facies and congenital heart defect | Barrow & Fitzsimmons (1984) |
| 3. Short limbs, narrow chest, advanced bone age, spondylometaphyseal abnormalities | Beemer et al (1985) |
| 4. Lethal short-limbed dwarfism with severe spondylocostal dysostosis, cardiovascular and urogenital malformations and a Dandy-Walker cyst | Moermann et al (1985) |
| 5. Short-limbed dwarfism with pulmonary hypoplasia (affected siblings) | McAlister et al (1985) |
| 6. Pyknoachondrogenesis; gross limb shortening and skeletal sclerosis | Camera et al (1986) |
| 7. Achondrogenesis, new type | Kozlowski et al (1986) |

## Non-lethal short-limbed dwarfism in the newborn

Several dwarfing skeletal dysplasias which present in the newborn are compatible with a prolonged survival or even a normal life span. These conditions have been reviewed in detail in Chapters 1 and 2, but for the sake of completion, they are listed in Table 3.2.

**Table 3.2** Non-lethal short-limbed dwarfism in the newborn

|  | Inheritance |
| --- | --- |
| Achondroplasia | AD |
| Chondroectodermal dysplasia (Ellis van Creveld) | AR |
| Mesomelic dysplasia (various types) | AD/AR |
| Rhizomelic dysplasia (various types) | AD/AR |
| Spondyloepiphyseal dysplasia congenita | AD |
| Spondylometaphyseal dysplasia | AD/AR |
| Metatropic dysplasia | AR? |
| Kniest syndrome | AD? |
| Diastrophic dysplasia | AR |
| Parastremmatic dysplasia | AD? |
| Opsismodysplasia | AR? |
| Kyphomelic dysplasia | XL? AR? |
| Grebe chondrodysplasia | AR |

REFERENCES

**Introduction**

Camera G, Mastroiacovo P 1982 Birth prevalence of skeletal dysplasias in the Italian Multicentric Monitoring System for Birth Defects. In: Papadatos C J, Bartsocas C S (eds) Skeletal Dysplasias. Liss, New York, p 441–449

Connor J M, Connor R, Sweet E M, Gibson A, Patrick W, McNay M B, Redford D 1985 Lethal neonatal chondrodysplasias in the West of Scotland 1970–1983 with a description of a thanatophoric, dysplasia-like, autosomal recessive disorder. American Journal of Medical Genetics 22: 243–253

Cremin B, Beighton P 1978 Bone Dysplasias of Infancy. Springer-Verlag, Berlin

Elejalde B R, Elejalde M M 1986 The prenatal growth of the human body determined by the measurement of bones and organs by ultrasonography. American Journal of Medical Genetics 24: 575–598

Filley R A, Golbus M S, Carey J C, Hall J G 1981 Short-limbed dwarfism, ultrasonographic diagnosis by mensuration of fetal femoral length. Radiology 138: 653–656

Filley R A, Golbus M S 1982 Ultrasonography of the normal and pathologic fetal skeleton. Radiologic Clinics of North America 20: 311–323

Hobbins J C, Mahoney M J 1980 The diagnosis of skeletal dysplasias with ultrasound. In: Sanders R C, James A E (eds) The Principles and Practice of Ultrasonography in Obstetrics and Gynecology 2nd ed. Appleton-Century-Crofts, New York

Hobbins J C, Bracken M B, Mahoney M J 1982 Diagnosis of fetal skeletal dysplasias with ultrasound. American Journal of Obstetrics and Gynecology 142: 306–312

Kurtz A B, Wapner R J 1983 Ultrasonographic diagnosis of second-trimester skeletal dysplasias: A prospective analysis in a high-risk population. Journal of Ultrasound in Medicine 2: 99–105

Lachman R, Hall J G 1979 The radiographic prenatal diagnosis of the generalized bone dysplasias and other skeletal abnormalities. Birth Defects: Original Article Series, XV, 5A, 3–24

Maroteaux P, Stanescu V, Stanescu R 1976 The lethal chondrodysplasias. Clinical Orthopaedics and Related Research 114: 31

Muller L M, Cremin B J 1985 Ultrasonic demonstration of fetal skeletal dysplasia. South African Medical Journal 67: 222

Orioli I M, Castilla E E, Barbosa-Neto J G 1986 The birth prevalence rates for the skeletal dysplasias. Journal of Medical Genetics 23: 328–332

Rimoin D L, Silberberg R, Hollister D W 1976 Chondro-osseous pathology in the chondrodystrophies. Clinical Orthopaedics and Related Research 114: 137–151

Sillence D O, Horton W A, Rimoin D L 1979 Morphologic studies in the skeletal dysplasias. American Journal of Pathology 96: 813–860

Sillence D O, Rimoin D L, Lachman R 1978 Neonatal dwarfism. Pediatric Clinics of North America 25: 453–483

Stanescu V, Stanescu R, Maroteaux P 1977 Morphological and biochemical studies of epiphyseal cartilage in dyschondroplasias. Archives Françaises de Pédiatrie 34(Suppl 3): 1–80

### Achondrogenesis

Benacerraf B, Osathanondh R, Bieber F R 1984 Achondrogenesis type I: ultrasound diagnosis in utero. Journal of Clinical Ultrasound 12: 357–359

Chen H, Liu C T, Yang S S 1981 Achondrogenesis: A review with special consideration of achondrogenesis type II (Langer-Saldino). American Journal of Medical Genetics 10: 379–394

Connor J M, Connor R, Sweet E M, Gibson A, Patrick W, McNay M B, Redford D 1985 Lethal neonatal chondrodysplasias in the West of Scotland 1970–1983 with a description of a thanatophoric, dysplasia-like, autosomal recessive disorder. American Journal of Medical Genetics 22: 243–253

Glenn L W, Teng S S 1985 In utero sonographic diagnosis of achondrogenesis. Journal of Clinical Ultrasound 13: 195–198

Golbus M S, Hall B D, Filley R A, Pozkanzer L B 1977 Prenatal diagnosis of achondrogenesis. Journal of Pediatrics 91: 464–466

Graham D, Tracey J, Winn K, Corson V, Sanders R C 1983 Early second trimester sonographic diagnosis of achondrogenesis. Journal of Clinical Ultrasound 11: 336–338

Houston C S, Awen C F, Kent H P 1972 Fatal neonatal dwarfism. Journal of the Canadian Association of Radiologists 23: 45

Maroteaux P, Stanescu V, Stanescu R 1976 The lethal chondrodysplasias. Clinical Orthopaedics and Related Research 114: 31

Muller L M, Cremin B J 1985 Ultrasonic demonstration of fetal skeletal dysplasia. South African Medical Journal 67: 222–226

Ornoy A, Sekeles E, Smith 1976 Achondrogenesis type I in three sibling fetuses. Scanning and transmission electron microscopic studies. American Journal of Pathology 82/1: 71

Scott C I 1972 In Steinberg A G, Bearn A G (eds) Progress in Medical Genetics, Vol. 8, Ch. 7. Grune and Stratton, New York

Smith W L, Breitweiser T D, Dinno N 1981 In utero diagnosis of achondrogenesis Type I. Clinical Genetics 19: 51–54

Verma I C, Bharagava S, Agarwal A 1975 An autosomal recessive form of lethal chondrodystrophy with severe thoracic narrowing, rhizoacromelic type of micromelia, polydactyly and genital anomalies. Birth Defects: Original Article Series 11/6: 167

Whitley C B, Gorlin R J 1983 Achondrogenesis: New nosology with evidence of genetic heterogeneity. Radiology 148: 693–698

Wiedemann H R, Remagen W, Hienz H A 1974 Achondrogenesis within the scope of connately manifested generalised skeletal dysplasias. Zeitschrift für Kinderheilkunde 116/4: 223

Yang S S, Heidelberger K P, Bernstein J 1976 Intracytoplasmic inclusion bodies in the chondrocytes of type I lethal achondrogenesis. Human Pathology 7: 667–673

Yang S S, Brough A J, Garewal G S, Bernstein J 1974 Two types of heritable lethal achondrogenesis. Journal of Paediatrics 85/6: 796–801

### Thanatophoric dysplasia

Beetham F G, Reeves J S 1984 Early ultrasound diagnosis of thanatophoric dwarfism. Journal of Clinical Ultrasound 12: 43–44

Bergstrom K, Gustavon K H, Jorulf H 1972 Thanatophoric dwarfism: diagnosis in utero. Australasian Radiology 16: 155

Bouvet J P, Maroteaux P, Feingold J 1974 Genetic study of thanatophoric dwarfism. Annals of Human Genetics 17/3: 181

Camera G, Dodero D. de Pascale S 1984 Prenatal diagnosis of thanatophoric dysplasia at 24 weeks. American Journal of Medical Genetics 18: 39–43

Cremin B J, Shaff M I 1977 Ultrasonic diagnosis of thanatophoric dwarfism. Radiology 124: 479–480

Elejalde B R, Elejalde M M 1985 Thanatophoric dysplasia: Fetal manifestations and prenatal diagnosis. American Journal of Medical Genetics 22: 669–683

Feingold M, Miller D, Bull M J 1973 The demise of a syndrome? Syndrome Identification 1: 21

Harris R, Patton J T 1971 Achondroplasia and thanatophoric dwarfism in the newborn. Clinical Genetics 2: 61

Hodach R J, Viseskul C, Gilbert E F et al 1975 Studies of malformation syndrome in man. XXXVI: The Pfeiffer syndrome, association with kleeblattschadel and multiple visceral anomalies: Case report and review. Zeitschrift fur Kinderheilkunde 119/2: 87

Horton W A, Harris D J, Collins D L 1983 Thanatophoric dysplasia of identical twins. American Journal of Medical Genetics 17: 703–706

Kozlowski K, Warren P S, Fisher C C 1985 Cloverleaf skull with generalised bone dysplasia. Report of a case with short review of the literature. Pediatric Radiology 15/6: 412–414

Kremens B, Kemperdick H, Borchard F, Liebert U G 1982 Thanatophoric dysplasia with cloverleaf skull: Case report and review of the literature. European Journal of Pediatrics 139: 298–303

Langer L O Jr, Spranger J W, Greinacher I, Herdman R C 1969 Thanatophoric dwarfism. A condition confused with achondroplasia in the neonate, with brief comments on achondrogenesis and homozygous achondroplasia. Radiology 92: 285

Mahoney B S, Filly R A, Callen P W, Golbus M S 1985 Thanatophoric dwarfism with the cloverleaf skull: A specific antenatal sonographic diagnosis. Journal of Ultrasound in Medicine 4: 151–154

Maroteaux P, Lamy M, Robert J M 1967 Le nanisme thanatophore. Presse Médicale 75: 2519

Maroteaux P, Stanescu V, Stanescu R 1976 The lethal chondrodysplasias. Clinical Orthopaedics and Related Research 114: 31

Moir D H, Kozlowski K 1976 Long survival in thanatophoric dwarfism. Pediatric Radiology 5: 123

O'Malley B P, Parker R, Saphyakhafon P, Qizilbach A H 1972 Thanatophoric dwarfism. Journal of the Canadian Association of Radiologists 23: 62

Ornoy A, Adomian G E, Eteson D J, Burgeson R E, Rimoin D L 1985 The role of mesenchyme-like tissue in the pathogenesis of thanatophoric dysplasia. American Journal of Medical Genetics 21: 613–630

Partington M W, Gonzales-Crussi F, Khakee S G, Wollin D G 1971 Cloverleaf skull and thanatophoric dwarfism. Report of four cases, two in the same sibship. Archives of Disease in Childhood 46: 656

Pena S D J, Goòdman H O 1973 The genetics of thanatophoric dwarfism. Pediatrics 51: 104

Pilz E, Swoboda W 1975 Cloverleaf skull syndrome. Pädiatrishe Fortbildungskurse fur die Praxis 16/2: 275

Rimoin D L, Silberberg R, Hollister D W 1976 Chondro-osseous pathology in the chondrodystrophies. Clinical Orthopaedics 114: 137–152

Salvo A F 1981 Short communications: In utero diagnosis of Kleeblattschadel (cloverleaf skull). Prenatal Diagnosis : 141–145

Sato D, Hosokawa Y, Nakamura Y, Mukae T, Nakashima T, Komatsu Y, Kabashima S 1981 Thanatophoric dysplasia of identical twins. Acta Pathologica Japonica 31: 895–902

Serville F, Carles D, Maroteaux P 1984 Letter to the editor: Thanatophoric dysplasia of identical twins. American Journal of Medical Genetics 17: 703–706

Temtamy S A, Shoukry A S, Fayad I, El Meligy M R 1975 Limb malformations in the cloverleaf skull anomaly. Birth Defects: Original Article Series 11/2: 247

Thompson B H, Parmley T H 1971 Obstetric features of thanatophoric dwarfism. American Journal of Obstetrics and Gynaecology 109: 396

Widdig K, Steinhoff R, Guenther H 1974 The cloverleaf skull syndrome. Zentralblan fur allgemeine Pathologie und pathologische Anatomie 118/4: 358
Young R S, Pocharzevsky R, Leonicas J C, Wexley I B, Ratney H 1973 Thanatophoric dwarfism and cloverleaf skull (Kleeblattschädel). Radiology 106: 401


**Homozygous achondroplasia**
Hall J G, Dorst J P, Taybi H, Scott C, Langer L O, McKusick V A 1969 Two probable cases of homozygosity for the achondroplasia gene. Birth Defects 5: 24–34
Kozlowski K, Masel J, Morris L, Kunze D 1978 Neonatal death dwarfism (a further report). ROFO 129: 626–633
Kozlowski K, Masel J, Morris L, Ryan J H, Collins E, van Vliet P, Woolnough H 1977 Neonatal death dwarfism (report of 17 cases). Australasian Radiology 21: 164–183
Murdoch J L, Walker B A, Hall J G, Abbey H, Smith K K, McKusick V A 1970 Achondroplasia: A genetic and statistical survey. Annals of Human Genetics 33: 227–244
Pauli R M, Conroy M M, Langer L O Jnr, McLone D G, Naidich T, Franciosi R, Ratner I M, Copps S C 1983 Homozygous achondroplasia with survival beyond infancy. American Journal of Medical Genetics 16: 459–473
Rogovits N, Weissenbacher G, Zweymuller E 1972 Homozygote Achondroplasie und thantophorer Zwergwuchs — Pranatal diagnosticzierbare Skelettstorunger. Geburtschilfe und Frauenheilkunde 32: 184–191


**Asphyxiating thoracic dysplasia (Jeune)**
Elejalde R F, Elejalde M M, Pansch D 1985 Prenatal diagnosis of Jeune syndrome. American Journal of Medical Genetics 21: 433–438
Schinzel A, Savoldelli G, Briner J, Schubiger G 1985 Prenatal sonographic diagnosis of Jeune syndrome. Radiology 154: 777–778
Turkel S B, Diehl E J, Richmond J A 1985 Necropsy findings in neonatal asphyxiating thoracic dystrophy. Journal of Medical Genetics 22: 112–118


**Short-rib syndromes**
Beemer F A, Langer L O, Klep-de Pater J et al 1983 A new short rib syndrome: Report of two cases. American Journal of Medical Genetics 14: 115–123
Bernstein R, Isdale J, Pinto M, Zaaijman J D, Jenkins T 1985 Short rib-polydactyly syndrome: A single or heterogeneous entity? A re-evaluation prompted by four new cases. Journal of Medical Genetics 22: 46–53
Bidot-Lopez P, Ablow R C, Ogden J A, Mahoney M J 1978 A case of short rib polydactyly. Pediatrics 61: 427–432
Cherstvoy E D, Lurie E W, Shved I A, Lazjuk G I, Ostrowskaya T I, Usoev S S 1980 Difficulties in classification of the short rib-polydactyly syndromes. European Journal of Pediatrics 133: 57–61
Gembruch U, Hansmann M, Fodisch H J 1985 Early prenatal diagnosis of short rib-polydactyly syndrome type I (Majewski) by ultrasound in a case at risk. Prenatal Diagnosis 5: 357–362
Majewski F, Pfeiffer R A, Lenz W, Muller R, Feil G, Seiler R 1971 Polysyndaktylie, verkurzte Gliedmassen und Genitalfehbildungen: Kennzeichen eines selbstandigen Syndromes? Zeitschrift fur Kinderheilkunde 111: 118–138
Mavel A, Mabille J P, Halfon D, Nivelon-Chevallier A, Feldmann J P, Barthelet J, Kamp A 1982 Syndrome 'polydactylie et cotes courtes' type Majewski. A propos d'un cas recent diagnostique in utero. Revue Francaise de Gynecologies et Obstetrique 77: 595–599
Motegi T, Kusunoki M, Nishi T, Hamada T, Sato N, Imamura T, Mohri N 1979 Short rib-polydactyly syndrome, Majewski type, in two male siblings. Human Genetics 49: 269–275
Naumoff P, Young L W, Mazer J, Amortegui A J 1977 Short rib-polydactyly type 3. Radiology 122: 443–447
Piepkorn M, Karp L E, Hickok D, Wiegenstein L, Hall J G 1977 A lethal neonatal dwarfing condition with short ribs, polydactyly, cranial synostosis, cleft palate and severe ossification defect. Teratology 16: 345–358

Richardson M M, Beaudet A L, Wagner M L, Malini S, Rosenberg H S, Lucci J A 1977
Prenatal diagnosis of recurrence of Saldino-Noonan dwarfism. Journal of Pediatrics
919: 467–471
Saldino R M, Noonan C D 1972 Severe thoracic dystrophy with striking micromelia,
abnormal osseous development, including the spine and multiple visceral anomalies.
American Journal of Roentgenology 114: 257
Sillence D O 1980 Invited editorial comment: Non-Majewski short rib-polydactyly
syndrome. American Journal of Medical Genetics 27: 223–229
Spranger J W, Grimm B, Weller M, Weissenbecher G, Hermann J, Gilbert E, Krepler R
1974 Short rib-polydactyly (SRP) syndromes, types Majewski and Saldino-Noonan.
Zeitschrift für Kinderheilkunde 116: 73–94
Tollager-Carsen K, Benzie R J 1984 Fetoscopy in prenatal diagnosis of the Majewski and
the Saldino-Noonan types of the short rib-polydactyly syndromes. Clinical Genetics
26: 56–60
Walley V M, Coates C F, Gilbert J J, Valentine G H, Davies E M 1983 Brief clinical
report: Short rib-polydactyly syndrome, Majewski type. American Journal of Medical
Genetics 14: 445–452
Yang S S, Lin C S, Saadi A A, Nongia B S, Bernstein J 1980 Short rib-polydactyly
syndrome, type 3, with chondrocytic inclusions. American Journal of Medical Genetics
7: 205–213

### Chondrodysplasia punctata, rhizomelic form

Connor J M, Connor R A C, Sweet E M, Gibson A, Patrick W, McNay M B, Redford D
1985 Lethal neonatal chondrodysplasias in the west of Scotland 1970–1983 with a
description of a thanatophoric, dysplasia-like, autosomal recessive disorder, Glasgow
variant. American Journal of Medical Genetics 22: 243–253
Fraser F C, Scriver J B 1954 A hereditary factor in chondrodystrophia calcificans congenita.
New England Journal of Medicine 250: 272
Gilbert E F, Opitz J M, Spranger J W, Langer L O, Wolfson J J, Visekul C 1976
Chondrodysplasia punctata — rhizomelic form. Pathologic and radiologic studies of three
infants. European Journal of Pediatrics 123: 89
Heselson N G, Cremin B J, Beighton P 1978 Lethal chondrodysplasia punctata. Clinical
Radiology 29: 675–684
Kaufmann H J, Mahboubi S, Spackman T J, Capitano M A, Kirkpatrick J 1976 Tracheal
stenosis as a complication of chondrodysplasia punctata. Annals of Radiology 19: 203
Melnick J C 1952 Chondrodystrophia calcificans congenita. American Journal of Diseases of
Children 110: 218
Mosekilde E 1952 Stippled epiphysis in the newborn. Acta radiologica 37: 291
Pauli M P, Madden J D, Kranzler K J, Culpepper W, Port R 1976 Warfarin therapy
initiated during pregnancy and phenotypic chondrodysplasia punctata. Journal of
Paediatrics 88: 506

### Campomelic dysplasia

Balcar I, Bieber F R 1983 Sonographic and radiologic findings in campomelic dysplasia.
American Journal of Roentgenology 141: 481–482
Becker M H, Finegold M, Genieser N B 1975 Campomelic dwarfism. Birth Defects:
Original Article Series 11/6: 113
Bianchine J W, Rismberg H M, Kanderian S S, Harrison H E 1971 Campomelic dwarfism.
Lancet i: 1017
Cremin B J, Orsmond G, Beighton P 1973 Autosomal recessive inheritance in campomelic
dwarfism (letter). Lancet i: 488
Eliachar E, Baux S, Maroteaux P 1975 Congenital curvature of the long bones: A new case.
Semaine des Hôpitaux de Paris 51: 161
Gardner L I, Assemany S R, Neu R L 1971 Syndrome of multiple osseous defects with
pretibial dimples. Lancet ii: 98
Hall B D, Spranger J W 1980 Campomelic dysplasia. American Journal of Diseases of
Children 134: 285–289
Hoefnugel D, Wurster-Hill D H, Dupree W B, Bernirschke K, Fuld G L 1978
Camptomelic dwarfism associated with XY gonadal dysgenesis and chromosome
anomalies. Clinical Genetics 13: 489–499

Houston C S, Opitz J M, Spranger J W, MacPherson R I, Reed M H, Gilbert E F, Herrmann J, Schinzel A 1983 The campomelic syndrome: Report of 17 cases and follow-up on the currently 17-year-old boy first reported by Maroteaux et al in 1971. American Journal of Medical Genetics 15: 3–38

Hovmoller M L, Osuna A, Eklof O, Fredga K, Hjerpe A, Lindsten J, Ritzen M, Stanescu V, Svenningsen N 1977 Camptomelic dwarfism. A genetically determined mesenchymal disorder combined with sex reversal. Hereditas 86: 51–62

Khajavi A, Lachman R S, Rimoin D L, Shimke R N, Dorsrt J P, Ebbin A J, Handmaker S, Perreault 1976 Heterogeneity in the campomelic syndromes: Long and short bone varieties. Birth Defects: Original Article Series 10/6: 93

Maroteaux P, Spranger J, Opitz J M, Kucera J, Lowry R B, Schimke R N, Kagan S M 1971 Le syndrome campomelique. Presse Médicale 79: 1157–1162

Rolland M, Dupie Y, Berges J-D et al 1975 Le syndrome camptomelique. A propos d'une observation. Pédiatrie 30: 860–861

Shafai T, Schwartz L 1976 Campomelic syndrome in siblings. Journal of Pediatrics 89: 512–513

Slater C P, Ross J, Nelson M M, Coetzee E J 1985 The campomelic syndrome: Prenatal ultrasound investigations. South African Medical Journal 67: 863–866

Spranger J, Langer L O, Maroteaux P 1970 Increasing frequency of a syndrome of multiple osseous defects? Lancet ii: 716

Stuve A, Wiedemann H R 1971 Congenital bowing of the long bones in two sisters (Letter). Lancet i: 495

Thurmon T F, Defraites E B, Anderson E E 1973 Familial camptomelic dwarfism. Journal of Pediatrics 83: 841

Weiner D S, Benfield G, Robinson H 1976 Camptomelic dwarfism. Report of a case and review of the salient features. Clinical Orthopaedics and Related Research 116: 29

Winter R, Rosenkranz W, Hofmann H, Zierler H, Becker H, Borkenstein M 1985 Prenatal diagnosis of campomelic dysplasia by ultrasonography. Prenatal Diagnosis 5: 1–8

## Osteogenesis imperefecta

Aylsworth A S, Seeds J W, Guilford W B, Burns C B, Washburn D B 1984 Prenatal diagnosis of a severe deforming type of osteogenesis imperfecta. American Journal of Medical Genetics 19: 707–714

Bonadio J, Byers P H 1985 Subtle structural alterations in the chains of type I procollagen produce osteogenesis imperfecta type II. Nature 316: 363–366

Braga S, Passage E 1981 Congenital osteogenesis imperfecta in three sibs. Human Genetics 58: 441–443

Brown B S 1984 The prenatal ultrasonographic diagnosis of osteogenesis imperfecta lethalis. Journal of the Canadian Association of Radiologists 35: 63–66

Chawla S 1964 Intrauterine osteogenesis imperfecta in four siblings. British Medical Journal 5475: 99–101

Elejalde B R, Elejalde M M 1983 Prenatal diagnosis of perinatally lethal osteogenesis imperfecta. American Journal of Medical Genetics 14: 353–359

Goldfarb A A, Ford D 1954 Osteogenesis imperfecta congenita in consecutive siblings. Journal of Pediatrics 44: 264–268

Heller R H, Winn K J, Heller R M 1975 The prenatal diagnosis of osteogenesis imperfecta congenita. American Journal of Obstetrics and Gynecology 121: 572–573

Milsom I, Mattsson L A, Dahlen-Nilsson J 1982 Antenatal diagnosis of osteogenesis imperfecta by real time ultrasound. Two case reports. British Journal of Radiology 55: 310–312

Ojita S, Kamei T, Masumoto M, Shimamoto T, Shimura K, Kawamuia T, Sujawa T 1976 Prenatal diagnosis of osteogenesis imperfecta congenita by means of fetography. European Journal of Pediatrics 123: 170–186

Patel Z M, Shah H L, Madon P F, Ambani L M 1983 Prenatal diagnosis of lethal osteogenesis imperfecta by ultrasonography. Prenatal Diagnosis 3: 261–263

Sillence D O, Barlow K K, Garber A P, Hall J G, Rimoin D L 1984 Osteogenesis imperfecta Type II: Delineation of the phenotype with reference to genetic heterogeneity. American Journal of Medical Genetics 17: 407–423

Shapiro J E, Byers P H, Levin L S, Barsh G S, Golstain P 1982 Prenatal diagnosis of lethal perinatal osteogenesis imperfecta congenita. Journal of Pediatrics 100: 127–133

Spranger J, Cremin B, Beighton P 1982 Osteogenesis imperfecta congenita. Pediatric
    Radiology 12: 21–27
Stephens J D, Filly R A, Callen P W, Golbus M S 1983 Prenatal diagnosis of osteogenesis
    imperfecta type II by real-time ultrasound. Human Genetics 64: 191–193
Wilson M G 1974 Congenital osteogenesis imperfecta. Birth Defects (Original Article Series)
    12: 296–298

**Hypophosphatasia lethalis**
Benzie R, Doran T A, Escoffery W, Gardner H A, Hoar D I, Hunter A, Malone R,
    Miskin M, Rudd N L 1976 Prenatal diagnosis of hypophosphatasia. Birth Defects
    (Original Article Series) 12: 271
Burck U, Kaitila I I, Goebel H H, Hoikka V, Palotie L, Vanneuville F J, Pahnke V 1982
    Clinical, radiological, morphological and biochemical data on fetal congenital lethal
    hypophosphatasia. In: Skeletal Dysplasias. Papadatos C J, Bartsocas C S (eds) Liss, New
    York, p 149–154
Clark P J, Pryse-Davies J, Sandler M, Blau K, Rattenbury J M, Pooley P J 1976 Prenatal
    diagnosis of hypophosphatasia. Lancet 1: 306
Rudd N L, Miskin M, Hoar D I, Benzie R, Doran T A 1976 Prenatal diagnosis of
    hypophosphatasia. New England Journal of Medicine 295: 146
Shin-Buehring Y S, Santer R, Osang M, Schaub J 1981 Properties of human alkaline
    phosphatase and prenatal diagnosis of hypophosphatasia. Journal of Inherited Metabolic
    Diseases 4: 125
Wladimiroff J W, Niermeijer M F, van der Harten J J et al 1985 Early prenatal diagnosis
    of congenital hypophosphatasia: Case report. Prenatal Diagnosis 5: 47–52

**Fibrochondrogenesis**
Eteson D J, Adomian G E, Ornay A, Koide T, Sugiura Y, Calabro A, Lungarotti S,
    Mastroiacovo, Lachman R S, Rimoin D L 1984 Fibrochondrogenesis: Radiologic and
    histologic studies. American Journal of Medical Genetics 19: 277–290
Lazzaroni-Fossati F 1979 La fibrodiscondrogenesi. Minerva Pediatrica 31: 1273–1280
Lazzaroni-Fossati F, Stanescu V, Stanescu R, Serra G, Magliono P, Maroteaux P 1978 La
    fibrochondrogenese. Archives Françaises de Pédiatrie 35: 1096–1104
Whitley C B, Langer L O, Ophoven J, Gilbert E F, Gonzalex C H, Mammel M, Coleman
    M, Rosemberg S, Rodriques J, Sibley R, Horton W A, Opitz J, Gorlin R J 1984
    Fibrochondrogenesis: Lethal, autosomal recessive chondrodysplasia with distinctive
    cartilage histopathology. American Journal of Medical Genetics 19: 265–275

**Hypochondrogenesis**
Borochowitz Z, Ornoy A, Lachman R, Rimoin D 1986 Achondrogenesis II —
    hypochondrogenesis: Variability versus heterogeneity. American Journal of Medical
    Genetics 24: 273–288
Hendriks G, Hoefsloot F, Kramer P, Van Haelst U 1983 Hypochondrogenesis: An
    additional case. European Journal of Pediatrics 140: 278–281
Maroteaux P, Stanescu V, Stanescu R 1982 Four recently described osteochondrodysplasias.
    Skeletal Dysplasias. Liss, New York, p 345–350
Maroteaux P, Stanescu V, Stanescu R 1983 Hypochondrogenesis. European Journal of
    Pediatrics 141: 14–22

**Atelosteogenesis**
Kozlowski K, Bateson E M 1984 Atelosteogenesis. Fortschritte der Rontgenstrahlen
    140: 224–225
Kozlowski K, Tsuruta T, Kameda Y, Kan A, Leslie G 1981 New forms of neonatal death
    dwarfism: Report of three cases. Pediatric Radiology 10: 155–160
Maroteaux P, Spranger J, Stanescu V, Le Marec B, Pfeiffer R A, Beighton P, Mattei J F
    1982 Atelosteogenesis. American Journal of Medical Genetics 13: 715–25
Optiz J 1983 Editorial comment. American Journal of Medical Genetics 15: 629
Rimoin D L, Sillence D O, Lachmann R S, Jenkins T, Riccardi V 1980 Giant cell
    chondrodysplasia: A second case of a rare lethal newborn skeletal dysplasia. American
    Journal of Human Genetics 125: 32
Sillence D O, Lachman R S, Jenkins T, Riccardi V M, Rimoin D L 1982
    Spondylohumerofemoral hypoplasia (giant cell chondrodysplasia): A neonatally lethal

short-limbed skeletal dysplasia. American Journal of Medical Genetics 13: 7–14
Yang S S, Roskamp J, Liu C T, Frates R, Singer D B 1983 Two lethal chondrodysplasias
    with giant chondrocytes. American Journal of Medical Genetics 15: 615–625

**Dyssegmental dysplasia**
Bueno M, Argemi J, Maroteaux P 1984 Dysplasie dyssegmentaire. A propos de 2 cas
    familiaux d'evolution lethale. Archives Françaises de Pédiatrie 41: 269
Dinno N D, Shearer L, Weisskopf B 1976 Chondrodysplastic dwarfism, cleft palate and
    micrognathia in a neonate: A new syndrome? European Journal of Pediatrics 123: 39
Fasanelli S, Kozlowski K, Reiter S, Sillence D 1985 Dyssegmental dysplasia. Skeletal
    Radiology 14: 173–177
Goodlin R C, Lowe E W 1974 Unexplained hydramnios associated with thanatophoric
    dwarfism. American Journal of Obstetrics & Gynecology 118: 873
Gruhn J G, Gorlin R J, Langer L O 1978 Dyssegmental dwarfism. A lethal anisospondylic
    camptomicromelic dwarfism. American Journal of Diseases of Children 132: 382
Handmaker S D, Campbell J A, Robinson L D, Chinwach O, Gorlin R J 1979
    Dyssegmental dwarfism: A new syndrome of lethal dwarfism. Birth Defects 8: 79
Langer L O, Gonzales-Ramos M, Chen H, Espiritu C E, Courtney N W, Opitz J M 1976
    A severe infantile micromelic chondrodysplasia which resembles Kniest disease. European
    Journal of Pediatrics 123: 29
Miething R, Stover B, Tuengerthal S, Winterling D, Svejcar J 1981 Dyssegmentaler
    Zwergwuchs Bericht uber zwei Falle. Radiologe 21: 190
Rolland J C, Laugier J, Grenier B, Desbuquois G 1972 Nanisme chondrodystrophique et
    division palatine chez un nouveau-né. Annales de Pédiatrie (Paris) 19: 139
Silverman F N 1969 Discussion on the relation between stippled epiphyses and the
    multiplex form of epiphyseal dysplasia. Birth Defects: Original Article Series 5/4: 68
Svejcar J 1983 Chemical abnormalities in connective tissue of osteodysplasty of Melnick-
    Needles and dyssegmental dwarfism. Clinical Genetics 23: 369–375

**Boomerang dysplasia**
Kozlowski K, Sillence D, Cortis-Jones R, Osborn R 1985 Boomerang dysplasia. British
    Journal of Radiology 58: 369–371
Kozlowski K, Tsuruta T, Kameda Y, Kan A, Leslie G 1981 New forms of neonatal death
    dwarfism: Report of 3 cases. Pediatric Radiology 10: 155–160
Tenconi R, Kozlowski K, Largaiolli 1983 Boomerang dysplasia. Fortschritte auf dem
    Gebiete der Roentgenstrahlen 138: 378–380

**De la Chapelle dysplasia**
De la Chapelle A, Maroteaux P, Havu N, Granroth G 1972 Une rare dysplasie osseuse
    lethale de transmission récessive autosomique. Archives Francaises Pédiatrie 29: 759–770
Whitley C B, Burke B A, Granroth G, Gorlin R J 1986 De la Chapelle dysplasia. American
    Journal of Medical Genetics 25: 29–39

**Schneckenbecken dysplasia**
Borochowitz Z, Jones K L, Silbey R, Adomian G, Lachman R, Rimoin D L 1986 A
    distinct lethal neonatal chondrodysplasia with snail-like pelvis: Schneckenbecken
    dysplasia. American Journal of Medical Genetics 25: 47–59
Chemke J, Graff G, Lancet M 1971 Familial thanatophoric dwarfism. Lancet 1: 1358
Graff G, Chemke J, Lancet M 1972 Familial recurring thanatophoric dwarfism. Obstetrics
    and Gynecology 39: 515–520
Knowles S, Winter R, Rimoin D 1986 A new category of lethal short-limbed dwarfism.
    American Journal of Medical Genetics 25: 41–46
Laxova R, O'Hara P T, Ridler M A C, Timothy J A D 1973 Family with probable
    achondrogenesis and lipid inclusions in fibroblasts. Archives of Disease in Childhood
    48: 212–216

**Miscellaneous, new, potentially lethal, neonatal dwarfing dysplasias**
Barrow M, Fitzsimmons J S 1984 A new syndrome: Short limbs, abnormal facial
    appearance and congenital heart defect. American Journal of Medical Genetics
    18: 431–433

Beemer F A, Kramer P, van der Harten H J, Gerards L J 1985 A new syndrome of dwarfism, neonatal death, narrow chest, spondylometaphyseal abnormalities and advanced bone age. American Journal of Medical Genetics 20: 555–558

Camera G, Giordano F, Mastroiacovo P 1986 Pyknoachondrogenesis: An association of skeletal defects resembling achondrogenesis with generalized bone sclerosis. A new condition? Clinical Genetics 30: 335–337

Colavita N, Kozlowski K 1984 Neonatal death dwarfism: A new form. Pediatric Radiology 14: 451–452

Kozlowski K, Tsuruta T, Taki N, Tsunoda I, Ozawa H, Hasegawa T, Sillence D 1986 A new type of achondrogenesis. Pediatric Radiology 16: 430–432

McAlister W H, Crane J P, Bucy R P, Craig R B 1985 A new neonatal short limbed dwarfism. Skeletal Radiology 13: 271–275

Moermann P, Vandenberghe K, Fryns J P, Haspeslagh M, Lauweryns J M 1985 A new lethal chondrodysplasia with spondylocostal dysostosis, multiple internal anomalies and Dandy-Walker cyst. Clinical Genetics 27: 160–164

# 4

# Disorders with disorganised development of cartilage and fibrous tissue

The conditions in this category share the feature of disorganised development of cartilage and fibrous tissue. Some are transmitted in a Mendelian fashion while others are apparently non-genetic. In view of the underlying embryological relationships and occasional overlap in clinical manifestations, the latter conditions have also been included in this chapter.

1. Dysplasia epiphysealis hemimelica
2. Multiple cartilagenous exostoses (diaphyseal aclasis)
3. Enchondromatosis (Ollier)
   a. Metachondromatosis
   b. Upington bone disease
   c. Spondyloenchondrodysplasia
4. Enchondromatosis with haemangiomata (Maffucci)
5. Neurofibromatosis (von Recklinghausen)
6. Fibrous dysplasia
   a. Monostotic and polyostotic fibrous dysplasia
   b. Fibromatoses
   c. Syndromic fibromatosis
   d. Craniofacial fibrous dysplasia
   e. Cherubism

## DYSPLASIA EPIPHYSEALIS HEMIMELICA

Dysplasia epiphysealis hemimelica (DEH) is a rare localised disorder of cartilage which was described by Trevor* (1950) and discussed in detail by Fairbank (1956). Kettelkamp et al (1966) presented 15 cases and reviewed the literature and a further 15 were added by Azouz et al (1985).

Pain and swelling are presenting features and diagnostic confirmation is obtained by recognition of the characteristic radiographic changes. Areas of sclerosis may be evident in the epiphyses of a single segment of a limb.

* David Trevor, whose name is sometimes applied to DEH, was a consultant at the Charing Cross and Royal National Orthopaedic Hospitals, London.

The lateral aspect of the lower end of the tibia and the corresponding bones of the tarsus and metatarsus are the sites of predilection, although the knee joint may also be involved. Onset usually occurs during childhood and growth of the affected bones may be disturbed. The disorder becomes quiescent following epiphyseal fusion but residual deformity persists. The orthopaedic implications of DEH have been discussed by Wolfgang & Heath (1976) and Fasting & Bjerkreim (1976). The changes are almost always confined to one side of a leg but, in exceptional cases, all limbs may be affected (Wiedemann et al 1981).

Apart from a report by Hensinger et al (1974) of a family in which DEH was seemingly transmitted as an irregular dominant, all patients mentioned in the literature have been sporadic. In a study of nine new cases and a long-term follow-up of seven others, there was no evidence to indicate any genetic factor in the aetiology of DEH (Connor et al 1983).

## MULTIPLE CARTILAGINOUS EXOSTOSES (Diaphyseal Aclasis)

Diaphyseal aclasis is one of the most common inherited skeletal disorders. A collection of 1100 cases was reported more than 60 years ago (Stocks & Barrington 1925) and since that time other large series have been published (Krooth et al 1961, Murken 1963, Solomon 1963).

### Clinical and radiographic features

Multiple bony swellings make their appearance in infancy and increase both in number and in size until growth ceases. The ends of the long bones, the pelvis and shoulder girdle are most commonly involved, while the skull and spine are usually spared. Infrequently, there is severe skeletal deformity and growth may be impaired. Problems arise from pressure upon tendons, nerves and blood vessels. Pelvic distortion may preclude normal parturition. The most important complication is malignant degeneration, as sarcoma is said to develop in adulthood in 10–20% of patients (McGuire & Reinert 1985). However, this figure may be an exaggeration of the true situation, and a risk estimate of about 3% may be realistic (Gordon et al 1981). There is considerable variation in the degree to which members of the same kindred are affected but, in general, the disorder is more severe in males than in females.

Radiological examination invariably reveals many more bony lesions than are clinically evident. The juxta-epiphyseal regions are the site of predilection in childhood and later in life the exostoses migrate towards the diaphyses. Irregular expansion of the metaphyses often interferes with normal bone development, resulting in malalignment of joints and limb deformity. A 'Madelung' configuration of the forearm is a common consequence of this disturbance of growth (Wood et al 1985).

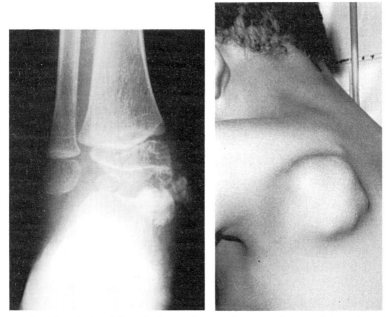

**Fig. 4.1** (left) Dysplasia epiphysealis hemimelica: involvement of the medial portion of the distal tibial epiphysis. (From Connor et al 1983 Journal of Bone and Joint Surgery 65B: 350–354.)

**Fig. 4.2** (right) Diaphyseal aclasis: a bony mass over the left scapula.

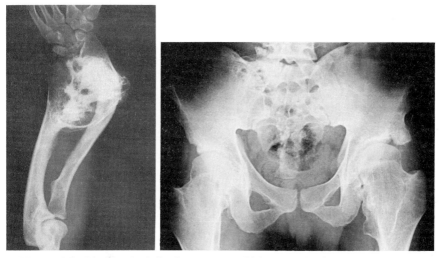

**Fig. 4.3** (left) Diaphyseal aclasis: the exostoses, which are usually found at the ends of the long bones, may interfere with growth and lead to deformity.

**Fig. 4.4** (right) Diaphyseal aclasis: the exostoses may be very numerous. Involvement of the pelvis may prevent normal childbearing.

*Genetics*

Autosomal dominant inheritance is well documented. An apparent excess of affected males led Harris (1948) to postulate that phenotypic expression might be influenced by a sex-limited modifying gene. However, when full radiographic studies have been carried out in kindreds, asymptomatic affected females have been detected and the equal sex ratio has been re-established. A single instance of non-penetrance was recognised in a six-generation family in which 35 persons had diaphyseal aclasis (Crandell et al 1984).

About 30% of affected individuals apparently represent new mutations of the gene. In his investigation in Germany, Murken (1963) calculated that the mutation rate lay between six and nine per million. Diaphyseal aclasis has a wide geographic distribution but it is particularly common on the Pacific Island of Guam. It has been established that the prevalence is about 1 in 1000 in the islanders (Krooth et al 1961).

As diaphyseal aclasis is relatively common and readily recognisable, it would lend itself to linkage studies. Investigations of this type yielded negative results two decades ago (Scholz & Murken 1963) but as new gene markers become available, repetition and augmentation of these studies might be worthwhile. Diaphyseal aclasis is amongst the few dominant disorders in which a possible homozygote has been encountered. Giedion et al.(1975) studied two brothers who were the offspring of affected parents and suggested that the severe and precocious manifestations in these boys might be indicative of homozygosity.

Solomon (1963) identified a kindred in which eight individuals in three generations had multiple exostoses which were virtually confined to the bones of the hands. As the intrafamilial manifestations of diaphyseal aclasis are usually very variable, it is possible that this particular condition is a unique genetic entity which differs from the common form of the disorder.

Exostoses are present in the Langer-Giedion form of acrodysplasia which is also known as trichorhinophalangeal dysplasia type II. This disorder is reviewed in Chapter 10.

## ENCHONDROMATOSIS (Ollier)*

Enchondromatosis (Ollier disease) is much less common than diaphyseal aclasis, with which it is sometimes confused. However, there is little similarity in the manifestations of these disorders and diagnostic distinction is not difficult.

---

* Louis Xavier Edouard Ollier (1830–1900) was a founder of French orthopaedic surgery. He spent his career at Lyons, where he developed successful new operative techniques for the management of war wounds.

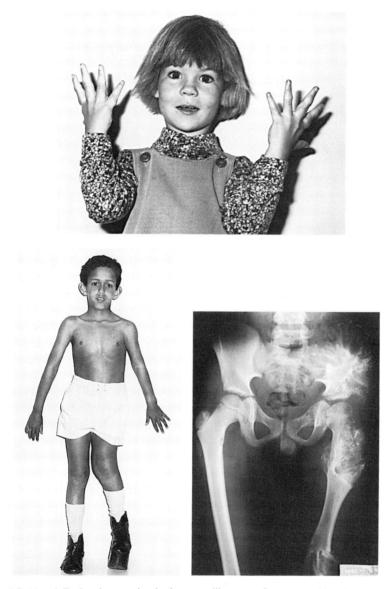

**Fig. 4.5** (above) Enchondromatosis: the bony swellings are often most evident in the digits.

**Fig. 4.6** (left) Enchondromatosis: the enchondromata may interfere with growth, producing marked disparity in limb length.

**Fig. 4.7** (right) Enchondromatosis: radiograph of the boy depicted in Figure 4.6. In the pelvis, the fan-like lesions radiate towards the iliac crest. The upper part of the shaft of the femur is expanded and distorted.

*Clinical and radiographic features*

Multiple bony swellings, particularly of the digits, are the presenting feature of Ollier disease. These enchondromatous lesions become static or regress after puberty. Localised disturbance of growth by enchondromata in the metaphyseal regions of long bones may produce deformity and limb asymmetry. Pathological fracture is a well recognised complication and the development of malignancy has been reported (Cannon & Sweetnam 1985). Precise figures are not available but this latter problem seems to be uncommon. Radiographically, the enchondromata appear as multiple, lucent defects. The digits and tubular bones are predominantly involved, while the spine and skull are spared. The orthopaedic management of a series of 21 patients with Ollier disease was reviewed by Shapiro (1982).

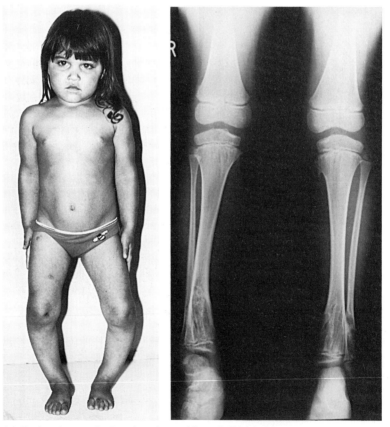

**Fig. 4.8** Enchondromatosis: bow legs due to bilateral tibial involvement.

**Fig. 4.9** Enchondromatosis: radiograph of the girl depicted in Figure 4.8. Symmetrical lesions in the lower tibial metaphyses.

*Genetics*

Enchondromatosis is usually considered to be non-genetic and the majority of reports have been concerned with sporadic individuals. Sibs have been reported by Lamy et al (1954), while Rossberg (1959) described a brother and sister with the condition, whose paternal grandfather had also been affected. McKusick (1983) suggested that these observations were compatible with autosomal dominant inheritance with reduced penetrance.

## Metachondromatosis

Metachondromatosis is an innocuous disorder in which multiple exostoses and enchondromata occur together. The exostoses are usually confined to the digits and frequently regress spontaneously. The enchondromata are found in the iliac crests and metaphyses of the long bones and are irregular or striated. Apart from minor deformity at the site of the lesions, affected persons are asymptomatic. Growth and stature are not compromised and there have been no reports of malignant degeneration.

The condition was delineated by Maroteaux (1971) when he studied two affected families and a further example of generation to generation transmission was reported by Giedion et al (1975), Kozlowski & Scougall (1975) suggested that metachondromatosis often presented as diaphyseal aclasis; this contention was born out in a survey of 91 families in Oregon with multiple exostoses amongst whom Beals (1982) identified three kindreds with metachondromatosis. Autosomal dominant inheritance was confirmed after Hinkel et al (1984) and Bassett & Cowell (1985) published pedigrees in which members of three generations were affected.

## Upington bone disease

A distinct disorder in which multiple enchondromata and ecchondromata were associated with Perthes-like changes in the femoral heads has been observed in 13 members of three generations of an Afrikaner kindred (Schweitzer et al 1971). The authors reported the condition under the title 'Upington disease: a familial dyschondrodysplasia'. This geographic appellation pertained to the patients' origin in a rural area in South Africa. In this kindred the pattern of transmission was consistent with autosomal dominant inheritance.

## Spondyloenchondrodysplasia

Spondyloenchondrodysplasia comprises multiple enchondromata and platyspondyly. The initial report concerned Iraqi-Jewish brothers with short stature (Schorr et al 1976). Their unaffected parents were consanguineous and autosomal recessive inheritance seemed likely. Further cases have been described by Spranger et al (1978) and Chagnon et al (1985) and the question of heterogeneity has arisen.

## ENCHONDROMATOSIS WITH HAEMANGIOMATA (Maffucci)*

A patient with enchondromatosis and haemangiomata was described by Maffucci in the last century and the syndromic relationship of these anomalies was firmly established by Carleton et al (1942). At that time about 20 cases had been reported. This total had risen to 62 when the literature was reviewed two decades later (Anderson 1965), while Gutman et al (1978) were able to identify more than 100 reported cases.

### Clinical and radiographic features

The components of the Maffucci syndrome are single or multiple enchondromata and cutaneous angiomatous lesions. The radiographic changes resemble those of Ollier disease although bone lesions do not have the same predilection for the digits. The skeletal anomalies appear during childhood and progress until growth ceases. Limb deformity of significant degree is present in at least 50% of cases. Cavernous or capillary haemangiomata of the skin are often evident at birth, although their appearance is sometimes delayed until infancy. These lesions may be widespread, with angiomatous involvement of internal organs (Allen 1978).

The most important consequence of the Maffucci syndrome is chondrosarcoma, which arises in about 17% of patients (Sun et al 1985). It has been suggested that the predisposition to neoplasia is not limited to the skeletal system and that there is an increased risk of malignancy at other sites (Lowell & Mathog 1979). Haemangiomata of the tongue occur in about 1% of affected persons (Laskaris & Skouteris 1984) and pose special problems in management (Ma & Leung 1984).

### Genetics

The Maffucci syndrome must be distinguished from the Klippel-Trenaunay-Weber syndrome, in which haemangiomata are associated with bone hypertrophy, in the absence of enchondromata. Vaz & Turner (1986) pointed out that the Maffucci syndrome and Ollier disease shared a propensity for the development of mesenchymal ovarian tumours and suggested that these disorders might be part of a spectrum rather than separate entities. A report by Phelan et al (1986) concerning a child with enchondromatosis, haemangiomata, soft tissue calcifications and hemihypertrophy has further complicated the nosological situation. Uncomplicated hemihypertrophy also has features in common with these disorders and in the individual patient accurate categorisation can be difficult (see Ch. 19).

---

* Angelo Maffucci (1847–1903) was an eminent Italian pathologist. He achieved recognition for his research into tuberculosis and became professor of anatomical pathology at the University of Pisa.

The essential features of this group of disorders are summarised in Table 4.1.

**Table 4.1**

| Disorder | Manifestations | Inheritance |
|---|---|---|
| Diaphyseal aclasis | Multiple exostoses | AD |
| Ollier disease | Multiple enchondromata | Non-genetic |
| Metachondromatosis | Multiple exostoses and enchondromata | AD |
| Upington bone disease | Enchondromata, ecchondromata and Perthes-like changes in the femoral heads | AD |
| Spondyloenchondrodysplasia | Platyspondyly with enchondromata | AR? |
| Maffucci syndrome | Enchondromata and haemangiomata | Non-genetic |
| Klippel-Trenaunay-Weber syndrome | Bone and soft tissue hypertrophy and haemangiomata | Non-genetic |
| Hemihypertrophy | Unilateral skeletal and soft tissue overgrowth. No haemangiomata | Non-genetic |

## NEUROFIBROMATOSIS

Neurofibromatosis, or von Recklinghausen* disease, is a well known disorder which is inevitably encountered at some stage in every medical career — frequently during viva voce examination for a specialist qualification! The clinical manifestations can be very severe and for many years it was thought that the 'elephant man' of the London hospital had this disorder. A diagnosis of the proteus syndrome is more probable (see Ch. 19). A recent monograph devoted to neurofibromatosis contains a full account of the phenotypic manifestations, natural history and pathogenesis (Riccardi & Eichner 1986).

### Clinical and radiographic features

Multiple pedunculated and sessile dermal tumours may be very numerous, while café-au-lait pigmented macules are invariably present. (The smooth outline of these patches has been likened to the 'coast of California' in distinction to the irregular 'coast of Maine' configuration of the macules in polyostotic fibrous dysplasia. As these conditions bear a resemblance to each other, this geographic concept is of practical value.) Plexiform neuromata occasionally develop into pendulous lesions, while neuromata on periph-

---

* Frederick Daniel von Recklinghausen (1833–1916) was an outstanding German pathologist during the last three decades of the 19th century. He was a pupil of the famous Professor Rudolf Virchow at the Pathological Institute, Berlin and ultimately became Rector of the University of Strasburg, Alsace.

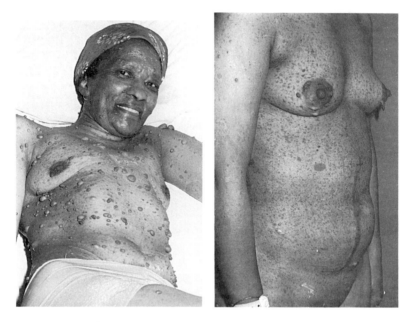

Fig. 4.10 (left) Neurofibromatosis: dermal tumours may cover the whole body. Phenotypic expression of this autosomal dominant gene is extremely variable.

Fig. 4.11 (right) Neurofibromatosis: the café-au-lait macule which is present over the right costal margin has the typical smooth 'coast of California' configuration.

eral nerves or in the spine sometimes cause neural compression. An acoustic neuroma may result in deafness. Phaeochromocytoma and other endocrine tumours occur and malignant degneration is not uncommon. Skeletal manifestations include vertebral abnormalities, rib fusion, pseudoarthrosis of the tibia and hypertrophy of a limb or digit. In a Scandinavian series of 3209 patients with scoliosis 3% had neurofibromatosis (Rezaian 1976). Apter et al (1975) have emphasised that the rate of development and the extent of the abnormalities is very variable, even in members of the same kindred.

Lisch nodules represent an important diagnostic indicator as these punctate pigmented hamartomatous lesions of the iris are pathognomonic for neurofibromatosis. They are best seen by slit-lamp examination, although they may be visible to the naked eye or with an ophthalmoscope. They were first documented by Waardenburg (1918) and received their name following a report by Lisch (1937). In affected persons the number of Lisch nodules increases with age and with the severity of skin lesions (Lewis & Riccardi 1981). A perspective of the diagnostic value of Lisch nodules can be gleaned from the fact that they were detected in 73% of a series of 30 patients with neurofibromatosis (Zehavi et al 1986).

Macromelanosomes, or giant pigment granules, are a useful morphologic marker in the diagnosis of neurofibromatosis. These bodies can be seen on

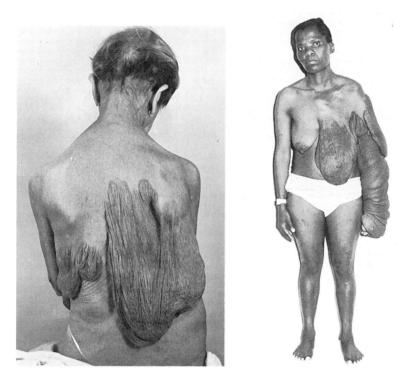

**Fig. 4.12** (left) Neurofibromatosis: pendulous dermal lesions.

**Fig. 4.13** (right) Neurofibromatosis: massive overgrowth of the soft tissues.

routine microscopy of skin biopsy specimens and although not pathognomonic, their presence in a potentially affected person would be highly
significant (Jimbow et al 1973). It is of considerable interest that in a series
of 10 patients of Indian stock, none had macromelanosomes (Slater et al
1986). This observation might be indicative of heterogeneity.

*Genetics*

Neurofibromatosis is one of the commonest autosomal dominant disorders.
The genetic basis was extensively reviewed by Crowe et al (1956). These
authors estimated that about 50% of patients represented new gene mutations and that the mutation rate was approximately $1 \times 10^{-4}$. If this figure
is accurate then neurofibromatosis bears the dubious distinction of having
the highest mutation rate of any clinically important genetic condition.
However, in view of the notorious variability of expression of the gene, it
is likely that neurofibromatosis is under-diagnosed in parents of sporadic
cases and that the quoted mutation rate is erroneously high. In family
studies, skipped generations are not unusual. Nevertheless, minor stigmata

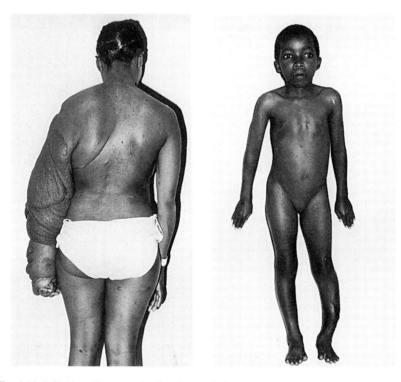

**Fig. 4.14** (left) Neurofibromatosis: the characteristic dermal macules and neurofibromata are evident on the trunk and limbs.

**Fig. 4.15** (right) Neurofibromatosis: pseudoarthrosis of the left tibia causing localised deformity. The pigmented macules on the truck are indicative that neurofibromatosis is the underlying disorder.

are often recognisable on close examination of ostensibly normal obligatory carriers of the gene. In this context it is of practical importance that the dermal manifestations of neurofibromatosis are progressive and that they may not become evident until mid-childhood.

If new mutation in neurofibromatosis is indeed a common event, the question of possible association with advanced parental age arises. In a formal analysis of 187 affected persons with normal parents, Riccardi et al (1984) calculated mean paternal and maternal ages of 32.8 years and 27.4 years, respectively. Both ages were significantly greater than those of control populations.

Ansari & Nagamani (1976) described a woman in whom neurofibromata increased in size in late pregnancy. The authors considered that the risk of this complication, together with the dominant mode of transmission of the gene, would justify abortion. Other complications which have been reported in pregnancy include sarcoma (Ginsburg et al 1981), severe hyper-

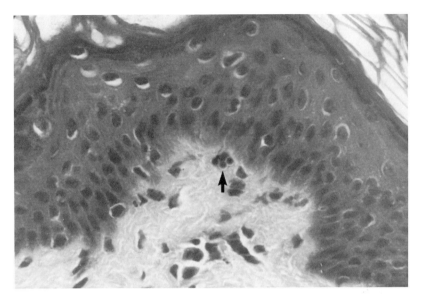

**Fig. 4.16** Neurofibromatosis: giant pigment granules are evident in the basal layer and papillary dermis (arrowed). These macromelanosomes are an important diagnostic indicator. (From Slater et al 1986 American Journal of Dermatopathology 8(4): (284–289.)

tension (Edwards et al 1983) and exacerbation of maternal dermal stigmata (Miller & Hall 1978, Jarvis & Crompton 1978).

For the purposes of genetic counselling and prognostication, the individual with minimal clinical manifestations poses considerable problems. It is generally assumed that the diagnosis can be confirmed or refuted on a basis of criteria such as the presence of a minimum number of café-au-lait macules which must be of a certain size. However, as the phenotype is very variable, a rigid approach will inevitably lead to diagnostic errors and inappropriate counselling.

The term 'central neurofibromatosis' has come into use for the association of acoustic neuromata with characteristic dermal lesions. This disorder is far less common than classical or peripheral neurofibromatosis and independent syndromic status is not fully established. The mode of inheritance is autosomal dominant (Huson & Thrush 1985).

Atypical cases of neurofibromatosis are not uncommon and it is sometimes difficult to reach a firm diagnosis. In addition, there may well be poorly delineated entities which resemble neurofibromatosis. For instance, Bradley et al (1974) reported a mother and two daughters who shared the features of neurofibromatosis, congenital deafness due to neuronal degeneration, partial albinism and a defect of the iris. The condition in this kindred could well be a 'private' syndrome.

The association of neurofibromatosis and the Noonan syndrome (stunted stature, webbed neck, learning disability) has recently attracted attention.

Four unrelated patients with the stigmata of both disorders were documented by Allanson et al (1985) and these authors suggested that this disorder might be a distinct entity. In a further report concerning three affected children, Kaplan & Rosenblatt (1985) emphasised the unusual facial features in this disorder.

For some time it was thought that the gene for neurofibromatosis might be linked to Gc on chromosome 4 or to the secretor locus and myotonic dystrophy on chromosome 19. Linkage analysis in 15 families, using molecular techniques, has now indicated that the locus for neurofibromatosis is in the pericentromeric region of chromosome 17 (Barker et al 1987).

In view of the high frequency and clinical importance of neurofibromatosis it can be anticipated that molecular technology will be applied to the disorder. The development of a gene probe or closely linked restriction fragment polymorphism for the condition would resolve the problems of diagnostic confirmation and prenatal diagnosis and elucidate the question of heterogeneity. Nerve growth factor has been implicated in the pathogenesis of neurofibromatosis (Schenkein et al 1974, Siggers et al 1975, Fabricant & Todaro 1981) but molecular studies have failed to demonstrate any alteration in the nerve growth factor gene in affected persons (Darby et al 1985).

## FIBROUS DYSPLASIA

### Monostotic and polyostotic fibrous dysplasia

The association of fibrosis with cystic lesions of bone is well recognised. If the fibrocystic changes are limited to one bone, the term 'monostotic fibrous dysplasia' is employed, while 'polyostotic fibrous dysplasia' implies widespread skeletal involvement. In both disorders irregular patches of light brown macular (café-au-lait) pigmentation may be present, not necessarily in anatomical relationship with the affected bone. The condition may remain unrecognised until pathological fracture occurs but if bone involvement is extensive, deformity may be severe. The cystic radiolucent lesions may be small and discreet, or may occupy and distort the whole shaft of the affected bone. Sarcomatous changes are uncommon.

The condition in this group which has attracted the most attention is the McCune-Albright syndrome. This disorder was described by McCune (1936) under the title 'osteitis fibrosa cystica' and further delineated by Albright et al (1937) when the association of polyostotic dysplasia with café-au-lait cutaneous pigmentation and precocious puberty was emphasised. Females are predominantly affected and secondary sexual characteristics, including pubic hair, breast enlargement and menstruation appear during the first decade. In many patients the epiphyses close prematurely, resulting

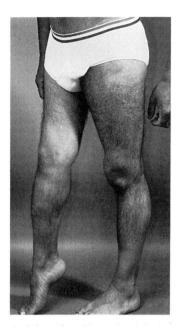

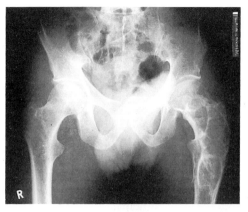

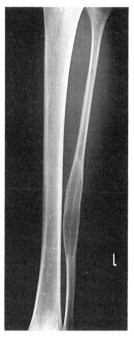

**Fig. 4.17** (above, left) Fibrous dysplasia. Shortening and deformity of the right leg, due to fibrous dysplasia of the femur and tibia.

**Fig. 4.18** (above, right) Fibrous dysplasia. Expansion and lucency of the upper portion of the left femur.

**Fig. 4.19** (right) Fibrous dysplasia. An isolated mon-ostotic fibrous lesion in the shaft of the fibula.

in failure to achieve normal adult height. About 3% of individuals in the polyostotic dysplasia group have the full McCune-Albright syndrome.

Lightner et al (1975) have suggested that the endocrinopathies in the McCune-Albright* syndrome are the result of hypothalamic dysfunction.

---

* Donovan James McCune (1902–1976) was a paediatrician at Columbia University, New York.

Fuller Albright (1900–1969) was a distinguished endocrinologist at the Massachusetts General Hospital, Harvard, USA. He made many original contributions and his name is now attached to several skeletal disorders.

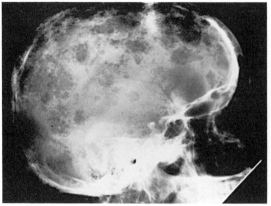

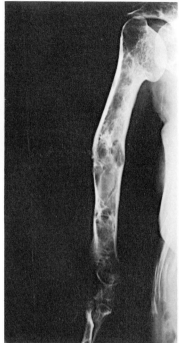

**Fig. 4.20** (left) Polyostotic fibrous dysplasia: skull radiograph of a severely affected girl.

**Fig. 4.21** (right) Polyostotic fibrous dysplasia: this girl had multiple bony lesions and areas of irregular café-au-lait dermal pigmentation. However, as sexual development was not precocious, an initial diagnosis of the McCune-Albright could not be substantiated.

The development of acromegaly in a patient with polyostotic fibrous dysplasia and a pituitary abnormality was reported by Harris (1985) and the pathogenesis of precocious puberty in the condition has been reviewed by Bost et al (1985). The role of radiographic procedures, together with ultrasound and computerised tomographic screening in the evaluation of endocrine involvement has been reviewed by Rieth et al (1984).

In the Paris Nomenclature the term 'Jaffe-Lichtenstein syndrome' is applied to polyostotic fibrous dysplasia without skin pigmentation or sexual precocity. Strictly speaking, this is a misnomer as the original review article concerning 23 cases published by Lichtenstein & Jaffe* (1942) made special mention of cutaneous involvement and precocious puberty. The nosological situation is further complicated by the fact that both authors had written earlier reports of disorders involving bone fibrosis. Although the use of these eponyms is open to question, their retention is probably warranted in order to avoid further confusion.

---

* Henry Jaffe (1896–1979) was the doyen of North American bone pathologists. He was director of laboratories at the Hospital for Joint Diseases, New York, for 40 years and established a formidable reputation.

Louis Lichtenstein (1906–1977) was a pathologist at the Hospital for Joint Diseases during his collaboration with Jaffe. He subsequently occupied academic posts in San Francisco and Los Angeles.

A simple classification of non-genetic monostotic and polyostotic fibrous dysplasia is given below:

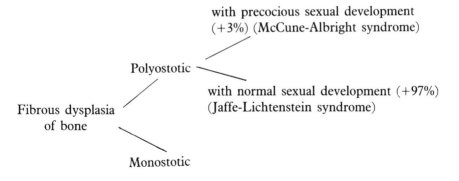

Fibrous dysplasia of bone

Polyostotic
- with precocious sexual development (+3%) (McCune-Albright syndrome)
- with normal sexual development (+97%) (Jaffe-Lichtenstein syndrome)

Monostotic

The vast majority of persons with monostotic or polyostotic fibrous dysplasia are sporadic and there is very little evidence for any genetic aetiology. However, there has been a report of a Mexican family in which several members had variable osseous and cutaneous manifestations of polyostotic fibrous dysplasia (Alvarez-Arratia et al 1983). The authors suggested that the condition is this kindred was inherited as an autosomal dominant trait.

Happle (1986) observed that the skin changes in the McCune-Albright syndrome followed the embryonic lines of Blaschko. On this basis he advanced the fascinating hypothesis that the syndrome is the consequence of mosaicism for a mutant autosomal dominant gene which causes early loss of the zygote in the non-mosaic situation. This concept would explain the asymmetrical distribution of the bony lesions and the variable endocrine involvement.

### Fibromatoses

This category contains disorders in which localised or generalised proliferation of fibrous tissue is the major feature. The nomenclature is very confused and these syndromes themselves are poorly delineated. Some of these disorders, notably aggressive fibromatosis (Griffiths et al 1983) are apparently non-genetic, whilst others seem to be heritable (Young & Fortt 1981).

In multiple fibromatosis, metaphyseal lesions are associated with fibrous soft tissue nodules (Stout 1954). The bone changes resemble those of Ollier disease (Schlangen 1976). Using the designation 'congenital generalised fibromatosis' Heiple et al (1972) reviewed the features of 24 patients. In some, bone involvement was minimal or absent. It is not clear if this variability is indicative of heterogeneity. Baird & Worth (1976) reported two sets of affected sibs in a consanguineous kindred, pointed out that the disorder had never been recorded in a parent of a child with the condition and postulated that inheritance was autosomal recessive.

## Syndromic fibromatosis

Excessive fibrosis at specific anatomical sites underlies several well-defined syndromes. Although the skeleton is not necessarily involved in these disorders, they are included for the sake of completion.

*Léri\* pleonosteosis* is a rare disorder in which moderate shortening of stature and broadening of the thumbs is associated with flexion contractures of the digits and extremities. Excessive fibrosis is present in the fascia, joint capsules, tendons and ligaments (Watson-Jones 1949). There have been several familial instances and inheritance is probably autosomal dominant (Léri 1922, Rukavina et al 1959).

*Dupuytren\*\* contracture* is a common disorder in which fibrosis of the palmar fascia causes progressive flexion deformities of the fingers, notably the fourth and fifth digits. There is a predominance of affected males and the genetic background is uncertain. There have been several reports of familial aggregation (Ling 1963, Lygonis 1966, Maza & Goodman 1968, Matthews 1979) and autosomal dominant inheritance with partial sex-limitation and frequent non-penetrance seems likely.

*Peyronie\*\*\* disease* presents in middle age as curvature of the penis, due to fibrous induration. The deformity is painless, slowly progressive and is most evident during erection. Familial aggregation has been recorded and there is an association with Dupuytren contracture (Nyberg et al 1982). The mode of inheritance is uncertain; an anonymous contributor to the Catalogue of Mendelian disorders, speaking — perhaps — from personal experience, suggested that Peyronie disease was sex-linked with reduced penetrance! (McKusick 1983). Wilma Bias and her colleagues studied three families with the disorder, recorded affected males in three generations and took the standpoint that the condition was a male-limited autosomal dominant trait (Bias et al 1982).

## Craniofacial fibrous dysplasia

Fibrous dysplasia of the bones of the skull, especially the mandible and maxilla, is an uncommon but well-recognised problem in otological and

---

\* Andre Léri (1875–1930) was a Parisian physician who published many articles on diverse topics. He is mainly remembered for his research into skeletal disorders.

\*\* Guillaume Dupuytren (1777–1825) was the pre-eminent French surgeon in the early years of the 19th century. He had a forceful personality and was known to his colleagues as the 'Napoleon of surgery'.

\*\*\*Francois de la Peyronie (1678–1747) was surgeon to King Louis XIV and a founder of the Royal Academy of Surgery, Paris.

dental practice. Several non-genetic conditions present in this way; syndromic boundries are ill-defined and their independent status is uncertain. Large series of cases have been reported by Zimmerman et al (1958), Waldron & Giansanti (1973) and Ramsey et al (1975). Some forms of craniofacial dysplasia are comparatively benign while others cause severe deformity and pose problems in management (Zanini et al 1985).

Using the term 'fibro-osseous dysplasia of the jaw' Chatterjee & Mazumder (1967) depicted a father and two sons who had gross facial abnormalities. This condition seems to be an autosomal dominant trait. Familial craniofacial polyostotic dysplasia was reported by Reitzik & Lownie (1975). A milder disorder involving the mandible, congenital monostotic fibrous dysplasia, is inherited as an autosomal recessive (El Deeb et al 1979).

## Cherubism

The descriptive designation 'cherubism' is applied to a rare disorder in which expansion of the maxilla and mandible produce an 'angelic' or 'cherubic' facies. This appearance is the result of multiple cystic changes in the jaw bones and is accompanied by hyperplasia of the submandibular lymph nodes. Multilocular cysts may also be present in the anterior ends of the ribs. Cherubism appears during the first year of life and progresses until the skeleton is mature. The hereditary nature of cherubism was discussed by Anderson & McClendon (1962) who concluded that the disorder was inherited as an autosomal dominant with varying expression. Jones (1965) concurred with this opinion. Subsequently, generation to generation transmission was reported by Salzano & Ebling (1966), Khosla & Korobkin (1970), Wayman (1978) and Wechsler & Grellet (1984). An impressive pedigree containing 20 cherubs in a single family was published by Peters (1979). Autosomal dominant inheritance was confirmed and penetrance was stated to be 80%.

An autosomal recessive form of cherubism in which fibrous dysplasia of the axillae was associated with mental retardation, epilepsy, gingival fibromatosis, hypertrichosis and stunted growth was recognised in a consanguineous Brazilian family by Pinn-Neto et al (1986). Of the four affected children, three also had juvenile rheumatoid arthritis, which the authors regarded as a syndromic component. The condition, without articular involvement, had previously been described in two siblings from a consanguineous union by Ramon et al (1967).

REFERENCES

**Dysplasia epiphysealis hemimelica**
Azouz E H, Slomic A M, Marton D, Rigault P, Finidori G 1985 The variable manifestations of dysplasia epiphysealis hemimelica. Pediatric Radiology 15: 44–49
Connor J M, Horan F T, Beighton P 1983 Dysplasia epiphysealis hemimelica: A clinical and genetic study. Journal of Bone and Joint Surgery 65B: 351–354

Fairbank T J 1956 Dysplasia epiphysialis hemimelica (tarso-epiphysial aclasis). Journal of Bone and Joint Surgery 38B: 257

Fasting O J, Bjerkreim I 1976 Dysplasia epiphysealis hemimelica. Acta Medica Scandinavica 47/2: 217

Hensinger R N, Cowell H R, Ramsey P L, Leopold R G 1974 Familial dysplasia epiphysealis hemimelica associated with chondromas and osteochondromas: Report of a kindred with variable presentations. Journal of Bone and Joint Surgery 56A: 1513–1516

Kettlekamp D B, Campbell C J, Bonfiglio M 1966 Dysplasia epiphysialis hemimelica: A report of 15 cases and a review of the literature. Journal of Bone and Joint Surgery 48A: 746

Trevor D 1950 Tarso-epiphyseal aclasis: A congenital error of epiphyseal development. Journal of Bone and Joint Surgery 32B: 204

Wiedemann H R, Mann M, von Kreudenstein P S 1981 Dysplasia epiphysealis hemimelica — Trevor disease: severe manifestations in a child. European Journal of Pediatrics 136: 311–316

Wolfgang G L, Heath R D 1976 Dysplasia epiphysealis hemimelica: A case report. Clinical Orthopaedics and Related Research 116: 32

### Multiple cartilaginous exostoses

Crandall B F, Field L L, Sparkes R S, Spence M A 1984 Hereditary multiple exostoses: Report of a family. Clinical Orthopaedics 190: 217–219

Giedion A, Kesztler R, Muggiasca F 1975 The widening spectrum of multiple cartilaginous exostosis (MCE). Paediatric Radiology 3/2: 93

Gordon S L, Buchanan J R, Ladda R L 1981 Hereditary multiple exostoses: Report of a kindred. Journal of Medical Genetics 18: 428–430

Harris H 1948 A sex-limiting modifying gene in diaphyseal aclasis. Annals of Eugenics 14: 165

Krooth R S, Macklin M A P, Hillbish T F 1961 Diaphyseal aclasis (multiple exostoses) on Guam. American Journal of Human Genetics 13: 340

McGuire R A, Reinert C M 1985 Hereditary multiple exostosis. Orthopaedic Review 14: 675–680

Murken J D 1963 Uber muiltiple cartilaginare Exostosen. Zeitschrift für menschliche Vererbungs- und Konstitutionslehre 36: 469

Scholz W, Murken J D 1963 Koppelungsuntersuchungen bei Familien mit multiplen cartilaeginaeren Exostosen. Zeitschrift für menschliche Vererbungs und Konstitutionslehre 37: 178

Solomon L 1963 Hereditary multiple exostosis. Journal of Bone and Joint Surgery 45B: 292

Solomon L 1964 Hereditary multiple exostosis. American Journal of Human Genetics 16: 351

Stocks P, Barrington A 1925 Hereditary disorders of bone development. Treasury of Human Inheritance Vol 3, part 1. Cambridge University Press, London

Wood V E, Sauser D, Mudge D 1985 The treatment of hereditary multiple exostosis of the upper extremity. Journal of Hand Surgery 10: 505–513

### Enchondromatosis (Ollier)

Cannon S R, Sweetnam D R 1985 Multiple chondrosarcomas in dyschondroplasia (Ollier's disease). Cancer 55: 836–840

Lamy M, Aussannaire M, Jammet M L, Nezelop C 1954 Trois cas de maladie d'Ollier dans une fratrie. Bulletins et Memoires de la Societé Médicale des Hôpitaux de Paris 53: 1491

McKusick V A 1983 Mendelian Inheritance in Man, 6th edn. Johns Hopkins Press, Baltimore

Rossberg A 1959 Zur Erblichkeit der Knochenchondromatose. Fortschritte auf dem Gebiete der Rotgenstrahlen und der Nuklearmedizin 90/1: 138

Shapiro F 1982 Ollier's disease. An assessment of angular deformity, shortening and pathological fracture in 21 patients. Journal of Bone and Joint Surgery 64A: 95–103

### Metachondromatosis

Bassett G S, Cowell H R 1985 Metachondromatosis. Journal of Bone and Joint Surgery 67: 811–814

Beals R K 1982 Metachondromatosis. Clinical Orthopaedics and Related Research 169: 167–170

Giedion A, Kesztler R, Muggiasca F 1975 The widening spectrum of multiple cartilaginous exostosis (MCE). Paediatric Radiology 3/2: 93

Hinkel G K, Rupprecht E, Harzer W 1984 Metachondromatosis: Report of another family with 4 patients. Helvetica Paediatrica Acta 39: 5–6

Kozlowski K, Scougal J S 1975 Metachondromatosis: Report of a case in a 6-year-old boy. Australian Paediatric Journal 11/1: 42

Maroteaux P 1971 La metachondromatose. Zeitschrift fur Kindereilkunde 109: 246

*Upington Bone Disease*

Schweitzer G, Jones B, Timme A 1971 Upington disease: A familial dyschondroplasia. South African Medical Journal 45: 994

*Spondyloenchondrodysplasia*

Chagnon S, Lacert P, Blery M 1985 Spondylo-enchondrodysplasia. Journal of Radiology 66: 75–77

Schorr S, Legum C, Ochshorn M 1976 Spondyloenchondrodysplasia. Enchondromatosis with severe platyspondyly in two brothers. Radiology 118/1: 133

Spranger J, Kemperdieck H, Bakowski H, Opitz J M 1978 Two peculiar types of enchondromatosis. Pediatric Radiology 7: 215–219

**Enchondromatosis with haemangiomata (Maffucci)**

Allen B R 1978 Maffucci's syndrome. British Journal of Dermatology 99(Suppl 16): 31–33

Anderson I F 1965 Maffucci's syndrome: report of a case, with a review of the literature. South African Medical Journal 39: 1066

Carleton A, Elkington J S C, Breenfield J G, Robb-Smith A H T, 1942 Maffucci's syndrome (dyschondroplasia with haemangiomata). Quarterly Journal of Medicine 2: 203

Gutman E, McCutcheon S, Garber P 1978 Enchondromatosis with hemangiomas (Maffucci's syndrome). Southern Medical Journal 71: 466–467

Laskaris G, Skouteris C 1984 Maffucci's syndrome: Report of a case with oral hemangiomas. Oral Surgery Oral Medicine Oral Pathology 57(3): 263–266

Lowell S H, Mathog R H 1979 Head and neck manifestations of Maffucci's syndrome. Archives of Otolaryngology 105: 427–430

Ma G F, Leung P C 1984 The management of the soft-tissue haemangiomatous manifestations of Maffucci's syndrome. British Journal of Plastic Surgery 37(4): 615–618

Phelan E M D, Carty H M, Kalos S 1986 Generalised enchondromatosis associated with haemangiomas, soft-tissue calcifications and hemihypertrophy. British Journal of Radiology 59/697: 69–74

Sun T C, Swee R G, Shives T C, Unni K K 1985 Chondrosarcoma in Maffucci's syndrome. Journal of Bone and Joint Surgery 67: 1214–1218

Vaz R M, Turner C 1985 Ollier disease (enchondromatosis) associated with ovarian juvenile granulosa cell tumor and precocious pseudopuberty. Journal of Pediatrics 108: 945–947

**Neurofibromatosis**

Allanson J E, Hall J G, van Allen M I 1985 Noonan phenotype associated with neurofibromatosis. American Journal of Medical Genetics 21/3: 457–462

Ansari A H, Nagamani M 1976 Pregnancy and neurofibromatosis (von Recklinghausen's disease). Obstetrics and Gynecology 47/1: 25–29

Apter N, Chemke J, Hurwitz N, Levin S 1975 Neonatal neurofibromatosis: unusual manifestations with malignant clinical course. Clinical Genetics 7/5: 388

Barker D, Wright E, Nguyen K et al 1987 Gene for von Recklinghausen neurofibromatosis is in the pericentromeric region of chromosome 17. Science 236: 1100–1102

Bradley W G, Richardson J, Frew I J C 1974 The familial association of neurofibromatosis, peroneal muscular atrophy, congenital deafness, partial albinism, and Axenfeld's defect. Brain 97/3: 521

Crowe F W, Schull W J, Neel J V 1956 A Clinical, Pathological and Genetic Study of Multiple Neurofibromatosis. Thomas, Springfield

Dunn B G, Farrell R E, Riccardi V M 1985 A genetic linkage study in 15 families of individuals with von Recklinghausen neurofibromatosis. American Journal of Medical Genetics 22: 403–407

Edwards J N, Fooks M, Davey D A 1983 Neurofibromatosis and severe hypertension in pregnancy. British Journal of Obstetrics and Gynaecology 90: 528–531

Fabricant R N, Todaro G J 1981 Increased serum levels of nerve growth factor in von Recklinghausen's disease. Archives of Neurology 38: 401–405

Ginsburg D S, Hernandex E, Johnson J W 1981 Sarcoma complicating von Recklinghausen's disease in pregnancy. Obstetrics and Gynecology 58: 385–387

Huson S M, Thrush D C 1985 Central neurofibromatosis. Quarterly Journal of Medicine 55/218: 213–24

Jarvis G J, Crompton A C 1978 Neurofibromatosis and pregnancy. British Journal of Obstetrics and Gynaecology 85: 844–846

Jimbow K, Szabo G, Fitzpatrick T B 1973 Ultrastructure of giant pigment granules (macromelanosomes) in the cutaneous pigmented macules of neurofibromatosis. Journal of Investigative Dermatology 61: 300–309

Kaplan P, Rosenblatt B 1985 A distinctive facial appearance in neurofibromatosis von Recklinghausen. American Journal of Medical Genetics 21/3: 463–470

Lewis R A, Riccardi V M 1981 von Recklinghausen neurofibromatosis. Incidence of iris hamartoma. Ophthalmology 88: 348–354

Lisch K 1937 Ueber Beteiligung der Augen, insbesondere das Vorkommen von Irisknoetchen bei der Neurofibromatose (Recklinghausen). Zeilschrift für Augenheilkunde 93: 137–143

Miller M, Hall J G 1978 Possible maternal effects on severity of neurofibromatosis. Lancet 2: 1071–1073

Rezaian S M 1976 The incidence of scoliosis due to neurofibromatosis. Acta Orthopaedica Scandinavica 47: 534

Riccardi V M, Dobson C E, Chakraborty R, Bontke C 1984 The pathophysiology of neurofibromatosis. American Journal of Medical Genetics 18: 169–176

Riccardi V M, Eichner J E 1986 Neurofibromatosis: Phenotype, Natural History and Pathogenesis. Johns Hopkins Press, Baltimore

Schenkein I, Bueker E D, Helson L, Axelrod F, Dancis J 1974 Increased nerve growth factor stimulating activity in disseminated neurofibromatosis. New England Journal of Medicine 290: 613–614

Siggers D C, Boyer S M, Eldridge R 1975 Nerve growth factor in disseminated neurofibromatosis. New England Journal of Medicine 292: 1134

Slater C, Hayes M, Saxe N, Temple-Camp C, Beighton P 1986 Macromelanosomes in the early diagnosis of neurofibromatosis. American Journal of Dermatopathology 8(4): 284–289

Waardenburg P J 1918 Heterochrome en Melanosis. Nederlands Tijdschrift voor Geneeskunde 2: 1453–1455

Zehavi C, Romano A, Goodman R M 1986 Iris (Lisch) nodules in neurofibromatosis. Clinical Genetics 29: 51–55

## Fibrous dysplasia

*Monostotic and polyostotic fibrous dysplasia*

Albright F, Bulter A M, Hampton A O, Smith P 1937 Syndrome characterised by osteitis fibrosa disseminata areas of pigmentation and endocrine dysfunction, with precocious puberty in females: A report of five cases. New England Journal of Medicine 216: 727

Alvarez-Arratia M C, Rivas F, Avila-Abundis A, Hernandex A, Nazara Z, Lopez C, Castillo A, Cantu J M 1983 A probable monogenic form of polyostotic fibrous dysplasia. Clinical Genetics 24: 132–139

Bost M, Andrini P, Jean D 1985 Pseudo precocious puberty in the McCune-Albright syndrome. Pediatrie 40: 55–60

Happle R 1986 The McCune-Albright syndrome: A lethal gene surviving by mosaicism. Clinical Genetics 29: 321–324

Harris R I 1985 Polyostotic fibrous dysplasia with acromegaly. American Journal of Medicine 78: 539–542

Lichtenstein L, Jaffe H L 1942 Fibrous dysplasia of bone. Archives of Pathology 33: 77

Lightner E S, Penny R, Fraser S D 1975 Growth hormone excess and sexual precocity in polyostotic fibrous dysplasia (McCune-Albright syndrome): Evidence for abnormal hypothalamic function. Journal of Pediatrics 87/61: 922

McCune D J 1936 Osteitis fibrosa cystica. American Journal of Diseases of Children 52: 745

Rieth K G, Comite F, Shawker T H, Cutler G B 1984 Pituitary and ovarian abnormalities demonstrated by CT and ultrasound in children with features of the McCune-Albright syndrome. Radiology 153: 389–393

*Fibromatoses*
Baird P A, Worth A J 1976 Congenital generalised fibromatosis: an autosomal recessive condition? Clinical Genetics 9: 488
Griffiths H J, Robinson K, Bonfiglio A 1983 Aggressive fibromatosis. Skeletal Radiology 9: 179–184
Heiple K G, Perrin E, Aikawa M 1972 Congenital generalised fibromatosis. A case limited to osseous lesions. Journal of Bone and Joint Surgery 54A: 663
Schlangen J T 1976 Congenital generalised fibromatosis. A case report with roentgen manifestations of the skeleton. Clinical Radiology 45: 18
Stout A P 1984 Juvenile fibromatoses. Cancer 7: 953
Young I D, Fortt R W 1981 Familial fibromatosis. Clinical Genetics 20: 211–216

*Syndromic fibromatosis*
*Léri pleonosteosis*
Léri A 1922 Dystrophie osseuse generalisée congénitale et héréditaire: La pléonostéose familiale. Presse Medicale 30: 13
Rukavina J G, Falls H F, Holt J F, Block W G 1959 Leri's pleonosteosis: A study of a family with a review of the literature. Journal of Bone and Joint Surgery 41A: 397
Watson-Jones R 1949 Léri's pleonosteosis, carpal tunnel compression of the medial nerves and Morton's metatarsalgia. Journal of Bone and Joint Surgery 31: 397

*Dupuytren contracture*
Ling R S M 1963 The genetic factor in Dupuytren's disease. Journal of Bone and Joint Surgery 45(b): 709–718
Lygonis C S 1966 Familial Dupuytren's contracture. Hereditas 56: 142–143
Matthews P 1979 Familial Dupuytren's contracture with predominantly female expression. British Journal of Plastic Surgery 32: 120–123
Maza R K, Goodman R M 1968 A family with Dupuytren's contracture. Journal of Heredity 59: 155–156

*Peyronie disease*
Bias W B, Walsh P C, Nyberg L M 1981 Peyronie's syndrome: A newly recognised autosomal dominant trait. American Journal of Human Genetics 33: 131
McKusick V A 1983 Mendelian Inheritance in Man. 6th edn. Johns Hopkins Press, Baltimore
Nyberg L M, Bias W B, Hochberg M C, Walsh P C 1982 Identification of an inherited form of Peyronie's disease with autosomal dominant inheritance and association with Dupuytren's contracture and histocompatibility B7 cross-reacting antigens. Journal of Urology 128: 48–51

*Craniofacial fibrous dysplasia*
Chatterjee S K, Mazumder J K 1967 Massive fibro-osseous dysplasia of the jaws in two generations. British Journal of Surgery 20: 648
El Deeb M, Waite D E, Gorlin R J 1979 Congenital monostotic fibrous dysplasia. A new possibly autosomal recessive disorder. Journal of Oral Surgery 37: 520–525
Ramsey H E, Strong E W, Frazell E L 1975 Fibrous dysplasia of the craniofacial bones. American Journal of Surgery 116: 542–547
Reitzik M, Lownie J F 1975 Familial polyostotic fibrous dysplasia. Oral Surgery 40/6: 769
Waldron C A, Giansanti J S 1973 Benign fibro-osseous lesions of the jaws: A clinical-radiologic-histologic review of 65 cases. Fibrous dysplasia of the jaws. Oral Surgery 35: 190
Zanini S A, Paillakis J M, Migowski W 1985 Craniofacial fibrous dysplasia. Annals of Plastic Surgery 14(4): 378–382
Zimmerman D C, Dahlin D C, Stafne E C 1958 Fibrous dysplasia of the maxilla and mandible. Oral Surgery 111: 55–68

*Cherubism*

Anderson D E, McClendon J L 1962 Cherubism—hereditary fibrous dysplasia of the jaws. 1. Genetic considerations. Oral Surgery 15: Supl. 2

Jones W A 1965 Cherubism: a thumbnail sketch of its diagnosis and a conservative method of treatment. Oral Surgery 20: 648

Khosla V M, Korobkin M 1970 Cherubism. American Journal of Diseases of Children 120: 458

Peters W J 1979 Cherubism: A study of 20 cases from one family. Oral Surgery, Oral Medicine, Oral Pathology 47: 307–311

Pina-Neto J M, Moreno A F, Silva L R, Velludo M A, Petean E B, Ribeiro M V, Athayde L, Voltarelli J C 1986 Cherubism, gingival fibromatosis, epilepsy and mental deficiency (Ramon syndrome) with juvenile rheumatoid arthritis. American Journal of Medical Genetics 25: 433–441

Ramon Y, Berman W, Bubis J J 1967 Gingival fibromatosis combined with cherubism. Oral Surgery 24: 436–448

Salzano F M, Ebling H 1966 Cherubism in a Brazilian kindred. Acta Geneticae Medicae et gemellologiae 15: 296

Wayman J B 1978 Cherubism: A report on three cases. British Journal of Oral Surgery 16: 47–56

Wechsler J, Grellet M 1984 A case of familially associated cherubism, centro-osseous giant cell granulomas and giant cell tumors of the jaws. Revue de Stomatologie et de Chirurgie Maxillo-Faciale 85: 337–340

# 5

# Disorders with diminished bone density

The conditions in this category share the common feature of increased radiolucency of the skeleton. The majority of them are heterogeneous and further subdivision can be foreseen.

1. Osteogenesis imperfecta
   a. OI type I
   b. OI type II
   c. OI type III
   d. OI type IV
   e. Sporadic OI
2. Juvenile idiopathic osteoporosis
3. Osteoporosis — pseudoglioma syndrome
4. Idiopathic osteolyses
   a. Osteolysis, phalangeal type
   b. Hajdu-Cheney syndrome
   c. Osteolysis, tarsocarpal type
   d. Osteolysis, nephropathic type
   e. Winchester syndrome
   f. Multicentric osteolysis
5. Hypophosphatasia
6. Vitamin D-resistant rickets
7. Pseudohypoparathyroidism

## OSTEOGENESIS IMPERFECTA

Osteogenesis imperfecta (OI) is one of the most common and best known of the inherited disorders of the skeleton. OI has been described and discussed by many authors during the last 150 years and the eponyms 'Lobstein', 'Vrolik' and 'van der Hoeve' are still associated with the condition. A tentative retrospective diagnosis of OI has been made in a variety of historic circumstances. Perhaps the earliest and best documented case is that of an affected Egyptian mummy (Gray 1970).

OI is currently the focus of considerable attention and rapid progress is being made in the elucidation of the basic biochemical and molecular

135

**Fig. 5.1** Osteogenesis imperfecta: a kindred in which several members have autosomal dominant osteogenesis imperfecta. The orthopaedic sequelae of multiple fractures are very obvious.

defects. The vast quantity of published material now defies succint analysis, but essential information can be found in a monograph entitled 'The Brittle Bone Syndrome' by Smith et al (1983).

*Nosology*

OI has been subdivided into a severe 'congenita' form, where stigmata are present at birth and a mild 'tarda' form, where problems arise in childhood. These are not distinct genetic entities, however, and although there is some practical value in this format, the issue of syndromic specificity is avoided. In the same way, radiographic subdivision into 'thick bone' and 'thin bone' types does not represent precise syndromic categorisation.

Impetus to the understanding of OI was provided by Sillence et al (1979) when subdivision into four major types was proposed on a basis of clinical and genealogical characteristics. This 'Sillence classification' rapidly gained general acceptance and has provided a framework for advances in the understanding of the disorder. Modification and further subgrouping has been introduced (Sillence 1981) but it is evident that the limit of clinical and genetic categorisation has been reached. There is continuing debate concerning the value of some of the diagnostic parameters and the existence of certain subcategories of the condition and it is highly probable that the

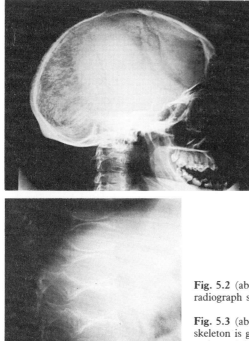

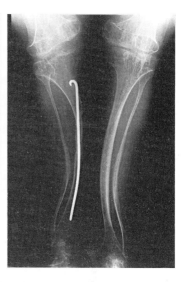

**Fig. 5.2** (above left) Osteogenesis imperfecta: skull radiograph showing multiple Wormian bones.

**Fig. 5.3** (above right) Osteogenesis imperfecta: the skeleton is gracile and undermineralised. In this patient, deformity of the tibia has been corrected by means of an intramedullary nail.

**Fig. 5.4** (left) Osteogenesis imperfecta: in some patients the vertebral bodies become flattened and biconcave.

nosology will change in the future. The current classification is shown below:

*OI type I* Osseous fragility (variable from minimal to
     moderately severe), distinctly blue sclerae,       AD
     presenile hearing loss         (heterogeneous)
  Subgroup A: normal teeth
  Subgroup B: with dentinogenesis imperfecta
  Subgroup C: blue sclerae, Wormian bones,
       dentinogenesis imperfecta, few
       fractures

*OI type II* Lethal perinatal OI. Extremely severe
     osseous fragility, stillbirth or death in the
     newborn
  Subgroup A: Radiographs show broad,
       crumpled long bones and broad,       AD
       beaded ribs        new mutation

> Subgroup B: Radiographs show broad,
>             crumpled long bones without
>             significant beading of the ribs                    AR?
> Subgroup C: Radiographs show thin, fractured
>             long bones and thin, beaded ribs                   AR?

*OI type III* Moderate to severe osseous fragility,
             normal sclerae, variable but severe
             deformity of long bones and spine,
             stunted stature. Generally non-lethal in
             the newborn                                         AR

*OI type IV* Osseous fragility with normal sclerae,
             severe deformity of long bones and spine           AD

When abnormalities at the biochemical or molecular level began to be elucidated, some researchers anticipated that the clinical subclassification would be confirmed by laboratory findings. Unfortunately this has not turned out to be the case. For instance, although a significant proportion

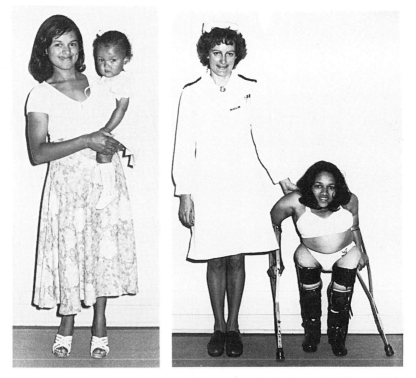

**Fig. 5.5** (left) Osteogenesis imperfecta: a mother and baby daughter with OI type I. They have experienced few fractures but their sclerae are impressively blue.

**Fig. 5.6** (right) Osteogenesis imperfecta type III: a young woman who suffered numerous fractures and now has severe dwarfing and deformity. Several sibs are affected, but her parents are normal. (From Horan F T, Beighton P 1975 Clinical Genetics 8: 107.)

of families with OI type I have mutations of the pro-alpha 1(1) collagen gene (Tsipouras et al 1984, Byers & Bonadio 1985), in others, linkage with the pro-alpha 2(1) gene has been established (Grobler-Rabie et al 1985, Wallis et al 1986, Sykes et al 1986).

Apart from the obvious clinical heterogeneity OI is proving to be extremely heterogeneous at the level of the gene (Prockop et al 1984, Byers & Bonadio 1985, Smith 1986). The outstanding nosological problem will be the reconciliation of clinical and molecular classifications!

## OI type I

### Clinical and radiographic features

Bone fragility is the cardinal feature of OI type I. This propensity is variable and some patients suffer many fractures, while others have few problems. The fractures heal rapidly but there may be residual deformity.

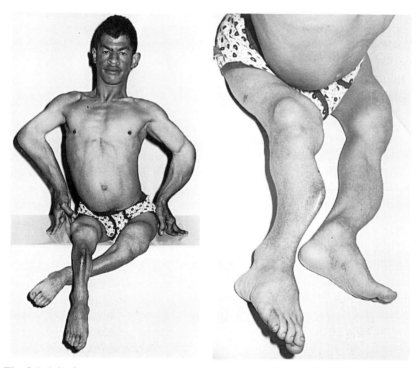

**Fig. 5.7** (left) Osteogenesis imperfecta: a young man with OI type III. He is dwarfed, with considerable deformity due to fracturing and bending of the long bones. His clinical appearance is very different from that of the persons with OI type I depicted in Figure 5.5. (From Beighton et al 1983 South African Medical Journal 64: 565–568.)

**Fig. 5.8** (right) Osteogenesis imperfecta type III. Anterior bowing produces a 'sabre tibia' configuration.

Abundant callus at a fracture site may simulate osteosarcoma and there have been erroneous diagnoses and needless amputations (Rutkowski et al 1979). A wide bi-temporal diameter, blueness of the sclera and pearly grey discolouration of the teeth (dentinogenesis imperfecta, DI) are variable concomitants. Deafness supervenes in adulthood in about 20% of cases. The severity of the condition is very variable and some patients are little troubled, while others are crippled and deformed (Moorfield & Miller 1980, Beighton et al 1983, Shapiro 1985). Multiple fractures and intracranial bleeding occasionally cause stillbirth or death in the neonatal period and for this reason, Caesarian section is the preferential method of delivery for a potentially affected infant (Roberts & Solomons 1975).

Radiographically the skeletal appearances range from normality to chaos. In general, the bones are gracile and porotic, with deformity consequent upon repeated fractures and mechanical forces. The vertebrae are sometimes flattened and bi-concave, but significant spinal malalignment is unusual. Wormian bones are present in the cranial sutures; these are evident in virtually every affected person, although they may be obliterated by sclerosis in adulthood (Cremin et al 1982). The characteristic radiographic features permit diagnosis in the majority of affected persons but it must be emphasised that in mild cases the skeleton can be radiographically normal.

*Genetics*

The existence of a common autosomal dominant form of OI was firmly established after Bell (1928) and Fuss (1935) analysed numerous pedigrees. Many reports of generation to generation transmission followed, culminating with a kindred in which 68 persons in four generations had the disorder (Quisling et al 1979). OI-I is notoriously variable in expression and complete non-penetrance has been reported. This variability is intrafamilial as well as interfamilial.

Sporadic individuals with a phenotype which is consistent with OI-I are usually regarded as representing new mutations of the faulty gene. Support for this concept was advanced by Carothers et al (1986) when they demonstrated that the mean paternal age in these circumstances was significantly higher than that of population controls.

It has been suggested that OI type I can be split into subgroups A and B on the basis of the presence or absence of dentinogenesis imperfecta (DI) (Levin et al 1980, Patterson et al 1983). There are problems, however, with the use of DI as a diagnostic discriminant. DI certainly breeds true in some families but in individual patients it is not unusual to find that some teeth are affected whereas others are apparently normal.

Another subcategory, OI-IC, has also been proposed. This corresponds to an autosomal dominant syndrome of blue sclerae, Wormian bones and DI, with a paucity of fractures (Beighton 1981) which was identified in 20 persons of three generations of a Cape Town family. As mentioned previously, it is probably pointless to attempt further subdivision on a basis of phenotypic features. Indeed it is likely that OI-I is very heterogeneous and that virtually every affected family may turn out to have a different defect at the molecular level.

## OI Type II

This form of OI largely corresponds to the category 'OI congenita' which enjoyed wide acceptance in the era of innocence and simplicity which preceded the molecular revolution. Further details of the genetic implications of OI-II are given in Chapter 3.

### Clinical and radiographic features

OI type II presents in the newborn with bowed limbs, caput membranaceum and variable scleral blueing. Radiographically, numerous fractures are present in the ribs and long bones, the calvarium is poorly ossified and Wormian bones are evident in the cranial sutures. In some affected infants the tubular bones and ribs are broad and crumpled, while in others these bones are gracile.

### Genetics

OI-II is usually sporadic and although autosomal recessive inheritance was initially thought likely, the paucity of affected siblings and the lack of parental consanguinity mitigated against this hypothesis. It is now accepted that the majority of affected infants represent new dominant mutations, although rare autosomal recessive forms may exist. The current speculative subdivision into OI-II A, B and C, based upon radiological appearances, embraces this concept (Sillence et al 1984, Byers et al 1984). At the molecular level various abnormalities of pro-alpha 1(1) and pro-alpha 2(1) collagen genes have been recognised and it is evident that there is great heterogeneity (Prockop et al 1986, Byers & Bonadio 1985, Smith 1985).

## OI Type III

OI type III is a rare but well-defined entity in which dwarfism, severe deformity, white sclerae and autosomal recessive inheritance are the major characteristics. In a reivew of 345 pedigrees of families with OI, Sillence et al (1986) could identify only seven kindreds with OI-III. However, the condition is common in South Africa (Beighton & Versfeld 1985) and in Zimbabwe (Viljoen & Beighton 1987) and more than 80 affected Blacks have been documented.

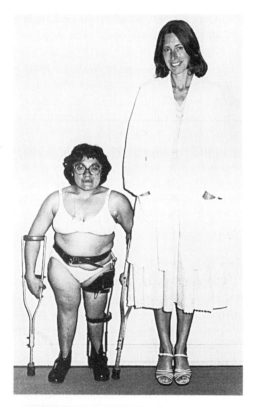

**Fig. 5.9** (left) Osteogenesis imperfecta: A severely affected young woman with blue sclera, disturbance of growth and the sequelae of multiple fractures. She is the only member of her family with the disorder, and the mode of inheritance is unknown. This form of OI is unclassifiable.

**Fig. 5.10** (below) Dentinogenesis imperfecta. The teeth are opalescent, with a purple-brown discolouration. (From: Beighton et al 1981 Journal of Medical Genetics 18: 124–128.)

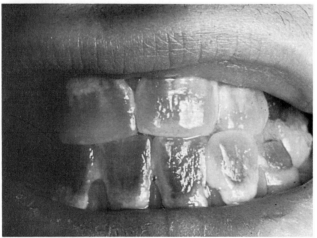

*Clinical and radiographic features*

The affected newborn presents with limb deformity and multiple fractures and at this stage differentiation from OI-II may be impossible. Fracturing continues in childhood, tubular bones bend due to mechanical forces and growth is disturbed. Due to these processes, affected persons are dwarfed, with malalignment of the limbs. The tibiae are often bowed anteriorly, producing the classical 'sabre tibia' appearance. The trunk and neck are short and the bitemporal diameter is wide. Whiteness of the sclerae is an important diagnostic feature. Many patients survive into adulthood but physical handicap is usually severe.

Radiographically the skeleton is porotic and gracile and the sequelae of numerous fractures are usually present. The tubular bones are bowed and Wormian bones are visible in the cranial sutures. Protrusio acetabulae and vertebral biconcavity are prominent features. In a review of the radiographic features of 16 children with the OI-III, Versfeld et al (1985) observed that the pedicles of lumbar vertebrae were markedly elongated; this feature may prove to be important in diagnostic categorisation.

*Genetics*

There have been several instances of affected siblings with normal parents and of parental consanguinity. In a kindred reported by Horan & Beighton (1975) two brothers had married two sisters; both couples produced several affected offspring. Autosomal recessive inheritance was firmly established following formal segregation analysis of six affected families (Sillence et al 1986).

The reasons for the high prevalence in the Black population of Southern Africa and Zimbabwe are obscure (Beighton & Versfeld 1985, Viljoen & Beighton 1987) but it is possible that the unaffected heterozygote has a biological advantage of some kind in the tropical environment. The extent of geographic distribution of OI-III in Africa is unknown but there is a report of the disorder in Nigeria (Adeyokunnu 1982). An inbred Jewish isolate, the Mosabites of Ghardia, Southern Algeria, also have a form of OI-III but their condition is probably a separate genetic entity (Kaplan & Baldino 1953).

There have been few biochemical studies of OI III. The most detailed was undertaken on a patient with a homozygous 4 base pair deletion in the pro-alpha 2(1) collagen gene who had originally been reported by Nicholls et al (1984). The mutated pro-alpha 2(1) chains were not incorporated into the type 1 collagen helix, which consisted exclusively of pro-alpha 1(1) chains (Dickson et al 1984, Pihlajaniemi et al 1984). Linkage studies have incriminated mutations in the pro-alpha 2(1) collagen gene (Tsipouras et al 1984, Sykes et al 1986) but the issues of heterogeneity and nosological status remain contentious. The investigation of further cases of OI III from different populations should throw light on possible heterogeneity.

## OI Type IV

The diagnostic criteria for OI-IV are autosomal dominant inheritance, white sclerae, stunted stature and severe fractures and deformity. Only a few

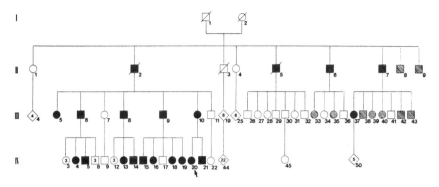

**Fig. 5.11** Pedigree of a kindred with autosomal dominant inheritance of Wormian bones and blue sclerae. This disorder is now regarded as a form of OI type I. (From: Beighton P 1981 Journal of Medical Genetic 18: 124–128.)

affected families have been documented. Although OI-IV is listed in the current classification there is debate concerning its existence as a distinct syndromic entity. Some experts claim that the disorder is fairly common, while others deny ever having seen a case. It is difficult to know whether this dissension reflects inconsistent geographical distribution of the disorder, or subjective differences in diagnostic appraisal. To a large extent this issue resolves around the assessment of scleral colour and it is difficult to know whether the problem is in the eye of the patient or the beholder!

Skin fibroblasts from a patient diagnosed as having mild OI type IV were shown to synthesise type I collagen molecules with alterations in their triple helical structure and thermal stability (Wenstrup et al 1986). This over-modification, which had previously been associated with the lethal OI-II phenotype, is thought to be the consequence of a mutation which alters the integrity of the type I collagen helix. The increased post-translational modification is thus a secondary effect and does not indicate molecular homogeneity. For this reason, although overmodification is of value as a marker for the presence of subtle mutations within the collagen helix (Byers & Bonadio 1985), it is of little value in the resolution of nosological problems in OI.

## Sporadic OI

Sporadic unclassifiable patients with a phenotype which is broadly compat-ible with OI-IV are not uncommon (Wynne-Davies & Gormley 1981). At present the underlying genetic mechanisms are unknown but these may include new mutation, autosomal recessive inheritance and genetic compounds. In view of the precedents for heterogeneity in OI, it is likely that these sporadic persons have a variety of separate conditions which differ in their basic biomolecular defects.

## JUVENILE IDIOPATHIC OSTEOPOROSIS

Juvenile idiopathic osteoporosis is a rare disease of uncertain aetiology. The condition was delineated by Dent & Friedman (1965) and reviewed by Dent (1969). Teotia et al (1979) described four affected children and tabulated the manifestations of 27 others whom they culled from the literature. In a study of six children with idiopathic osteoporosis, Smith (1980) gave details of comprehensive histological and biochemical investigations. The diagnosis is usually reached after the exclusion of renal, gastrointestinal, endocrine or other metabolic disorders which lead to decreased skeletal density. At a clinical level, confusion with osteogenesis imperfecta is a common problem.

### Clinical and radiographic features

Onset is in the prepubescent period. Bone pain, or fracture on minor trauma are the presenting features. The course is self-limiting, with remission within 5 years. However, there may be residual stunting and skeletal deformity. During the active phase, serum calcium concentrations are persistently low and intestinal absorption of calcium is impaired.

Radiographically, the skeleton is demineralised, and bowing of the tubular bones and concavity of the vertebral bodies may be seen. Unlike osteogenesis imperfecta there are no Wormian bones in the skull.

### Genetics

All the case description of juvenile idiopathic osteoporosis have concerned sporadic individuals and there is no evidence to indicate a genetic aetiology. The condition described by Jackson (1958) under the title 'Osteoporosis of unknown cause in younger people', affects adults in the third decade. It is probably a separate entity.

## OSTEOPOROSIS — PSEUDOGLIOMA SYNDROME

Bianchine & Murdoch (1969) reported a boy with osteoporosis, in whom both eyes had been enucleated in infancy for pseudoglioma. Subsequently, Bianchine et al (1972) described two more kindreds and concluded that the osteoporosis-pseudoglioma syndrome was a specific disorder which was inherited as an autosomal recessive. The total of reported patients reached 24 when Frontali et al (1985) published details of three affected siblings and identified a further 21 patients in eight families in the literature.

### Clinical and radiographic features

The major manifestations are in the skeleton and eyes. Osteoporosis is present from early childhood and leads to frequent fractures, limb bowing,

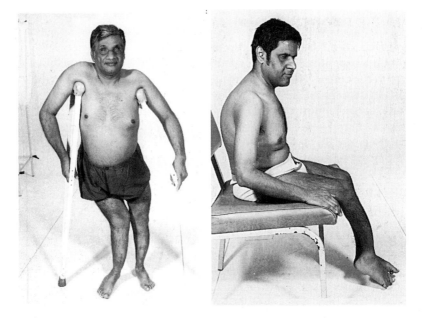

**Fig. 5.12** (left) Osteoporosis-pseudoglioma syndrome. A male, blind from birth, with anterior bowing of the tibiae and valgus deformity of the right knee. (From Beighton et al 1985 Clinical Genetics 25: 69–75.)

**Fig. 5.13** (right) Osteoporosis-pseudoglioma syndrome. The brother of the person depicted in Figure 5.12. He was blind from birth in the left eye and defective vision in the right eye was extinguished by the age of 15 yrs. He sustained several fractures during childhood and is now severely handicapped. (From Beighton et al 1985 Clinical Genetics 25: 69–75.)

spinal malalignment and disturbance of growth. Blindness from the time of birth is usual, although there have been a few instances of later onset or preservation of residual vision. The structural changes in the eye are variable and include microphthalmia, microcornea, iris atrophy, posterior synechiae and cataracts. In two instances when the eye was enucleated after a misdiagnosis of glioma, a mass of pseudogliomatous tissue was found in the posterior chamber (Bianchine & Murdoch 1969, Briard & Frezal 1976). Ligamentous laxity and mental retardation are inconsistent concomitants.

Radiographically the skeleton is gracile and porotic, with bowing of the tubular bones, biconcavity of vertebral bodies and the sequelae of old fractures. The metaphyses are expanded and lucent, the femoral necks are usually long, and Wormian bones are present in the cranial sutures. These changes are indistinguishable from those of osteogenesis imperfecta.

*Genetics*

Autosomal recessive inheritance is beyond doubt. Parental consanguinity

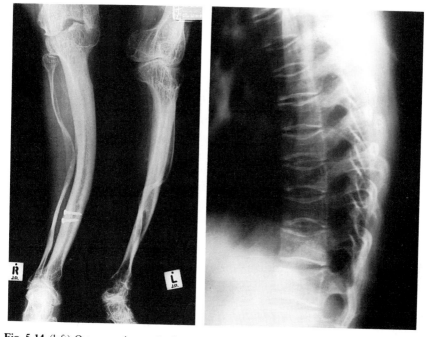

**Fig. 5.14** (left) Osteoporosis-pseudoglioma syndrome. Radiograph of the lower legs of the patient depicted in Figure 5.12. The tibiae are bowed, with cortical thickening and lucency of the periarticular regions. (From Beighton et al 1985 Clinical Genetics 25: 69–75.)

**Fig. 5.15** (right) Osteoporosis-pseudoglioma syndrome. The vertebral bodies are porotic and biconcave. (From Beighton et al 1985 Clinical genetics 25: 69–75.)

was present in the family studied by Sauvegrain et al (1981) and suspected in a kindred documented by Bartsocas et al (1982). Other descriptions of affected siblings include those of Saraux et al (1967, 1969) and Neuhauser et al (1976). The majority of these reports have emanated from Mediterranean countries, notably France, Italy, Greece and Tunisia but the condition has also been encountered in a consanguineous South African kindred of Indian stock, initially misdiagnosed as having osteogenesis imperfecta (Beighton et al 1985).

## IDIOPATHIC OSTEOLYSES

The idiopathic osteolyses are a group of unusual disorders, in which spontaneous skeletal rarefaction progresses to partial or complete disappearance of the affected bones. These conditons are classified according to the anatomical distribution of the areas which are involved. Although rare, they are very heterogeneous in terms of their clinical manifestations and genetic background. The idiopathic osteolyses can be listed under the following headings:

1. Osteolysis: phalangeal type
2. Hajdu-Cheney syndrome
3. Osteolysis: tarsocarpal type
4. Osteolysis: nephropathic type
5. Winchester syndrome
6. Multicentric osteolysis

Osteolysis may be a component of a large number of genetic and acquired disorders, such as pycnodysostosis, hyperparathyroidism, ainhum, rheumatoid arthritis, Raynaud's disease, Sudeck arthropathy and polyvinyl-chloride poisoning. Similarly, bone reabsorption due to neurological damage is found in many conditions, including diabetes mellitus, leprosy and meningomyelocele. By definition, these secondary forms of osteolysis are excluded from the idiopathic category. Zugibe et al (1974) emphasised that 'osteolysis' is a radiological concept and that the underlying process may be either abnormal bone destruction or defective primary bone formation. The biochemical basis of acro-osteolysis has been discussed by Brown et al (1976).

## Osteolysis — phalangeal type

The majority of case reports have mentioned onset in childhood, with pain and swelling in the affected fingers. Osteolysis develops when the inflammatory process becomes quiescent. Bony collapse follows, with subsequent deformity and contracture of the digits.

A kindred in which the disorder was transmitted through four generations was described by Lamy & Maroteaux (1961) and another in which inheritance followed the same autosomal dominant pattern was investigated by Cazalis (1982). Families in which inheritance was probably autosomal recessive have been reported by Giaccai (1952) and Hozay (1953). Another distinctive autosomal recessive form of distal osteolysis was documented by Petit & Fryns (1986). These authors described two adult siblings, from a consanguineous family, in whom progressive destruction of the phalanges was associated with mental retardation and stunted stature. These patients had an unusual facial appearance, with maxillary hypoplasia, relative exophthalmos, a broad nasal tip and loss of the teeth. These facial features are also present in persons with the dominantly inherited carpotarsal and multicentric forms of osteolysis and although possibly providing a clue to the underlying pathogenesis, they are not of value as diagnostic discriminants.

## Hajdu*-Cheney syndrome

A syndrome in which phalangeal osteolysis is associated with generalised osteoporosis, absence of frontal sinuses, deafness, hypoplasia of the ramus of the mandible, articular laxity and Wormian bones in the skull was recognised in a mother and her four children by Cheney (1965). Further cases were reported by Dorst & McKusick (1969) and Herrmann et al (1973). These latter authors reviewed the features of the disorder, mentioned an earlier report by Hajdu & Kauntze (1948) and proposed the term 'arthro-dento-osteodysplasia (Hajdu-Cheney syndrome)'. The evolution of the disease process was discussed in detail by Silverman et al (1974) when they described a long-term follow-up on the cases of Hajdu & Kauntze (1948) and Herrmann et al (1973). The phenotypic features of an affected boy and 13 other previously reported patients were tabulated by Zugibe et al (1974) and two additional patients were documented by Weleber & Beals (1976). Van den Houten et al (1985) reported an affected mother and son, together with a sporadic patient with consanguineous parents. It seems probable that the syndrome is genetically heterogeneous but so far phenotypic differentiation has not been accomplished. About 20 cases have been reported.

## Osteolysis, tarsocarpal type

Acro-osteolysis, which primarily involved the carpus and tarsus, was recognised in three generations of a kindred by Gluck & Miller (1972). Dominant inheritance was also a feature of the disorder in the kindreds reported by Thieffry & Sorrel-Dejerine (1958) and Kohler et al (1973). The condition in this latter family is probably a separate entity, as the affected individuals had the additional stigmata of frontal bossing, pes cavus, micrognathia and a Marfanoid habitus. In a review of the literature Beals & Bird (1975) identified 14 patients with tarsocarpal acro-osteolysis. Of these, nine were sporadic and five had affected kin.

## Osteolysis, nephropathic type

Autosomal dominant inheritance of tarsocarpal acro-osteolysis in association with nephropathy was reported by Shurtleff et al (1964). Sporadic patients were described by Torg & Steel (1968), Macpherson et al (1973) and Tuncbilek et al (1985).

## Winchester syndrome

The eponym 'Winchester' is applied to a syndrome in which osteolysis of

---

* Nicholas Hajdu qualified in medicine at the University of Prague in 1934 and became consultant radiologist at St George's Hosptial, London in 1949.

the tarsus and carpus is associated with short stature, severe joint contractures, corneal opacities, a coarse facies and generalised osteoporosis.

This disorder was first recognised in two sisters in a consanguineous Puerto Rican kindred (Winchester et al 1969). By means of electron microscopic studies of corneal biopsy material, Brown & Kuwabara (1970) demonstrated that the condition was a storage disease. The autosomal recessive mode of inheritance was confirmed by Hollister et al (1974) when they investigated three patients in two consanguineous sibships.

**Multicentric osteolysis**

There are problems in the subclassification of multicentric osteolysis and, indeed, with the other forms of osteolysis. In the Paris Nomenclature the Hajdu-Cheney and Winchester syndromes are listed under this heading together with the Torg form (vide infra). Other classifications have been proposed by Sage & Allen (1974) and Hardegger et al (1985) and at present there is no nosological unanimity. Gorham & Stout (1955) reviewed the features of 24 children in whom widespread osteolysis had developed. No genetic basis was established. This condition is associated with angiomatosis and the osteolysis arises at the site of the vascular abnormality in the bone. The course is benign and the disorder is self-limiting.

Torg et al (1969) reported three affected sibs who had consanguineous parents. A second family in which four out of 12 siblings were affected was documented by Sauvegrain et al (1981).

Evidence for genetic heterogeneity was provided by Canun et al (1976) following their studies of a mother and her two children, in whom generalised osteolysis was apparently transmitted as a dominant trait. The daughter experienced severe episodes of joint pain during childhood and later developed generalised osteolysis and contractures. As the mother and son were less severely affected, it is evident that phenotypic expression of the abnormal gene is very variable. An affected father and son with this disorder were documented by Whyte et al (1978).

## HYPOPHOSPHATASIA

The hypophosphatasias are a heterogeneous group of conditions with clinical and radiographic stigmata which resemble those of dietary rickets, in conjunction with a low level of serum alkaline phosphatase. These disorders are classified according to the age of onset and the severity of manifestations into varieties such as 'congenita', 'juvenile', 'tarda' and 'adult'. However, there is considerable overlap and intermediate types have been reported. The congenita form is lethal while in the tarda variety survival is usual. The antenatal diagnosis of the severe congenita form of hypoplasia is reviewed in Chapter 3.

*Clinical and radiographic features*

In hypophosphatasia congenita grossly defective calvarial ossification produces a caput membranaceum and the affected neonate usually dies as a result of respiratory distress or intracranial bleeding. The limbs are shortened and deformed, and clinical differentiation from the other lethal skeletal dysplasias may be difficult. Radiographically, the skeleton is poorly ossified, with marked metaphyseal irregularities. Changes in the tarda form are similar but less severe. Multiple fractures and spontaneous bowing distort the shafts of the long bones. The manifestations ameliorate during early childhood, but the patient may be left with short stature and bone deformities. Premature fusion of the sutures leads to craniostenosis and the teeth are often lost prematurely. In all forms of hypophosphatasia, serum levels of alkaline phosphatase are reduced and the urinary excretion of phosphoethanolamine is increased. Hypercalcaemia is sometimes present. The biochemical abnormalities have been discussed by Gorodischer et al (1976).

*Genetics*

The heterozygote can be identified with ease and certainty (Rathbun et al 1961). Recognition is based upon the demonstration of diminished serum concentrations of alkaline phosphatase and increased urinary excretion of phosphoethanolamine. These heterozygotes often lose their permanent teeth at an early age (Pimstone et al 1966). Parental consanguinity has been a feature of several cases (Svejcar & Walther 1975), and there is little doubt that inheritance is autosomal recessive.

Controversy exists as to whether the various types of hypophosphatasia are the result of great variation in clinical expression of the same underlying genetic defect, or whether they are truly separate genetic entities. MacPherson et al (1972) have reported mild and severe cases within the same kindred, thus lending support to the former contention. On the other hand, Mehes et al (1972) found consistent clinical manifestations in affected children in an inbred Hungarian community. This observation seems to confirm the existence of a distinct 'juvenile' form of hypophosphatasia. In further studies in the same kindred, Rubecz et al (1974) found 4 homozygotes and 25 heterozygotes amongst the 53 individuals in whom biochemical studies were undertaken.

Benzie et al (1976) reported the antenatal detection of hypophosphatasia in the 19th week of pregnancy by ultrasonographic demonstration of defective cranial ossification in one instance, and by radiographic identification of generalised skeletal undermineralisation in a second. These authors suggested that assessment of the activity of alkaline phosphatase in cultured fibroblasts might be of diagnostic value.

A 'rickety' form of hypophosphatasia, which was present in a man and his two sons, led Silverman (1962) to postulate that inheritance might be autosomal dominant. The possibility of pseudodominance in this kindred, due to unusually marked expression in the heterozygote, was discounted by the fact that the patient's wife had a normal serum alkaline phosphatase level. A very mild dominant type of hypophosphatasia, characterised principally by premature loss of primary teeth, was identified in female members of three generations of a kindred by Danovitch et al (1968).

Whyte et al (1979) reviewed the literature and documented a family with apparent autosomal dominant transmission in which males were more severely affected than females. Weinstein & Whyte (1981) discussed the possibility of heterogeneity in the autosomal dominant adult form of hypophosphatasia and Whyte et al (1982) recorded variable expression in this disorder.

Some idea of the relative prevalence of the various clinical categories of hypophosphatasia can be obtained from a multicentre radiographic analysis of 24 cases which was undertaken by Kozlowski et al (1976). Three were considered to have the neonatal lethal form. The authors emphasised that the diagnostic situation was not clear-cut and that in a sporadic case it is not always possible to distinguish between the dominant and recessive forms of the condition.

In a review of the literature Terheggen & Wischermann (1984) identified 278 cases of hypophosphatasia. Of these 49 had the lethal congenital form, 94 had a severe early infantile form, 112 had a benign juvenile form and 23 had an innocuous adult form. Inheritance was autosomal recessive in all four types and, in addition, there was evidence for autosomal dominant transmission in some instances of the adult form.


## VITAMIN D-RESISTANT RICKETS

Vitamin D-resistant rickets, or familial hypophosphataemia, is clinically and radiographically very similar to other forms of dietary and metabolic rickets. Following delineation of the condition by Albright et al (1937) there have been reports of many affected kindreds.


### Clinical and radiographic features

Bony changes become evident during infancy, usually when walking commences. Bowing of the legs is the commonest presentation but knock-knees or a combined genu valgum and genu varum 'windswept' configuration is sometimes encountered. Expansion of the bone ends and costochondral beading may develop. Hypophosphataemia differs from dietary rickets in that tetany and muscular weakness do not occur. The clinical course is very variable, but the stature of adults is usually diminished, with some residual bone deformity. The serum concentration of phosphorus is consistently low and in the active phase the serum alkaline phosphatase level may be moderately raised. The nature of the basic abnormality is not

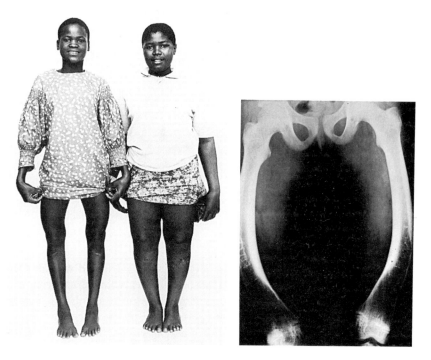

**Fig. 5.16** (left) Vitamin D-resistant rickets: affected sisters with femoral bowing. Bilateral tibial osteotomy has been employed to straighten the shins. These girls are of Zulu stock, a group in whom the condition reaches a relatively high prevalence.

**Fig. 5.17** (right) Vitamin D-resistant rickets: femoral bowing and demineralisation. Secondary cortical buttressing is evident on the medial side of the femora.

fully understood, and opinion is divided as to whether calcium, phosphorus or vitamin D absorption, excretion or metabolism are at fault.

Investigations in mice with X-linked hypophosphataemia indicate that a low renal threshold for phosphate reabsorption is important in the pathogenesis but it is uncertain if this represents the primary defect (Brunette 1985).

Radiographically, the skeleton is demineralised, particularly in the juxta-metaphyseal regions. The metaphyses are cupped, irregular and uneven. The shafts of the long bones of the legs become bowed, with secondary cortical buttressing in their concavities. Ligamentous calcification may develop in adulthood (Polisson et al 1985).

*Genetics*

Vitamin D-resistant rickets is one of the few clinically important X-linked dominant conditions. In this mode of inheritance, a female with the disorder would be expected to produce equal numbers of affected and

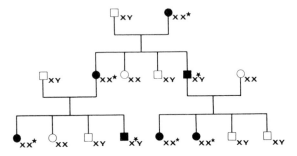

**Fig. 5.18** Vitamin D-resistant rickets: a hypothetical pedigree showing X-linked dominant inheritance. A female with the condition is at a theoretical risk of having equal numbers of affected and unaffected sons and daughters. Conversely, all the daughters and none of the sons of an affected male will inherit the condition. *Key:* □ normal male; ○ normal female; ■ affected male; ● affected female; X, normal X chromosome; X*, X chromosome bearing the abnormal gene; Y, Y chromosome.

unaffected sons and daughters, while an affected male would transmit the condition to all his daughters but none of his sons. Clinical expression is very variable within any kindred and in some instances, the only phenotypic manifestation is a persistently lowered serum phosphorus concentration.

The X-linked dominant mode of transmission was confirmed by Graham et al (1959) following a genetic analysis of five affected kindreds. Subsequently, Burnett et al (1964) described the pedigrees of 24 patients.

There have been a few reports of kindreds with autosomal dominant inheritance of hypophosphataemia (Harrison et al 1966, Pak et al 1972). The fact that expression of the abnormal gene may be limited to a low level of serum phosphorus complicates the issue. However, in these particular families autosomal dominant inheritance seems to have been confirmed. The kindreds with generation to generation transmission of a similar disorder were subsequently reported by Scriver et al (1977) under the designation 'hypophosphataemic non-rachitic bone disease'.

Pseudo vitamin D-resistant rickets or vitamin D-dependent rickets is a severe disorder in which gross skeletal dysplasia is associated with hypophosphataemia, marked hypocalcaemia and mild aminoaciduria. This condition differs from the common form of hypophosphataemic rickets in that tetany and myopathy occur, while the response to vitamin D therapy is usually good. Patients with normal consanguineous parents have been reported by Dent et al (1968) and Arnaud et al (1970), and affected sibs have been described by Birtwell et al (1970). There is little doubt that pseudo vitamin D-resistant rickets is inherited as an autosomal recessive.

A non-hyperaminoaciduric form of vitamin D-dependent rickets was identified in two Mexican siblings by Cantu (1974). In a similar report, Stamp & Baker (1976) described a brother and sister with severe bowing of the legs, craniostenosis and nerve deafness, in association with persistent hypophosphataemia. Perry & Stamp (1978) subsequently described adult

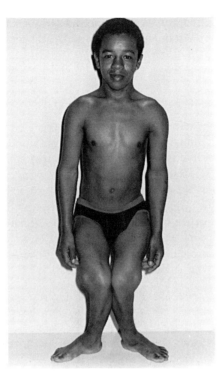

**Fig. 5.19** Vitamin D-resistant rickets: a young man with severe deformity of the legs. The configuration of limb bowing is non-specific.

brothers with the same disorder and emphasised the features of gross osteosclerosis, extra-skeletal calcification and osteomalacia. As the unaffected parents were consanguineous in both families, it is likely that inheritance of this entity was autosomal recessive.

## PSEUDOHYPOPARATHYROIDISM

In a report of three patients, Albright et al (1942) used the term 'pseudo-hypoparathyroidism' for the syndrome in which hypocalcaemia was associated with a characteristic facies and habitus in the absence of other evidence of hypoparathyroidism. Descriptions then followed of individuals with identical clinical stigmata, in whom serum calcium concentrations were normal. This latter condition, which was considered to be a separate entity, was designated 'pseudo-pseudohypoparathyroidism'. It was later recognised that the hypocalcaemia may fluctuate, and it is now accepted that these disorders are manifestations of the same abnormal gene. For this reason, to the great relief of the semantic purists, the repetitious term 'pseudo-pseudohypoparathyroidism' has been discarded. In view of the confusion which still persists, 'Albright's hereditary osteodystrophy' is a useful alternative common designation. The manifestations and pathogenesis of the disorder were reviewed by Fitch (1982).

### Clinical and radiographic features

Moderate shortness of stature, obesity and a 'moon face' are associated with some degree of shortening of the tubular bones of the extremities. The bone changes are very variable but the fourth metacarpal is most often affected. Hypocalcaemia, which is resistant to treatment with parathormone, may lead to cataracts, mental deficiency, tetany and ectopic calcification. Endocrine involvement may include primary hyperparathyroidism (Sasaki et al 1985) and oligomenorrhea due to hypogonadism (Halal et al 1985).

Radiographically, cone-shaped epiphyses may be identified in the phalanges. Other changes include thickening of the calvarium and calcification in the basal ganglia and subcutaneous tissues.

### Genetics

The problems concerning the relationship of pseudohypoparathyroidism and pseudo-pseudohypoparathyroidism were finally settled when both of these disorders were recognised within the same kindred (Mann, et al 1962, Brito Suarenz et al 1975). Biochemical confirmation was forthcoming when Levine et al (1985) documented siblings with the former condition and demonstrated that their mother had the latter.

There are several recorded instances of generation to generation transmission, and an excess of affected females has led to a contention that the disorder may be an X-linked dominant trait. Against this is the fact that descriptions of affected fathers producing affected sons were given by Weinberg & Stone (1972) and Brito Suarez et al (1975). The situation remains confused but it is possible that the disorder is heterogeneous, one form being an X-linked dominant and the other an autosomal dominant trait. Alternative explanations would be incomplete ascertainment or partial sex limitation.

Defective activity of erythrocyte N-protein, which promotes coupling of stimulatory hormone receptors and catalytic cyclase has been demonstrated in a series of affected persons (Farfel et al 1981). Inheritance of the phenotype plus this abnormality was autosomal dominant in some families. The normality of N-protein in other kindreds might be indicative of heterogeneity.

Cederbaum & Lippe (1973) reviewed the genetics of pseudohypoparathyroidism and reported an anomalous kindred in which inheritance was apparently autosomal recessive. Using the designation pseudohypoparathyroidism type II, Drezner et al (1973) reported a male infant with hypocalcaemia and other endocrine abnormalities. The genetic status of this condition and its relationship to the usual form of pseudohypothyroidism is uncertain.

## REFERENCES

**Osteogenesis imperfecta**

Adeyokunnu A A 1982 Spectrum of bone dysplasias in African children: Ibadan Nigerian experience. In: Papadatos C J, Bartsocas C S (eds) Skeletal dysplasias. Liss, New York p 427–439

Beighton P 1981 Familial dentinogenesis imperfecta, blue sclerae and Wormian bones without fractures: Another type of osteogenesis imperfecta? Journal of Medical Genetics 18: 124–128

Beighton P, Versfeld G A 1985 On the paradoxically high relative prevalence of osteogenesis imperfecta type III in the Black population of South Africa. Clinical Genetics 27: 398–401

Beighton P, Spranger J, Versfeld G 1983 Skeletal complications in OI: Review of 153 South African patients. South African Medical Journal 64: 565–568

Bell J 1928 Blue sclerotics and fragility of bone. Treasury of Human Inheritance, Vol 2, part III. Cambridge University Press, Cambridge

Byers P H, Bonadio J F 1985 The molecular basis of clinical heterogeneity in osteogenesis imperfecta. In: Lloyd J, Scriver C R (eds) Metabolic and Genetic Disease in Pediatrics. Butterworth, London

Byers P H, Bonadio J F, Steinmann B 1984 Osteogenesis imperfecta: Update and perspective. American Journal of Medical Genetics 17: 429–435

Carothers A D, McAllion S J, Paterson C R 1986 Risk of dominant mutation in older fathers: Evidence from osteogenesis imperfecta. Journal of Medical Genetics 23: 227–230

Cremin B, Goodman H, Spranger J, Beighton P 1982 Wormian bones in osteogenesis imperfecta and other disorders. Skeletal Radiology 8: 35–38

Dickson L A, Pihlajaniemi T, Deak S, Pope F M, Nicholls A, Prockop D J, Myers J C 1984 Nuclease S1 mapping of a homozygous mutation in the carboxy-propeptide region of the pro-alpha 2(1) collagen gene in a patient with osteogenesis imperfecta. Proceedings of the National Academy of Science 81: 4524–4528

Fuss H 1935 Die erbliche Osteopsathyrosis. Deutsche Zeitschrift fur Chirurgie 245: 279

Gray P H K 1970 A case of osteogenesis imperfecta, associated with dentinogenesis imperfecta, dating from antiquity. Clinical Radiology 21: 106

Grobler-Rabie A F, Wallis G, Brebner D K, Beighton P, Bester A J, Mathew C G P 1985 Detection of a high frequency RsaI polymorphism in the human pro-alpha 2(1) collagen gene which is linked to an autosomal dominant form of osteogenesis imperfecta. EMBO Journal 4: 1745–1748

Horan F, Beighton P 1975 Autosomal recessive inheritance of osteogenesis imperfecta. Clinical Genetics 8/2: 107

Kaplan M, Baldino C 1953 Dysplasie periostale paraissant familiale et transmise suivant le mode mendelien récessif. Archives Françaises de Pédiatrie 10: 943

Levin L S, Brady J M, Melnick M 1980 Scanning electron microscopy of teeth in autosomal dominant osteogenesis imperfecta: Support for genetic heterogeneity. American Journal of Medical Genetics 5: 189–199

Moorfield W G, Miller G R 1980 Aftermath of osteogenesis imperfecta: The disease in adulthood. Journal of Bone and Joint Surgery 62: 113–119

Nicholls A C, Osse G, Schloon H G, Lenard H G, Deak S, Myers J C, Prockop D J, Weigel W R, Fryer P, Pope F M 1984 The clinical features of homozygous alpha 2(1) collagen deficient osteogenesis imperfecta. Journal of Medical Genetics 21: 257–262

Patterson C R, McAllion S, Miller R 1983 Heterogeneity of osteogenesis imperfecta type I. Journal of Medical Genetics 20: 203–205

Pihlajamieni T, Dickson L A, Pope M, Korhonen V R, Nicholls A, Prockop D J, Myers J C 1984 Osteogenesis imperfecta: cloning of a pro-alpha 2(1) collagen gene with a frameshift mutation. Journal of Biological Chemistry 259: 12941–12944

Prockop D J, Kivirikko K I 1984 Heritable diseases of collagen. New England Journal of Medicine 311: 376–386

Quisling R W, Moore G R, Jahrsdoerfer R A, Cantrell R W 1979 Osteogenesis imperfecta: A study of 160 family members. Archives of Otolaryngology 105: 207–211

Roberts J M, Solomons C C 1975 Management of pregnancy in osteogenesis imperfecta: New perspectives. Obstetrics and Gynecology 45: 168

Rutkowski R, Resnick P, McMaster J H 1979 Osteosarcoma occurring in osteogenesis imperfecta. A case report. Journal of Bone and Joint Surgery 61: 606–608

Shapiro F 1985 Consequences of an osteogenesis imperfecta diagnosis for survival and ambulation. Journal of Pediatric Orthopedics 5: 456–462

Sillence D O 1981 Osteogenesis imperfecta. An expanding panorama of variants. Clinical Orthopaedics 159: 11–25

Sillence D O, Senn A, Danks D M 1979 Genetic heterogeneity in osteogenesis imperfecta. Journal of Medical Genetics 16: 101–116

Sillence D O, Barlow K K, Garber A P et al 1984 Osteogenesis imperfecta type II. Delineation of the phenotype with reference to genetic heterogeneity. American Journal of Medical Genetics 17: 407–423

Sillence D O, Barlow K K, Cole W G, Dietrich S, Garber A P, Rimoin D L 1986 Osteogenesis imperfecta type III. Delineation of the phenotype with reference to genetic heterogeneity. American Journal of Medical Genetics 23: 821–832

Smith R 1986 The molecular genetics of collagen disorders. Clinical Science 71: 129–135

Smith R, Francis M J O, Houghton G R 1983 The Brittle Bone syndrome: Osteogenesis imperfecta. Butterworth, London

Sykes B, Ogilvie D, Wordsworth P, Anderson J, Jones N 1986 Osteogenesis imperfecta is linked to both type I collagen structural genes. Lancet ii: 69–72

Tsipouras P, Borresen A L, Dickson L A, Berg K, Prockop D J, Ramirez F 1984 Molecular heterogeneity in the mild autosomal dominant forms of osteogenesis imperfecta. American Journal of Human Genetics 36: 1172–1179

Versfeld G A, Beighton P H, Katz K, Solomon A 1985 Costovertebral anomalies in osteogenesis imperfecta. Journal of Bone and Joint Surgery 67: 602–604

Viljoen D, Beighton P 1987 Osteogenesis imperfecta type III: An ancient mutation in Africa? American Journal of Medical Genetics 27, 4: 907–912

Wallis G, Beighton P, Boyd C, Mathew C G 1986 Mutations linked to the pro alpha 2(1) collagen gene are responsible for several cases of osteogenesis imperfecta type I. Journal of Medical Genetics 23: 411–416

Wenstrup R J, Hunter A G W, Byers P H 1986 Osteogenesis imperfecta type IV: evidence of abnormal triple helical structure of type I collagen. Human Genetics 74: 47–53

Wynne-Davis R, Gormley J 1981 Clinical and genetic patterns in osteogenesis imperfecta. Clinical Orthopaedics and Related Research 159: 26–35

## Juvenile idiopathic osteoporosis

Dent C E 1969 Idiopathic juvenile osteoporosis (IJO). Birth Defects: Original Article Series 5/4: 134

Dent C E, Friedman M 1965 Idiopathic juvenile osteoporosis. Quarterly Journal of Medicine 34: 177

Jackson W P U 1958 Osteoporosis of unknown cause in younger people. Journal of Bone and Joint Surgery 40B: 420

Smith R 1980 Idiopathic osteoporosis in the young. Journal of Bone and Joint Surgery 62B/4: 417–427

Teotia M, Teotia S P S, Singh R K 1979 Idiopathic juvenile osteoporosis. American Journal of Diseases of Children, 133: 894–900

## Osteoporosis — pseudoglioma syndrome

Bartsocas C S, Zeis P M, Elia M, Papadatos C J 1982 Syndrome of osteoporosis with pseudoglioma. Annales de Génétique 25: 61–61

Beighton P, Winship I, Behari D 1985 The ocular form of osteogenesis imperfecta: a new autosomal recessive syndrome. Clinical Genetics 28: 69–75

Bianchine J W, Murdoch J L 1969 Juvenile osteoporosis in a boy with bilateral enucleation of the eyes for pseudoglioma. The Clinical Delineation of Birth Defects, National Foundation, New York, p 225

Bianchine J W, Briard-Gullemot M L, Maroteaux P, Frezal J, Harrison H E 1972 Generalised osteoporosis with bilateral pseudoglioma — an autosomal recessive disorder of connective tissue: report of three families — review of the literature. American Journal of Human Genetics 24: 34A

Briard M L, Frezal J 1976 Le pseudogliome bilateral avec osteoporose generalizée, une affection récessive autosomique. Journal de Génetique Humaine 24: 665–674

Frontali M, Stomea C, Dallapiccola B 1985 Osteopetrosis — pseudoglioma syndrome. American Journal of Medical Genetics 22: 35–47

Neuhauser G, Kaveggia E G, Opitz J M 1976 Autosomal recessive syndrome of pseudogliomatous blindness, osteoporosis and mild mental retardation. Clinical Genetics 9: 324–332

Saraux H, Miller H, Mawas J, Mawas E, Prepin F 1969 La dysplasie hyaloidorétinienne (pseudogliome) à hérédité récessive autosomale. Annales Oculaire 202: 1131–1137

Saraux H, Frézal J, Roy C, Aron J J, Hayat B, Lamy M 1967 Pseudogliome et fragilité osseuse hereditaire à transmission autosomale récessive. Annales Oculaire 200: 1241–1252

Sauvegrain J, Dufier J L, Vacher H, Charlot J C, Le Ho'ang Phuc, Haye C 1981 Dégenerescence hyaloido-rétienne avec ostéoporose et fragilité osseuse. Journal de Radiologie 62: 537–543

**Preamble** **Idiopathic osteolyses**
Brown D M, Bradford D S, Gorlin R J 1976 The acro-osteolysis syndrome: morphologic and biochemical studies. Journal of Pediatrics 88/41: 573

Zugibe F T, Herrmann J, Opitz J M, Gilbert E F, McMillan G 1974 Arthro-dento-osteodysplasia: A genetic 'acro-osteolysis' syndrome. Birth Defects: Original Article Series 10/5: 145

*Osteolysis — phalangeal type*
Cazalis P 1982 Acro-osteolyse de la main. Annales de Radiologie 25: 337–340

Giaccai L 1952 Familial and sporadic neurogenic acro-osteolysis. Acta Radiologica 38: 17

Hozay J 1953 Sur une dystrophie familiale particuliere (inhibition precoce de la croissance et osteolyse non mutilante acrales avec dysmorphie faciale). Revista de Neurologia Clinica 89: 245

Lamy M, Maroteaux P 1961 Acro-osteolyse dominante. Archives Françaises de Pédiatrie 18: 693

Petit P, Fryns J-P 1986 Distal osteolysis, short stature, mental retardation, and characteristic facial appearance: delineation of an autosomal recessive subtype of essential osteolysis. American Journal of Medical Genetics 25: 537–541

*Hajdu-Cheney syndrome*
Cheney W D 1965 Acro-osteolysis. American Journal of Roentgenology, Radium Therapy and Nuclear Medicine 94: 595

Dorst J P, McKusick V A 1969 Acro-osteolysis (Cheney syndrome). The Clinical Delineation of Birth Defects, National Foundation, New York, p 215

Hajdu N, Kauntze R 1948 Cranioskeletal dysplasia. British Journal of Radiology 21: 42

Herrmann J, Zugibe F T, Gilbert E F, Opitz J M 1973 Arthro-dento-osteo dysplasia (Hajdu–Cheney syndrome). Zeitschrift für Kinderheilkunde 11: 1

Silverman F N, Dorst J P, Hajdu N 1974 Acro-osteolysis (Hajdu-Cheney syndrome). Birth Defects: Original Article Series 10/12: 106

Van den Houten B R, Ten Kate L P, Gerding J C 1985 The Hajdu-Cheney syndrome. A review of the literature and report of three cases. International Journal of Oral Surgery 14: 113–125

Weleber R G, Beals R K 1976 The Hajdu-Cheney syndrome: Report of two cases and review of the literature. Journal of Pediatrics 88: 243–249

Zugibe F T, Herrmann J, Opitz J M, Gilbert E F, McMillan G 1974 Arthro-dento-osteodysplasia: a genetic 'acro-osteolysis syndrome. Birth defects: Original Article Series 10/6: 145

*Osteolysis, tarsocarpal type*
Beals R K, Bird C B 1975 Carpal and tarsal osteolysis. Birth Defects: Original Article Series 11/6: 107

Gluck J, Miller J J 1972 Familial osteolysis of the carpal and tarsal bones. Pediatrics
    81: 506
Kohler E, Babbitt D, Huizenga B, Good T A 1973 Hereditary osteolysis. Radiology 108: 99
Thieffry S, Sorrel-Dejerine J 1958 Forme speciale d'ostéolyse essentielle héréditaire et
    familiale à stabilisation spontanée, survenant dans l'enfance. Presse Médicale 66: 1858

*Osteolysis, nephropathic type*
Macpherson R I, Walker R D, Kowall M H 1973 Essential osteolysis with nephropathy.
    Journal of the Canadian Association of Radiologists 24: 98
Shurtleff D B, Sprakes R S, Clawson K, Guntheroth W G, Mottet N K 1964 Hereditary
    osteolysis with hypertension and nephropathy. Journal of American Medical Association
    188: 363
Torg J S, Steel H H 1968 Essential osteolysis with nephropathy: a review of the literature
    and case report of an unusual syndrome. Journal of Bone and Joint Surgery 50: 1629
Tuncbilek E, Besim Am Bakkaloglu A et al 1985 Carpo-tarsal osteolysis. Pediatric
    Radiology 15: 255-258

*Winchester syndrome*
Brown S I, Kuwabara T 1970 Peripheral corneal opacification and skeletal deformities: a
    newly recognized acid mucopolysaccharidosis simulating rheumatoid arthritis. Archives of
    Ophthalmology 83: 667
Hollister D W, Rimoin D L, Lachman R S, Cohen A H, Reed W B, Westin G W 1974
    The Winchester syndrome: A nonlysosomal connective tissue disease. Journal of
    Pediatrics 84: 701
Winchester P, Grossman H, Lim W N, Danes B S 1969 A new acid mucopolysaccharidosis
    with skeletal deformities simulating rheumatoid arthritis. American Journal of
    Roentgenology 106: 121

*Multicentric osteolysis*
Canun S, Torres P, Del Castillo V, Carnevale A 1976 Hereditary osteolysis with dominant
    transmission. Excerpta medica: Fifth International Congress of Human Genetics, Mexico,
    p 63.
Gorham L W, Stout A P 1955 Massive osteolysis. Journal of Bone and Joint Surgery
    37A: 985
Hardegger F, Simpson L A, Segmueller G 1985 The syndrome of idiopathic osteolysis.
    Journal of Bone and Joint Surgery 67: 89–93
Sage M R, Allen P W 1974 Massive osteolysis. Journal of Bone and Joint Surgery 56B: 130
Sauvegrain J, Gaussin G, Blondet P, Legendre H, Challe J Y, D'Aboville M 1981
    Ostéolyse multicentrique à transmission récessive. Annales de Radiologie 24: 638–642
Torg J S, de George A M, Kirkpatrick J A, Trujillo M M 1969 Hereditary multicentric
    osteolysis with recessive transmission: a new syndrome. Journal of Pediatrics 75: 243
Whyte M P, Murphy W A, Kleerekoper M, Teitelbaum S L, Avioli L V 1978 Idiopathic
    multicentric osteolysis: Report of an affected father and son. Arthritis and Rheumatism
    21: 367–376

**Hypophosphatasia**
Benzie R, Doran T A, Escoffery W, Gardner H A, Hoar D I, Hunter A, Malone R,
    Miskin M, Rudd N L 1976 Prenatal diagnosis of hypophosphatasia. Birth Defects:
    Original Article Series 12/6: 271
Danovitch S H, Baer P N, Laster L 1968 Intestinal alkaline phosphatase activity in familial
    hypophosphatasia. New England Journal of Medicine 278: 1253
Gorodischer R, Davidson R G, Mosovich L L, Yaffe S J 1976 Hypophosphatasia: a
    developmental anomaly of alkaline phosphatase? Pediatric Research 10/7: 650
Kozlowski K, Sutcliffe J, Barylak A, Harrington G, Kemperdick H, Nolte K 1976
    Hypophosphatasia — a review of 24 cases. Pediatric Radiology 5: 103
Macpherson R I, Krocker M, Houston C S 1972 Hypophosphatasia. Journal de l'Association
    Canadienne des Radiologistes 23: 16
Mehes K, Klujiber L, Lassu G, Kajtar P 1972 Hypophosphatasia: screening and family
    investigations in an endogamous Hungarian village. Clinical Genetics 3: 60

Pimstone B, Eisenberg E, Silverman S 1966 Hypophosphatasia: genetic and dental studies. Annals of Internal Medicine 65: 722

Rathbun J C, MacDonald J W, Robinson H M C, Wanklin J M 1961 Hypophosphatasia: a genetic study. Archives of Disease in Childhood 36: 540

Rubecz I, Mehes K, Klujber L, Bozzay L, Weisenbach J, Fenyvasi J 1974 Hypophosphatasia: screening and family investigation. Clinical Genetics 6: 155

Silverman J L 1962 Apparent dominant inheritance of hypophosphatasia. Archives of Internal Medicine 110: 191

Svejcar J, Walther A 1975 The diagnosis of the early infantile form of hypophosphatasia tarda. Humangenetika 28/1: 49

Terheggen H G, Wischermann A 1984 Congenital hypophosphatasia. Monatsschrift Kinderheilkunde 132: 512–522

Weinstein R S, Whyte M P 1981 Heterogeneity of adult hypophosphatasia: Report of severe and mild cases. Archives of Internal Medicine 141: 727–731

Whyte M P, Vrabel L A, Schwartz T D 1982 Adult hypophosphatasia: Generalized deficiency of alkaline phosphatase activity demonstrated with cultured skin fibroblasts. Clinical Research 30: 557A

Whyte M P, Teitelbaum S L, Murphy W A, Avioli L V 1979 Adult hypophosphatasia: Clinical, laboratory and genetic investigation of a large kindred with review of the literature. Medicine 58: 329–347

**Vitamin D-resistant rickets**

Albright F, Butler A M, Bloomberg E 1937 Rickets resistant to vitamin D therapy. American Journal of Diseases of Children 54: 529

Arnaud C D, Maijer R, Reade T, Scriver C R, Whelan D T 1970 Vitamin D dependency: an inherited post-natal syndrome with secondary hyperparathyroidism. Pediatrics 46: 871

Birtwell W M, Magsamen B F, Fenn P A, Torg J S, Tourtellotte C D, Martin 1970 An unusual hereditary osteomalacic disease: pseudo-vitamin D deficiency. Journal of Bone and Joint Surgery 52A: 1222

Brunette M G 1985 The X-linked hypophosphatemic vitamin D resistant rickets: Old and new concepts. International Journal of Pediatric Nephrology 6: 55–62

Burnett C H, Dent C E, Harper C, Warland B J 1964 Vitamin D-resistant rickets. Analysis of 24 pedigrees with hereditary and sporadic cases. American Journal of Medicine 36: 222

Cantu J M 1974 Autosomal recessive nonhyperaminoaciduric vitamin D-dependent rickets. Birth Defects: Original Article Series 10/4: 294

Dent C E, Friedman M, Watson L 1968 Hereditary pseudo-vitamin D deficiency rickets. Journal of Bone and Joint Surgery 50B: 708

Graham J B, McFalls V W, Winters R W 1959 Familial hypophosphataemia with vitamin D-resistant rickets, II. Three additional families of the sex-linked dominant type with a genetic analysis of five such families. American Journal of Human Genetics 11: 311

Harrison H E, Harrison H C, Lifshitz F, Johnson A D 1966 Growth disturbance in hereditary hypophosphataemia. American Journal of Diseases of Children 112: 290

Pak C Y C, Deluca H F, Bartter F C, Henneman D H, Frame B, Simopoulos A, Delea C S 1972 Treatment of vitamin D-resistant rickets with 25-hydroxycholecalciferol. Archives of Internal Medicine 129: 894

Perry W, Stamp T C B 1978 Hereditary hypophosphataemic rickets with autosomal recessive inheritance and severe osteosclerosis. Journal of Bone and Joint Surgery 60: 430–434

Polisson R P, Martinez S, Khoury M et al 1985 Calcification of entheses associated with X-linked hypophosphatemic osteomalacia. New England Journal of Medicine 313: 1–6

Scriver C R, MacDonald W, Reade T M, Glorieux F H, Nogrady B 1977 Hypophosphatemic nonrachitic bone disease: An entity distinct from X-linked hypophosphatemia in renal defect, bone involvement and inheritance. American Journal of Medical Genetics 1: 101–117

Stamp T C, Baker L R I 1976 Recessive hypophosphataemic rickets and possible aetiology of the Vitamin D-resistant syndrome. Archives of Disease in Childhood 51: 360

**Pseudohypoparathyroidism**

Albright F, Burnett C H, Smith P H, Parson W 1942 Pseudohypoparathyroidism: example of 'Seabright-bantam syndrome'; a report of three cases. Endocrinology 30: 922

Brito Suarez M, Herendez C, de la Rosa J 1975 Pseudohypoparathyroidism: three familial cases of Albright's hereditary osteodystrophy. Revista Española de Reumatismo y Enfermedades Osteoarticulares 18/2: 99

Cederbaum S D, Lippe B M 1973 Probable autosomal recessive inheritance in a family with Albright's hereditary osteodystrophy and an evaluation of the genetics of the disorder. American Journal of Human Genetics 25: 638

Drezner M, Neelson F A, Lebovitz H E 1973 Pseudohypoparathyroidism type II: a possible defect in the reception of the cyclic AMP signal. New England Journal of Medicine 289: 1056

Farfel Z, Brothers V M, Brickman A S, Conte F, Neer R, Bourne H R 1981 Pseudohypoparathyroidism: Inheritance of deficient receptor-cyclase coupling activity. Proceedings of National Acadamy of Science 78: 3098–3102

Fitch N 1982 Albright's hereditary osteodystrophy: A review. American Journal of Medical Genetics 11: 11–29

Halal F, Van Dop C, Lord J 1985 Differential diagnosis in young women with oligomenorrhea and the pseudo-pseudohypoparathyroidism variant of Albright's hereditary osteodystrophy. American Journal of Medical Genetics 21: 551–568

Levine M A, Jap T-S, Hung W 1985 Infantile hypothyroidism in two sibs: An unusual presentation of pseudohypoparathyroidism type Ia. Journal of Pediatrics 107: 919–922

Mann J B, Alterman S, Hill A G 1962 Albright's hereditary osteodystrophy comprising pseudohypoparathyroidism and pseudo-pseudohypoparathyroidism, with a report of two cases representing the complete syndrome occurring in successive generations. Annals of Internal Medicine 56: 315

Sasaki H, Tsutsu N, Asano T, Yamamoto T, Kikuchi M, Okumura M 1985 Coexisting primary hyperparathyroidism and Albright's hereditary osteodystrophy — an unusual association. Postgraduate Medical Journal 61: 153–155

Weinberg A G, Stone R T 1972 Autosomal dominant inheritance in Albright's hereditary osteodystrophy. Journal of Pediatrics 79: 996

# 6

# Osteoscleroses

The osteoscleroses share the feature of increased skeletal density, with little or no disturbance of the bony contours. The various forms of osteopetrosis are the commonest and most important osteoscleroses, but a few other disorders are conventionally included in this general category.

1. Osteopetrosis — benign or tarda form, AD
2. Osteopetrosis — malignant or congenita form, AR
3. Osteopetrosis — intermediate form, AR
4. Osteopetrosis with renal tubular acidosis
5. Pycnodysostosis
6. Osteomesopyknosis.

The terms 'osteopetrosis' and 'Albers-Schönberg* disease' are often used loosely and erroneously for any of the numerous osteoscleroses, craniotubular dysplasias and craniotubular hyperostoses. In the strict sense, the designation 'osteopetrosis' is applicable only to the specific autosomal dominant and autosomal recessive conditions which are described in this section.

The nomenclature of the osteopetroses has been a source of confusion for many years. In the first half of this century terminological problems were compounded by the use of the designations 'marble bones' (Schulze 1921) and 'osteosclerosis fragilis generalisata' (Laurell & Wallgren 1920). The situation was improved when Karshner (1926) proposed the term 'osteopetrosis' and clarified when McPeak (1936) recognised the existence of malignant and benign forms of the condition. The nosology and implications of the osteoscleroses have been reviewed by Beighton et al (1977) and Horan & Beighton (1978) and formed the subject of a monograph entitled 'Sclerosing Bone Dysplasias' (Beighton & Cremin 1980).

Genetic osteopetrosis occurs in laboratory animals and experiments involving rodents have opened up promising lines of research into the patho-

---

* Heinrich Albers-Schönberg [1865–1921] was the doyen of German radiologists at the beginning of the 20th century. Like many of his contemporaries he suffered severely from the effects of radiation.

genesis of this group of disorders (Cotton et al 1976, Loutit & Sansom 1976, Marks 1976). On a basis of investigations of osteosclerotic mice, Seifert & Marks (1985) concluded that osteopetrosis was the result of reduced bone resorption due to decreased osteoclastic activity. The histo-pathological changes in chondro-osseous tissue from patients with various forms of hyperostotic bone dysplasia have been reviewed by Kaitila & Rimoin (1976). The histological appearances are indicative of disparity in the basic defects in these conditions. Milgram & Jasty (1982) undertook a detailed study of material from 21 persons with osteopetrosis, ranging in age from 2 months to 78 years. They pointed out that the abnormally dense tissue was composed of lamellar bone and calcified cartilage, and correlated the radiographic and pathological changes.

In addition to the genetic osteoscleroses, increased skeletal density is a prominent feature of renal osteodystrophy and parathyroid dysfunction (Resnick 1981, Hall et al 1981). Physiological bony sclerosis of the newborn can also mimic osteopetrosis but regression rather than progression resolves this diagnostic problem in the early months of life. Localised bone sclerosis in infancy can result from idiopathic hypercalcaemia, lead poisoning and hypothyroidism (McCarthy 1968).

## OSTEOPETROSIS — BENIGN OR TARDA FORM, AD

Albers-Schönberg (1904) gave his name to posterity when he described a 26 year-old-male in whom an unusual increase in bone density had been detected radiologically. In a subsequent report Lorey & Reye (1923) employed the eponym 'Albers-Schönberg disease'. This individual was eventually lost to medical authors when he died at the age of 49. As the mother of Albers-Schönberg's original patient was affected, it is reasonable to reserve this eponym for the dominant form of the disorder. This form of osteopetrosis is comparatively common, with a wide ethnic and geographic distribution, and more than 400 cases have now been reported.

### Clinical and radiographic features

Affected individuals may remain totally asymptomatic and the diagnosis is often reached by chance when radiographs are taken for some unrelated purpose. The facies, physique, mentality and lifespan are normal and general health is unimpaired. A mild anaemia is an infrequent complication. In a proportion of patients, the presenting feature is facial palsy or deafness, consequent upon cranial nerve compression by bony overgrowth. Pathologi-cal fractures due to bone fragility may occur, tooth extraction is sometimes difficult and osteomyelitis of the mandible occasionally develops.

Radiologically, bone sclerosis becomes increasingly apparent as childhood progresses. Sclerotic foci, termed 'endobones' or 'bones within bone', are

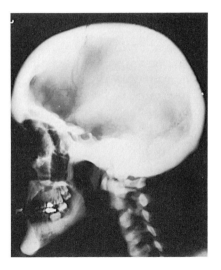

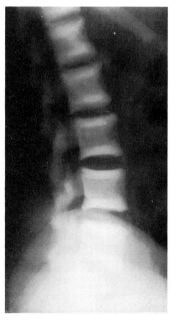

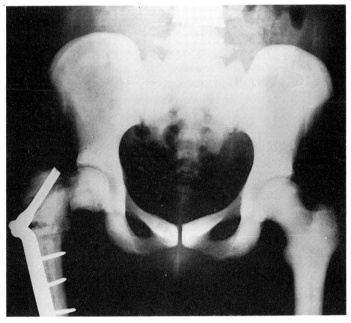

**Fig. 6.1** (top left) Osteopetrosis—AD type: skull radiograph showing widening and increased density in the base and calvarium. (From Beighton P, Horan F T, Hamersma H 1977 Postgraduate Medical Journal 53: 507.)

**Fig. 6.2** (above) Osteopetrosis — AD type: radiograph showing increased skeletal density without disturbance in bone contours. Fractures of the upper femoral region have been stabilised by fixation. In some patients the bones are fragile, while others have no problems of this type.

**Fig. 6.3** (top right) osteopetrosis — AD type: the 'rugger jersey' spine. Sclerosis of the end plates of the vertebral bodies produces a banded appearance. (From Beighton P, Horan F T, Hamersma H 1977 Postgraduate Medical Journal 53: 507.)

a striking feature. These changes usually disappear by the end of the second decade. Bone involvement is widespread but certain regions, particularly the extremities, are sometimes spared. The calvarium is dense and the sinuses may be obliterated. In the spine, thickening of the vertebral end plates gives rise to the characteristic 'rugger jersey' appearance. (The jersey worn by rugby players traditionally carries transverse bands as this configuration produces an illusion of increased body bulk, thereby disconcerting the opposition!)

*Genetics*

Johnston et al (1968) reviewed 19 kindreds with 85 affected individuals and demonstrated that inheritance was autosomal dominant. There is considerable interfamilial variation and there is little doubt that the condition is heterogeneous. Although the manifestations are usually consistent within a particular kindred, there have been reports of anomalous situations, in which there has been disparity in the degree to which members of successive generations have been affected (Thomson 1949). The gene may occasionally be non-penetrant and skipped generations have been described. Johnston et al (1968) concluded that dominant osteopetrosis was a single entity with variable manifestations. However, as case reports accumulate, the evidence lends support to the concept of heterogeneity. The dominant form of osteopetrosis has been recognised antenatally in a fetus during X-ray pelvimetry in late pregnancy. The diagnosis was confirmed in the mother at the same examination (Delahaye et al 1976).

## OSTEOPETROSIS — MALIGNANT OR CONGENITA FORM, AR

The autosomal recessive type of osteopetrosis is much less common than the autosomal dominant form. Although the radiographic changes are similar in these entities, their clinical features differ, and distinction is usually not difficult.

*Clinical and radiographic features*

The manifestations are evident during infancy. Bony overgrowth is associated with marrow dysfunction and presenting symptoms include failure to thrive, spontaneous bruising, abnormal bleeding and anaemia. The teeth become carious and hepatosplenomegaly develops. The bones are fragile and pathological fractures are a frequent complication. Palsies of the optic, oculomotor and facial nerves may occur in the later stages. Death from overwhelming infection or haemorrhage usually takes place in the first decade.

Generalised bone sclerosis is the predominant radiological feature. Penetrated films of the tubular bones reveal transverse bands in the meta-

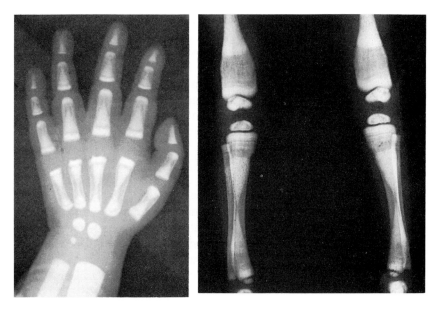

**Fig. 6.4** (left) Osteopetrosis — AR type: endobones or 'bones within a bone' are evident in the metacarpals and phalanges.

**Fig. 6.5** (right) Osteopetrosis — AR type: the lower ends of the femora have a club-like configuration. Transverse bands and longitudinal striations are seen in their shafts.

physeal regions and longitudinal striations in the shafts. The vertebrae show the classical 'rugger jersey' appearance and endobones are evident in the axial skeleton. As the condition progresses, the proximal humerus and distal femur develop a flask-shaped configuration. The skull becomes progressively thickened, with encroachment upon the foramina of the cranial nerves.

The management of this potentially lethal form of osteopetrosis has been revolutionised by bone marrow transplantation (Coccia et al 1980, Kadota & Smithson 1984). Initial results have been very promising although, as yet, the long-term outcome is uncertain.

*Genetics*

Evidence for autosomal recessive inheritance includes affected sibs with normal parents in an inbred kindred (Enell & Pehrson 1958) and parental consanguinity (Tips & Lynch 1962). The abnormal gene is present in relatively high frequency in Costa Rica where 26 affected children in 12 families have been encountered by Loria-Cortes et al (1977). The genes for one form of autosomal recessive osteopetrosis in the mouse have been assigned to chromosome 12 (Marks & Lane 1976). As yet, there is no comparable human information.

Severe osteopetrosis was recognised in utero in two pregnancies from a consanguineous union of Moroccan parents. In the second, the diagnosis was confirmed by ultrasonography at 14 weeks of gestation. The affected siblings, a boy and a girl, had multiple fractures and detailed histological studies demonstrated a marked reduction in osteoclasts (Khazen et al 1986). The authors suggested that this disorder might be a new entity for which they proposed the title 'autosomal recessive lethal osteopetrosis'.

## OSTEOPETROSIS — INTERMEDIATE FORM, AR

The intermediate form of osteopetrosis was recognised in four affected adult South Africans of mixed ancestry by Beighton et al (1979). The clinical manifestations were more severe than those of the conventional AD type of osteopetrosis; in particular, osteomyelitis of the jaw and fractures due to bone fragility were significant complications, while hepatosplenomegaly and a bleeding diathesis were variable features. Radiographically, generalised bone sclerosis was associated with modelling defects in the long bones which were reminiscent of the congenita form of osteopetrosis.

It is very probable that intermediate osteopetrosis is inherited as an autosomal recessive trait as the four original patients included a brother and sister whose parents were clinically, radiographically and biochemically normal. Further evidence for recessive inheritance was provided by Kaibara et al (1982) when they reported two Japanese brothers with unaffected, nonconsanguineous parents.

A mild form of osteopetrosis was identified in two brothers with normal parents by Horton et al (1980). The authors demonstrated unusual histological features and suggested that this condition was yet another distinct entity, with either autosomal recessive or X-linked inheritance. In a comprehensive review of the literature Kahler et al (1984) were able to recognise 18 persons in 10 families with intermediate or mild autosomal recessive osteopetrosis. Interfamilial phenotypic variation might be indicative of further heterogeneity.

## OSTEOPETROSIS WITH RENAL TUBULAR ACIDOSIS, AR

A unique form of osteopetrosis with renal tubular acidosis was recognised by Guibaud et al (1972), Sly et al (1972) and Vainsel et al (1972) and the development of cerebral calcification in this disorder was subsequently documented by Ohlsson et al (1980) and Whyte et al (1980). More than 20 affected persons have been reported and autosomal recessive inheritance has been confirmed. The condition has been reviewed in detail by Sly et al (1985).

*Clinical and radiographic features*

The disorder manifests in early childhood with failure to thrive and stunted growth plus weakness and hypotonia due to renal tubular acidosis. Anaemia is mild or absent and the serious neurological and visceral complications of infantile osteopetrosis do not develop. Other variable features are involvement of the optic and auditory nerves, a fracturing tendency and dental abnormalities. The majority of affected children have learning difficulties and some degree of mental retardation. Activity of red cell carbonic anhydrase II is defective (Sly et al 1983).

The radiographic changes resemble those of classical osteopetrosis, but differ in that they tend to regress in later childhood. Intracranial calcification in the basal ganglia and periventricular regions is a consistent finding.

*Genetics*

Autosomal recessive inheritance has been firmly established on the basis of affected siblings with normal parents and several instances of parental consanguinity. The condition has a wide geographic distribution and in addition to families in the USA, Belgium and France, it has been recognised in Algeria, Saudi Arabia and Kuwait. Heterozygotes are clinically and radiographically normal, but their status can be accurately determined by enzymatic investigation of blood specimens (Tashian et al 1984). There is some evidence that the disorder may be heterogeneous (Sly et al 1985).

## PYCNODYSOSTOSIS

Maroteaux & Lamy (1962) defined the characteristics of pycnodysostosis and established it as an entity in its own right. Previously, the predominant clinical feature of shortness of stature led to confusion with other types of dwarfism, while the generalised skeletal sclerosis and clavicular hypoplasia prompted some authors to regard the condition as a form of osteopetrosis or cleidocranial dysplasia (Palmer & Thomas 1958). However, the distinctive clinical and radiographic features permit accurate diagnosis. Recognition at a histological level is also possible, as ultrastructural studies of cartilage have revealed abnormal inclusions in the chondrocytes (Stanescu et al 1975). The kinetics of calcium metabolism in pycnodysostosis have been discussed by Cabrejas et al (1976).

*Clinical and radiographic features*

Individuals with pycnodysostosis have small faces with a hooked nose, receding chin and carious misplaced teeth. The cranium bulges and the anterior fontanelle remains patent. The terminal phalanges are short, with dysplasia of the fingernails. Bony fragility predisposes to spontaneous fracture (Roth 1976). Other less consistent skeletal changes include narrowing of the thorax and spinal deformity. Adult height does not usually exceed 150 cm.

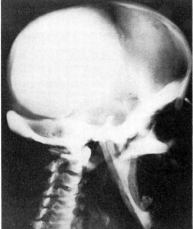

**Fig. 6.6** (left) Pycnodysostosis: skull radiograph showing increased density of the base, wide fontanelles, hypoplasia of the sinuses and an obtuse angle to the mandibular ramus.

**Fig. 6.7** (below, left) Pycnodysostosis: hand radiograph from an affected child. Although the bones are dense, their outlines are undisturbed. The terminal phalanges are shortened and irregular.

**Fig. 6.8** (below, right) Pycnodysostosis: the affected child has a small lower jaw, a hooked nose and a bulging cranium.

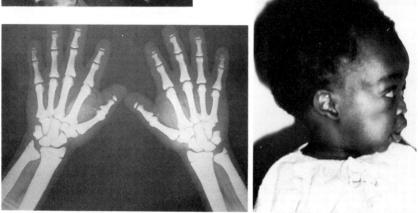

The impressionist painter Toulouse-Lautrec is thought to have had pycnodysostosis (Maroteaux & Lamy 1965). Indeed, his appearance and medical history serve as a useful 'aide-memoire' to the manifestations of the disorder. It is well-known that he was of short stature, and the 'stove-pipe' hat which he habitually wore might have covered a patent fontanelle. Similarly, his beard may have been grown to conceal a receding chin. The stick which he carried is a reminder of bone fragility; he suffered two femoral fractures in childhood as a result of minor trauma. Finally, the fact that his aristocratic parents were first cousins is in keeping with the autosomal recessive inheritance of the condition.

Bone sclerosis becomes radiographically apparent in childhood and increases throughout the years of growth. Skeletal modelling and bony contours are undisturbed and neither striations nor endobones are seen. The calvarium is not particularly dense but patency of the fontanelles and the presence of multiple Wormian bones can usually be demonstrated. The facial bones and paranasal sinuses are hypoplastic and the angle of the

mandible is obtuse. The terminal phalanges are shortened, with distal irregularity, similar to that encountered in acro-osteolysis. The clavicles may be
gracile, with underdevelopment of their lateral portions (Maroteaux &
Fauré 1973).

*Genetics*

Pycnodysostosis is inherited as an autosomal recessive. Following a review
of the literature Sedano et al (1968) estimated that about 30% of patients
were the offspring of consanguineous unions. The majority of case descriptions have been from Europe and the USA but the condition has been
encountered in Southern Africa (Palmer 1960, Wolpowitz & Matison 1974),
Portugal (Meneses de Almeida 1972), Japan (Sugiura et al 1974, Kawahara
et al 1977), India (Diwan & Gogate 1974), Israel (Roth 1976), Indonesia
(Srivastava et al 1978) and from Morocco (Bennani-Smires et al 1984). More
than 50 affected kindreds have now been recorded.

An affected individual with a deletion of the short arm of a G group chromosome, probably chromosome 22, led Elmore et al (1966) to speculate that
the abnormal gene might be located at that particular chromosomal site.
This finding has not been confirmed.

An atypical form of pycnodysostosis was reported by Kozlowski & Yu
(1972) when they documented an Australian child with the additional
features of hepatosplenomegaly, anaemia and rickets. In a further report
entitled 'craniomandibular dermatodysostosis' Danks et al (1974) described
a boy with another condition which bore some resemblance to pycnodysostosis. It is probable, however, that both conditions are separate entities.

## OSTEOMESOPYCNOSIS

Simon et al (1979) published an account of a young woman with backache,
in association with sclerosis of the spine, pelvis and proximal portions of
the long bones. The authors proposed the title 'axial osteosclerosis' for this
condition and as the patient's father and two brothers were similarly
affected, they suggested that inheritance was autosomal dominant. Maroteaux (1980) recognised the condition in five members of four separate
families, emphasised that the usual presentation was lumbar pain in early
childhood and introduced the title 'osteomesopycnosis'. Stoll et al (1981)
reported another affected kindred and confirmed that the condition was
transmitted as an autosomal dominant trait. The early reports all emanated
from France but families with the condition were subsequently recognised
in North America by Whyte et al (1981) and Proschek et al (1985). As
osteomesopycnosis is comparatively innocuous, and in view of the autosomal dominant mode of inheritance, it seems likely that previously undiagnosed cases will be increasingly recognised in the future.

## REFERENCES

### Preamble

Beighton P, Cremin B J 1980 Sclerosing bone dysplasias. Springer-Verlag, Berlin

Beighton P, Horan F, Hamersma H 1977 A review of the osteopetroses. Postgraduate Medical Journal 53: 507–515

Cotton W R, Williams G A, Hargis G K, Gaines J F 1976 Parathyroid hormone as a possible causal factor in osteopetrosis of the TL rat. Endocrinology 99/3: 872

Hall F M, Segall-Blank M, Genant H K, Kolb F O, Hawes L E 1981 Pseudohypoparathyroidism presenting as renal osteodystrophy. Skeletal Radiology 6: 43–46

Horan F T, Beighton P H 1978 'Osteopetrosis' in the Fairbank Collection. Journal of Bone and Joint Surgery 60B: 53–55

Kaitila I, Rimoin DL L 1976 Histologic heterogeneity in the hyperostotic bone dysplasias. Birth Defects: Original Article Series 12/6: 71

Karshner R G 1926 Osteopetrosis. American Journal of Roentgenology, Radium Therapy and Nuclear Medicine 16: 405

Laurell H, Wallgren A 1920 Untersuchungen uber einen Fall einer eigenartigen Skeletterkrankung (osteosclerosis fragilis generalisata). Upsala Läkareförenings Föhrandlingar 25: 309

Loutit J F, Sansom J M 1976 Osteopetrosis of microphthalmic mice — a defect of the hematopoietic stem cell? Calcified Tissue Research 20/3: 251

McCarthy J M T 1968 Bone sclerosis in infancy. Postgraduate Medical Journal 44: 908–916

McPeak C N 1936 Osteopetrosis. Report of eight cases occurring in three generations of one family. American Journal of Roentgenology, Radium Therapy and Nuclear Medicine 36: 816

Marks S C 1976 Osteopetrosis in the IA rat cured by spleen cells from a normal littermate. American Journal of Anatomy 146/3: 331

Milgram J W, Jasty M 1982 Osteopetrosis. Journal of Bone and Joint Surgery 64: 912–929

Resnick D 1981 The 'rugger jersey' vertebral body. Arthritis and Rheumatism 24: 1191–1194

Schulze F 1921 Das Wesen des Krankheitsbildes der 'Marmorknochen' (Albers-Schönberg). Archiv für Klinische Chirugie 118: 411

Seifert M F, Marks S C Jr 1985 Morphological evidence of reduced bone resorption in the osteosclerotic (oc) mouse. American Journal of Anatomy 172: 141–153

### Osteopetrosis, AD

Albers-Schönberg H 1904 Röntgenbilder einer seltenen Knochenerkrankung. Münchener Medizinische Wochenschrift 51: 365

Delahaye R P, Metges P J, Anglade J P, Malmexat X, Pascal-Suisse P 1976 Découverte simultanée d'une ostéopetrose chez la mère et le foetus à l'occasion d'une radiopelvimetrie. Journal de Radiologie 57/4: 359

Johnston C C Jr, Lavy N, Lord T, Vellios F, Merritt A D, Deiss W P Jr 1968 Osteopetrosis. A clinical, genetic, metabolic, and morphologic study of the dominantly inherited, benign form. Medicine 47: 149

Lorey A, Reye B 1923 Über Marmorknochen (Albers-Schönbergsche krankheit). Fortschritte auf dem Gebiete der Röntgenstrahlen und der Nukearmedizin 30: 35

Thomson J 1949 Osteopetrosis in successive generations. Archives of Disease in Childhood 24: 143

### Osteopetrosis — malignant or congenita form AR

Coccia P F, Krivit W, Cervenka J, Clawson C, Kersey J H, Kim T H, Nesbit M E, Ramsay N K, Warkentin P I, Teitelbaum S L, Kahn A J, Brown D M 1980 Successful bone-marrow transplantation for infantile malignant osteopetrosis. New England Journal of Medicine 302: 701–708

Enell H, Pehrson M 1959 Studies on osteopetrosis. I. Clinical report of three cases with genetic considerations. Acta Paediatrica 47: 279

Kadota R P, Smithson W A 1984 Bone marrow transplantation for diseases of childhood. Mayo Clinic Proceedings 59(3): 171–184

Khazen N, Faverly D, Vamos E, Van Regemorter N, Flament-Durand J, Carton B, Cremer-Perlmutter N 1986 Lethal osteopetrosis with multiple fractures in utero. American Journal of Medical Genetics 23: 811–819

Loria-Cortes R, Quesada-Calvo E, Cordero-Chaverri C 1977 Osteopetrosis in children: a report of 26 cases. Journal of Pediatrics 91: 43

Marks S C, Lane P W 1976 Osteopetrosis: a new recessive mutation on chromosome 12 of the mouse. Journal of Heredity 67: 11

Tips R L, Lynch H T 1962 Malignant congenital osteopetrosis resulting from a consanguineous marriage. Acta Paediatrica 47: 279

### Osteopetrosis — intermediate form, AR

Beighton P, Hamersma H, Cremin B J 1979 Osteopetrosis in South Africa. South African Medical Journal 55: 659–665

Horton W A, Schimke R N, Iyama T 1980 Osteopetrosis: further heterogeneity. Journal of Pediatrics 97: 580–585

Kahler S G, Burns J A, Aylsworth A S 1984 A mild autosomal recessive form of osteopetrosis. American Journal of Medical Genetics 17: 451–464

Kaibara N, Katsuki I, Hotokebuchi T, Takagishi K 1982 Intermediate form of osteopetrosis with recessive inheritance. Skeletal Radiology 9: 47–51

### Osteopetrosis with renal tubular acidosis, AR

Guibaud P, Labre F, Freycon M T, Genoud J 1972 Ostéopétrose et acidose rénale tubulaire. Deux cas de cette association dans une fratrie. Archives Françaises de Pédiatrie 29: 269

Ohlsson A, Stark G, Sakati N 1980 Marble bone disease: recessive osteopetrosis, renal tubular acidosis and cerebral calcification in three Saudi Arabian families. Developments in Medicine and Child Neurology 22: 72–96

Sly W S, Lang R, Avioli L, Haddad J, Lubowitz H, McAlister W 1972 Recessive osteopetrosis: a new clinical phenotype. American Journal of Human Genetics 24: 34

Sly W S, Hewett-Emmett D, Whyte M P, Yu, Y L, Tashian R E 1983 Carbonic anhydrase II deficiency identified as the primary defect in the autosomal recessive syndrome of osteopetrosis with renal tubular acidosis and cerebral calcification. Proceedings of National Academy of Science USA 80: 2752–2756

Sly W S, Whyte M P, Sundaram V, Tashian R E, Hewett-Emmett D, Guibaud P, Vainsel M, Baluarte J, Gruskin A, Al-Mosawi M, Sakati N, Ohlsson A 1985 Carbonic anhydrase II deficiency in 12 families with the autosomal recessive syndrome of osteopetrosis with renal tubular acidosis and cerebral calcification. New England Journal of Medicine 313: 139–145

Tashian R E, Hewett-Emmett D, Dodgson S J, Forester R E II, Sly W S 1984 The value of inherited deficiences of human carbonic anhydrase isozymes in understanding their cellular roles. Annals of New York Academy of Science 429: 262–275

Vainsel M, Fondu P, Cadranel S, Rocmans C, Gepts W 1972 Osteopetrosis associated with proximal and distal tubular acidosis. Acta Pediatrica Scandinavica 61: 429–434

Whyte M P, Murphy W A, Fallon M D et al 1980 Osteopetrosis, renal tubular acidosis and basal ganglia calcification in three sisters. American Journal of Medicine 69: 64–74

### Pycnodysostosis

Bennani-Smires C, Rhjoti el Alamy N, Bouchareb N 1984 Pyknodysostosis. Typical and atypical features and report on 7 cases. Journal de Radiologie 65/10: 689–695

Cabrejas M L, Fromm G A, Roca J F 1976 Pycnodysostosis. Some aspects concerning kinetics of calcium metabolism and bone pathology. American Journal of Medical Science 271/2: 215

Danks D M, Mayne V, Wettenhall H N B, Hall R K 1974 Craniomandibular dermatodysostosis. Birth Defects: Original Article Series 10/12: 99

Diwan R V, Gogate A N 1974 Pycnodysostosis (first report of a family from India). Indian Journal of Radiology 28: 268

Elmore S M, Nance W E, McGee B J, Engel-de Montmollin M, Engel E 1966 Pycnodysostosis, with a familial chromosome anomaly. American Journal of Medicine 40: 273

Kawahara K, Nishikiori M, Imai K, Ksihi K, Fujiki Y 1977 Radiographic observations of pycnodysostosis: report of a case. Oral Surgery 44: 476

Kozlowski K, Yu J S 1972 Pycnodysostosis. A variant form with visceral manifestations. Archives of Disease in Childhood 47: 804

Maroteaux P, Fauré C 1973 Pycnodysostosis. Presse Médicale 4: 403

Maroteaux P, Lamy M 1965 The malady of Toulouse-Lautrec. Journal of the American Medical Association 191: 715

Maroteaux P, Lamy M 1962 La pycnodysostose. Presse Médicale 70: 999

Meneses de Almeida L 1972 Contribution à l'étude génétique de la pycnodysostose. Annales de Génétique 15: 99–101

Palmer P E S 1960 Osteopetrosis with multiple epiphyseal dysplasia. British Journal of Radiology 33: 455

Palmer P E S, Thomas J E P 1958 Osteopetrosis with unusual changes in the skull and digits. British Journal of Radiology 31: 705

Roth V G 1976 Pycnodysostosis presenting with bilateral subtrochanteric fractures: case report. Clinical Orthopaedics and Related Research 117: 247

Sedano H D, Gorlin R J, Anderson V E 1968 Pycnodysostosis. Clinical and genetic considerations. American Journal of Diseases of Children 116: 70

Srivastava K K, Bhattacharya A K, Galatius-Jensen F, Tamaela L A, Borgstein A, Kozlowski K 1978 Pycnodysostosis: report of 4 cases. Australasian Radiology 22: 70

Stanescu R, Stanescu V, Maroteaux P 1975 Ultrastructural abnormalities of chondrocytes in pycnodysostosis. Nouveautés Médicales 4/37: 247

Sugiura Y, Yamado Y, Koh 1974 Pycnodysostosis in Japan. Report of six cases and a review of the Japanese literature. Birth Defects: Original Article Series 10/12: 78

Wolpowitz A, Matisson A 1974 A comparative study of pycnodysostosis, cleidocranial dysostosis, osteopetrosis and acro-osteolysis. South African Medical Journal 48: 1011

**Osteomesopycnosis**

Maroteaux P 1980 L'ostéomesopycnose. Archives Françaises de Pédiatrie 37: 153–157

Proschek R, Labelle H, Bard C, Marton D 1985 Osteomesopycnosis: case report. Journal of Bone and Joint Surgery 67: 652–653

Simon D, Cazalis P, Dryll A, Roland R, de Vernejoul M, Ryckewaert A 1979 Une ostéosclerose axiale de transmission dominante autosomique: nouvelle entité? Rêvue du Rhumatisme 46: 375–382

Stoll C G, Dominique C, Dreyfus J 1981 Brief clinical report: osteomesopycnosis: an autosomal dominant osteosclerosis. American Journal of Medical Genetics 8: 349–353

Whyte M P, Fallon M D, Murphy W A, Teitelbaum S L 1981 Axial osteomalacia. Clinical, laboratory and genetic investigation of an affected mother and son. American Journal of Medicine 71: 1041–1049

# 7

# Craniotubular dysplasias

The craniotubular dysplasias are a group of disorders in which abnormal modelling of the skeleton is the predominant feature (Table 7.1). Increased radiological density of bone may be present and if the cranium is involved, complications include facial distortion and cranial nerve compression.

1. Metaphyseal dysplasia (Pyle)
2. Craniometaphyseal dysplasia
3. Craniodiaphyseal dysplasia
4. Frontometaphyseal dysplasia
5. Osteodysplasty (Melnick-Needles)
6. Dysosteosclerosis
7. Tubular stenosis (Kenny-Caffey).

Table 7.1 The genetic status and frequency of the craniotubular dysplasias

|  | Inheritance | Approximate number of reported cases |
|---|---|---|
| Metaphyseal dysplasia (Pyle) | AR | 20 |
| Craniometaphyseal dysplasia | AD | 60 |
|  | AR | 10 |
| Frontometaphyseal dysplasia | XL? AD? | 30 |
| Osteodysplasty | AD? | 45 |
|  | AR form? |  |
| Dysosteosclerosis | XL | 12 |
| Tubular stenosis | AD | 5 |

## METAPHYSEAL DYSPLASIA (PYLE*)

Metaphyseal dysplasia or Pyle disease is a rare autosomal recessive disorder which is often the subject of semantic confusion with the craniometaphyseal

* Edwin Pyle 1891–1961 was an orthopaedic surgeon at the Waterbury hospital, Connecticutt, USA.

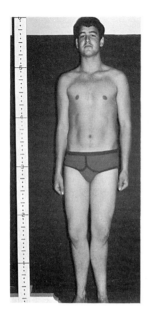

Fig. 7.1

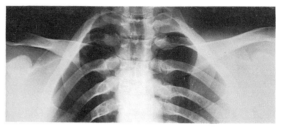

Fig. 7.2

Fig. 7.3

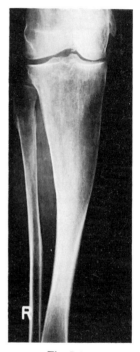

Fig. 7.4

**Fig. 7.1** (above, left) Pyle disease: the only clinical manifestations are widening of the lower regions of the thighs and medial ends of the clavicles. The innocuous nature of the disorder is evident from the fact that this young man is a keen rugby player. (From Raad & Beighton 1978 Clinical Genetics 14: 251–256.)

**Fig. 7.2** (above, right) Pyle disease: gross widening of the medial ends of the clavicles. (From Heselson et al 1979 British Journal of Radiology 52: 431–440.)

**Fig. 7.3** (right) Pyle disease: the lower end of the femur is grossly expanded. In spite of the dramatic radiographic changes, there are few clinical manifestations. (From Raad & Beighton 1978 Clinical Genetics 14: 251–256)

**Fig. 7.4** (far right) Pyle disease: the tubular bones are undermodelled, and their cortices are thin. (From Raad & Beighton 1978 Clinical Genetics 14: 251–256.)

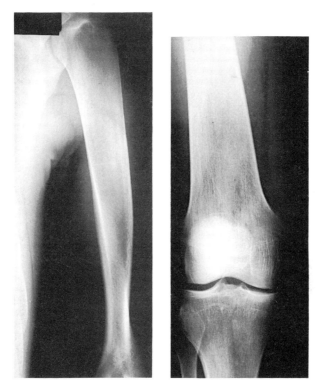

**Fig. 7.5** (left) Pyle disease: undertubulation of the proximal two-thirds of the humerus. (From Heselson et al 1979 British Journal of Radiology 52: 431–440.)

**Fig. 7.6** (right) Pyle disease: the lower femur of an obligate heterozygote, showing mild widening. (From Raad & Beighton 1978 Clinical Genetics 14: 251–256.)

dysplasias (vide infra). However, the very marked disturbance in bone modelling and the lack of cranial sclerosis serves to distinguish Pyle disease from these conditions. About 20 cases have been recorded.

*Clinical and radiographic features*

Valgus deformity of the knees may be the only obvious abnormality, but muscular weakness, scoliosis, and bone fragility are sometimes present. In contrast to the mild clinical stigmata, the radiographic changes are striking. The tubular bones of the legs show gross 'Erlenmeyer flask' flaring, particularly in the distal portions of the femora. The long bones of the arms are also undermodelled and the cortices are generally thin. The skull is virtually normal, apart from a supraorbital prominence. The bones of the pelvis and thoracic cage are expanded. The manifestations have been reviewed by Gorlin et al (1970) and Heselson et al (1979). Dental involvement was emphasised by Pazhayattil (1986).

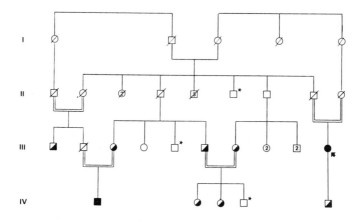

**Fig. 7.7** Pyle disease: the pedigree of an affected family. Several obligate heterozygotes have minor but definite bone widening, most evident in the lower femoral regions. (From Raad & Beighton 1978 Clinical Genetics 14: 251–256.)
*Key*: ■ male, affected; ○ female, clinically and radiographically normal; ● female, affected; ◪ male, clinically normal, minor radiographic changes in femora; ◑ female, clinically normal, minor radiographic changes in femora; / deceased.

*Genetics*

Although the dramatic radiographic changes are unmistakable, the clinical features may be mild, and Pyle disease is probably underdiagnosed. The evidence indicates that the disorder is inherited as an autosomal recessive. Affected sibs featured in the original report of Pyle (1931) and other sets were mentioned by Bakwin & Krida (1937), Hermel et al (1953) and Feld et al (1955). Parental consanguinity was present in the kindred described by Daniel (1960) and Mabille et al (1973). An affected male and female from a consanguineous Afrikaner kindred had relatives in whom minor widening of the distal femora was probably indicative of heterozygosity (Raad & Beighton 1978).

Temtamy et al (1974) reported two sisters who had the clinical and radiological stigmata of Pyle disease, together with dermal lesions and optic atrophy. As the unaffected parents were consanguineous, the authors speculated that this entity was inherited as an autosomal recessive.

## CRANIOMETAPHYSEAL DYSPLASIA

Craniometaphyseal dysplasia (CMD) was delineated by Jackson et al (1954) and more than 60 cases of the autosomal dominant form have been reported. The autosomal recessive variety is much less common and much more severe. The eponym 'Pyle' has been applied erroneously to both forms in the past thereby engendering much confusion.

*Clinical and radiographic features*

Paranasal bossing develops during infancy and progressive expansion and thickening of the skull and mandible distort the jaw and face. These changes, which are very variable in degree, become static in the third decade. Paradoxically, the paranasal bossing diminishes with the passage

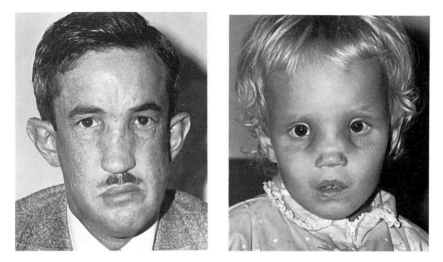

**Fig. 7.8** (left) Craniometaphyseal dysplasia — AD type: an adult male with enlargement and asymmetry of the mandible. Deafness and facial palsy were additional problems.

**Fig. 7.9** (right) Craniometaphyseal dysplasia — AD type: the daughter of the patient shown in Figure 7.8. Paranasal bossing is a prominent feature. Curiously, this abnormality regresses in later life.

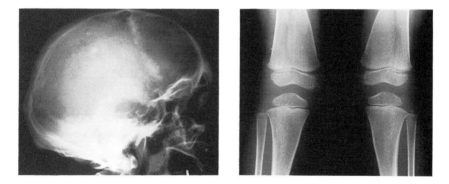

**Fig. 7.10** (left) Craniometaphyseal dysplasia — AD type: a skull radiograph from the male shown in Figure 7.8. The base is sclerotic. (From Beighton et al 1979 Clinical Genetics 15: 252–258)

**Fig. 7.11** (right) Craniometaphyseal dysplasia — AD type: radiograph of the knees of the child shown in Figure 7.9. The lower femoral metaphyses have a characteristic club-shaped configuration. (From Spiro et al 1975 South African Medical Journal 49: 839–842.)

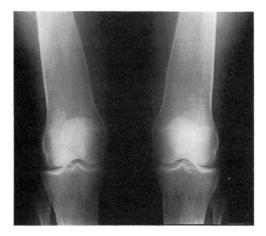

**Fig. 7.12** Craniometaphyseal dysplasia: the femora of an affected adult, showing metaphyseal widening. (From Beighton et al 1979 Clinical Genetics 15: 252–258.)

of time. Bone encroachment leads to entrapment and dysfunction of the cranial nerves and some degree of facial palsy and deafness is usually present. Dental problems arise from malocclusion of the jaws and partial obliteration of the sinuses predisposes to recurrent nasorespiratory infection. The bones are not fragile and pathological fractures do not occur. Intelligence, height, general health and life span are normal.

The radiographic changes are age-related, usually becoming evident by the age of 5. The main feature in the skull is sclerosis, which is maximal in the base, although the cranium is always involved to some degree. The long bones have widened metaphyses and normal diaphyses, presenting a club-shaped configuration, particularly at the lower end of the femur. These changes are much less severe than those encountered in Pyle disease. Minor degrees of expansion and cortical thinning are evident in the ribs and clavicles, while the spine and pelvis are uninvolved. The radiographic features of CMD have been reviewed by Spiro et al (1975).

*Genetics*

Cases in successive generations can be recognised in the reports of Spranger et al (1965), Lejeune et al (1966), Gladney & Monteleone (1970) and Stool & Caruso (1973). The author has personal knowledge of a kindred with branches in England and South Africa, in which there are 15 affected persons in five generations. Male to male transmission has occurred in this family, and the pedigree is entirely consistent with autosomal dominant inheritance (Beighton et al 1979).

The autosomal recessive form is a rarity, and fewer than 10 cases have

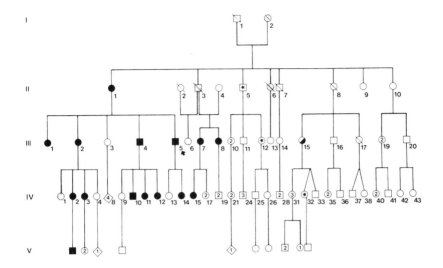

**Fig. 7.13** Craniometaphyseal dysplasia: the pedigree of an affected family. Autosomal dominant transmission is apparent. (From Beighton et al 1979 Clinical Genetics 15: 252–258.)

been recorded. The recessive mode of inheritance is evident from the reports of Lehmann (1957) and Millard et al (1967) who described affected sibs born to normal parents, and Lievre & Fischgold (1956) who mentioned parental consanguinity. The stigmata are of much greater severity and earlier onset than those of the dominant form. In particular, cranial nerve entrapment is universal and occurs at an early age. Facial distortion may be gross, hence the archaic descriptive term 'leontiasis ossea' which pertained to this disorder. All reports have concerned children or young adults and the long-term prognosis is unknown.

## CRANIODIAPHYSEAL DYSPLASIA

Craniodiaphyseal dysplasia was delineated by Joseph et al (1958). As with the recessive form of craniometaphyseal dysplasia, earlier reports appeared in the literature under the designation 'leontiasis ossea' (Gemmell 1935).

### Clinical and radiographic features

Overgrowth of the skull results in grotesque deformation of the face with entrapment of the facial, auditory and optic nerves. The importance of nasal and lacrimal obstruction was emphasised by Kaitila et al (1975). Radiographic changes are maximal in the skull and mandible, where massive hyperostosis and sclerosis are evident. The ribs and clavicles are widened and the tubular bones are undermodelled. There is no metaphyseal flaring and the shape of the long bones had been likened to that of a policeman's truncheon. The evolution of the radiographic changes has been reviewed by Tucker et al (1976).

*Genetics*

The disorder is recognisable in a description of a brother and sister with gross facial deformity (de Souza 1927). An affected child with consanguineous normal parents was reported by Halliday (1949) and isolated cases have been described by Joseph et al (1958) and Stransky et al (1962). It is reasonable to assume that inheritance is autosomal recessive. Macpherson (1974) reported three patients with widely disparate clinical and radiogical features and emphasised the probable heterogeneity of craniodiaphyseal dysplasia. A mother and infant son with a severe progressive dominant form of the disorder were documented by Schaefer et al (1986).

## FRONTOMETAPHYSEAL DYSPLASIA

Frontometaphyseal dysplasia (FMD) was delineated by Gorlin & Cohen (1969) and about 30 cases have now been described.

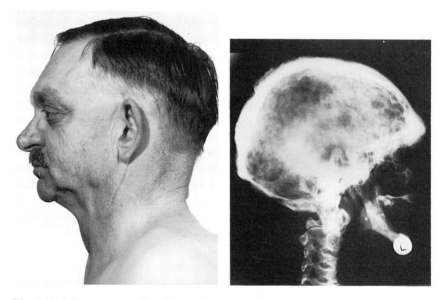

**Fig. 7.14** (left) Frontometaphyseal dysplasia: this patient presented with deafness at the age of 40. The supraorbital region is prominent and the mandible is constricted anteriorly. (From Beighton & Hamersma 1980 Journal of Medical Genetics 17: 53–56.)

**Fig. 7.15** (right) Frontometaphyseal dysplasia: skull radiograph showing massive thickening of the frontal region and patchy sclerosis throughout the skull. (From Beighton & Hamersma 1980 Journal of Medical Genetics 17: 53–56.)

*Clinical and radiographic features*

A prominent supraorbital ridge, which resembles a knight's visor, is the outstanding feature. In several published photographs the mandible appears to be hypoplastic, with anterior constriction. Dental anomalies are common, and deafness may develop in adulthood. Progressive contractures in the digits may simulate rheumatoid arthritis. General health is good and height is normal.

Radiographically, overgrowth of the supraorbital region is very marked. Sclerosis of the cranial vault is of mild degree, and may be patchy. The vertebral bodies are dysplastic but not sclerotic. The iliac crests are abruptly

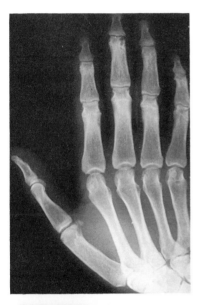

**Fig. 7.16** (left) Frontometaphyseal dysplasia: the tubular bones are undermodelled. (From Beighton & Hamersma 1980 Journal of Medical Genetics 17: 53–56.)

**Fig. 7.17** (below) FMD. The iliac crests are flared, the pelvic inlet is distorted and the femoral heads are dysplastic. (From Beighton & Hamersma 1980 Journal of Medical Genetics 17: 53–56.)

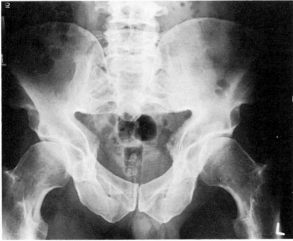

flared and the pelvic inlet is distorted. The femoral capital epiphyses are flattened with expansion of the femoral heads and a coxa valga deformity. Changes in the tubular bones are unremarkable. However, the bones of the fingers are undermodelled, and erosions and fusions may be present in the carpus.

*Genetics*

The majority of reports have concerned sporadic cases (Gorlin & Cohen 1969; Holt et al 1972; Danks et al 1972; Sauvegrain et al 1975). There was doubt concerning the mode of transmission of the disorder until Weiss et al (1975) described an affected mother and son and proposed that transmission was dominant. Kassner et al (1977) produced further evidence to support this contention. However, Jarvis & Jenkins (1975) had encountered two mentally defective males who were born of different fathers to the same unaffected mother. This raises the question as to whether FMD is heterogeneous, existing in autosomal dominant and X-linked forms.

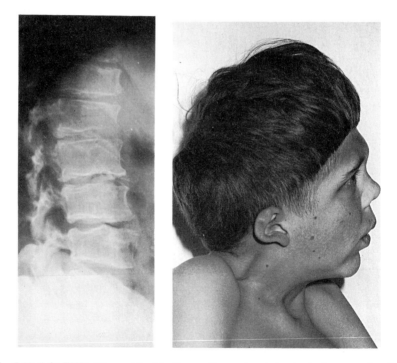

**Fig. 7.18** (left) FMD. The vertebral bodies are flattened and dysplastic, but not sclerotic. (From Beighton & Hamersma 1980 Journal of Medical Genetics 17: 53–56.)

**Fig. 7.19** (right) FMD. A boy aged 10 years with frontal prominence and a hypoplastic mandible. The initial misdiagnosis of osteodysplasty was subsequently amended to FMD.

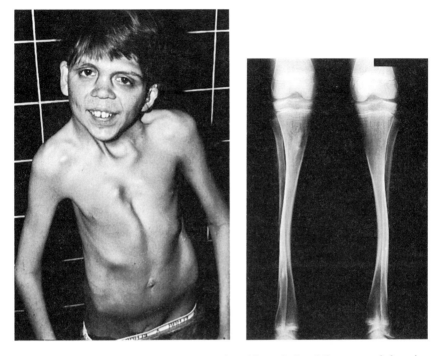

**Fig. 7.20** (left) FMD. The affected boy is deaf, and has spinal malalignment and thoracic asymmetry.

**Fig. 7.21** (right) FMD. The bones of the shins have a wavy configuration.

Gorlin & Winter (1980) reviewed all published cases and wherever possible, updated existing pedigrees. They concluded that FMD was X-linked, with severe manifestations in affected males, and minor but variable changes in heterozygous females. In an independent analysis, Beighton & Hamersma (1980) reached a similar conclusion. They also observed that FMD had many features in common with osteodysplasty (vide infra) and suggested that some reported cases, including one from their own series (Sellars & Beighton 1978) might have been incorrectly diagnosed. As virtually all published cases of osteodysplasty have been females, it is possible that there is a homogeneous X-linked syndrome, in which severely affected males are labelled 'FMD' and the mildly affected females designated 'osteodysplasty'.

Metachromasia has been detected in cultured fibroblasts from a patient with FMD (Danks et al 1972). If similar changes are present in amniotic fluid cells, antenatal diagnosis may be possible. As yet, there are no reports of attempts at this procedure.

## OSTEODYSPLASTY

About 25 individuals with osteodysplasty have been reported. These include members of two kindreds investigated by Melnick & Needles (1966) and sporadic cases described by Coste et al (1968), Wendler & Kellerer (1975) and Stoll et al (1976). The manifestations of the condition were reviewed by Leiber et al (1975). The designation 'osteodysplasty' pertains to the generalised nature of the skeletal dysplasia and it is intended to convey the meaning of 'badly formed'.

### Clinical and radiographic features

The forehead is prominent and the mandible is small. Variable skeletal malformations include kyphoscoliosis, genu valgum and shortening of the distal phalanges. Radiographically, irregular ribbon-like constrictions of the ribs and tubular bones are a striking feature. The pelvis is distorted and coxa valga is present. The base of the skull may be thickened and patchy areas of sclerosis are seen in the cortices of the long bones.

### Genetics

Transmission through four generations of a kindred and three generations of another was reported by Melnick & Needles (1966). It initially seemed likely that osteodysplasty was inherited as an autosomal dominant trait but Gorlin & Knier (1982) subsequently determined that several putatively affected males were actually normal. Following a thorough analysis of the literature, they identified a total of 23 patients in 15 pedigrees, all of whom were female. There was no advanced paternal age, which might have indicated new mutation, and the authors speculated that osteodysplasty might be either X-linked or an autosomal dominant trait which was lethal in males. A relationship with frontometaphyseal dysplasia is also possible (see previous section) although this matter remains unresolved.

   In the context of male lethality, von Oeyen et al (1982) reported an affected woman who gave birth to a son with osteodysplasty plus an omphalocele and hypoplastic kidneys; this child died in early infancy. Similarly, Theander & Ekberg (1981) reported a pregnant female with osteodysplasty in whom radiographic studies revealed fetal malformation. After delivery, abnormalities of the abdominal wall, bowel and eyes were evident together with soft tissue calcification and skeletal changes.

   A severe type of osteodysplasty, which is lethal in infancy, has been encountered in two Albanian sibs and a Polish girl. Danks et al (1974) designated this condition 'a precocious form of osteodysplasty', and suggested that inheritance was autosomal recessive. A similar mode of inheritance was postulated by ter Haar et al (1982) when they reported three individuals with the skeletal changes of osteodysplasty, one of whom also had bilateral glaucoma while the other two had congenital cardiac defects.

Two brothers with osteodysplasty and mental retardation were reported by Ruvalcaba et al (1971); the condition was partially expressed in two female maternal cousins. It seems very probable that these entities are different from classical osteodysplasty.

Svejcar (1983) has undertaken biochemical studies of skeletal material from a patient with osteodysplasty and demonstrated that the collagen content is increased. The basic defect, however, remains unknown.

## DYSOSTEOSCLEROSIS

Dysosteosclerosis is a rare disorder which was delineated by Spranger et al (1968). The manifestations have been reviewed by Liesti et al (1975) and Houston et al (1978).

### Clinical and radiographic features

Published reports have concerned children with short stature, fragile bones and enamel hypoplasia. Overgrowth of the skull may lead to cranial nerve compression. The clinical status of the affected adult is unknown. Sclerosis of the skull and axial skeleton, together with platyspondyly, are the major radiographic features. The long bones may be bowed, with metaphyseal expansion and epiphyseal sclerosis.

### Genetics

Affected sibs with consanguineous parents are recognisable in the case reports of Ellis (1934) and Field (1939). Spranger et al (1968) and Houston et al (1978) also recorded parental consanguinity. Of the 12 reported patients, 11 have been males (Stehr 1941, Kaitila & Rimoin 1976, Pascual-Castroviejo et al 1977). The only female thought to be affected was reported by Roy et al (1968); this girl had additional neurological and dermatological manifestations and it is possible that she had a different disorder. On the basis of existing evidence, it seems probable that dysosteosclerosis is X-linked.

## TUBULAR STENOSIS

Kenny & Linnarelli (1966) reported a mother and son with proportionate dwarfism, low birth weight, delayed closure of the anterior fontanelle, narrowing of the tubular bones, hypocalcaemia and tetany. The radiographic changes in these patients were subsequently reviewed by Caffey (1967) and the alternative designation 'Kenny-Caffey syndrome' came into use. The diameter of the shafts of the tubular bones was reduced, while the metaphyses were relatively flared. The external contours of these bones were irregular and the cortices were thickened. Narrowing of the medullary cavities was emphasised by Frech & McAlister (1968).

Larsen et al (1985) undertook extensive biochemical investigations in a sporadic adult male with the syndrome and reported that treatment with vitamin D and calcium resulted in normocalcaemia.

In addition to the family documented by Kenny & Linnarelli (1966), generation to generation transmission has been recorded by Segond et al (1973). Autosomal dominant inheritance is probable.

## REFERENCES

**Metaphyseal dysplasia**

Bakwin H, Krida A 1937 Familial metaphyseal dysplasia. American Journal of Diseases of Children 53: 1521

Daniel A 1960 Pyle's disease. Indian Journal of Radiology 14: 126

Feld H, Switzer R A, Dexter M W, Langer E W 1955 Familial metaphyseal dysplasia. Radiology 65: 206

Gorlin R J, Koszalka M F, Spranger J 1970 Pyle's disease (familial metaphyseal dysplasia). Journal of Bone and Joint Surgery 52A: 347

Hermel M B, Gershon-Cohen J, Jones D T 1953 Familial metaphyseal dysplasia. American Journal of Roentgenology, Radium Therapy and Nuclear Medicine 70: 413

Heselson N G, Raad M S, Hamersma H, Cremin B J, Beighton P 1979 The radiological manifestations of metaphyseal dysplasia (Pyle disease). British Journal of Radiology 52/618: 431–440

Mabille J-P, Benoit J-P, Castera D 1973 Dysplasie metaphysaire de Pyle. Annals of Radiology (Paris) ·16/11: 273

Pazhayattil S 1986 Pyle's Disease (Familial Metaphyseal Dysplasia). Saudi Medical Journal 4: 409–411

Pyle E 1931 Case of unusual bone development. Journal of Bone and Joint Surgery 13: 874

Raad M S, Beighton P 1978 Autosomal recessive inheritance of metaphyseal dysplasia (Pyle disease). Clinical Genetics 14: 251–256

Temtamy S A, El-Meligy M R, Badrawy H A, Meguid M S A, Safwat H M 1974 Metaphyseal dysplasia, anetoderma and optic atrophy: an autosomal recessive syndrome. Birth Defects: Original Article Series 10/12: 61

**Craniometaphyseal dysplasia**

Beighton P, Hamersma H, Horan F 1979 Craniometaphyseal dysplasia — variability of expression within a large family. Clinical Genetics 15: 252–258

Gladney J H, Monteleone P L 1970 Metaphyseal dysplasia. Lancet 2: 44

Jackson W P U, Albright F, Drewry G, Hanelin J, Rubin M L 1954 Metaphyseal dysplasia, epiphyseal dysplasia, diaphyseal dysplasia and related conditions. Archives of Internal Medicine 94/6: 871

Lehmann E C H 1957 Familial osteodystrophy of the skull and face. Journal of Bone and Joint Surgery 39B: 313

Lejeune E, Anjou A, Bouvier M, Robert J, Vauzelle J L, Jeanneret J 1966 Dysplasie cranio-metaphysaire. Revue du Rhumatisme et des Maladies osteo-articulares 33: 714

Lievre J A, Fischgold H 1956 Leontiasis ossea chez l'enfant (osteopetrose partielle probable). Presse Médicale 64: 763

Millard D R, Maisels D D, Batstone J H F, Yates B W 1967 Craniofacial surgery in craniometaphyseal dysplasia. American Journal of Surgery 113: 615

Spiro P C, Hamersma H, Beighton P 1975 Radiology of the autosomal dominant form of craniometaphyseal dysplasia. South African Medical Journal 49: 839

Spranger J, Paulsen K, Lehmann W 1965 Die kraniometaphysaere Dysplasie (Pyle). Zeitschrift für Kinderheilkunde 93: 64

Stool S E, Caruso V G 1973 Cranial metaphyseal dysplasia. Archives of Otolaryngology 97: 410

**Craniodiaphyseal dysplasia**

de Souza O 1927 Leontiasis ossea. Case reports. Porto Alegre Faculty of Medicine 13: 47

Gemmell J H 1935 Leontiasis ossea: a clinical and roentgenological entity. Report of a case. Radiology 25: 723

Halliday J 1949 Rare case of bone dystrophy. British Journal of Surgery 37: 52

Joseph R, Lefebvre J, Guy E, Job J C 1958 Dysplasia cranio-diaphysaire progressive. Ses relations avec la dysplasie diaphysaire progressive de Camurati-Engelmann. Annals of Radiology 1: 477

Kaitila I, Stewart R E. Landow E, Lachman R, Rimoin D L 1975 Craniodiaphyseal dysplasia. Birth Defects 11/6: 359

Macpherson R I 1974 Craniodiaphyseal dysplasia, a disease or group of diseases? Journal of the Canadian Association of Radiologists 25: 22

Stransky E, Mabilangan L, Lara R T 1962 On Paget's disease with leontiasis ossea and hypothyreosis, starting in early childhood. Annals of Paediatrics 199: 399

Schaefer B, Stein S, Oshman D, Rennert O, Thurnau G, Wall J, Bodensteiner J, Brown O 1986 Dominantly inherited craniodiaphyseal dysplasia: A new craniotubular dysplasia. Clinical Genetics 30: 381–391

Tucker A S, Klein L, Antony G J 1976 Craniodiaphyseal dysplasia: Evolution over a 5 year period. Skeletal Radiology 1: 47

**Frontometaphyseal dysplasia**

Beighton P, Hamersma H 1980 Frontometaphyseal dysplasia: autosomal dominant or x-linked? Journal of Medical Genetics 17/1: 53–56

Danks D M, Mayne C, Hall R K, McKinnon M C 1972 Frontometaphyseal dysplasia. A progressive disease of bone and connective tissue. American Journal of Diseases of Children 123: 254

Gorlin R J, Cohen M M 1969 Frontometaphyseal dysplasia. A new syndrome. American Journal of Diseases of Children 118: 487

Gorlin R J, Winter R B 1980 Frontometaphyseal dysplasia — Evidence for x-linked inheritance. American Journal of Medical Genetics 5: 81–84

Holt J F, Thompson G R, Arenberg I K 1972 Fronto-metaphyseal dysplasia. Radiology Clinics of North America 10: 225

Jarvis G A, Jenkins E C 1975 In: Bergsma (ed) Syndrome Identification, Volume 3, Number 1. The National Foundation — March of Dimes, New York, p 18

Kassner E G, Haller J O, Reddy V H, Mitarotundo A, Katz I 1977 Frontometaphyseal dysplasia: evidence for autosomal dominant inheritance. American Journal of Roentgenology 127: 927

Sauvegrain J, Lombard M, Garel L, Truscelli D 1975 Frontometaphyseal dysplasia. Annals of Radiology 18/2: 155

Sellars S, Beighton P 1978 Deafness in Osteodysplasty of Melnick and Needles. Archives of Otolaryngology 104: 225–227

Weiss L, Reynolds W A, Szymanowski R T 1975 Familial frontometaphyseal dysplasia: evidence for dominant inheritance. Birth Defects: Original Article Series 11/5: 55

**Osteodysplasty**

Coste F, Maroteaux P, Chouraki L 1968 Osteodysplasty (Melnick and Needles' syndrome). Report of a case. Annals of Rheumatic Diseases 27: 360

Danks D M, Mayne C, Kozlowski K 1974 A precocious autosomal recessive type of osteodysplasty. In: The Clinical Delineation of Birth Defects, Number 19. Williams and Wilkins, Baltimore

Gorlin R J, Knier J 1982 X-linked or autosomal dominant, lethal in the male, inheritance of the Melnick-Needles (osteodysplasty) syndrome? A reappraisal. American Journal of Medical Genetics 13/4: 465–467

Leiber B, Olbrich G, Moelter N, Walther A 1975 Melnick–Needles syndrome. Monatsschrift für Kinderheilkunde 123/9: 178

Melnick J C, Needles C F 1966 An undiagnosed bone dysplasia. A family study of four generations and three generations. American Journal of Roentgenology, Radium Therapy and Nuclear Medicine 97: 39

Ruvalcaba R H, Reichert A, Smith D W 1971 A new familial syndrome with osseous dysplasia and mental deficiency. Journal of Pediatrics 79: 450

Stoll C L, Levy J M, Gardea A, Weil J 1976 L'ostéodysplastie. Pédiatrie 31/2: 195

Svejcar J 1983 Biochemical abnormalities in connective tissue of osteodysplasty of Melnick-Needles and dyssegmental dwarfism. Clinical Genetics 23: 369–375

ter Haar B, Hamel B, Hendriks J, de Jager J 1982 Melnick-Needles syndrome: indication for an autosomal recessive form. American Journal of Medical Genetics 13/4: 469–477

Theander G, Ekberg O 1981 Congenital malformations associated with maternal osteodysplasty. A new malformation complex. Acta Radiol. Ser. Diagn. 22/3B: 369–377

von Oeyen P, Holmes L B, Trelstad R L, Griscom N T 1982 Omphalocele and multiple severe congenital anomalies associated with osteodysplasty (Melnick-Needles syndrome). American Journal of Medical Genetics 13/4: 453–463

Wendler H, Kellerer K 1975 Osteodysplastic syndrome (Melnick and Needles). Fortschritte auf dem Gebiete der Röntgenstrahlen und der Nuklearmedizin 122/4: 309

### Dysosteosclerosis

Ellis R W B 1934 Osteopetrosis (marble bones: Albers-Schönberg's disease: osteosclerosis fragilis generalisata: congenital osteosclerosis). Proceedings of the Royal Society of Medicine 27: 1563

Field C E 1939 Albers-Schönberg disease. An atypical case. Proceedings of the Royal Society of Medicine 32: 320

Houston C S, Gerrard J W, Ives E J 1978 Dysosteosclerosis. American Journal of Radiology 130: 988

Kaitila I, Rimoin D L 1976 Histologic heterogeneity in the hyperostotic bone dysplasias. Birth Defects 12/6: 71

Liesti J, Kaitila I, Lachman R S, Asch M J, Rimoin D L 1975 Dysosteosclerosis. Birth Defects: Original Article Series 11/6: 349

Pascual-Castroviejo I, Casas-Fernandez C, Lopez-Martin V, Martinez-Bermeio A 1977 X-linked dysosteosclerosis. Four familial cases. European Journal of Pediatrics 126: 127

Roy C, Maroteaux P, Kremp L, Courtrecuise V, Alagille D 1968 Un nouveau syndrome osseux avec anomalies cutanées et troubles neurologiques. Archives Françaises de Pédiatrie 25: 985

Spranger J W, Albrecht C, Rohwedder H J, Wiedemann H R 1968 Die Hysosteosklerose: eine Sonderform der generalisierten Osteosklerose. Fortschritte auf dem Gebiete der Röntgenstrahlen und der Nuklearmedizin 109: 504

Stehr L 1941 Pathogenese und Klinik der Osteosklerosen. Archiv für orthopädische und Unfall-Chirurgie mit besonderer Berucksichtigung der Frakturenlehre und der orthopädisch-Chirurgischen Technik 41: 156

### Tubular stenosis

Caffey J P 1967 Congenital stenosis of medullary spaces in tubular bones and calvaria in two proportionate dwarfs, mother and son, coupled with transitory hypocalcaemic tetany. American Journal of Roentgenology, Radium Therapy and Nuclear Medicine 100: 1

Frech R S, McAlister W H 1968 Medullary stenosis of the tubular bones associated with hypocalcemic convulsions and short stature. Radiology 91: 457

Kenny F M, Linarelli L 1966 Dwarfism and cortical thickening of tubular bones. Transient hypocalcaemia in a mother and son. American Journal of Diseases of Children 111: 201

Larsen J L, Kivlin J, Odell W D 1985 Unusual case of short stature. American Journal of Medicine 78/6 pt 1: 1025–1032

Segond P, Menkes C J, Maroteaux P, Braun S, Delbarre F 1973 Le retrecissement du canal medullaire des os à transmission dominante. Nouvelle Presse Médicale 2: 2728

# 8

# Craniotubular hyperostoses

Overgrowth of bone, which leads to alteration of contours and increase in radiological density of the skeleton, is the predominant feature of the craniotubular hyperostoses. In this group of conditions, hyperostosis is present in the skull and other regions in varying degrees and combinations. As with the craniotubular dysplasias, confusion with the osteopetroses is a recurring theme.

1. Endosteal hyperostosis AR, Van Buchem type
2. Endosteal hyperostosis AD, Worth type
3. Sclerosteosis
4. Oculodento-osseous dysplasia
5. Diaphyseal dysplasia (Camurati-Engelmann)
6. Infantile cortical hyperostosis (Caffey)
7. Osteoectasia
8. Osteitis deformans (Paget)

Table 8.1  Genetic status and frequency of the craniotubular hyperostoses

|  | Inheritance | Approx. no of cases |
|---|---|---|
| 1. Endosteal hyperostosis (van Buchem) | AR | 20 |
| 2. Endosteal hyperostosis (Worth) | AD | 35 |
| 3. Sclerosteosis | AR | 60 |
| 4. Oculodento-osseous dysplasia | AD | 50 |
| 5. Diaphyseal dysplasia (Camurati-Engelmann) | AD | 100 |
| 6. Infantile cortical hyperostosis (Caffey) | AD? | 120 |
| 7. Osteoectasia | AR | 30 |
| 8. Osteitis deformans (Paget) | ? | 1000 + |

## ENDOSTEAL HYPEROSTOSIS, Van Buchem* type

Using the term 'hyperostosis corticalis generalisata familiaris' van Buchem et al (1955) described two sibs in Holland with cranial sclerosis and

* Van Buchem (1898–1979) was the Professor of internal medicine at Groningen, Holland, when he published the account of the condition which bears his name.

191

widening of the diaphyses of the long bones. Further reports followed (van Buchem et al 1962, van Buchem 1971) and the accumulated information was eventually published as a monograph (van Buchem et al 1976). About 20 cases of van Buchem disease have been reported and the eponymous designation is widely accepted.

There are semantic problems with the term 'endosteal hyperostosis' as an autosomal dominant, or 'Worth' form is also recognised. In addition, sclerosteosis is sometimes categorised as a subdivision of endosteal hyperostosis. This latter format is employed to denote possible syndromic homogeneity but it has the disadvantage of engendering confusion. The nosological situation has been reviewed in detail by Eastman & Bixler (1977).

*Clinical and radiographic features*

Overgrowth and distortion of the mandible and brow become evident during the latter part of the first decade. Subsequently, entrapment of the cranial nerves leads to facial palsy and deafness. Two sibs mentioned in the original case report were mentally defective, but other patients have been of normal intelligence. The disorder is progressive, and optic nerve involve-

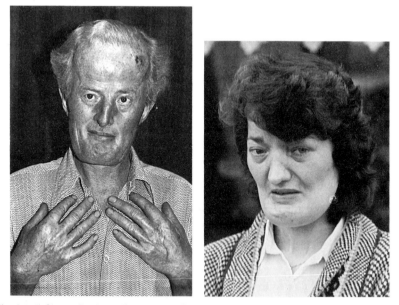

**Fig. 8.1** (left) van Buchem disease: mandibular overgrowth of moderate degree. In distinction to sclerosteosis (vide infra) the digits are normal. (From Beighton et al 1984 Clinical Genetics 25: 275–181.)

**Fig. 8.2** (right) van Buchem disease: bilateral facial palsy and marked mandibular overgrowth. (From Beighton et al 1984 Clinical Genetics 25: 275–181.)

ment may be a late complication. However, the lifespan is not compromised, stature is normal and the bones are not fragile.

Widening and sclerosis of the calvarium, cranial base and mandible are the major radiographic features. Endosteal thickening is present in the diaphyses of the tubular bones. The external configuration of these bones is relatively undisturbed.

*Genetics*

The disorder described by van Buchem et al (1955, 1962) is undoubtedly autosomal recessive. None of these Dutch patients had affected parents or offspring. Parental consanguinity was present in one kindred, and a pair of sibs and a set of dizygous twins were encountered in other families.

Several of the affected persons lived on the island of Urk in the Zuider Zee and the constraints imposed by this geographic isolation presumably influenced mating patterns and the emergence of recessive disease. The island is now connected to the mainland with a causeway and it can be foreseen that inbreeding will diminish and the incidence of affected homozygotes will decline.

Apart from the Dutch patients with van Buchem disease, the only other genuine cases which are recognisable in the literature are four Scots siblings (Dixon et al 1982). It is not known whether this family had any ancestral links with Holland. Van Buchem disease is similar to sclerosteosis, which is present in Afrikaners of Dutch stock and it is possible that these disorders share the same basic genetic defect (Beighton et al 1984). This issue is discussed on page 197.

ENDOSTEAL HYPEROSTOSIS, Worth* type, AD

The Worth, or autosomal dominant type of endosteal hyperostosis is distinct from van Buchem disease, the only link between these entities being common terminology. The difficulty arose when Worth & Wollin (1966) reported generation to generation transmission of a condition which they termed 'hyperostosis corticalis generalisata congenita'. Maroteaux et al (1971) recognised that this autosomal dominant disorder was different from the autosomal recessive van Buchem form, and Spranger et al (1974) used the term 'endosteal hyperostosis' to embrace both types. As case reports accumulated Gelman (1977) and Gorlin & Glass (1977) introduced the alternative designation 'autosomal dominant osteosclerosis' in order to emphasise syndromic identity. Unfortunately, this title is also used for a different condition which was delineated by Stanescu et al (1963; see Ch. 9), but the addition of the eponym 'Worth' clarifies this issue. About 35 cases have been reported.

---

* Worth was a senior radiologist in Vancouver, Canada, until his retirement a few years ago.

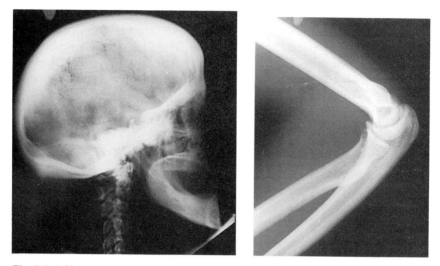

**Fig. 8.3** (left) Endosteal hyperostosis, Worth type: skull radiograph showing moderate widening and sclerosis of the calvarium, base and mandible.

**Fig. 8.4** (right) Endosteal hyperostosis, Worth type: the diaphyses of the long bones are sclerotic.

*Clinical and radiological features*

The Worth type of endosteal hyperostosis is a mild disorder, in which progressive overgrowth of the mandible may be the only clinical manifestation. Intelligence and stature are normal, cranial nerve involvement is unusual, and intracranial pressure does not become elevated. The bones are resistant to fracturing. Torus palatinus and mandibular asymmetry are associated with malocclusion and dental problems.

The radiographic changes are very similar to those of the benign autosomal dominant form of osteopetrosis and differentiation between these conditions can be difficult. In both disorders the skeleton shows increased density, without disturbance of the bony contours. In the Worth form of endosteal hyperostosis, in distinction to osteopetrosis, the vertebral bodies do not have the 'rugger jersey' appearance, which is produced by endplate sclerosis, although the pedicles are dense. Hyperostosis of the mandible in the former condition is another distinguishing feature. In the mildly affected person the diagnosis is sometimes made by chance, following radiological studies for some unrelated purpose.

*Genetics*

Autosomal dominant inheritance is well established. In addition to the reports of familial transmission by Worth & Wollin (1966), Maroteaux et al (1971) and Scott & Gautby (1974), 14 affected persons in four generations were reported by Beals (1976). Owen (1976) recognised the disorder in six members of three generations of a British family who had previously been studied by Dyson (1972). Other families in which inheritance was autosomal dominant were reported by Lapresle et al (1976), Vayssairat et al (1976), Gelman (1977) and Gorlin & Glass (1977).

In view of the innocuous nature of this condition and its wide geographic distribution it is possible that it is fairly common.

## SCLEROSTEOSIS

Truswell (1958) recognised the existence of this disorder when he described six patients in an article entitled 'Osteopetrosis with syndactyly; a morphological variant of Albers-Schönberg disease'. The designation 'sclerosteosis' is now in general use, although it does not accurately describe the manifestations of the condition. About 60 cases have been reported, the majority among the Afrikaner community of South Africa (Beighton et al 1976, Beighton & Hamersma 1979). Other known patients are a sibship in New

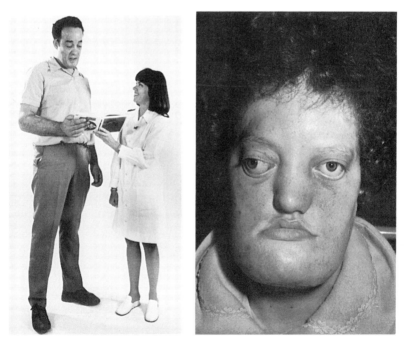

**Fig. 8.5** (left) Sclerosteosis: tall stature is a notable feature. This young man is 214 cm in height. (From Epstein et al 1979 South African Medical Journal 55: 1105–1110.)

**Fig. 8.6** (right) Sclerosteosis: progressive bone thickening leads to distortion of the face and jaws, with proptosis, facial palsy and deafness. (From Beighton et al 1977 Clinical Genetics 11: 1.)

York (Higinbotham & Alexander 1941) a woman in Switzerland (Pietruschka 1958), a kindred of mixed ancestry in Maryland, USA (Kelley & Lawlah 1946, Witkop 1965), a girl in Japan (Sugiura & Yasuhara 1975) and a consanguineous family in Brazil (Alves et al 1982).

*Clinical and radiographic features*

Progressive overgrowth and sclerosis of the skeleton, particularly the skull, develop in early childhood. Height and weight are often excessive. Indeed, a patient known to the author was a heavyweight boxing champion at the age of 14. The consistency of his bones undoubtedly facilitated his pugilistic activities. Deafness and facial palsy due to cranial nerve entrapment may be a presenting feature. Distortion of the facies, which is apparent by the age of 10, eventually becomes very severe. In adulthood elevation of intracranial pressure may cause headache. Several adults have died suddenly from impaction of the brain stem in the foramen magnum. Cutaneous or bony syndactyly of the second and third fingers serves to distinguish sclerosteosis from the other disorders in this group. The terminal phalanges are deviated radially, with dystrophy of the finger nails. The bones are resistant to trauma and pathological fractures do not occur.

Without prophylactic craniectomy survival beyond middle age is unusual, although an affected male aged 77 years has been documented (Barnard et al 1980).

Gross widening and sclerosis of the skull is the predominant radiographic feature. Hypertrophy of the mandible and frontal regions leads to relative mid-facial hypoplasia. The vertebral bodies are spared although their pedicles are dense. The pelvic bones are sclerotic but their contours are normal. The cortices of the long bones are sclerosed and hyperostotic. The radiographic features of sclerosteosis have been reviewed by Beighton et al (1976).

Epstein et al (1979) were unable to demonstrate any significant endocrine dysfunction in these affected adults. Following comprehensive histological studies Stein et al (1983) suggested that the condition might be a disorder of osteoblast hyperactivity. As yet, however, the nature of the basic defect is unknown.

*Genetics*

Sclerosteosis is inherited as an autosomal recessive. Consanguinity was present in 5 of 15 Afrikaner kindreds, into which a total of 25 affected individuals had been born (Beighton et al 1977). There were several sets of sibs in this series but the parents were all normal. In this community the minimum prevalence of sclerosteosis is 1: 75 000, with a gene frequency of 0.0035. At least 1 in every 140 Afrikaners is a carrier of the gene and there are about 10 000 clinically normal heterozygotes in South Africa.

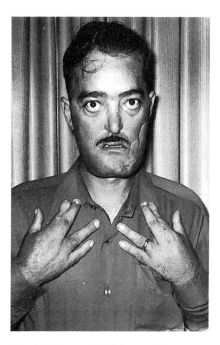

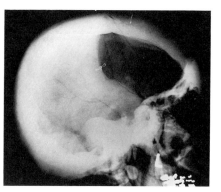

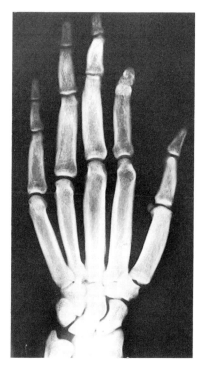

**Fig. 8.7** (above, left) Sclerosteosis: gross asymmetrical mandibular overgrowth with dental malalignment. Syndactyly of the 2nd and 3rd fingers of each hand is a characteristic feature of the condition. (From Beighton et al 1984 Clinical Genetics 25: 175–181.)

**Fig. 8.8** (above, right) Sclerosteosis: the calvarium is sclerotic and hyperostotic. In this patient, craniotomy has been undertaken to relieve raised intracranial pressure. (From Beighton P, Durr L, Hamersma H 1976 Annals of Internal Medicine 84: 393.)

**Fig. 8.9** (right) Sclerosteosis: the tubular bones are undermodelled, with thickening of their cortices. (From Beighton P, Cremin B J, Hamersma H 1976 British Journal of Radiology 49: 934.)

Sclerosteosis has similar manifestations to van Buchem disease, differing by virtue of greater severity and the additional component of syndactyly. However, a mild example of sclerosteosis, lacking syndactyly, and a severely affected person with van Buchem disease would be indistinguishable from each other. The progenitors of the Afrikaner community left Holland about 300 years ago and as the overwhelming majority of individuals with van Buchem disease are Dutch, it seems likely that there is some fundamental genetic link between these entities. It is possible that they are

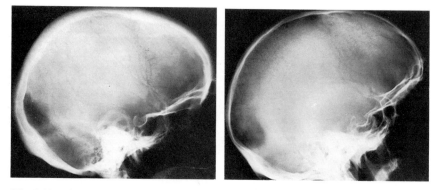

**Fig. 8.10** (left) Sclerosteosis; the clinically normal heterozygote may have a minor degree of calvarial thickening.

**Fig. 8.11** (right) Sclerosteosis; a normal skull, for comparison with Figure 8.10.

both the consequence of homozygosity for the same mutant allelles, with the phenotype in the Afrikaners being modified by epistatic influences (Beighton et al 1984). At present, this concept is speculative and the elucidation of this fascinating problem awaits the application of molecular techniques. The Dutch origins of the sclerosteosis gene were further emphasised by Alves et al (1982) in their description of an affected kindred of mixed ancestry in the State of Bahia, Brazil, when they pointed out that this region had been occupied by Hollanders during the 17th century.

Calvarial thickening has been recognised in lateral skull radiographs from a number of obligatory heterozygotes. This finding has implications for the genetic counselling of the potentially heterozygous relatives of known patients. At present, however, there are no firm objective methods for detecting the asymptomatic carriers of the faulty gene.

## OCULODENTO-OSSEOUS DYSPLASIA

Oculodento-osseous dysplasia (ODOD) was delineated by Gorlin et al (1963) under the title 'oculodentodigital dysplasia' and Reisner et al (1969) were able to assemble data concerning 43 affected persons. The title of the condition appeared in the 1970 version of the Paris Nomenclature as 'oculodento-osseous dysplasia (ODOD)' and in the 1983 revision it was subdivided into mild autosomal dominant and severe autosomal recessive types. This format has facilitated differentiation from the oculodentodigital syndrome type II (O'Rourk & Bravos 1969) in which preaxial polydactyly and the absence of defects of bone modelling are distinguishing features. About 50 cases of the mild autosomal dominant form of ODOD have been recorded, but there are less than 10 reports of the severe autosomal recessive form.

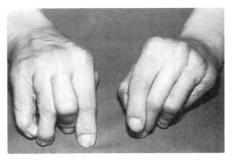

**Fig. 8.12** (left) ODOD: A young man with mandibular overgrowth, a narrow nose and microphthalmia with microcornea. (From Barnard et al 1980 South African Medical Journal 59: 758–762.)

**Fig. 8.13** (right) ODOD: The 5th fingers are flexed and incurved, due to hypoplasia of the middle phalanx. Soft tissue syndactyly of the 4th and 5th digits has been corrected surgically.

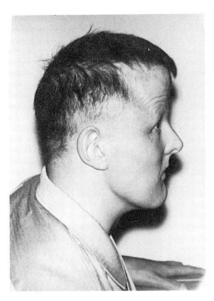

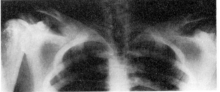

**Fig. 8.14** (left) ODOD: An affected relative of the patient depicted in Figure 8.12. He is blind, with defective dental enamel, and fine, sparse hair. The alae nasi are hypoplastic. (From Beighton et al 1979 Clinical Genetics 16: 169–177.)

**Fig. 8.15** (above) ODOD: The clavicles show gross medial expansion and have a 'mutton chop' configuration. Patchy sclerosis is present in the ribs and humeri. (From Beighton et al 1979 Clinical Genetics 16: 169–177.)

## Clinical and radiographic features

The abnormalities in the mild and severe forms of ODOD have a similar anatomical distribution but they differ greatly in severity. The main stigmata are microphthalmia, fine sparse hair and a narrow nose. Ophthalmological involvement has been emphasised by Sugar (1978), Judisch et al (1979) and Gutierrez Diaz et al (1982). Glaucoma is a frequent complication

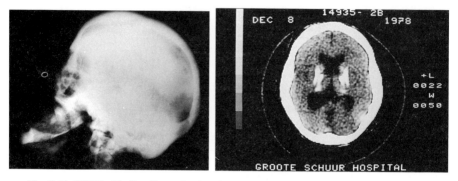

**Fig. 8.16** ODOD: Skull radiograph, showing marked hyperostosis and sclerosis. (From Beighton et al 1979 Clinical Genetics 16: 169–177.)

**Fig. 8.17** ODOD: Skull, computerised tomographic scan. The calvarium is widened and the ventricles show mild dilatation. Calcification of the basal ganglia is an important diagnostic feature. (From Barnard et al 1980 South African Medical Journal 59: 758–762.)

and hypoplasia of the enamel of the teeth is a constant feature. Cutaneous syndactyly with flexion and ulnar deviation of the fourth and fifth fingers are additional components of this syndrome. Undermodelling of the tubular bones, clavicles, ribs and mandible is radiographically evident. In the severe form, cranial hyperostosis and mandibular overgrowth are prominent features, while the clavicles are expanded into a 'mutton chop' configuration. Calcification of the basal ganglia is an important diagnostic feature (Barnard et al 1981).

*Genetics*

Autosomal dominant inheritance of the mild form of ODOD can be inferred from the reports of Mohr (1939), Duggan & Hassard (1961) and Pfeiffer et al (1968). In particular, Littlewood & Lewis (1963) described father to son transmission, while Rajic & de Veber (1966) reported seven individuals in three generations of a family. The severe form has been described in affected siblings with normal parents (Gillespie 1964) and cousins born into an Afrikaner kindred in South Africa in which a pair of brothers had married a pair of sisters (Beighton et al 1979). In a second Afrikaner family the mother of a young affected male with severe stigmata had partial soft tissue syndactyly of the fourth and fifth fingers and second and third toes. It is possible that these features represented minor phenotypic manifestations in an obligate heterozygote. Further evidence for autosomal recessive inheritance was produced by Traboulsi et al (1986) when they reported a young Lebanese girl with ODOD who had been born to unaffected consanguineous parents.

## DIAPHYSEAL DYSPLASIA (Camurati-Engelmann*)

Diaphyseal dysplasia is comparatively well known and more than 100 cases have been reported. Although the eponym 'Camurati-Engelmann' disease is often used, McKusick (1975) contended that the disorder described by Camurati was a different condition and that the designation 'Engelmann disease' is more accurate. This issue is unresolved and the conjoined eponym has been retained in the Paris Nomenclature.

### Clinical and radiographic features

Diaphyseal dysplasia presents in mid-childhood with muscular pain, weakness and wasting, typically in the legs. The condition is self-limiting and generally resolves by the age of 30. Cranial nerve compression and raised intracranial pressure are occasional complications. The manifestations are variable and some patients are severely handicapped, while others are virtually asymptomatic. There is no specific treatment but Lindstrom (1974) reported that steroid therapy alleviated muscle pain and improved exercise tolerance in an affected 16 year old boy. A similar rapid clinical and radiographic response to steroids in an adolescent Pakistani girl was obtained by Minford et al (1981) and in a series of 12 affected persons by Naveh et al (1985).

Marked thickening of the cortices of the leg bones is the predominant feature. The medullary canals are narrowed and the external bone contours are irregular. The changes are diaphyseal and the metaphyses and epiphyses remain uninvolved. The long bones of the arms may be affected but the extremities and the axial skeleton are usually spared. Infrequently, the skull is involved, with calvarial widening and basal sclerosis. As with the clinical features, the radiological changes are very variable (Sparkes & Graham 1972, Sty et al 1978). Patients with severe cranial involvement may be difficult to differentiate from individuals with mild forms of craniodiaphyseal dysplasia but, in general, the clinical course and mode of inheritance permit firm categorisation.

### Genetics

Autosomal dominant inheritance is well established and kindreds with cases in successive generations have been reported by Girdany (1959), Ramon & Buchner (1966), Allen et al (1970), Hundley & Wilson (1973) and Crisp & Brenton (1982). Thurmon & Jackson (1976) encountered dominantly transmitted Camurati-Engelmann disease in the St Landry Mulattoes, a

---

* Mario Camurati, 1896–1948, qualified in Bologna and became a senior orthopaedic surgeon in Ancona, Italy.

Guido Engelmann, born in 1876, trained in Berlin and practised orthopaedic surgery in Vienna.

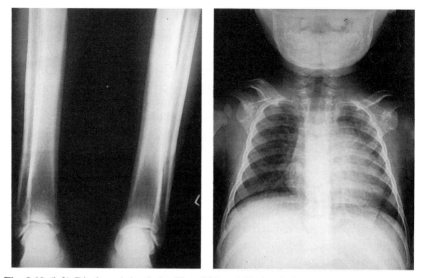

**Fig. 8.18** (left) Diaphyseal dysplasia: The tibial and fibular shafts are hyperostotic with narrowing of the medullary canals and irregularity of the external bony contours.

**Fig. 8.19** (right) Infantile cortical hyperostosis: the mandible and right clavicle are expanded and have a characteristic 'ground glass' appearance.

social isolate in Louisiana. In the largest family yet reported, Naveh et al (1984) identified 13 affected persons in three generations and confirmed autosomal dominant transmission.

There is marked intrafamilial variation in severity. In some instances, phenotypic expression has been confined to minor radiological changes and complete non-penetrance of the gene has been recorded (Sparkes & Graham 1972). The apparent excess of sporadic individuals who might otherwise be assumed to represent new mutations, can be explained on a basis of this phenotypic inconsistency.

A Norwegian family in which members of three generations had a condition which resembled Camurati-Engelmann disease was reported by Koller et al (1979). The affected persons had the additional features of a fracturing tendency and icthyosis and it seems likely that this disorder is an autonomous entity.

## INFANTILE CORTICAL HYPEROSTOSIS (Caffey)*

Infantile cortical hyperostosis, or Caffey disease, is an unusual but relatively well known disorder. Following the first report by Caffey & Silverman (1945), the features have been reviewed by Sherman & Hellyer (1950),

* John Caffey (1895–1978) was an eminent paediatric radiologist in the USA.

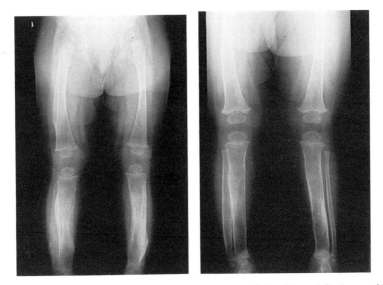

**Fig. 8.20** (left) Infantile cortical hyperostosis: the cortices of the tibia and fibula are wide and irregular.

**Fig. 8.21** (right) Infantile cortical hyperostosis: the patient depicted in Figure 8.20, 3 months later. The condition is quiescent and the bone changes are resolving.

Caffey (1957), Holman (1962), Rademacher et al (1975), Finsterbush & Rang (1975), Cremin (1979) and Landthaler et al (1984).

### Clinical and radiographic features

The condition presents with pain, swelling and inflammation of a localised area, often the mandible, shoulder girdle or limb, and the acute episode is accompanied by pyrexia, leucocytosis and a raised erythrocyte sedimentation rate. Spontaneous resolution takes place within a few weeks, but relapse is not unusual. Infantile cortical hyperostosis most frequently occurs before the age of 6 months, and it has been recognised radiologically in the fetus. There have been a few reports of presentation in later childhood, as in a 12 year old Moroccan girl studied by Gillet et al (1974).

Radiographically, the cortices of affected bones are widened and sclerotic, with irregular contours. With remission of the illness the bones regain their normal appearance.

### Genetics

The localisation and acute inflammatory nature of infantile cortical hyperostosis is most unlike a genetic disorder. Nevertheless, there have been several reports which strongly suggest autosomal dominant inheritance

(Gerrard et al 1961, van Buskirk et al 1961, Holman 1962, Bull & Feingold 1974). A most impressive French Canadian kindred with 34 affected persons in a family with Caffey disease was documented by Maclachlan et al (1984). The sibs described by Clemett & Williams (1963) might be taken as evidence for autosomal recessive inheritance. However, non-penetrance of the gene in a parent could also explain this anomalous observation. These reports of familial aggregation are too numerous to be ignored and there is little doubt that there is a strong genetic element in the pathogenesis of infantile cortical hyperostosis. In an account of 11 cases in two generations of a kindred, Fráňa & Sekanina (1976) aptly commented that the unfavourable genetic outlook was balanced by the benign character of the disease.

Yousefzadeh et al (1979) recorded the birth of affected cousins on the same day in different centres and speculated that environmental factors might be operative against a background of genetic predisposition. In a review of the literature, which revealed 104 affected persons, with 25 instances of familial aggregation, Saul et al (1982) made the interesting observation that ostensibly sporadic cases were becoming infrequent. They also noted that the familial type of Caffey disease had earlier onset, less mandibular involvement and more involvement of the bones of the legs. Caffey disease has been recognised in a fetus, following radiographic and ultrasonographic investigation of hydramnios (Labrune et al 1983).

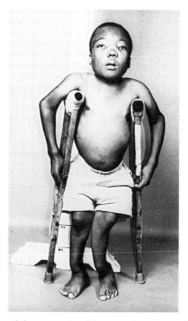

**Fig. 8.22** Osteoectasia: An adolescent male with stunted stature, enlargement of the skull and limb deformity. (From Saxe & Beighton 1982 Clinical and Experimental Dermatology 7(6): 605–609.)

## OSTEOECTASIA WITH HYPERPHOSPHATASIA

Virtually every case report concerning this disorder has appeared under a different designation, and there is therefore considerable semantic confusion. Amongst the titles which have been employed are endosteal hyperostosis, juvenile Paget disease, hyperostosis corticalis deformans, chronic congenital idiopathic hyperphosphataemia, congenital hyperphosphatasia and osteoectasia with hyperphosphatasia. The latter designation is used in the Paris Nomenclature and enjoys general acceptance. About 30 cases have now been reported; all have been children and a significant proportion have originated on the Caribbean island of Puerto Rico.

*Clinical and radiographic features*

Osteoectasia has many features in common with the usual adult form of Paget disease, the main difference being the precocious onset and the severity. This serious progressive disorder usually begins in infancy, with swelling of the long bones and enlargement of the skull. The limbs become bowed, and the deformity may be compounded by repeated pathological fractures. Stature is diminished. Deafness and optic atrophy are late complications and death may be the consequence of vascular involvement. Angioid streaks in the retina and persistently elevated levels of serum alkaline phosphatase are reminiscent of Paget disease. The indurated yellow 'cobblestone' flexural skin lesions which characterise pseudo-xanthoma elasticum have been recognised in osteoectasia (Mitsudo 1971, Fretzin 1975, Saxe & Beighton 1982). As angioid retinal streaks are also a feature of the former condition, it is possible that there is a common pathogenic mechanism.

Radiographically, the skeletal changes are asymmetrical, generalised demineralisation being the major feature. In the skull the calvarium is thickened with patchy areas of sclerosis, while the facial bones are spared. Platyspondyly and protrusio acetabulae may develop. Involvement of the tubular bones is severe and their width is greatly increased, with bowing and lack of modelling.

*Genetics*

Osteoectasia is inherited as an autosomal recessive. Affected sisters with consanguineous parents were reported by Swoboda (1958) and Bakwin et al (1964), while other sets of sibs were mentioned by Marshall (1962), Stemmermann (1966), Eyring & Eisenberg (1968) and Thompson et al (1969). In a review of the literature, Caffey (1973) mentioned seven affected children in two families in Puerto Rico. Geographic distribution is very wide and other patients have been reported from Greece (Choremis et al (1958), Brazil (Fanconi et al 1964), Britain (Woodhouse et al 1972), India (Desai et al 1973), Egypt (Temtamy et al 1974), the USA (Fretzin 1975), Guatemala

(Whalen et al 1977), Italy (Duillo et al 1977) and South Africa (Saxe & Beighton 1982).

Minor skeletal changes which were recognised on radiographic skeletal survey of the mother of the sibs described by Bakwin et al (1964) could represent phenotypic expression in the heterozygote. Caffey (1973) speculated that this woman might eventually develop the manifestations of the adult form of Paget disease.

There have been no reports of attempts at antenatal diagnosis of osteoectasia. However, as alkaline phosphatase levels are raised the condition might be recognisable in the fetus by determination of the activity of this enzyme in cultured amniotic fluid cells.

## OSTEITIS DEFORMANS (PAGET)*

Although more than a century has elapsed since osteitis deformans was delineated, the pathogenesis is still obscure. Indeed, in spite of the fact that Paget disease is comparatively common, there is still speculation as to whether or not the condition has a genetic basis.

### Clinical and radiographic features

Individuals with the characteristic radiographic bone changes may remain totally asymptomatic. The prevalence increases with advancing age, and Paget disease is seldom encountered before middle life. Nager (1975) estimated that 13% of the population over the age of 40 have the disorder. The abnormalities may be localised to one bone, or widely distributed throughout the skeleton. The skull, axial skeleton and proximal long bones are the sites of predilection. Bone pain is sometimes intractable and deafness, skeletal deformity, spontaneous fractures and osteosarcomatous changes are late complications. In severe cases, the vascular component acts as an arteriovenous shunt and cardiac failure may supervene. The serum alkaline phosphatase level is consistently raised. The management of Paget disease has been improved by the introduction of calcitonin therapy and in adequately treated patients, symptomatic remission can be induced.

Radiographically, bone involvement may be localised or widespread. Radiolucent areas appear in the early stages of the disease, followed by sclerosis, expansion and distortion of the bones. These changes sometimes mimic neoplastic metastases.

### Genetics

The familial aggregation of Paget disease is well documented; nevertheless, the mode of genetic transmission has not yet been firmly established. The

---

* Sir James Paget, 1814–1899, was the senior surgeon at St Bartholomew's Hospital and president of the Royal College of Surgeons, London. He had a brilliant career and became surgeon to Queen Victoria.

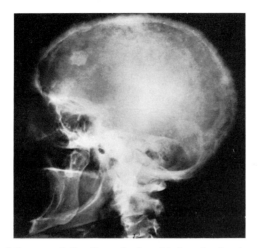

**Fig. 8.23** Osteitis deformans: skull radiograph of an elderly male, showing patchy sclerosis.

late onset often precludes the investigation of consecutive generations. The great variability of the manifestations and their potential for clinical silence makes any family study dependent upon radiology, with or without serum alkaline phosphatase determinations.

It is generally accepted that Paget disease results from osteoblast malfunction. The observation of intracellular virus-like bodies has prompted the suggestion that the condition results from a chronic slow virus infection, perhaps dependent upon the operation of additional genetic and environmental factors (Hosking 1981). In the latter context, Solomon (1979) drew attention to a champion billiards player who developed Paget disease in his digits at sites which corresponded to the alignment of his cue!

There are considerable geographic and ethnic discrepancies in the distribution of Paget disease. In his large scale study in Australia, Barry (1969) observed that the condition was virtually confined to immigrants from Britain. Paget disease is rare in indigenous African populations, although isolated cases have been reported from Senegal, Nigeria and South Africa (van Meerdervoort & Richter 1976). Simon et al (1975) investigated the HLA antigenic status of 46 patients without finding any association with Paget disease.

Ashley-Montagu (1949) proposed that inheritance was X-linked, with clinical expression in a proportion of the female heterozygotes. However, this hypothesis is untenable, as male to male transmission was documented when Gutman & Kasabach (1963) analysed a series of 116 cases. McKusick (1972) reviewed the situation at length, published a collection of pedigrees of kindreds containing more than one affected member and concluded that Paget disease is inherited as an autosomal dominant, with varying clinical expression. Since Paget disease is predominantly a disorder of old age, it is unlikely that a patient would seek genetic counselling!

REFERENCES

**Endosteal Hyperostosis, Van Buchem**

Beighton P, Barnard A, Hamersma H, van der Wouden A 1984 The syndromic status of sclerosteosis and van Buchem disease. Clinical Genetics 25: 175–181

Buchem F S P van 1971 Hyperostosis corticalis generalisata. Eight new cases. Acta Medica Scandinavica 189: 257

Buchem F S P van, Hadders H N, Ubbens R 1955 Hyperostosis corticalis generalisata familiaris. Acta Radiologica 44: 109

Buchem F S P van, Prick J J G, Jaspar H H J 1976 Hyperostosis Corticalis Generalisata Familiaris (Van Buchem's Disease). Excerpta Medica, Amsterdam

Buchem F S P van, Hadders H N, Hansen J F, Woldring M G 1962 Hyperostosis corticalis generalisata. Report of seven cases. American Journal of Medicine 33: 387

Dixon J M, Cuill R E, Gamble P 1982 Two cases of Van Buchem's disease. Journal of Neurology, Neurosurgery and Psychiatry 45: 913–918

Eastman J R, Bixler D 1977 Generalised cortical hyperostosis (van Buchem disease): Nosologic considerations. Radiology 125: 297–301


**Endosteal Hyperostosis, Worth**

Beals R K 1976 Endosteal hyperostosis. Journal of Bone and Joint Surgery 58: 1172

Dyson D P 1972 Van Buchem's disease (Hyperostosis corticalis generalisata familiaris). British Journal of Oral Surgery 9: 237

Gelman M I 1977 Autosomal dominant osteosclerosis. Radiology 125: 289

Gorlin R J, Glass L 1977 Autosomal dominant osteosclerosis. Radiology 125: 547

Lapresle J, Maroteaux P, Kuffer R, Said G, Meyer O 1976 Hyperostose corticale generalise dominante avec atteinte multiple des nerfs cranies. Nouvelle Presse Médicale 5/40: 2703

Maroteaux P, Fontaine G, Scharfman W, Farriaux J P 1971 L'hyperostose corticale generalisée à transmission dominante. Archives Françaises de Pédiatrie 28: 685–698

Owen R H 1976 Van Buchem's disease. British Journal of Radiology 49: 126

Scott W C, Gautby T H T 1974 Hyperostosis corticalis generalisata familiaris. British Journal of Radiology 47: 500

Spranger J W, Langer L O, Wiedemann H R 1974 Bone Dysplasias. An Atlas of Constitutional Disorders of Skeletal Development. Saunders, Philadelphia, p 331–334

Vayssairat M, Prier A, Meisel C, Camus J P, Grellet J 1976 New cases of familial generalized cortical hyperostosis with dominant transmission (Worth's type). Journal de Radiologie 57: 719–724

Worth H M, Wollin D G 1966 Hyperostosis corticalis generalisata congenita. Journal of the Canadian Association of Radiologists 17: 67


**Sclerosteosis**

Alves A F, Rubim J L, Cardoso L, Rabelo M M 1982 Sclerosteosis: A marker of Dutch ancestry? Brazil Journal of Genetics 4: 825–834

Barnard A, Hamersma H, Kretzmar J H, Beighton P 1980 Sclerosteosis in old age. South African Medical Journal 58: 401–403

Beighton P, Hamersma H 1979 Sclerosteosis in South Africa. South African Medical Journal 55: 783–788

Beighton P, Cremin B, Hamersma H 1976 The radiology of sclerosteosis. British Journal of Radiology 49: 934

Beighton P, Durr L, Hamersma H 1976 The clinical features of sclerosteosis. A review of the manifestations in twenty-five affected individuals. Annals of Internal Medicine 84/4: 393

Beighton P, Davidson J, Durr L, Hamersma H 1977 Sclerosteosis — an autosomal recessive disorder. Clinical Genetics 11: 1

Buchem F S P van, Hadders H N, Hansen J F, Woldring M G 1962 Hyperostosis corticalis generalisata. Report of seven cases. American Journal of Medicine 33: 387

Epstein S, Hamersma H, Beighton P 1979 Endocrine function in sclerosteosis. South African Medical Journal 55: 1105–1110

Higinbotham N L, Alexander S F 1941 Osteopetrosis, four cases in one family. American Journal of Surgery 53: 444

Kelley C H, Lawlah J W 1946 Albers-Schönberg disease — A family survey. Radiology 47: 507

Pietruschka G 1958 Weitere Mitteilungen uber die Marmorknochenkrankheit (Albers-Schonbergsche Krankheit) nebst Bemerkungen zur Differentialdiagnose. Klinische Monatsblatter für Augenheilkunde 132: 509

Stein S A, Witkop C, Hill S, Fallon M D, Viernstein L, Gucer G, McKeever P, Long D, Altman J, Miller N R, Teitelbaum S L, Schlesinger S 1983 Sclerosteosis: Neurogenetic and pathophysiologic analysis of an American kinship. Neurology 33: 267–277

Sugiura Y, Yasuhara T 1975 Sclerosteosis: a case report. Journal of Bone and Joint Surgery 57/2: 273

Truswell A S 1958 Osteopetrosis with syndactyly. A morphological variant of Albers-Schönberg disease. Journal of Bone and Joint Surgery 40: 208

Witkop C J 1965 Genetic disease of the oral cavity. In: Tiecke R W (ed) Oral Pathology, McGraw-Hill, New York

**Oculodento-osseous dysplasia**

Barnard A, Hamersma H, de Villiers J C, Beighton P 1981 Intracranial calcification in oculodento-osseous dysplasia. South African Medical Journal 59: 758–762

Beighton P, Hamersma H, Raad M 1979 Oculodento-osseous dysplasia: Heterogeneity or variable expression? Clinical Genetics 16: 169

Duggan J W, Hassard D T 1961 Familial microphthalmos. Transactions of the Canadian Ophthalmology Society 24: 210–215

Gillespie F D 1964 Hereditary dysplasia oculodentodigitalis. Archives of Ophthalmology 71: 187–193

Gorlin R J, Meskin L H, St Geme J W 1963 Oculodentodigital dysplasia. Journal of Pediatrics 63: 69–73

Gutierrez Diaz A, Alonso M J, Borda M 1982 Oculodentodigital dysplasia. Ophthalmic Paediatrics and Genetics 1: 227–232

Judisch G F, Martin-Casals A, Hanson J W, Olin W H 1979 Oculodentodigital dysplasia: Four new cases and a literature review. Archives of Ophthalmology 97: 878–884

Littlewood J M, Lewis G M 1963 The Holmes-Adie syndrome in a boy with acute juvenile rheumatism and bilateral syndactyly. Archives of Disease in Childhood 38: 193–197

Mohr O L 1939 Dominant acrocephalo-syndactyly. Hereditas (Lund) 25: 193–197

O'Rourk T R, Bravos A 1969 Oculodento-digital syndrome II. Birth Defects: Original Article Series Vol V, 2: 226

Pfeiffer R A, Epstein H, Junemann G 1968 Oculodento-digitale dysplasie. Klinische Monatsblatter für Augenheilkunde 152: 247–252

Rajic D S, De Veber L L 1966 Hereditary oculodento-osseous dysplasia. Annales of Radiology 9: 224–229

Reisner S H, Kott B, Bornstein B, Salinger H, Kaplan I, Gorlin R J 1969 Oculodentodigital dysplasia. American Journal of Diseases of Children 118: 600–606

Sugar H S 1978 Oculodentodigital dysplasia syndrome with angle-closure glaucoma. American Journal of Ophthalmology 86: 36–41

Traboulsi E I, Faris B M, Der Kaloustian V M 1986 Persistent hyperplastic primary vitreous and recessive oculo-dento-osseous dysplasia. American Journal of Medical Genetics 24: 95–100

**Diaphyseal dysplasia (Camurati-Engelmann)**

Allen D T, Saunders A M, Northway W H, Williams G F, Schafer I A 1970 Corticosteroids in the treatment of Engelmann's disease: progressive diaphyseal dysplasia. Pediatrics 46: 523

Crisp A J, Brenton D P 1982 Engelmann's disease of bone — a systemic disorder? Annals of the Rheumatic Diseases 41: 183–188

Girdany B R 1959 Engelmann's disease (progressive diaphyseal dysplasia — a non-progressive familial form of muscular dystrophy with characteristic bone changes). Clinical Orthopaedics and Related Research 14: 102

Hundley J D, Wilson F C 1973 Progressive diaphyseal dysplasia. Review of the literature and report of seven cases in one family. Journal of Bone and Joint Surgery 55A: 461

Koller M E, Maurseth K, Haneberg B, Aarskog D 1979 A familial syndrome of diaphyseal cortical thickening of the long bones, bowed legs, tendency to fracture and icthyosis. Pediatric Radiology 8: 179–182

Lindstrom J A 1974 Diaphyseal dysplasia (Engelmann) treated with corticosteroids. Birth Defects: Original Article Series 10/12: 504

McKusick V A 1975 Engelmann disease (progressive diaphyseal dysplasia) Mendelian Inheritance in Man, 4th Edn, Johns Hopkins University Press, Baltimore, p 95

Minford A M, Hardy G J, Forsythe W I, Fitton J M, Rowe V L 1981 Engelmann's disease and the effect of corticosteroids. Journal of Bone and Joint Surgery : 597–600

Naveh Y, Kaftori J K, Alon U, Ben-David J, Berant M 1984 Progressive diaphyseal dysplasia: Genetics and clinical and radiologic manifestations. Pediatrics 74: 399–405

Naveh Y, Alon U, Kaftori J K, Berant M 1985 Progressive diaphyseal dysplasia: Evaluation of corticosteroid therapy. Pediatrics 75: 321–323

Ramon Y, Buchner A 1966 Camurati-Engelmann's disease affecting the jaws. Oral Surgery 22: 592

Sparkes R S, Graham C B 1972 Camurati-Engelmann's disease. Genetics and clinical manifestations, with a review of the literature. Journal of Medical Genetics 9: 73

Thurmon T H, Jackson J 1976 Tumoral calcinosis and Engelmann disease. Birth Defects: Original Article Series 12/5: 321

**Infantile cortical hyperostosis (Caffey disease)**

Bull M J, Feingold M 1974 Autosomal dominant inheritance of Caffey disease. In: Bergsma D (ed) Skeletal Dysplasia. The National Foundation; pub. Symposia Specialists, Miami, p 139

Caffey J 1957 Infantile cortical hyperostosis: a review of the clinical and radiographic features. Proceedings of the Royal Society of Medicine 50: 347

Caffey J, Silverman W A 1945 Infantile cortical hyperostosis: preliminary report on a new syndrome. American Journal of Roentgenology, Radium Therapy and Nuclear Medicine 54: 1

Clemett A R, Williams J H 1963 The familial occurrence of infantile cortical hyperostosis. Radiology 80: 409

Cremin B J 1979 Caffey's disease in Cape Town. South African Medical Journal 55: 377

Finsterbush A, Rang M 1975 Infantile cortical hyperostosis. Acta Orthopaedica Scandinavica 46: 727–736

Fráňa L, Sekanina M 1976 Infantile cortical hyperostosis. Archives of Disease in Childhood 51: 589

Gerrard J, Holman G H, Gorman A A, Morrow I H 1961 Infantile cortical hyperostosis: an inquiry into its familial aspects. American Journal of Roentgenology, Radium Therapy and Nuclear Medicine 85: 613

Gillet J, Imani F, Benzakour M, Guignard J 1974 Infantile cortical hyperostosis (Caffey's disease); two cases. Annals of Radiology 17/7: 707

Holman G H 1962 Infantile cortical hyperostosis: a review. Quarterly Review of Pediatrics 17: 24

Labrune M, Guedj G, Vial M et al 1983 Maladie de Caffey à debut antenatal. Archives Françaises de Pédiatrie 40: 39–43

Landthaler G, Loizeau A, Tron P, Mallet E, Le Dosseur P, De Menibus C H 1984 Caffey's disease in a mother and her 2 children. Archives Françaises de Pédiatrie 41: 275–278

Mclachlan A K, Gerrard J W, Houston C S, Ives E J 1984 Familial infantile cortical hyperosteosis in a large Canadian family. Canadian Medical Association Journal 130: 1172–1174

Rademacher K H, Grossmann I, Wildner G P 1975 Infantile cortical hyperostosis (IKH). Short survey and own observations. Radiological Diagnosis 16/4: 585

Saul R A, Lee W H, Stevenson R E 1982 Caffey's disease revisited. Further evidence for autosomal dominant inheritance with incomplete penetrance. American Journal of Diseases of Children 136: 55–60

Sherman M S, Hellyer D T 1950 Infantile cortical hyperostosis. Review of the literature and report of five cases. American Journal of Roentgenology, Radium Therapy and Nuclear Medicine 63: 212

Van Buskirk F W, Tampas J P, Peterson O S 1961 Infantile cortical hyperostosis: an inquiry into its familial aspects. American Journal of Roentgenology, Radium Therapy and Nuclear Medicine 85: 613

Yousefzadeh D K, Brosnan P, Jackson J H 1979 Infantile cortical hyperostosis, Caffey's disease, involving two cousins. Skeletal Radiology 4: 141–147

## Osteoectasia with hyperphosphatasia

Bakwin H, Golden A, Fox S 1964 Familial osteoctasia with macrocranium. American Journal of Roentgenology, Radium Therapy and Nuclear Medicine 91: 609

Caffey J 1973 Familial hyperphosphatasemia with ateliosis and hypermetabolism of growing membranous bone: review of the clinical, radiographic and chemical features. In: Kaufmann H J (ed) Intrinsic diseases of bones vol 4. Karger, Basel, p 438

Choremis C, Yannakos D, Papadatos C, Baroutsou E 1958 Osteotis deformans (Paget's disease) in an 11 year old boy. Helvetica Paediatrica Acta 13: 185

Desai M P, Joshi N C, Shah K N 1973 Chronic idiopathic hyperphosphatasia in an Indian child. American Journal of Diseases of Children 126: 626

Eyring E J, Eisenberg E 1968 Congenital hyperphasphatasia. Journal of Bone and Joint Surgery 50A: 1099

Fanconi G, Moreira G, Uehlinger E, Giedion A 1964 Osteochalasia desmalis familiaris. Helvetica Paediatrica Acta 19: 279

Fretzin D F 1975 Pseudoxanthoma elasticum in hyperphosphatasia. Archives of Dermatology 111: 271

Marshall W C 1962 Chronic progressive osteopathy with hyperphosphatasia. Proceedings of Royal Society of Medicine 55: 238

Mitsudo S M 1971 Chronic idiopathic hyperphosphatasia associated with pseudoxanthoma elasticum. Journal of Bone and Joint Surgery 53: 303

Saxe N, Beighton P 1982 Cutaneous manifestations of osteoectasia. Clinical and Experimental Dermatology 7: 605–609

Stemmermann G N 1966 An histologic and histochemical study of familial osteoectasia (chronic idiopathic hyperphosphatasia). American Journal of Pathology 48: 641

Swoboda W 1958 Hyperostosis corticalis deformans juvenilis. Helvetica Paediatrica Acta 13: 292

Temtamy S A, El-Meligy M, Salem S, Osman N 1974 Hyperphosphatasia in an Egyptian child. Birth Defects: Original Article Series 10/12: 196

Thompson R C, Gaul G E, Horwitz S J, Schenk R K 1969 Hereditary hyperphosphatasia. Study of 3 siblings. American Journal of Medicine 47: 209

Whalen J P, Horwith M, Krook L, Macintyre I, M Mena E, Viteri F, Torun B, Nunex E A 1977 Calcitonin treatment in hereditary bone dysplasia with hyperphosphatesemia: A radiographic and histologic study of bone. American Journal of Roentgenology 129: 29

Woodhouse N J, Fisher M T, Sigurdsson G, Joplin G F, Macintyre I 1972 Paget's disease in a 5 year old. Acute response to human calcitonin. British Medical Journal 4: 267

## Osteitis deformans (Paget)

Ashley-Montagu M F 1949 Paget's disease (osteitis deformans) and heredity. American Journal of Human Genetics 1: 94

Barry H C 1969 Paget's Disease of Bone. Livingstone, London, p 14

Gutman A B, Kasabach H 1963 Paget's disease (osteitis deformans). Analysis of 116 cases. American Journal of Medical Science 191: 361

Hosking D J 1981 Paget's disease of bone. British Medical Journal 28: 686–688

McKusick V A 1972 Heritable Disorders of Connective Tissue 4th edn. Mosby, St Louis

Nager G T 1975 Paget's disease of the temporal bone. Annals of Otolaryngology 84/4: 22

Simon L, Blotman F, Seignalet J, Claustre J 1975 The etiology of Paget's bone disease. Revue du Rhumatisme et des Maladies Osteo-articulares 42/10: 535

Solomon L R 1979 Billiard-player's fingers: An unusual case of Paget's disease of bone. British Medical Journal 1: 931

Van Meerdervoort H F P, Richter G G 1976 Paget's disease of bone in South African Blacks. South African Medical Journal 50: 1897

# 9

# Miscellaneous sclerosing and hyperostotic disorders

The conditions in this group are uncommon disorders in which clinical silence often contrasts with striking radiological manifestations.

1. Osteopathia striata
2. Osteopoikilosis
3. Melorheostosis
4. Pachydermoperiostosis
5. Osteosclerosis, Stanescu type
6. Osteosclerosis, distal type
7. Weismann-Netter-Stuhl syndrome
8. Other sclerosing and hyperostotic disorders
   a. Calvarial hyperostosis
   b. Familial osteodysplasia
   c. Schwarz-Lélek syndrome
   d. Sclerotic bone-dentine dysplasia syndrome
   e. Osteopetrosis with familial paraplegia
   f. Osteosclerosis with mental retardation
   g. Osteosclerosis with bamboo hair

## OSTEOPATHIA STRIATA

The designation 'osteopathia striata' was used by Fairbank (1935) following the initial case description by Voorhoeve (1924). The term pertains to multiple lines of increased density which are radiologically apparent in the long bones and the pelvis. Striations of this type are a component of several conditions, including the osteopetroses, osteopoikilosis and focal dermal hypoplasia. In addition, osteopathia striata with cranial sclerosis is a specific bone disorder, which is inherited as an autosomal dominant trait.

*Clinical and radiographic features*

Individuals with isolated osteopathia striata are usually asymptomatic. The radiographic abnormalities in Voorhoeve's two original patients were unchanged when Fermin (1962) re-examined them three decades later.

212

Radiographically, multiple parallel lines of sclerosis run along the shafts of the tubular bones. In the ilia, the striations have a fan-shaped configuration. The contours of the bones are undisturbed.

In the osteopathia striata-cranial sclerosis syndrome (OS-CS), the additional component of increased density and overgrowth of the calvarium and base causes variable but usually mild facial distortion and cranial nerve palsies (Bloor 1954, Jones & Mulcahy 1968, Walker 1969). Height, habitus and intelligence are normal, but increased head circumference, frontal bossing and deafness may be present. Pectus excavatum, and spinal malalignment are occasional features.

*Genetics*

Osteopathia striata was observed by Voorhoeve (1924) in a father and his daughter. In the kindred reported by Rucker & Alfidi (1964) under the designation 'Fairbank disease', members of three generations were affected. It is possible that isolated osteopathia striata can be inherited as an autosomal dominant, but in view of the variability of the OS-CS, it is likely that these disorders are both the consequence of the same faulty gene.

Autosomal dominant inheritance and syndromic identity of OS-CS was established by Horan & Beighton (1978) following a review of four affected families. Other familial instances were subsequently reported from Scotland by Jones (1979) and from Japan by Nakamura et al (1985).

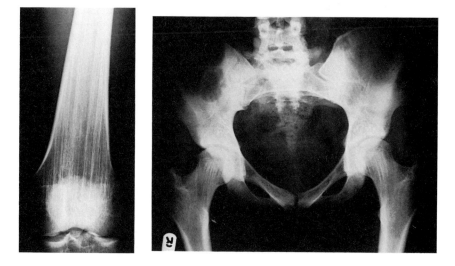

**Fig. 9.1** (left) Osteopathia striata: radiograph of the lower end of the femur of a young woman, showing multiple parallel lines of sclerosis. The base of the skull was sclerotic and the patient was deaf, due to entrapment of the auditory nerves.

**Fig. 9.2** (right) Osteopathia striata: striae are evident in the femoral necks and in the ilia. Inheritance was autosomal dominant and members of three generations of the kindred had similar changes.

Winter et al (1980) encountered the disorder in three generations and recorded concomitant cleft palate and severe micrognathia. It is uncertain whether the range of phenotypic expression of OS-CS embraces these latter anomalies or whether they are aetiologically unrelated.

At present, about 20 affected persons in 10 families have been documented.

## OSTEOPOIKILOSIS

Osteopoikilosis or 'spotty bones' is an unusual but not uncommon condition. More than 300 cases have been reported (Szabo 1971). The diagnosis is usually reached fortuitously, following radiographic examination for an unrelated purpose. Alternatively, the dermal lesions may draw attention to the disorder. (Hence the Lancashire proverb attributed to Granny Peard — 'what's bred in the bone comes out in the flesh'.)

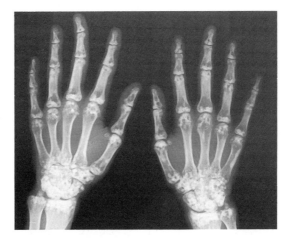

**Fig. 9.3** Osteopoikilosis: multiple sclerotic foci in the carpus and at the ends of the tubular bones.

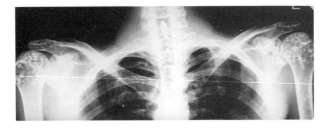

**Fig. 9.4** Osteopoikilosis: in this patient, the diagnosis was made by chance, after routine chest X-ray.

*Clinical and radiographic features*

Radiographically, small sclerotic foci are found throughout the skeleton. Many hundreds of these lesions may be present, and they tend to be congregated in the epiphyseal and metaphyseal regions of the tubular bones. Multiple sessile dermal naevi, known as 'dermatofibrosis lenticularis disseminata', are an associated feature.

These skin lesions were discussed in detail by Verbov & Graham (1986) who employed the eponym 'Buschke-Ollendorf' for the complete syndrome. The radiographic and pathological features of osteopoikilosis have been reviewed by Lagier et al (1984).

Osteopoikilosis is usually innocuous but a report of the development of osteosarcoma (Mindell et al 1978) may have sinister implications. Spinal canal stenosis (Weisz 1982) and the basal cell nevus syndrome (Blinder et al 1984) have been recognised in persons with osteopoikilosis. The significance of these inter-relationships is unknown.

*Genetics*

Reports of several large kindreds have been published and autosomal dominant inheritance is well established. Melnick (1959) described the disorder in 17 individuals in four generations, while father to son transmission was observed by Raque & Wood (1970). Clinical expression of the gene is variable and the dermal and osseous changes may occur together or separately in the same family (Berlin et al 1967). Schoenenberg (1975) observed the typical bone appearances in eight members of three generations of a kindred. Six also had the dermal manifestations.

## MELORHEOSTOSIS

Melorheostosis is a rare disorder which is characterised by irregular cortical thickening, principally of the tubular bones. The appellation, which pertains to the 'streaming' appearance of the bones, is derived from the Greek. Green et al (1962) differentiated melorheostosis and osteopetrosis. Campbell et al (1968) described 14 cases and reviewed the literature and in a further review (Beauvais et al 1977) were able to identify 200 reports. The condition has been recognised in a fibula from a prehistoric Alaskan Eskimo burial ground (Lester 1967).

*Clinical and radiographic features*

The majority of individuals with melorheostosis are asymptomatic. However, vague pains in the bones and joint swelling may be the presenting feature. The skin over affected bones sometimes becomes indurated and thickened, and soft tissue contractures may develop. Onset usually occurs

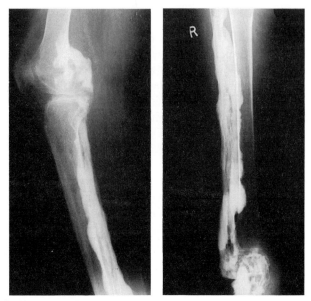

**Fig. 9.5** (left) Melorheostosis: lateral radiograph of the knee of an affected woman. The overlying skin was indurated and sclerotic.

**Fig. 9.6** (right) Melorheostosis: the radiographic changes have the appearance of melted wax flowing down the side of a candle.

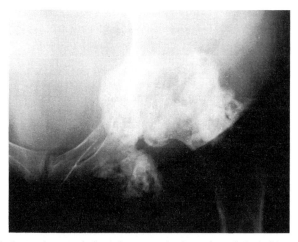

**Fig. 9.7** Melorheostosis: ectopic bone is present in the region of the ischium.

in early adulthood and progression is very slow (Murray 1951, Patrick 1969).

The radiographic manifestations are striking. The changes in the long bones have the appearance of melted wax flowing down the sides of a candle. The disorder is usually confined to a single bone or limb and limited to one side of the body. Ectopic bone occasionally forms in the adjacent soft tissues.

### Genetics

Murray (1978) documented 30 affected persons, pointed out that the skeletal changes often corresponded with the sclerotomes and postulated that the underlying pathological mechanism involved sensory nerve lesions.

There is no evidence that melorheostosis is a genetic disorder. As the condition is often clinically silent, family studies would have to be based upon skeletal surveys of relatives of known cases. So far, none have been reported.

Mixed-sclerosing-bone-dystrophy comprises features of melorheostosis, osteopoikilosis, osteopathia striata and osteopetrosis. In a comprehensive review, Whyte et al (1981) identified six cases, all sporadic, and discussed the nosology and interrelationships of this disorder. The genetic background, if any, and independent syndromic status remain uncertain.

## PACHYDERMOPERIOSTOSIS

Pachydermoperiostosis is a relatively common disorder in which clubbing of the digits is associated with thickening and hyperhydrosis of the skin and extremities, and oiliness and seborrhoea of the scalp.

### Clinical and radiographic features

The manifestations appear at puberty and are generally more severe in males. Radiographically, the cortices of the long bones are widened and

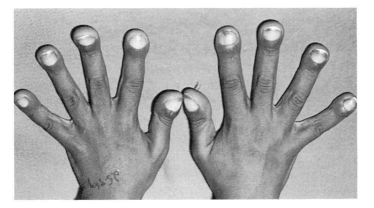

**Fig. 9.8** Familial digital clubbing: several members of this boy's kindred also had clubbing of the fingers and toes. This familial condition must be distinguished from the clubbing which develops in chronic cardiopulmonary disease.

sclerotic, with increased medullary trabeculation and periosteal thickening at the distal extremities. Clinical importance centres around the digital clubbing, which must be distinguished from the pulmonary osteoarthropathy of neoplastic or chronic cardiopulmonary disease. The individual with familial pachydermoperiostosis who was needlessly subjected to bronchoscopy and cardiac catheterisation would be in full agreement with this point!

*Genetics*

Clinical expression is very variable and individuals with the gene may manifest only certain components of the disorder. For this reason, family studies are difficult. Following a review of the literature, Rimoin (1965) concluded that pachydermoperiostosis was inherited as an autosomal dominant. However, there have been several instances of affected sibs with normal parents and parental consanguinity has been reported (Findlay & Oosthuizen 1951). It is therefore possible that there is an autosomal recessive variety of pachydermoperiostosis. McKusick (1975) has postulated that familial simple digital clubbing without skin changes might represent a separate genetic defect.

Digital clubbing, in association with widespread periosteal new bone formation in the shafts of the long bones and defects in cranial ossification, was encountered in two young Negro sisters by Currarino et al (1961). The parents and two sibs were normal, but the development of the condition in a third sister was subsequently reported by Chamberlain et al (1965). A further sporadic case was described by Cremin (1970). The cranial defects distinguish the disorder from the other familial digital clubbing syndromes. The mode of inheritance of this condition, which has been designated 'familial idiopathic osteoarthropathy of children', has not been elucidated.

Pachydermia in association with digital clubbing and hyperostosis of the frontal region of the skull and cortices of the tubular bones, has been termed the Touraine-Solenti-Golé syndrome. The disorder, which is a form of familial acromegaly, was reviewed by Gray & Steyn (1978).

## OSTEOSCLEROSIS, STANESCU* TYPE

Stanescu et al (1963) reported a French family in which 11 individuals had gross thickening of the cortices of the long bones, in association with craniofacial dysostosis. A sporadic patient in the USA with a similar phenotype was described by Hall (1974) and three additional affected children in the original kindred were documented by Maximilian et al (1981). The condition was listed in the 1983 Paris Nomenclature as 'Dominant Osteosclerosis,

---

* Victor and Ritta Stanescu were born in Rumania and collaborated at the Bucharest Institute of Endocrinology. They now hold senior research posts at the Hôpital des Enfants Malades, Paris.

Stanescu type', and an affected mother and daughter were reported from the Argentine under this designation by Dipierri & Guzman (1984).

The clinical manifestations include asymmetry of the skull, brachycephaly, premature closure of the fontanelles, proptosis and a hooked nose. Stature is stunted, and some affected persons have brachydactyly. The most striking radiological changes are sclerosis and hyperostosis of the diaphyses of the limb bones. The features are age-related and there is a wide range of phenotypic expression.

Although it is rare, the syndromic status of the condition seems to be secure. There is potential confusion, however, as Gorlin & Glass (1977) used the term 'autosomal dominant osteosclerosis' as a synonym for the dominant form of endosteal hyperostosis described by Worth & Wollin (1966).

## OSTEOSCLEROSIS, DISTAL TYPE

Osteosclerosis, distal type was documented in five persons in two generations of a South African kindred of mixed ancestry by Beighton et al (1980). The affected persons had clinically obvious widening of the distal part of their forearms and shins, but the condition was otherwise innocuous. General health was good, stature and intelligence were normal, and there were no systemic ramifications. Radiographically, the radii and ulnae were somewhat bowed, with marked cortical widening and sclerosis. Similar

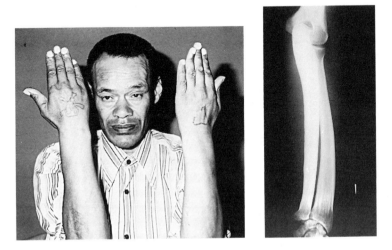

**Fig. 9.9** (above, left) Distal osteosclerosis: the distal portions of the forearms are widened, but there is no disability. (From Beighton et al 1980 Clinical Genetics 18: 298–304.)

**Fig. 9.10** (above, right) Distal osteosclerosis: the radius and ulna are curved, sclerotic and undermodelled. (From Beighton et al 1980 Clinical Genetics 18: 298–304.)

changes were present in the tibiae and fibulae. The skull showed mild sclerosis of the calvarium and base, and there was minor localised sclerosis in other regions of the skeleton, notably in the vertebral pedicles and the pelvis.

The condition was present in a father and four adult offspring and seems to be a 'private' syndrome which is inherited as an autosomal dominant trait.

## WEISMANN-NETTER*-STUHL SYNDROME

Using the designation 'toxopachyostèose diaphysaire tibio-péronière', Weismann-Netter & Stuhl (1954) documented the familial recurrence of anterior curvature of the tibia and fibula, stunted stature and variable mental retardation. About 20 cases have been reported, mainly in French publications, and the eponym is well established.

### Clinical and radiographic features

The condition presents with 'sabre shins' thus generating diagnostic confusion with the sequelae of dietary rickets and syphylis. The bowing is often asymmetrical, concomitant stunting of stature is variable and spinal malalignment sometimes occurs. Mental retardation may be severe but general health is unimpaired.

Radiographically, anterior bowing of the tibia and fibula predominates, but the other long bones may also be involved. The diaphyses and cortices are hyperostotic, but the external contours are undisturbed. The iliac bones are squared and the falx cerebri may be calcified.

### Genetics

Familial aggregation was reported by Weismann-Netter & Stuhl (1954), Weismann-Netter & Rouaux (1956), Krewer (1961), Azimi & Bryan (1974) and a history of involvement of five persons in three generations was elicited by Breuzard et al (1960). Other reported cases have been sporadic (Heully et al 1967, Hoefnagel 1969, Keats & Alavi 1970, Alavi & Keats 1973). The majority of published case descriptions have concerned African Negroes. The syndromic status of the Weismann-Netter-Stuhl syndrome is well established. Although the available evidence points to a genetic aetiology, the mode of transmission has not yet been elucidated.

---

* Robert Weismann-Netter (1894–1980) was a physician at the Beaujon Hospital, Paris. He shared an interest in skeletal dysplasias with his colleague, Dr L. Stuhl.

## OTHER SCLEROSING AND HYPEROSTOTIC DISORDERS

Skeletal sclerosis or overgrowth is a predominant feature of a number of rare disorders which have not yet achieved recognition as independant syndromic entities. Those which are adequately defined are briefly mentioned below:

### Calvarial hyperostosis

Calvarial hyperostosis was recognised in three related males by Pagon et al (1986). The affected persons had significant asymmetry of their skulls, with lateral frontal prominences. The changes were evident in infancy and increased with age, but apart from the unsightly appearance the bone overgrowth did not cause elevation of intracranial pressure or other complications. Radiographically, the frontoparietal regions were irregular and sclerotic but the skeleton was otherwise normal.

The three males had an uncle-nephew-second cousin relationship; the intervening relatives were all normal females and it is very possible that inheritance is X-linked.

### Familial osteodysplasia

The term 'familial osteodysplasia' was used by Buchignani et al (1972) when they described four siblings with prominence of the forehead, flattening of the nasal bridge, mid-facial hypoplasia, dental malocclusion and scoliosis. The long bones showed cortical hyperostosis and medullary narrowing and the calvarium, superior pubic rami and ribs were thin. Anderson et al (1972) published a separate review of the same patients. This disorder is distinct from the osteodysplasty of Melnick and Needles, and so far it seems to be a private syndrome. The mode of inheritance is uncertain.

### Schwarz-Lélek syndrome

The Schwarz-Lélek syndrome is characterised by severe cranial hyperostosis, with lateral bowing and metaphyseal widening of the femora. In the original case descriptions, Schwartz (1960) used the designation 'craniometaphyseal dysplasia' while Lélek (1961) termed the condition 'Camurati-Engelmann disease'. The independent identity of the condition was recognised by Gorlin et al (1969) when they employed the eponym 'Schwarz-Lélek'. The genetic basis of the disorder is unknown.

### Sclerotic bone-dentine dysplasia syndrome

Sclerosis and endosteal cortical thickening of the long bones and carpus in association with dental dentine dysplasia were identified in members of four generations of a Californian kindred by Morris & Augsburger (1977). Inheritance of this private syndrome is apparently autosomal dominant.

## Osteopetrosis with familial paraplegia

The development of paraplegia in middle age, in the absence of any evidence of spinal cord compression was documented in members of three generations of a kindred with osteopetrosis by Jacques et al (1975). It is uncertain whether this condition is an autonomous entity.

## Osteosclerosis with mental retardation

Hunter & MacPherson (1978) gave a detailed account of a young woman with progressive skeletal sclerosis and severe mental retardation. The bony abnormalities were recognised at the age of 3 years and resembled those of the autosomal recessive form of osteopetrosis. At the time of her death in early adulthood she had widespread changes in the skeleton. In particular, the calvarium was sclerotic and the facial bones and mandible were hyperostotic. The tubular bones were undermodelled and dense. This condition is unlike any other sclerosing bone dysplasia; the pathogenesis and genetic background are unknown.

## Osteosclerosis with bamboo hair

An infant with sclerosis of the axial skeleton, in association with 'bamboo' hair and ichthyosis was documented by Johnson et al (1978). The changes in the hair and skin were consistent with a diagnosis of the Netherton syndrome, and the authors raised the question as to whether skeletal involvement was a consistent but inconspicuous syndromic component, which had previously been overlooked.

The trichodento-osseous syndrome is a similar condition in which fuzzy hair is associated with dental enamel hypoplasia and generalised osseous sclerosis. Quattromani et al (1983) discussed the syndromic status and clinical heterogeneity of this disorder.

REFERENCES

**Osteopathia striata**
Bloor D U 1954 A case of osteopathia striata. Journal of Bone and Joint Surgery (Br) 36: 261
Fairbank H A T 1935 Generalised disorders of the skeleton. Proceedings of the Royal Society of Medicine 28: 1611
Fermin H E A 1962 Osteorhabdotose. Een voor het eerst door N. voorhoeve beschreven bijzondere vorm van osteopathia condensans disseminata. Nederlansch Tijdschrift voor Geneeskunde 106: 1188

Horan F T, Beighton P H 1978 Osteopathia striata with cranial sclerosis. An autosomal
dominant entity. Clinical Genetics 13: 201
Jones D N 1979 Hyperostosis generalisata with striations of the bones. A further report in
two related families. Clinical Radiology 30: 87
Jones M D, Mulcahy N D 1968 Osteopathia striata, osteopetrosis, and impaired hearing. A
case report. Archives of Otolaryngology 87: 116
Nakamura T, Yokomizo Y, Kanda S, Harada T, Naruse T 1985 Osteopathia striata with
cranial sclerosis affecting three family members. Skeletal Radiology 14: 267–269
Rucker T N, Alfidi R J 1964 A rare familial systematic affection of the skeleton: Fairbank's
disease. Radiology 82: 63
Voorhoeve N 1924 L'image radiologique non encore décrite d'une anomalie de squelette.
Acta radiologica 3: 407
Walker B A 1969 Osteopathia striata with cataracts and deafness. Clinical Delineation of
Birth Defects. National Foundation, New York, p 295
Winter R M, Crawfurd M d'A, Meire H B, Mitchell N 1980 Osteopathia striata with
cranial sclerosis: highly variable expression within a family including cleft palate in two
neonatal cases. Clinical Genetics 18: 462–474

## Osteopoikilosis

Berlin R, Hedensio B, Lilja B, Linder L 1967 Osteopoikilosis — a clinical and genetic
study. Acta Medica Scandinavica 181: 305
Blinder G, Barki Y, Pezt M, Bar-Ziv J 1984 Widespread osteolytic lesions of the long
bones in basal cell nevus syndrome. Skeletal Radiology 12: 196–198
Lagier R, Mbakop A, Bigler A 1984 Osteopoikilosis: A radiological and pathological study.
Skeletal Radiology 11: 161–168
Melnick J C 1959 Osteopathia condensans disseminata (osteopoikilosis). Study of a family of
four generations. American Journal of Roentgenology, Radium Therapy and Nuclear
Medicine 82: 229
Mindell E R, Northup C S, Douglass H O 1978 Osteosarcoma associated with
osteopoikilosis: Case report. Journal of Bone and Joint Surgery (Am) 60: 406
Raque C J, Wood M G 1970 Connective tissue naevus. Dermatofibrosis lenticularis
disseminata with osteopoikilosis. Archives of Dermatology 102: 290
Schoenenberg H 1975 Osteopoikilia with dermofibrosis lenticularis disseminata (Buschke
Ollendorf syndrome). Klinische Paediatrie 187/2: 123
Szabo A D 1971 Osteopoikilosis in a twin. Clinical Orthopaedics and Related Research
79: 156
Verbov J, Graham R 1986 Buschke-Ollendorf syndrome — Disseminated dermatofibrosis
with osteopoikilosis. Clinical and Experimental Dermatology 11/1: 17–26
Weisz G M 1982 Lumbar spinal canal stenosis in osteopoikilosis. Clinical Orthopaedics
166: 89–92

## Melorheostosis

Beauvais P, Faure C, Montagne J-P, Chigot P L, Maroteaux P 1977 Léri's melorheostosis:
Three pediatric cases and a review of the literature. Pediatric Radiology 6: 153
Campbell C J, Papademetriou T, Bonfiglio M 1968 Melorheostosis. A report of the clinical,
roentgenographic and pathological findings in fourteen cases. Journal of Bone and Joint
Surgery 50A: 1281
Green A E, Elwood W H, Collins J R 1962 Melorheostosis and osteopoikilosis, with a
review of the literature. American Journal of Radiology 87: 1096
Lester C W 1967 Melorheostosis in a prehistoric Alaskan skeleton. Journal of Bone and
Joint Surgery 49: 142
Murray R O 1951 Melorheostosis associated with congenital arteriovenous aneurysms.
Proceedings of the Royal Society of Medicine 44: 473
Murray R O 1978 Sclerotome theory of melorheostosis. Proceedings of the 5th Annual
Meeting of the International Skeletal Society. Boston. Case 26, p 9
Patrick J H 1969 Melorheostosis associated with arteriovenous aneurysm of the left arm and
trunk. Journal of Bone and Joint Surgery 51B: 126
Whyte M P, Murphy W A, Fallon M D, Hahn T J 1981 Mixed-Sclerosing-Bone-dystrophy:
Report of a case and review of the literature. Skeletal Radiology 6: 95–102

## Pachydermoperiostosis

Chamberlain D S, Whitaker J, Silverman F N 1965 Idiopathic osteoarthropathy and cranial defects in children. (Familial idiopathic osteoarthropathy). American Journal of Roentgenology, Radium Therapy and Nuclear Medicine 93: 408

Cremin B J 1970 Familial idiopathic osteoarthropathy of children: a case report and progress. British Journal of Radiology 43: 568

Currarino G, Tierney R C, Giesel R G, Weihl C 1961 Familial idiopathic osteoarthropathy. American Journal of Roentgenology, Radium Therapy and Nuclear Medicine 85: 633

Findlay G H, Oosthuizen W J 1951 Pachydermoperiostitis: syndrome of Touraine, Solente and Gole. South African Medical Journal 25: 747

Gray P I, Steyn A F 1978 Touraine-Solenti-Gole syndrome: A case report. South African Medical Journal 54: 1071

McKusick V A 1975 Mendelian Inheritance in Man 4th edn. Johns Hopkins University Press, Baltimore, p 246

Rimoin D L 1965 Pachydermoperiostosis (idiopathic clubbing and periostosis). Genetic and physiologic considerations. New England Journal of Medicine 272: 923

## Osteosclerosis, Stanescu type

Dipierri J E, Guzman J D 1984 A second family with autosomal dominant osteosclerosis-type Stanescu. American Journal of Medical Genetics 18: 13–18

Gorlin R J, Glass L 1977 Autosomal dominant osteosclerosis. Radiology 125: 289

Hall J G 1974 Craniofacial dysostosis — either Stanescu dysostosis or a new entity. Birth Defects: Original Article Series X/12: 521–523

Maximilian C, Dumitriu L, Ionitui D, Ispas I, Firu P, Ciovirnache M, Duca D 1981 Syndrome de dysostose cranio-faciale avec hyperplasie dyaphysaire. Journal de Genetique Humaine 29: 129–139

Stanescu V, Maximilian C, Poenaru S, Floria I, Stanescu R, Ionesco V, Ioanitiu D 1963 Syndrome héréditaire dominant, réunissant une dysostose craniofaciale de type particulier, une insuffisance de croissance d'aspect chondrodystrophique et un épaississessement massif de la corticale des os longs. Rev. Franc. Endocrinol. Clin. 4: 219–131

Worth H M, Wollin D G 1966 Hyperostosis corticalis generalisata congenita. Journal of the Canadian Association of Radiologists 17: 67

## Osteosclerosis, distal type

Beighton P, Macrae M, Kozlowski K 1980 Distal osteosclerosis. Clinical Genetics 18: 298–304

## Weismann-Netter-Stuhl syndrome

Alavi S M, Keats T E 1973 Toxypachyosteose diaphysaire tibio-péronière. Weismann-Netter syndrome. American Journal of Roentgenology, Radium Therapy & Nuclear Medicine 118: 314–317

Azimi F, Bryan P J 1974 Weismann-Netter-Stuhl syndrome (Toxopachyostéose diaphysaire tibiopéronière). British Journal of Radiology 47: 618–620

Breuzard J, Tixier P, Sallet J 1960 A propos des incurvations non rachitiques des membres inférieurs. Deux nouveaux cas de toxo-pachy-ostéose diaphysaire tibio-péronière observes chez l'adulte. Bulletin et Mémoires de la Societe Médicine des Hôpitaux de Paris 76: 165–170

Heuily F, Gaucher A, Gaucher P, Laurent J, Vautrin D 1967 Une curieuse association: Maladie de Weismann-Netter et Stuhl, maladie de Biermer goitre congenital. La Presse Médicale 75: 1577–1578

Hoefnagel D 1969 Malformation syndromes with mental deficiency. Birth defects: Original Article Series 5/2: 11–14

Keats T E, Alavi M S 1970 Toxopachyostéose diaphysaire tibio-péronière. (Weismann-Netter syndrome.) American Journal of Roentgenology, Radium Therapy and Nuclear Medicine 109: 568–574

Krewer B 1961 Dysmorphie jambière de Weismann-Netter (Toxo-pachy-ostéose diaphysaire tibio-péronière). Chez deux vrais jumeaux. La Presse Médicale 69: 419–420

Weismann-Netter R, Rouaux Y 1956 Tochopachyostéose diaphysaire tibio-péronière (R. Weismann-Netter et L. Stuhl). Chez deux soeurs. La Presse Médicale 64: 790

Weismann-Netter R, Stuhl L 1954 D'une ostéopathie congénitale éventuellement familiale. Surtout définie par l'incurvation antéro-postérieure et l'épaississement des deux os de la jambe (Toxopachyostéose diaphysaire tibio-péronière). La Presse Médicale 62: 1618–1622

**Other sclerosing and hyperostotic disorders**
*Calvarial hyperostosis*
Pagon R A, Beckwith J B, Ward B H 1986 Calvarial hyperostosis: a benign X-linked recessive disorder. Clinical Genetics 29: 73–78

*Familial osteodysplasia*
Anderson L G, Cook A J, Coccaro P J, Bosma J F 1972 Familial osteodysplasia. Journal of the American Medical Association 220: 1687
Buchignani J S, Cook A J, Anderson L G 1972 Roentgenographic findings in familial osteodysplasia. American Journal of Roentgenology 116: 602

*Schwarz-Lélek syndrome*
Gorlin R J, Spranger J, Kosalka M F 1969 Genetic craniotubular bone dysplasias and hyperostoses: A critical analysis. Birth Defects 5/4: 79
Lélek I 1961 Camurati-Engelmann'sche Erkraankung. Fortschritte Rontgenstrahlen 94: 702
Schwarz E 1960 Craniometaphyseal dysplasia. American Journal of Roentgenology 84: 461

*Sclerotic Bone-Dentine Dysplasia syndrome*
Morris M E, Augsburger R H 1977 Dentine dysplasia with sclerotic bone and skeletal anomalies inherited as an autosomal dominant trait. Oral Surgery 43: 267

*Osteopetrosis with familial paraplegia*
Jacques S, Garner J T, Johnson D, Sheldon C H 1975 Osteopetrosis associated with familial paraplegia: Report of a family. Paraplegia 13: 143

*Osteosclerosis with mental retardation*
Hunter A G W, Macpherson R I 1978 Mental retardation and osteosclerosis. American Journal of Medical Genetics 2: 267–273

*Osteosclerosis with Bamboo hair*
Johnson F, Flores C, Dodgson W B 1978 Case report 50. Skeletal Radiology 2: 185
Netherton E W 1958 A unique case of trichorrhexis nodosa. Archives of Dermatology 78: 483
Quattromani F, Shapiro S D, Young R S 1983 Clinical heterogeneity in the tricho-dento-osseous syndrome. Human Genetics 64/2: 116–121

# 10

# Miscellaneous skeletal disorders

Skeletal disorders of unknown pathogenesis which do not fit neatly into any nosologic category are reviewed in this chapter:

1. Cleidocranial dysplasia
   a. Spondylo-megaepiphyseal-metaphyseal dysplasia
   b. Bilateral glenoid dysplasia
2. Dyschondrosteosis
3. Hereditary arthro-ophthalmopathy (Stickler)
4. Larsen syndrome
5. Acrodysplasias
   a. Thiemann form
   b. Geleophysic dysplasia
   c. Saldino-Mainzer syndrome
6. Trichorhinophalangeal syndrome (Giedion)
7. Coffin-Lowry syndrome
8. Freeman-Sheldon syndrome
9. Nail-patella syndrome
10. Congenital bowing of long bones
    a. Blount disease
    b. Kyphomelic dysplasia
    c. Fuhrmann dysplasia
    d. St Helena genu valgum

## CLEIDOCRANIAL DYSPLASIA

Cleidocranial dysplasia, formerly known as cleidocranial dysostosis, is a well-defined condition in which maldevelopment of the clavicles is associated with mild shortness of stature and a characteristic facies. More than 500 cases have been reported. As many individuals with the disorder are totally asymptomatic, it is probable that cleidocranial dysplasia is even commoner than this figure suggests.

A skeleton of a woman who lived in the Pylos region of Greece during the 16th century BC, had absent clavicles and stunted stature which were suggestive of cleidocranial dysplasia (Bartsocas 1977). Another possible example of the condition in Ancient Greece was the ugly hero Thersites, described by Homer as being able to bring his shoulders together in front of his chest (Bartsocas 1973).

Photographs of the famous 19th century vaudeville artist 'Little Tich' are indicative that he had cleidocranial dysplasia. This diagnosis would account for his small stature and unusual mobility but his theatrical genius must be attributed to his personality rather than his physique.

**Fig. 10.1** Cleidocranial dysplasia: two brothers demonstrating the unusual mobility of their shoulder girdles. These boys are members of a large extended Cape Town family in which more than 500 individuals have the disorder.

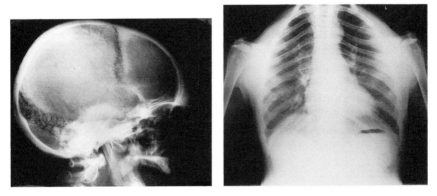

**Fig. 10.2** Cleidocranial dysplasia: skull radiograph of an affected adult showing widening of the sutures and multiple Wormian bones. (Wormian bones, which are found in a few other skeletal dysplasias, notably osteogenesis imperfecta, are named after Oluff Worm (1588–1654), professor of anatomy at the University of Copenhagen. Worm was known for his opposition to William Harvey's concepts concerning the circulation of the blood in the human body.)

**Fig. 10.3** (right) Cleidocranial dysplasia: chest radiograph shows absence of the clavicles.

*Clinical and radiographic features*

The forehead is broad and the parietal region is wide. Clavicular hypoplasia, which may be asymmetrical, permits undue mobility of the shoulder girdle. Many photographs have been published depicting patients in a classic pose, with their arms wrapped around their chests and their shoulders almost touching anteriorly. The only consistent complication is recurrent dislocation, chiefly of the shoulder, elbow and hip joints. Vertebral malalignment is a less frequent but more serious problem.

In skull radiographs patency of the fontanelles, widening of the sutures and multiple Wormian bones may be evident. The sphenoid is short, the foramen magnum is enlarged and the paranasal sinuses are absent. The clavicles may be totally absent, although the medial portions are often spared. The pubic and iliac bones are hypoplastic. The long bones are gracile and the distal phalanges are shortened (Fauré & Maroteaux 1973).

*Genetics*

Autosomal dominant inheritance is well established. Large series were reported by Lasker (1946) who studied 73 kindreds and Jarvis & Keats (1974) who reviewed the radiological features of 40 patients. A family with concordant monozygous twins and discordant dizygous twins was investigated by de Weerdt & Wildervanck (1973). Cleidocranial dysplasia is well known in Cape Town, where several hundred affected individuals had a common ancestor in an energetic polygamous immigrant of Chinese stock (Jackson 1951). This extended kindred are aware of the configuration of their skulls and use the family name to describe this feature, terming it the 'Arnold head'.

Herndon (1951) estimated that about 16% of patients are apparently sporadic. On this basis it appears that the mutation rate is high but variability of phenotypic expression and underdiagnosis could also account for this observation. Goodman et al (1973) described three severely affected individuals in two Iraqi-Jewish kindreds in Israel. As both sets of parents were consanguineous these authors postulated that there might be an autosomal recessive form of the disorder. It is therefore possible that a proportion of the apparently sporadic patients have inherited the condition as an autosomal recessive. Nevertheless, for the purposes of genetic counselling, it is reasonable to assume that the vast majority of patients have an autosomal dominant disorder. As cleidocranial dysplasia is comparatively innocuous, it is unlikely that many parents would wish to limit the size of their families because of their possesion of the gene.

The syndromic associations of congenital clavicular abnormalities have been reviewed in detail by Hall (1982). He identified conditions in which hypoplasia of the clavicles is a frequent component, as listed below:

1. Skeletal dysplasias
   a. Cleidocranial dysplasia, classic autosomal dominant type.
   b. Cleidocranial dysplasia, rare autosomal recessive type (Goodman et al 1975).
2. Skeletal dysostoses
   a. Congenital clavicular pseudoarthrosis
   b. Parietal foramina and lateral clavicular deficiency (Eckstein & Hoare 1963)
3. Chromosome disorders
   a. 11q partial trisomy
   b. 11q/22q partial trisomy
   c. 20p trisomy (Schinzel 1980)
4. Multiple congenital anomalies
   a. Dysplasia cleidofacialis (Kozlowski et al 1970)
   b. Digital-mandibular-clavicular hypoplasia (Yunis & Varon 1980)
   c. Microcephaly-micrognathia-contractural dwarfism (Bixler & Antley 1974)
   d. Imperforate anus-psoriasis-clavicle deficiency (Fukuda 1981)
   e. Focal dermal hypoplasia
   f. Coffin-Siris syndrome

Hall pointed out that clavicular dysgenesis sometimes develops postnatally in progeria, pycnodysostosis and in some forms of acro-osteolysis and that it may occur in conjunction with maldevelopment of the upper limbs, as in the Roberts, Holt-Oram and radial aplasia-thrombocytopenia syndromes. It is thus apparent that, although the presence of clavicular abnormalities is suggestive of cleidocranial dysplasia, several other conditions warrant consideration before the diagnosis is finally established.

## Spondylo-megaepiphyseal-metaphyseal dysplasia

Eight persons with a new bone dysplasia which resembled cleidocranial dysplasia were reported in the radiological literature by Silverman & Reiley (1985). In this disorder which was termed 'spondylo-megaepiphyseal-metaphyseal dysplasia', ossification of the vertebral bodies is defective, while the ossification centres in the long bones are enlarged. Inheritance is thought to be autosomal recessive.

## Bilateral glenoid dysplasia

Dysplasia of the glenoid and neck of the scapula are associated with variable changes in the acromion and coracoid in this uncommon and clinically innocuous condition. Kozlowski et al (1985) gave details of eight cases, reviewed the manifestations and differential diagnosis and alluded to an additional 29 affected persons described in the literature, including a series of nine cases reported by Petterson (1981). They mentioned adult siblings with the disorder, one of whom had produced two affected children and suggested that inheritance was autosomal dominant, with variable phenotypic expression.

## DYSCHONDROSTEOSIS

Dyschondrosteosis was delineated by Léri & Weill (1929) when they recognised the association of mild mesomelic dysplasia with a bilateral Madelung deformity of the forearm. This deformity, which carries the name of Otto Madelung, a German surgeon (1846–1926) consists of shortening and dorsilateral bowing of the shaft of the radius. The ulna is subluxed dorsally at the wrist and the bones of the carpus are wedged between the inclined articular surfaces of the deformed radius and protruding ulna. Seen from the side, the forearm and wrist have a 'dinner fork' configuration. The Madelung deformity may be the result of trauma or infection, or part of a generalised skeletal dysplasia, such as diaphyseal aclasis or multiple enchondromatosis. The deformity may also be present as a non-genetic isolated congenital abnormality, which seems to be confined to females (Golding & Blackburne 1976).

### Clinical and radiographic features

In dyschondrosteosis defective growth becomes evident in mid-childhood. The ultimate height of affected adults is usually between 135 and 170 cm (Kaitila et al 1976). This stunting is largely the result of symmetrical shortening of the tibia and fibula. The Madelung deformity of the forearm is often asymmetrical and the range of movement at the elbow joints is usually limited. In the lower limbs genu valgum is a common feature. The skeleton is otherwise normal.

### Genetics

There has been considerable controversy as to whether all individuals with a primary Madelung deformity have dyschondrosteosis (Langer 1965) or whether the disorders exist as separate entities (Felman & Kirkpatrick 1969, Kozlowski & Zychowicz 1971). Evidence in support of the latter contention was provided by Golding & Blackburne (1976) when they studied 26 individuals with primary Madelung deformity of the wrist and demonstrated that none had any additional radiographic stigmata of dyschondrosteosis. As no bony abnormalities were detected in 65 first degree relatives of seven of these patients, it is possible that the isolated primary Madelung deformity is non-genetic, although this remains uncertain.

Dyschondrosteosis is transmitted as a dominant trait (Henry & Thorn-

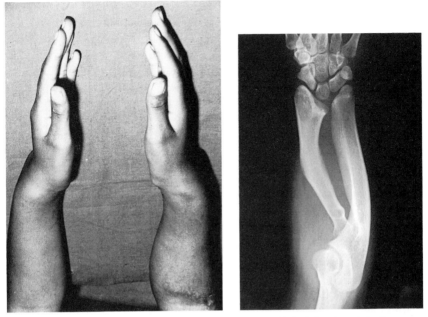

**Fig. 10.4** (left) Dyschondrosteosis: the Madelung or dinner fork deformity of the forearm.

**Fig. 10.5** (right) Dyschondrosteosis: in the Madelung deformity, the radius is shortened and bowed, and the bones of carpus are wedged between inclined articular surfaces.

burn 1967, Carter & Currey 1974, Beals & Lovrien 1976). The preponderance of females has led some authorities to believe that the condition may be an X-linked dominant. However, as females are often more severely affected than their male relatives, this discrepancy is also explicable in terms of bias of ascertainment. In a review of 13 affected persons from eight families Dawe et al (1982) concluded that inheritance was autosomal dominant, with about 50% penetrance. As mentioned in Chapter 1, the Langer form of mesomelic dwarfism may be the result of homozygosity of the dyschondrosteosis gene (Böök 1950, Silverman 1975, Espirutu et al 1975, Goldblatt et al 1987).

Funderburk et al (1976) described a kindred in which members of four generations had dyschondrosteosis and chronic nephritis. The authors speculated that two closely linked dominant genes might be responsible, although simple X-linked dominant inheritance could not be ruled out.

## HEREDITARY ARTHRO-OPHTHALMOPATHY (Stickler* syndrome)

At the turn of the century Dr C. H. Mayo, of the Mayo Clinic, examined a middle aged woman with eye complications and swollen joints. In the

* Gunnar Stickler is chairman of paediatrics at the Mayo Medical School, Minnesota, USA.

ensuing years other members of the family presented with the same stigmata and ultimately Stickler et al (1965) documented the results of a
comprehensive investigation of the kindred. Stickler & Pugh (1967)
expanded the diagnostic features of the syndrome, mentioning that deafness
could be a component. Ocular features are attracting increasing attention
(Blair et al 1979, Weingeish et al 1982) and it is becoming evident that a
high proportion of persons with the Stickler syndrome have mitral valve
prolapse (Liberfarb & Goldblatt 1986).

There is disagreement concerning the syndromic boundaries of the
condition and the true frequency is unknown. Some experts accept minimal
diagnostic criteria and believe that the disorder is comparatively common,
whereas others take a stringent approach and regard the syndrome as a
rarity.

*Clinical and radiographic features*

Enlargement of large joints, particularly the wrists, knees and ankles, may
be present at birth. Repeated episodes of acute arthritis precede degenerative arthropathy and physical activity may be considerably impaired by
middle age. Myopia and choroidoretinal abnormalities constitute the
ophthalmological facets of the syndrome. Retinal detachment sometimes
leads to blindness in childhood and painful secondary glaucoma may necessitate eventual enucleation of the eyeball. Conductive deafness, cleft palate
and structural abnormalities of the vertebrae are inconsistent features.

**Fig. 10.6** Stickler syndrome: a mother and three offspring with myopia, generalised
arthropathy and deafness. This family were initially thought to have an unusual form of
multiple epiphyseal dysplasia but the correct diagnosis is probably the Stickler syndrome.
(From Beighton et al 1978 Clinical Genetics 14: 173–177.)

Radiographically the epiphyses are dysplastic and in adulthood severe degenerative changes may be evident. The vertebral bodies may show some degree of flattening and irregularity.

*Genetics*

The sex distribution of patients and the transmission of the disorder through five generations of the Mayo Clinic kindred is indicative of autosomal dominant inheritance. This genetic mechanism was confirmed when Popkin & Polomeno (1974) recognised more than 20 cases in two large Canadian kindreds and by Liberfarb et al (1981) following a study of 22 affected families. The clinical stigmata are often very variable in the same family (Turner 1974, Herrmann et al 1975, Kozlowski & Turner 1975). This point was emphasised by Hall & Herrod (1975) when they compared the ocular, skeletal and orofacial features in members of three generations of a kindred. Conversely, in some families, expression can be fairly consistent; Vallat et al (1985) reported 11 severely affected persons in three French kindreds, all of whom had severe arthropathy, with deafness and visual problems.

The concept that the Stickler syndrome is an 'ice berg' trait has gained wide acceptance. According to this hypothesis the pleiotropic effects of the autosomal dominant gene can manifest as minor anomalies. In this way simple cleft palate or myopia in relatives of a person with the complete Stickler syndrome could be indicative of their possession of the faulty gene (Opitz et al 1972, Hall 1974). This situation has obvious significance in genetic counselling.

O'Donnell et al (1976) described a father and two sons with a spondyloepiphyseal type of skeletal dysplasia which was complicated by cataracts and neural deafness. The authors pointed out that although this disorder resembled Stickler syndrome, it was probably a separate entity. They speculated that their patients might have the Marshall syndrome, a condition in which involvement of the skeleton had not been previously recognised. This disorder is a separate entity and not a variant of the Stickler syndrome.

Kelly et al (1982) reported an infant with micrognathia, myopia and dumbell-shaped femora who was born into a family with the Stickler syndrome. They considered that these features were reminiscent of the Weissenbacher-Zweymueller syndrome and postulated that the two disorders were in fact variants of the same entity. It is now generally accepted that the Weissenbacher-Zweymueller syndrome is, indeed, the early stage of the Stickler syndrome.

From the foregoing, it is evident that the diagnosis of the Stickler syndrome is not always an easy matter and that it can be confused with other disorders. For instance, with hindsight, it seems that an Afrikaner kindred reported under the title 'dominant inheritance of multiple epiphyseal dysplasia, myopia and deafness' (Beighton et al 1978) should probably be regarded as having the Stickler syndrome.

## LARSEN* SYNDROME

The Larsen syndrome is an unusual disorder in which marked articular hypermobility and a flattened facies are associated with various skeletal abnormalities (Larsen et al 1950). It is likely that the condition exists in mild autosomal dominant and severe autosomal recessive forms. More than 100 cases have now been reported.

*Clinical and radiographic features*

Laxity of the joints is the cardinal feature of the Larsen syndrome. The knee joint is frequently unstable and dislocations may be recurrent. Equinovarus deformities of the feet are common and spinal malalignment may develop. The depressed nasal bridge and widely spaced eyes produce a 'dish face'. The fingers have a cylindrical configuration and the thumbs are spatulate. The manifestations are variable in degree and mildly affected individuals have few problems. However, those with severe involvement may have a disturbed gait, spinal deformity and cord compression. Structural cardiac malformations have been observed in a few patients (Kiel et al 1983). The orthopaedic complications in the Larsen syndrome have been discussed by Oki et al (1976), Michel et al (1976) and Habermann et al (1976).

The radiographic stigmata of the Larsen syndrome have been reviewed by Kozlowski et al (1974). The presence of an extra ossification centre in the calcaneum is an important diagnostic feature. This centre appears in infancy and fuses by the end of the first decade. The radial heads are often dislocated and the sequelae of repeated dislocation or subluxation may be evident in other joints. Vertebral anomalies are sometimes encountered in the cervical and dorsal regions. Supernumerary bones may be present in the carpus and the phalanges are usually misshapen.

*Genetics*

The heterogeneity of the Larsen syndrome has been emphasised by Maroteaux (1975). In keeping with autosomal recessive transmission Curtis & Fisher (1970) and Steel & Kohl (1972) reported affected sibs with normal parents. Strisciuglio et al (1983) gave details of a brother and sister with the syndrome in whom severe structural cardiac defects were also present. The unaffected parents were consanguineous and autosomal recessive inheritance is possible.

There have been several case descriptions which are consistent with autosomal dominant inheritance. Harris & Cullen (1971) reported an affected mother and daughter. The maternal grandfather had a similar

---

* Loren Larsen is chairman emeritus of the department of orthopaedic surgery at the Shriner's Hospital for Crippled Children, San Francisco, USA.

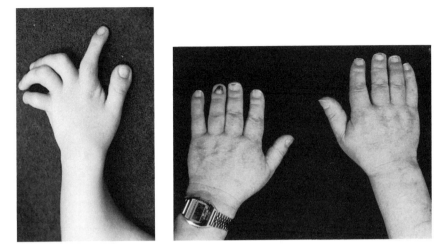

**Fig. 10.7** (left) Larsen syndrome: the thumb is spatulate and the fingers are hypermobile.

**Fig. 10.8** (right) Acrodysplasia: the digits are stubby. The patient also had stunted stature, retinitis pigmentosa and renal failure; the Saldino-Mainzer syndrome was diagnosed.

facies and probably also had the condition. Latta et al (1971) and Sugarman (1975) also described generation to generation transmission. Retrospectively, it is possible that McFarlane (1947) had encountered the dominant form of the Larsen syndrome when he reported a woman who, in three marriages, had produced children with bilateral dislocation of the knees. Orthopaedic complications can also involve the feet; Stanley & Seymour (1985) gave details of the foot problems in eight affected members of four generations of a family in Devon, England.

Piussan et al (1975) described six patients with short stature, hypermobility and diffuse skeletal sclerosis. Payet (1975) reported five sporadic children with the typical stigmata of the Larsen syndrome, in whom advanced bone age and osteoporosis were also present. The relationship of these disorders to the classical Larsen syndrome is uncertain but they are probably separate entities.

In the absence of affected kin genetic counselling in the Larsen syndrome is not easy. In the present state of knowledge the severity of the stigmata and the presence of cardiac and vertebral malformations are probably indicative of the autosomal recessive type, while mild manifestations are suggestive of the autosomal dominant form.

## ACRODYSPLASIAS

The acrodysplasias are a group of disorders in which the peripheral tubular bones are maldeveloped, so that the digits are shortened. Brailsford (1948) drew attention to these conditions and subsequently his name has been used in conjunction with the designation 'peripheral dysostosis' or 'epiphyso-metaphyseal acrodysplasia'. Giedion (1976) emphasised that acrodysplasia may be part of several major syndromes in which generalised skeletal changes overshadow the digital manifestations. The acrodysostosis group of conditions are further discussed in Ch. 12.

Peripheral dysostosis of clinically significant degree, in the absence of other associated features, has been inherited as an autosomal dominant in kindreds reported by Singleton et al (1960) and Bachman & Norman (1967). An autosomal recessive form of acrodysplasia was recognised by Prata et al (1984) in two brothers born to consanguineous parents. In addition to brachydactyly, these boys had scoliosis, spina bifida occulta and carpal synostosis. In the acrodysplasias, the phalangeal epiphyses have a cone-shaped configuration. However, epiphyses of this type are non-specific and they can also occur as unimportant isolated minor variants (Newcombe & Keats 1969).

*Thiemann form of acrodysplasia* is a rare disease in which soft tissue swelling of the proximal interphalangeal joints develops at puberty. Rubin-stein (1975) reviewed the topic and concluded that Thiemann disease was inherited as an autosomal dominant. Giedion (1976) disagreed, commenting that although the designation is frequently encountered in the early litera-ture, there is considerable doubt as to whether the condition exists as a specific entity.

*Geleophysic dysplasia* is a rare syndrome in which brachydactyly is associ-ated with small stature and cardiac valvular disease (Spranger et al 1971). The authors were impressed by their patients' jovial appearance and they coined the designation from the Greek 'geleos' or happy and 'physis' or nature. The autosomal recessive mode of inheritance was substantiated by reports of two affected sisters with normal parents (Koiffmann et al 1984) and three siblings with the disorder (Spranger et al 1984a). The latter authors suggested that the glycoprotein metabolism might be at fault. In a further article (Spranger et al 1984b) described a girl with a condition which resembled geleophysic dysplasia but differed by virtue of the mild-ness of the hand changes and the facial appearance. It is uncertain whether this form of acrofacial dysplasia represents a separate entity.

*Saldino-Mainzer syndrome* is yet another disorder which can be classified with the acrodysplasias. Affected persons have stubby digits, stunted stature and a facial resemblance to each other. Retinitis pigmentosa and renal abnormalities are additional components which have prompted the designation 'hereditary renal-retinal dysplasia' (Schimke 1969). Renal involvement, which may be life-threatening, has been emphasised in several reports (Saldino & Mainzer 1971, Diekmann et al 1977, Giedion 1979). Affected siblings described by Mainzer et al (1970) also had cerebellar ataxia, as had the single patient described by Popovic-Rolovic et al (1976).

Autosomal recessive inheritance is likely but because of the paucity of case reports, the condition has not yet been fully delineated. Indeed, it had not received the asterix of syndromic respectability in the 1983 edition of McKusick's catalogue, although it is listed in the 1983 version of the Paris Nomenclature.

## TRICHORHINOPHALANGEAL DYSPLASIA

Giedion (1966) delineated trichorhinophalangeal dysplasia (TRP) and since that time more than 100 cases have been reported. The disorder causes little disability and it is probably underdiagnosed.

### Clinical and radiographic features

Individuals with the condition are of short stature and have a bulbous pear-shaped nose and sparse hair. Expansion of the interphalangeal joints may lead to an erroneous diagnosis of rheumatoid arthritis. Prior to skeletal maturation, cone-shaped epiphyses are radiographically evident in the digits. In the adult the tubular bones of the hands are shortened and, in some instances, the articular surfaces may be indented. The femoral capital epiphyses are small and Perthes-like changes may supervene in the hip joints. Morris et al (1985) in a review of eight cases from five families in the Antipodes emphasised the practical importance of severe hip complications.

### Genetics

There have been several reports of generation to generation transmission and autosomal dominant inheritance is well established (Murdoch 1969, Beals 1973, Giedion et al 1973). In a review of the genetics of the condition

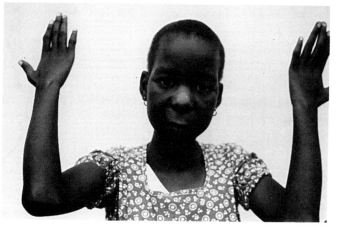

**Fig. 10.9** Trichorhinophalangeal dysplasia: a girl with short stature, expansion of the interphalangeal joints, sparse hair and a bulbous nose.

Weaver et al (1974) described three sibs who had inherited the disorder from their father. Giedion (1976) pointed out that in four kindreds a total of nine sibs with the condition had normal parents. Parental consanguinity was present in one instance. This author postulated that in these families inheritance might be recessive or, conversely, that the dominant gene may be non-penetrant.

Trichorhinophalangeal dysplasia type II, or the Langer-Giedion syndrome, is a separate entity in which the facial appearance is reminiscent of the classical form of the disorder. Additional features are microcephaly, mental retardation, loose joints and exostoses (Giedion 1976). There is disagreement as to whether TRP I and II are separate entities and syndromic boundaries are ill defined. Indeed, it has been suggested that exostoses represent the only consistent diagnostic discriminant at the clinical or radiographic level (Langer et al 1984, Buhler & Malik 1984).

Deletions in the long arm of chromosome 8 have been detected in about 20 patients with TRP type II (Buhler et al 1980, Pfeiffer 1980, Fryns et al 1979, 1981, 1983, Zabel & Baumann 1982, Turleau et al 1982, Zaletajev & Marincheva 1983, Wilson et al 1983, Fukushima et al 1983, Buhler et al 1983, Schwartz et al 1985). These deletions have been of variable size but the critical segment is believed to be situated in the region of band 23 (i.e. 8q23). Similar deletions in persons with TRP I have been recognised by Hamers et al (1983) and Goldblatt & Smart (1986).

There is little doubt that the deletion in chromosome 8 is a genuine component of TRP but the precise role of this abnormality in the pathogenesis of the condition is still uncertain. Not all patients have the deletion and the question of heterogeneity and separate syndromic status of TRP I and II are still unsettled. Perhaps the last word should remain with Andreas Giedion of Zurich who originally described the condition. In a personal communication he stated 'Whatever the final conclusion concerning the deletion, in my opinion the two conditions should be separated.'

## COFFIN-LOWRY* SYNDROME

The Coffin-Lowry syndrome is a rare disorder in which facial and digital abnormalities are associated with mental deficiency and skeletal deformities. In a review of 28 cases in eight kindreds, including eight patients in three unreported families, Temtamy et al (1975) demonstrated that the conditions reported independently by Coffin et al (1966) and Lowry et al (1971) were the same entity.

---

* Grange Coffin is a paediatrician in San Francisco, USA.
  Brian Lowry is professor of medical genetics at the University of Calgary, Canada.

*Clinical and radiographic features*

Intellectual impairment is the major problem in the syndrome. The face becomes progressively coarsened and the eyes have an antimongoloid slant. The hands are stubby, the fingers are hypermobile and the skin is extensible. The cervical vertebrae may be fused. Radiographically minor dysplastic changes are evident throughout the skeleton.

*Genetics*

The Coffin-Lowry syndrome has been transmitted from generation to generation in several families. Males are more severely affected than females, in whom the stigmata are inconsistent. It is possible that the condition is X-linked, with variable manifestations in the female heterozygote. However, as no affected male has reproduced, the issue remains unresolved. Temtamy et al (1975) have suggested that the syndrome might be transmitted as an autosomal dominant sex-influenced trait. Mattei et al (1981) reported affected sisters in a North African kindred. There was a hint of parental consanguinity and the possibility of autosomal recessive inheritance was raised.

In the absence of objective criteria, diagnosis can be difficult, especially in early life (Wilson & Kelly 1981). Membrane-limited intracytoplasmic inclusions have been identified in the skin and conjunctiva of one patient with the Coffin-Lowry syndrome. These findings may be of eventual significance in the elucidation of the mode of genetic transmission and in the development of techniques for antenatal diagnosis of the disorder.

Christian et al (1977) have described four male cousins with short stature, mental retardation and skeletal anomalies which were reminiscent of the Coffin-Lowry syndrome. The condition was X-linked and of five female obligatory carriers, three had fusions of the cervical vertebrae.

## FREEMAN-SHELDON* SYNDROME

Since the first report by Freeman & Sheldon (1938) more than 50 cases have been described. The descriptive designation 'whistling face syndrome' pertains to the shape of the small puckered mouth (Burian 1963). The title 'craniocarpotarsal dystrophy' which is sometimes employed, is neither specific nor accurate and could well be discarded.

*Clinical and radiographic features*

The face is immobile, with ptosis, strabismus, a long philtrum, microstomia

---

* Ernest Freeman (1900–1975) and Joseph Sheldon (1893–1972) occupied the posts of senior orthopaedic surgeon and senior physician at the Royal Wolverhampton Hospital, England. Their original report concerned two unrelated children whom they presented at the Royal Society of Medicine, London.

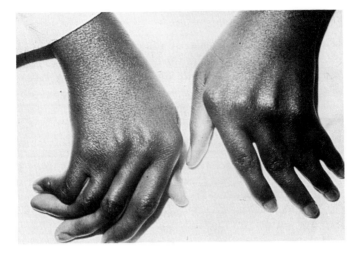

**Fig. 10.10** Freeman-Sheldon syndrome: ulnar deviation of the elongated fingers produces a windmill-vane configuration.

and a dimpled chin. The elongated fingers have ulnar deviation and a 'windmill vane' configuration. Skeletal changes are very variable. Talipes equinovarus and scoliosis may be severe and misdiagnosis of arthrogryposis is not unusual. Intelligence is normal but stature is sometimes reduced. The orthopaedic features of 28 patients have been analysed by Rinsky & Bleck (1976) and in a further study, Weinstein et al (1980) drew attention to the potential complication of severe scoliosis. Sauk et al (1974) reported that the facial musculature was electromyographically and histologically abnormal and in further studies Vanek et al (1986) found evidence of a myopathy of the congenital fibre disproportion type in two unrelated patients. These authors suggested that the muscle abnormalities represented the primary defect and that the skeletal deformities were a secondary manifestation.

*Genetics*

The Freeman-Sheldon syndrome was identified in seven individuals in four generations of a kindred by Jacquemain (1966) and in a father and son by Fraser et al (1970). Pedigree data is consistent with autosomal dominant inheritance. In a review of 24 cases MacLeod & Patriquin (1974) emphasised the phenotypic variability of the syndrome. For this reason diagnosis may be difficult in the sporadic individual. Indeed, this problem arose when the condition was recognised in newborn, non-identical twins (Kousseff et al 1982). As the parents and other members of the family were normal, these authors suggested that there might be an autosomal recessive form of the syndrome. Hall et al (1982) reached the same conclusion when they

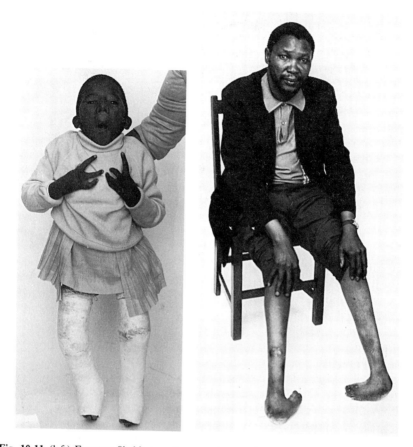

**Fig. 10.11** (left) Freeman-Sheldon syndrome: microstomia and myotonia of the facial muscles produce the typical whistling face. Abnormalities of the lower limbs have been corrected surgically.

**Fig. 10.12** (right) Freeman-Sheldon syndrome: the father of the patient shown in Fig. 10.11. Although the feet are severely deformed, the face is comparatively normal. Phenotypic expression of this autosomal dominant condition is notoriously variable.

recognised more than 18 apparently sporadic cases in the literature. Families in which autosomal recessive inheritance seems likely include those reported by Alves & Azevedo (1977), Fitzsimmons et al (1984) and Sanchez & Kaminker (1986). Phenotypic discriminants for the recognition and separation of the dominant and recessive forms of the Freeman-Sheldon syndrome have not yet been defined.

## NAIL-PATELLA SYNDROME

The nail-patella syndrome, or osteo-onychosdysostosis, is relatively common, occurring in about 1 in 50 000 newborn infants (Renwick & Izatt 1965). More than 250 cases have been described.

## Clinical and radiographic features

Dystrophy of the nails and hypoplasia or absence of the patellae are the most obvious manifestations. Extension of the elbow joints is sometimes limited. Many patients develop renal lesions, which are now recognised as an integral component of the syndrome. Structural abnormalities of the elbow joints and horn-shaped protuberances on the postero-lateral aspects of the ilia may be radiographically evident. Orthopaedic complications, especially involving malformations of the feet, may be more severe than generally recognised.

## Genetics

Large kindreds have been reported by many authors, including Beals & Eckhardt (1969), Aggarwal & Mittal (1970) and Bennett et al (1973) and autosomal dominant inheritance is well established. The gene is linked to the ABO blood group locus, with a recombination fraction of approximately 10% (Renwick & Lawler 1955, Renwick & Schulze 1965). Further linkage studies in which these findings were confirmed were undertaken in a large Corsican kindred by Serville et al (1974).

The renal lesions seem to aggregate in families and it is possible that there are nephropathic and non-nephropathic forms of the condition. Mckusick (1975) pointed out that as there are no obvious discrepancies in the linkage data, these two forms of the condition, if they indeed exist, may be allelic.

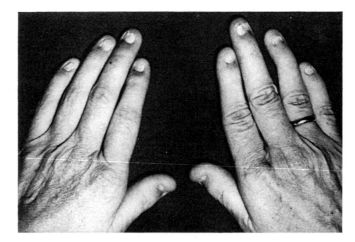

**Fig. 10.13** Nail-patella syndrome: hypoplasia of the finger nails.

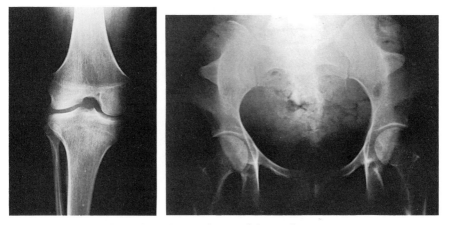

**Fig. 10.14** (left) Nail-patella syndrome: absence of the patella.

**Fig. 10.15** (right) Nail-patella syndrome: iliac horns.

An infant with a 48XXXY chromosome constitution, together with the stigmata of the nail-patella syndrome, was reported by Jansen et al (1976). The child's mother and grandfather also had the nail-patella syndrome and the authors concluded that the occurrence of the two syndromes in the child was fortuitous. In the same way, Gilula & Kantor (1975) encountered a family in which carcinoma of the colon coexisted with the nail-patella syndrome. These authors speculated that predisposition to colonic carcinoma might represent a previously undetected component of the nail-patella syndrome. However, as there have been no similar reports, this association is almost certainly a chance event.

Absence of the patella and aniridia have been observed in members of three generations of a kindred by Mirkinson & Mirkinson (1975). No other features of osteo-onychodysostosis were present and this condition is apparently a distinct autosomal dominant entity.

## CONGENITAL BOWING OF LONG BONES

Bowing of the long bones, particularly those of the lower limbs, may occur in the absence of any obvious metabolic disturbance. Thompson et al (1976) reviewed the problem of bowed limbs in the neonate and emphasised that this abnormality may be physiological or the presenting feature of a variety of disorders. Campomelic dwarfism, a potentially lethal condition in the newborn in which the long bones are bowed, has been discussed in Chapter 3.

Bow legs due to tibial torsion was recognised in eight individuals in four generations of a kindred by Blumel et al (1957). A second family was reported by Fitch (1974). Inheritance is apparently autosomal dominant and it is probable that this disorder is more common than is generally recog-

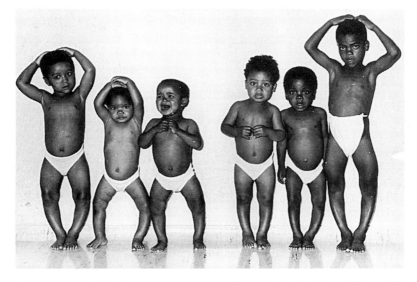

**Fig. 10.16** Limb bowing: there are many causes of limb bowing in infancy. These children have a variety of conditions, including dietary and metabolic rickets, metaphyseal dysplasia and Blount disease. (Courtesy of Mr R. A. de Méneaud, Cape Town.)

nised. A mother and daughter with mild self-limiting femoral bowing were documented by Kapar & van Vloten (1986). This condition seems to be a separate genetic entity.

A kindred with autosomal dominant inheritance of congenital tibial bowing in association with pseudoarthrosis and pectus excavatum was reported by Beals & Fraser (1975). These authors reviewed the causes of tibial bowing and concluded that this particular syndrome had not been previously reported. Newell & Durbin (1976) discussed the pathogenesis of congenital angulation and the relationship of this deformity to pseudo-arthrosis.

Kozlowski et al (1978) gave details of 10 patients with assorted forms of congenital bowing of the long bones and reviewed the literature concerning abnormalities of this type. They drew attention to the Caffey variety of generalised bowing, which rightly or wrongly was attributed to fetal compression and abnormal positioning. Hall & Spranger (1979, 1980) commented that many unclassifiable conditions present with congenital bowing of long bones.

### Blount* disease

The infantile form of Blount disease, or tibia vara, is characterised by bowing of the legs which develops during the second year of life. The

---

* Walter Blount, born 1900, is emeritus professor of orthopaedics at the Marquette Medical School, USA.

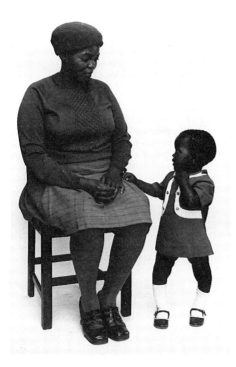

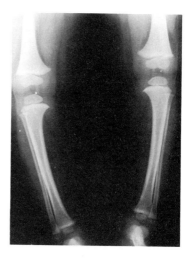

**Fig. 10.17** (left) Blount disease: bowing of the legs develops during the second year of life. The aetiology is unknown.

**Fig. 10.18** (above) Blount disease: beaking of the upper medial tibial metaphysis and buttressing of the cortex of the medial side of the tibial shaft.

medial aspect of the upper tibial epiphysis is primarily involved but changes are sometimes present in the adjacent tibial metaphysis and at the lower end of the femur (Blount 1937). Blount disease is particularly common in the African negro population of Southern Africa. However, the aetiology is unknown and in a survey of over 100 affected children there was no evidence to implicate any simple genetic mechanism (Bathfield & Beighton 1978).

Blount disease of late onset, which appears at puberty, is almost certainly a separate condition from the infantile form of Blount disease. An isolated report of late-onset Blount disease in a father and his two sons is suggestive of autosomal dominant inheritance (Tobin 1957) but the majority of cases of this heterogeneous disorder are sporadic.

## Kyphomelic dysplasia

Kyphomelic dysplasia has emerged as a clearly delineated form of congenital bowing of the long bones. The major changes are in the femora, which are short, broad and angulated. Other bones show lesser bowing, while the vertebrae are flattened. It is of diagnostic significance that affected infants bear a facial resemblance to each other. There are no systemic ramifications but dwarfism and deformity are severe and orthopaedic intervention is usually required.

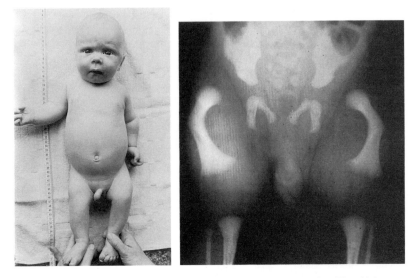

**Fig. 10.19** (left) Kyphomelic dysplasia. An affected child aged 3 months. The thighs are short and bowed.

**Fig. 10.20** (right) Kyphomelic dysplasia. The femora are bowed and shortened, with wide metaphyses, the pubic symphysis is underossified and segmentation of the sacrum is abnormal.

Affected children of unknown sex were reported by Khajavi et al (1976), while Hall & Spranger (1979) described brothers with the condition. MacLean et al (1983) gave details of an affected boy and another sporadic male child has been encountered by the author in Natal, South Africa. As yet, it is uncertain whether inheritance is autosomal recessive or X-linked.

## Fuhrmann dysplasia

Excessive bowing of the femora is a significant feature of Fuhrmann dysplasia, in which fibular hypoplasia and malformations of the feet and hands are additional components (Fuhrmann et al 1980). This condition occurred in three sons and a daughter in a Turkish-Arabian family. As the unaffected parents were members of a minority group and came from neighbouring villages, autosomal recessive inheritance is probable. The last of these affected offspring was diagnosed antenatally by ultrasound at the 19th week of gestation and the histological findings in the abortus were subsequently documented (Fuhrmann et al 1982).

## St Helena genu valgum

Members of several generations of a family on the island of St Helena in the South Atlantic Ocean had severe valgus deformity of the knees and variable, mild malalignment of the elbows and wrists. The knee deformity, which appears in early childhood, is progressive and leads to crippling. Hypoplasia of the lateral femoral and humeral condyles seems to be the primary defect. St Helena genu valgum is apparently a private syndrome, inherited as an autosomal dominant trait (Eickhoff & Beighton 1985, Beighton et al 1986).

REFERENCES

**Cleidocranial dysplasia**

Bartsocas C S 1973 Cleidocranial dysostosis in Homer. Archeia Ellin Paediatrica Etair 36: 107

Bartsocas C S 1977 Stature of Greeks of the Pylos area during the second millenium BC. Hippocrates (Athens) 2: 157

Bixler D, Antley R M 1974 Microcephalic dwarfism in sisters. Birth Defects: Original Article Series 10(7): 161–165

Eckstein H B, Hoare R D 1963 Congenital parietal 'foramina' associated with faulty ossification of the clavicles. British Journal of Radiology 36: 220–221

Fauré C, Maroteaux P 1973 Cleidocranial dysplasia. Progress in Pediatric Radiology 4: 211–237

Fukuda K 1981 Two siblings with cleidocranial dysplasia associated with atresia ani and psoriasis-like lesions: A new syndrome? European Journal of Pediatrics 136: 109–111

Goodman R M, Tadmor R, Zaritsky A, Becker S A 1975 Evidence for an autosomal recessive form of cleidocranial dysostosis. Clinical Genetics 8/1: 20

Hall B D 1982 Syndromes and situations associated with congenital clavicular hypoplasia or agenesis. In: Papadatos C J, Bartsocas C S (eds) Skeletal Dysplasias, Liss, New York p 279–288

Herndon C N 1951 Cleidocranial dysostosis. American Journal of Human Genetics 3: 314

Jackson W P U 1951 Osteo-dental dysplasia (cleido-cranial dysostosis) the 'Arnold head'. Acta Medica Scandinavica 139: 292

Jarvis L J, Keats T E 1974 Cleidocranial dysostosis, a review of 40 new cases. American Journal of Roentgenology, Radium Therapy and Nuclear Medicine 121: 5

Kozlowski K, Hanicks M. Zygulska-Machowa H 1970 Dysplasia cleido-facialis. Zeitschrift für Kinderheilkunde 108: 331–338

Lasker G W 1946 The inheritance of cleidocranial dysostosis. Human Biology 18: 103

Schinzel A 1980 Trisomy 20 pter — q11 in a malformed boy from a t(13;20) (p11, q11) translocation carrier mother. Human Genetics 53: 169–172

Weerdt C J de, Wildervanck L S 1973 A family with dysostosis cleido-cranialis in twins, with rare or never mentioned aspects in the relatives. Clinical Genetics 4: 490

Yunis E, Varon H 1980 Cleidocranial dysostosis, severe micrognathism, bilateral absence of thumbs and first metatarsal bones and distal aphalangia. American Journal of Diseases of Children 134: 649–653

*Spondylo-megaepiphyseal-metaphyseal dysplasia*

Silverman F N, Reiley M A 1985 Spondylo-megaepiphyseal-metaphyseal dysplasia: A new bone dysplasia resembling cleidocranial dysplasia. Radiology 156: 365–371

*Bilateral glenoid dysplasia*

Kozlowski K, Colavita N, Morris L, Little K E T 1985 Bilateral glenoid dysplasia: Report of eight cases. Australasian Radiology 29: 174–177

Petterson H 1981 Bilateral dysplasia of the neck of scapula and associated anomalies. Acta Radiologica Diagnosis 22: 81

**Dyschondrosteosis**

Beals R D, Lovrien E W 1976 Dyschondrosteosis and Madelung's deformity: report of three kindreds and review of the literature. Clinical Orthopaedics and Related Research 116: 24

Book J A 1950 A clinical and genetical study of disturbed skeletal growth. Hereditas 36: 161–180

Carter A R, Currey H L F 1974 Dyschondrosteosis (mesomelic dwarfism); A family study. British Journal of Radiology 47/562: 634

Dawe C, Wynne-Davies R, Fulford G E 1982 Clinical variation in dyschondrosteosis: A report on 13 individuals in 8 families. Journal of Bone and Joint Surgery 64: 377–381

Espiritu C, Chen H, Wooley P V 1975 Mesomelic dwarfism as the homozygous expression of dyschondrosteosis. American Journal of Diseases of Children 129: 375

Felman A H, Kirkpatrick J A 1969 Madelung's deformity; observations in 17 patients. Radiology 93: 1037

Funderburk S J, Smith L, Falk R E, Bergstein J M, Winter H 1976 A family with concurrent mesomelic shortening and hereditary nephritis. Birth Defects: Original Article Series 12/6: 47

Goldblatt J, Wallis C, Viljoen D, Beighton P 1987 Heterozygous manifestations of Langer mesomelic dysplasia. Clinical Genetics 31: 19–25

Golding J S R, Blackburne J S 1976 Madelung's disease of the wrist and dyschondrosteosis. Journal of Bone and Joint Surgery 58B: 350

Henry A, Thornburn M J 1967 Madelung's deformity. Journal of Bone and Joint Surgery 49B: 66

Kaitila I I, Leisti J T, Rimoin D L 1976 Mesomelic skeletal dysplasias. Clinical Orthopaedics and Related Research 114: 94

Kozlowski K, Zychowicz D 1971 Dyschondrosteosis. Acta Radiologica 11: 459

Langer L O 1965 Dyschondrosteosis: a hereditable bone dysplasia with characteristic roentgenographic features. American Journal of Roentgenology, Radium Therapy and Nuclear Medicine 95: 178

Léri A, Weill J 1929 Une affection congénitale et symetrique du developement osseux; la dyschondrostéose. Bulletins et Mémoires de la Société Médicale des Hôpitaux de Paris 53: 1491

Silverman F N 1975 Mesomelic dwarfism. In: Kaufman H J (ed) Progress in Pediatric Radiology, Vol. 4 Karger, Basel, p 456

## Hereditary arthro-ophthalmology (Stickler syndrome)

Beighton P, Goldberg L, Op't Hof J 1978 Dominant inheritance of multiple epiphyseal dysplasia, myopia and deafness. Clinical Genetics 14: 173–177

Blair N P, Albert D M, Liberfarb R M, Hirose T 1979 Hereditary progressive arthro-ophthalmopathy of Stickler. American Journal of Ophthalmology 88(5): 876–888

Hall J G 1974 Stickler syndrome presenting as a syndrome of cleft palate, myopia and blindness inherited as a dominant trait. Birth Defects: Original Article Series X 18: 157–171

Hall J G, Herrod H 1975 The Stickler syndrome presenting as a dominantly inherited cleft palate and blindness. Journal of Medical Genetics 1214: 397

Herrmann J, France T D, Spranger J W, Opitz J, Wiffler C 1975 The Stickler syndrome. Birth Defects: Original Article Series 112: 76–103

Kelly T E, Wells H H, Tuck K B 1982 The Weissenbacher-Zweymueller syndrome: Possible neonatal expression of the Stickler syndrome. American Journal of Medical Genetics 11: 113–119

Kozlowski K, Turner G 1975 Stickler syndrome — report of a second Australian family. Paediatric Radiology 3: 230

Liberfarb R M, Goldblatt A 1986 Prevalence of mitral valve prolapse in the Stickler syndrome. American Journal of Medical Genetics 24: 387–392

Liberfarb R M, Hirose T, Holmes L B 1981 The Wagner-Stickler syndrome: A study of 22 families. Journal of Pediatrics 99: 394–399

O'Donnell J J, Sirkin S, Hall B D 1976 Generalised osseous abnormalities in the Marshall syndrome. Birth Defects: Original Article Series 12/5: 299

Opitz J M, France T, Herrmann J, Spranger J W 1972 The Stickler syndrome. New England Journal of Medicine 286: 546–547

Popkin J S, Polemeno R C 1974 Stickler's syndrome (hereditary progressive artho-ophthalmopathy). Canadian Medicine Association Journal 111/10: 107

Stickler G B, Belau P G, Farrell F J, Jones J D, Pugh D G, Steinberg A G, Ward L E 1965 Hereditary progressive arthro-ophthalmopathy. Mayo Clinic Proceedings 40: 433

Stickler G B, Pugh D G 1967 Hereditary progressive arthro-ophthalmopathy II. Additional observation on vertebral abnormalities, a hearing defect, and a report of a similar case. Mayo Clinic Proceedings 42: 495

Turner G 1974 The Stickler syndrome in a family with the Pierre Robin syndrome and severe myopia. Australian Paediatric Journal 10/2: 103

Vallat M, Fritsch D, van Coppenolle F, Detre J, Moze M 1985 Stickler's syndrome or hereditary progressive arthro-ophthalmopathy. Journal Française d'Ophthalmologie 8: 301–307

Weingeish T A, Hermsen V, Hanson J W, Burnsted R M, Weinstein S L, Olin W H 1982 Ocular and systemic manifestations of Stickler's syndrome. Birth Defects: Original Article Series 18(6): 539–560

## Larsen syndrome

Curtis B H, Fisher R L 1970 Heritable congenital tibio-femoral subluxation. Journal of Bone and Joint Surgery 52A: 1104

Habermann E T, Sterling A, Dennis R I 1976 Larsen syndrome: a heritable disorder. Journal of Bone and Joint Surgery 58/4: 558

Harris R, Cullen C H 1971 Autosomal dominant inheritance in Larsen's syndrome. Clinical Genetics 2: 87

Kiel E A, Frias J L, Victoria B E 1983 Cardiovascular manifestations in the Larsen syndrome. Pediatrics 71: 942–946

Kozlowski K, Robertson F, Middleton R 1974 Radiographic findings in Larsen's syndrome. Australasian Radiology 18/3: 336

Larsen L J, Schottstaedt E R, Bost F D 1950 Multiple congenital dislocations associated with a characteristic facial abnormality. Journal of Paediatrics 37: 574

Latta R J, Graham C B, Aase J M, Scham S M, Smith D W 1971 Larsen's syndrome: a skeletal dysplasia with multiple joint dislocations and unusual facies. Journal of Paediatrics 78: 291

McFarlane A L 1947 A report of four cases of congenital genu recurvatum occurring in one family. British Journal of Surgery 34: 388

Maroteaux P 1975 Heterogeneity of Larsen's syndrome. Archives Françaises de Pédiatrie 32/7: 597

Michel L J, Hall J E, Watts H G 1976 Spinal instability in Larsen's syndrome: report of three cases. Journal of Bone and Joint Surgery 58/4: 562

Oki T, Terashima Y, Murachi S, Nogami H 1976 Clinical features and treatment of joint dislocations in Larsen's syndrome. Report of three cases in one family. Clinical Orthopaedics and Related Research 119: 206

Payet G 1975 Dwarfism with hyperlaxity, facial malformations and multiple dislocations. Larsen's syndrome. Archives Françaises de Pédiatrie 32/7: 601

Piussan C, Maroteaux P, Castroviejo I, Risbourg B 1975 Bone dysplasia with dwarfism and diffuse skeletal abnormalities. Archives Françaises de Pédiatrie 32/6: 541

Stanley D, Seymour N 1985 The Larsen syndrome occurring in four generations of one family. International Orthopaedics 8: 267–272

Steel H H, Kohl E J 1972 Multiple Congenital dislocations associated with other skeletal anomalies. (Larsen's syndrome) in three siblings. Journal of Bone and Joint Surgery 54A: 75

Strisciuglio P, Sebastio G, Andria G 1983 Severe cardiac anomalies in sibs with Larsen syndrome. Journal of Medical Genetics 20/6: 422–424

Sugarman G I 1975 The Larsen syndrome. Autosomal dominant form. Birth Defects: Original Article Series 11/2: 121

Ventrutoin Festa B, Sebastio L, Sebastio G, Catani L 1976 Larsen syndrome in two generations of an Italian family. Journal of Medical Genetics 13/6: 538

## Acrodysplasias

Bachman R K, Norman A P 1967 Hereditary peripheral dysostosis (three cases). Proceedings of the Royal Society of Medicine 60: 21

Brailsford J F 1948 The Radiology of Bones and Joints. Churchill, London

Giedion A 1976 Acrodysplasias: peripheral dysostosis, acrodysostosis and Thiemann's disease. Clinical Orthopaedics and Related Research 114: 107

Newcombe D S, Keats T E 1969 Roentgenographic manifestations of hereditary peripheral dysostosis. American Journal of Roentgenology, Radium Therapy and Nuclear Medicine 106: 178

Prata M F, Liberal M I, Goncalves V, Maroteaux P, Magalhaes J 1984 Acrodysplasia (hands and feet) with scoliosis by autosomal recessive transmission. Annales Génétique (Paris) 27: 233–236

Singleton E B, Daeschner C W, Teng C T 1960 Peripheral dysostosis. American Journal of Roentgenology, Radium Therapy and Nuclear Medicine 84: 499

*Thiemann form*

Giedion A 1976 Acrodysplasia: Peripheral dysostosis, acrodysostosis and Thiemann's disease. Clinical Orthopaedics and Related Research 114: 107

Rubinstein H M 1975 Thiemann's disease: A brief reminder. Arthritis and Rheumatism 18/4: 357

*Geleophysic dysplasia*

Koiffmann C P, Wajntal A, Ursich M J M, Pupo A A 1984 Brief clinical report: Familial recurrence of geleophysic dysplasia. American Journal of Medical Genetics 19: 483–486

Spranger J, Gilbert E F, Tuffli G A, Rossiter F P, Opitz J 1971 Geleophysic dwarfism: A 'focal' mucopolysaccharidosis? Lancet 2: 97–98

Spranger J, Gilbert E F, Arya S, Hoganson G M I, Opitz J M 1984a Geleophysic dysplasia. American Journal of Medical Genetics 19: 487–489

Spranger J, Gilbert E F, Flatz S, Burdelski M, Kallfelz H C 1984b Acrofacial dysplasia resembling geleophysic dysplasia. American Journal of Medical Genetics 19: 501–506

*Saldino-Mainzer syndrome*

Diekmann L, Louis C, Schulte-Kemna 1977 Familiare Nephropathie mit Retinitis Pigmentosa und peripherer Dysostose. Helvetica Paediatrica Acta 32: 375–382

Giedion A 1979 Phalangeal cone-shaped epiphysis of the hands (PhCSEH) and chronic renal disease. The conorenal syndromes. Pediatric Radiology 8: 32–38

Mainzer F, Saldino R M, Ozonoff M B, Minagi H 1970 Nephropathy associated with retinitis pigmentosa, cerebellar ataxia and skeletal abnormalities. American Journal of Medical Genetics 49: 556–562

Popovic-Rolovic M, Calic-Perisic N, Bunjevacki G, Negovanovic D 1976 Juvenile nephronophthisis associated with retinitis pigmentosa dystrophy, cerebellar ataxia and skeletal abnormalities. Archives of Disease in Childhood 51: 801–803

Saldino R M, Mainzer F 1971 Cone-shaped epiphyses (CSE) in siblings with hereditary renal disease and retinitis pigmentosa. Radiology 98: 39–45

Schimke R N 1969 Hereditary renal-retinal dysplasia. Annals of Internal Medicine 70: 735–744

**Trichorhinophalangeal dysplasia**

Beals R K 1973 Tricho-rhino-phalangeal dysplasia. Journal of Bone and Joint Surgery 55: 821

Buhler E M, Malik N J 1984 The Tricho-rhino-phalangeal syndrome(s): Chromosome 8 long arm deletion: Is there a shortest region of overlap between reported cases? TRP I and TRP II syndromes: Are they separate entities? American Journal of Medical Genetics 19: 113–119

Buhler E M, Buhler U K, Christen R 1983 Terminal or interstitial deletion in chromosome 8 long arm in Langer-Giedion syndrome (TRP II syndrome)? Human Genetics 64: 163–166

Buhler E M, Buhler U K, Stalder G R, Jani L, Jurik L P 1980 Chromosomal deletion and multiple cartilaginous exostoses. European Journal of Pediatrics 133, 163–166

Fryns J P, Logghe N, van Eygen M van den Berghe H 1979 Interstitial deletion of the long arm of chromosome 8, karyotype 46, XY del (8)(21). Human Genetics 48: 127–130

Fryns J P, Logghe N, van Eygen M, van den Berghe H 1981 Langer-Giedion syndrome and deletion of the long arm of chromosome 8. Human Genetics 58: 231–232

Fryns J P, Heremans G, Marien J, van den Berghe H 1983 Langer-Giedion syndrome and deletion of the long arm of chromosome 8. Confirmation of the critical segment to 8q23. Human Genetics 64: 194–195

Fukushima Y, Kuroki Y, Izawa T 1983 Two cases of the Langer-Giedion syndrome with the same interstitial deletion of the long arm of chromosome 8: 46,XY or XX del (8)(q23.8q24.13). Human Genetics 64: 90–93

Giedion A 1966 Das Tricho-rhino-phalangeale Syndrom. Helvetia Paediatrica Acta 21: 475

Giedion A 1976 Acrodysplasias: peripheral dysostosis, acrodysostosis and Thiemann's disease. Clinical Orthopaedics and Related Research 114: 107

Geidion A, Burdea M, Fruchter Z, Meloni T, Trosc V 1973 Autosomal dominant transmission of the tricho-rhino-phalangeal syndrome. Report of four unrelated families and a review of 60 cases. Helvetica Paediatrica Acta 28: 249

Goldblatt J, Smart R 1986 Tricho-rhino-phalangeal syndrome without exostoses, with an interstitial deletion of 8q23. Clinical Genetics 29: 434–438

Hamers A P, Jongbloet P, Peeters G, Geraedts J 1983 Micro-cytogenetics of chromosome 8q. Poster presented at the 8th International Chromosome Conference, Lubeck, September (Abstract 2–8)

Langer L O, Krassikoff N, Laxova R, Scheer-Williams M, Lutter L D Gorlin R J, Jennings C G, Day D W 1984 The Tricho-rhino-phalangeal syndrome with exostoses (or Langer-Giedion syndrome): Four additional patients without retardation and review of the literature. American Journal of Medical Genetics 19: 81–111

Morris L, Kozlowski K, McNaught P, Silink M 1985 Tricho-rhino-phalangeal syndrome I (Report of 8 cases). Australasian Radiology 229: 167–173

Murdoch J L 1969 Tricho-rhino-phalangeal dysplasia with possible autosomal dominant transmission. The Clinical Delineation of Birth Defects. National Foundation, New York, Ch 2, p 218

Pfeiffer R A 1980 Langer-Giedion syndrome and additional congenital malformations with interstitial deletion of the long arm of chromosome 8: 46,XY, del 18(q13–22). Clinical Genetics 18: 142–146

Schwartz S, Beisel J H, Panny S R, Cohen M M 1985 A complex rearrangement, in a case of Langer-Giedion syndrome. Clinical Genetics 27: 175–182

Turleau C, Chavin-Colin F, de Grouchy P, Maroteaux P, Rivera H 1982 Langer-Giedion syndrome with and without del 8q. Assignment of critical segment to 8q23. Human Genetics 62: 183–187

Weaver D D, Cohen M M, Smith D W 1974 The tricho-rhino-phalangeal syndrome. Journal of Medical Genetics 11/3: 312

Wilson W G, Wyandt H E, Shah H 1983 Interstitial deletion of 8q. American Journal of Diseases of Children 137: 444–448

Zabel B U, Baumann W A 1982 Langer-Giedion syndrome with interstitial 8q — deletion. American Journal of Medical Genetics 11: 353–358

Zaletajev D V, Marincheva G S 1983 Langer-Giedion syndrome in a child with complex structural aberration of chromosome 8. Human Genetics 63: 178–182

## Coffin-Lowry syndrome

Christian J C, Demyer W, Franken E A, Huff J S, Khairi S, Reed T 1977 X-linked skeletal dysplasia with mental retardation. Clinical Genetics 11/2: 128

Coffin G S, Siris E, Wegienka L C 1966 Mental retardation with osteocartilagenous anomalies. American Journal of Diseases of Children 112: 205

Lowry R B, Miller J R, Fraser F C 1971 A new dominant gene mental retardation syndrome: associated with small stature, tapering fingers, characteristic facies, and possible hydrocephalus. American Journal of Diseases of Children 121: 496

Mattei J F, Laframboise R, Rouault F, Giraud F 1981 Coffin-Lowry syndrome in sibs. American Journal of Medical Genetics 8: 315–319

Temtamy S A, Miller J D, Maumenee I 1975 The Coffin-Lowry syndrome: An inherited faciodigital mental retardation syndrome. Journal of Pediatrics 86/5: 724

Wilson W G, Kelly T E 1981 Early recognition of the Coffin-Lowry syndrome. American Journal of Medical Genetics 8: 215–220

## Freeman-Sheldon syndrome

Alves A F P, Azevedo E S 1977 Recessive form of Freeman-Sheldon syndrome or 'whistling face'. Journal of Medical Genetics 14: 139–141

Burian F 1963 The 'Whistling face' characteristic in a compound cranio-facio-corporal syndrome. British Journal of Plastic Surgery 16: 140–143

Fitzsimmons J S, Zaldua V, Chrispin A R 1984 Genetic heterogeneity in the Freeman-Sheldon syndrome: 2 adults with probable autosomal recessive inheritance. Journal of Medical Genetics 21: 364–368

Fraser F C, Pashayan H, Kadish M E 1970 Cranio-carpo-tarsal dysplasia. Report of a case in father and son. Journal of the American Medical Association 211: 1374

Freeman E A, Sheldon J H 1938 Cranio-carpo-tarsal dystrophy. An undescribed congenital malformation. Archives of Disease in Childhood 13: 277

Hall J G, Reed S D, Greene G 1982 The distal arthrogryposes: Delineation of new entities. Review and nosologic discussion. American Journal of Medical Genetics 11: 185–239

Jacquemain B 1966 Die angeborene Windmuehlenfluegelstellung als erbliche Kombinationsmissbildung. Zeitschrift für Orthopaedie und ihre Grenzgebiete 102: 146

Kousseff B G, McConnachie P, Hadro T A 1982 Autosomal recessive type of whistling face syndrome in twins. Pediatrics 69: 328–331

MacLeod P, Patriquin H 1974 The whistling face syndrome: cranio-carpo-tarsal dysplasia. Report of a case and a survey of the literature. Clinical Pediatrics 13/2: 184

Rinsky L A, Bleck E E 1976 Freeman-Sheldon ('whistling face') syndrome. Journal of Bone and Joint Surgery 58A: 148

Sanchez J M, Kaminker C P 1986 New evidence for genetic heterogeneity of the Freeman-Sheldon syndrome. American Journal of Medical Genetics 25: 507–511

Sauk J J, Delaney J R, Reaume C, Brandjord R, Witkop C J 1974 Electromyography of oral-facial musculature in craniocarpotarsal dysplasia (Freeman-Sheldon syndrome). Clinical Genetics 6: 132

Vanek J, Janda J, Amblerova V, Losan F 1986 Freeman-Sheldon syndrome: A disorder of congenital myopathic origin? Journal of Medical Genetics 23: 231–236

Weinstein A, Buchinger G, Braun A, von Bazan U B 1980 A family with whistling-face syndrome. Human Genetics 55: 177–189

**Nail-patella syndrome**

Aggarwal N D, Mittal R L 1970 Nail-patella syndrome. Journal of Bone and Joint Surgery 52B: 29

Beals R K, Eckhardt A L 1969 Hereditary onycho-osteodysplasia (nail-patella syndrome). Journal of Bone and Joint Surgery 51A: 505

Bennett W M, Musgrave J E, Campbell R A, Elliot D, Cox R, Brooks R E, Lovrien E W, Beals R K, Porter G A 1973 The nephropathy of the nail-patella syndrome: clinico-pathologic analysis of 11 kindreds. American Journal of Medicine 54: 304

Gilula L A, Kantor O S 1975 Familial colon carcinoma in nail-patella syndrome. American Journal of Roentgenology, Radium Therapy and Nuclear Medicine 123/4: 783

Jansen J, Hansen E, Holbolth N, Jacobson P, Mikkelsen 1976 48,XXXY Klinefelter syndrome and nail-patella syndrome in the same child. Clinical Genetics 9: 163

Mckusick V A 1975 Nail-patella syndrome. Mendelian Inheritance in Man, 4th edn, Johns Hopkins University Press, Baltimore, p 225

Mirkinson A E, Mirkinson N K 1975 A familial syndrome of aniridia and absence of the patella. Birth Defects: Original Article Series 11/5: 129

Renwick J H, Izatt M M 1965 Some genetical parameters of the nail-patella locus. Annals of Human Genetics 28: 369

Renwick J H, Lawler S D 1955 Genetic linkage between the ABO and nail-patella loci. Annals of Human Genetics 28: 312

Renwick J H, Schulze J 1965 Male and female recombination fractions for the nail-patella ABO linkage in man. Annals of Human Genetics 28: 379

Serville F, Verger P, Astruc J 1974 Osteo-onychodysostosis: a new family. Humangenetika 24/4: 333

**Congenital bowing of long bones**

Beals R K, Fraser W 1975 Familial congenital bowing of the tibia with pseudoarthrosis and pectus excavatum. Birth Defects: Original Article Series 11/6: 87

Blount W P 1937 Tibia vara: osteochondrosis deformans tibiae. Journal of Bone and Joint Surgery 19: 1

Blumel J, Eggers G W, Evans E B 1957 Eight cases of hereditary bilateral tibial torsion in four generations. Journal of Bone and Joint Surgery 39A: 1198

Fitch N 1974 Male-to-male transmission of tibial torsion. American Journal of Human Genetics 26: 662

Kapur S, Van Vloten A 1986 Isolated congenital bowed long bones. Clinical Genetics 29: 165–167

Kozlowski K, Butzler H O, Galatius-Jensen F, Tulloch A 1978 Syndromes of congenital bowing of the long bones. Pediatric Radiology 7: 40–48

Hall B D, Spranger J W 1979 Familial congenital bowing with short bones. Radiology 132: 611

Hall B D, Spranger J W 1980 Congenital bowing of the long bones: A review and phenotype analysis of 13 undiagnosed cases. European Journal of Pediatrics 133: 131

Newell R L M, Durbin F C 1976 The aetiology of congenital angulation of tubular bones with constriction of the medullary canal, and its relationship to congenital pseudoarthrosis. Journal of Bone and Joint Surgery 58B/4: 444

Thompson W, Oliphant M, Grossman H 1976 Bowed limbs in the neonate: significance and approach to diagnosis. Annales de Pédiatrie 5/1: 50

### Blount disease

Bathfield C A, Beighton P 1978 Blount disease. Clinical Orthopaedics 135: 29–33

Blount W P 1937 Tibia vara: Osteochondrosis deformans tibiae. Journal of Bone and Joint Surgery 19: 1

Tobin W J 1957 Familial osteochondritis dissecans with associated tibia vara. Journal of Bone and Joint Surgery 39: 1091

### Kyphomelic dysplasia

Hall B D, Spranger J W 1979 Familial congenital bowing with short bones. Radiology 132: 611

Khajavi A, Lachman R, Rimoin D, Schimke R N, Dorst J, Handmaker S, Ebbin A, Perreault G 1976 Heterogeneity in the campomelic syndromes. Long and short bone varieties. Radiology 120: 641

Maclean R N, Prater W K, Lozzio C B 1983 Skeletal dysplasia with short angulated femora (kyphomelic dysplasia). American Journal of Medical Genetics 14: 373

### Fuhrmann dysplasia

Fuhrmann W, Fuhrmann-Rieger A, de Sousa F 1980 Poly-, syn- and oligodactyly, aplasia or hypoplasia of fibula, hypoplasia of pelvis and bowing of femora in three sibs: A new autosomal recessive syndrome. European Journal of Pediatrics 133: 123

Fuhrmann W, Fuhrmann-Rieger A, Jovanovic V, Rehder H 1982 A new autosomal recessive skeletal dysplasia syndrome — Prenatal diagnosis and histopathology. In: Papadatos C J, Bartsocas C S (eds) Skeletal Dysplasias. Liss, New York, p 519–524

### St Helena genu valgum

Beighton P, Myers H S, Aldridge S J, Sedgewick J, Eickhoff S 1986 St Helena familial genu valgum. Clinical Genetics 30: 309–314

Eickhoff S, Beighton P 1985 Genetic disorders on the island of St Helena. South African Medical Journal 68: 475–478

# 11

## Mucopolysaccharidoses and other metabolic disorders

The mucopolysaccharidoses (MPS), mucolipidoses (ML) and related disorders are lysosomal storage diseases which are grouped together as heteroglycanoses. These conditions, in which skeletal involvement is a significant feature, are the consequence of disturbed metabolism of complex carbohydrates. In the majority the underlying enzymatic defect has been identified. New entities in this general category continue to be delineated and biochemical heterogeneity is becoming increasingly apparent. The nosology of these disorders has been reviewed by Spranger (1987).

A. *Mucopolysaccharidoses*
1. MPS I-H (Hurler syndrome)
2. MPS I-S (Scheie syndrome)
3. MPS II (Hunter syndrome)
4. MPS III (Sanfilippo syndrome)
5. MPS IV (Morquio syndrome)
6. MPS V (vacant)
7. MPS VI (Maroteaux-Lamy syndrome)
8. MPS VIII Beta-Glucuronidase deficiency
B. *Mucolipidoses and related disorders*
9. True glycoproteinoses
    a. Sialidosis
    b. Fucosidosis
    c. Mannosidosis
    d. Aspartylglycosaminuria
10. Heteroglycanoses with altered glycoprotein catabolism
    a. I-cell disease
    b. Pseudo-Hurler polydystrophy
    c. GM1 gangliosidoses
    d. Sialuria
    e. Multiple sulphatase deficiency
C. *Sphingolipidoses*
11. Gaucher disease

## A. MUCOPOLYSACCHARIDOSES

The mucopolysaccharidoses (MPS) are a group of conditions in which defective enzymatic activity leads to storage of incompletely degraded glycosaminoglycans. A coarse facies, short stature and skeletal dysplasia are the major features. Other variable manifestations include progressive intellectual impairment, hepatosplenomegaly, corneal clouding and infiltration of the cardiac valves. In recent years the MPS have attracted considerable medical interest and effort. Indeed, there is some truth in the contention that there are more research workers currently investigating the MPS than there are patients suffering from these conditions!

*Nosology*

The MPS provide an object lesson in the way in which nosology is influenced by the development of knowledge. In the early part of this century, affected children were considered to resemble the grotesque gargoyles that traditionally embellished medieval buildings. The disorders were then grouped together under the unfortunate designation 'gargoylism'. Later, when it was appreciated that hepatosplenomegaly was often present in association with abnormalities of the skeleton, the more acceptable term

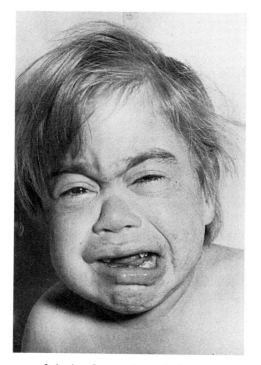

**Fig. 11.1** MPS: a coarse facies is a feature of several of the mucopolysaccharidoses. This child has MPS I.

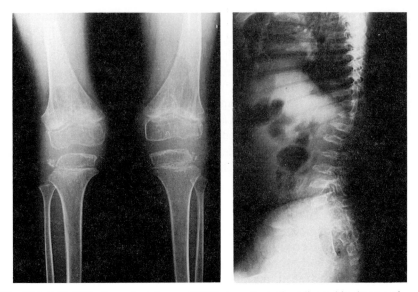

**Fig. 11.2** (left) MPS: the various forms of MPS share the basic radiographic changes of dysostosis multiplex. The epiphyses and metaphyses of the tubular bones are irregular.

**Fig. 11.3** (right) MPS: in dysostosis multiplex, the lumbar vertebrae are flattened, with anterior beaking.

'lipochondrodystrophy' was employed. The recognition of consistent patterns of clinical stigmata led to the concept of heterogeneity, and eponyms such as Hunter, Hurler or Morquio syndrome were applied to the various subdivisions.

The generic term 'mucopolysaccharidosis' came into being when it became apparent that excess mucopolysaccharides were excreted in the urine. Subsequently, abnormal quantities of urinary constituents such as dermatan, heparan and keratan sulphate were identified in various forms of the disorder. It was then evident that the clinical categories of the MPS conformed to these biochemical subtypes and the individual conditions were therefore given numeric designations as alternatives to their eponyms.

In the past few years, the situation has become increasingly complex. Histochemical studies of cultured fibroblasts and the demonstration of cross-correction of the metabolic defect in mixed fibroblast cultures have had an enormous impact. New MPS have been delineated, numerical categories have been subdivided, allelism has been recognised, genetic compounds have been encountered and basic enzymatic defects have been identified. The present state of affairs is summarised in Table 11.1.

*Radiographic features*

Spranger et al (1974) pointed out that the fundamental radiographic changes

**Table 11.1**

| Designation | Eponym | Enzymatic defect |
|---|---|---|
| MPS I-H | Hurler syndrome | Alpha-L-iduronidase |
| MPS I-S | Scheie syndrome | |
| MPS I-H/S | Hurler-Scheie compound | |
| Other forms | | |
| MPS II | Hunter syndrome (severe and mild forms) | L-iduronosulphate sulphatase |
| MPS III | Sanfilippo syndrome | |
| MPS III A | | Heparan N-sulphatase |
| MPS III B | | N-acetyl-alpha-D-glucosaminidase |
| MPS III C | | Acetyl CoA: alpha-glucosaminide-N-Acetyl transferase |
| MPS III D | | N-Ac-glucosamine-6-sulphate sulphatase |
| MPS IV | Morquio syndrome | |
| MPS IV A | (severe, intermediate and mild forms) | N-Ac-galactosamine-6-sulphite-sulphatase |
| MPS IV B | | Beta-galactosidase |
| MPS V | redesignated MPS I-S | |
| MPS VI | Maroteaux-Lamy syndrome (severe and mild forms) | N-Ac-galactosamine-4-sulphate sulphatase |
| MPS VII | Sly syndrome | Beta-glucuronidase |

were similar in each of the MPS and they proposed the non-specific term 'dysostosis multiplex' for these bone abnormalities. In dysostosis multiplex, the skeleton is osteoporotic. The skull shows calvarial thickening and a J-shaped pituitary fossa. The ribs are oar-shaped and the vertebrae are flat, with anterior beaking. The ilia are flared, and the acetabulae are dysplastic. The tubular bones are shortened, with defective diaphyseal modelling and irregularities of the metaphyses and epiphyses. As with many other skeletal dysplasias, the abnormalities in dysostosis multiplex are age-related. The individual MPS differ in the severity of skeletal involvement and recognition of characteristic features is often of diagnostic importance. The radiology of the MPS has been reviewed in detail by Grossman & Dorst (1973).

*General considerations*

With the exception of MPS II, the Hunter syndrome, which is X-linked, all forms of MPS are inherited as autosomal recessives.

Rational genetic, orthopaedic and surgical management is strongly influenced by the long-term prognosis, which varies greatly in the different forms of the condition. In turn, prognostication is dependent upon diagnostic precision. Accuracy will become even more important in the future if specific therapy becomes available. It is therefore imperative that every patient should be categorised at a biochemical level.

Diagnosis is not always an easy matter. Apart from the problem of change in radiographic appearances with advancing years, the urinary excretion of mucopolysaccharides may be age-related. The role of urinary screening tests for MPS has been discussed by Huang et al (1985). Diagnostic confirmation is based upon sophisticated enzymatic investigations on cultured cells. In view of the complexity of these methods and the comparative rarity of the MPS disorders, these tests are usually undertaken in centralised reference laboratories.

It is likely that further MPS remain to be delineated. Equally, a number of obscure skeletal dysplasias may well be the result of abnormalities of MPS metabolism in cartilage, in the absence of mucopolysacchariduria. The mucolipidoses (ML) and related disorders, which resemble the MPS, are distinct entities and they are classified separately.

In an attempt to replace the missing factor, usually a lysosomal enzyme, treatment with infusions of plasma has been employed in MPS I-H and MPS II (Hussels et al 1974) but it is now generally accepted that this approach does not have any practical value. In their review of the MPS, Pennock & Barnes (1976) pointed out that the discovery of the specific enzymatic defects may eventually provide the basis for therapy by enzyme replacement. This has not yet been achieved, although bone marrow transplantation has had some success. Transplantation of amniotic epithelial membranes has also been attempted in MPS I and II, but the results have been disappointing (Akle et al 1985).

Prenatal diagnosis is possible in the MPS disorders by means of enzymatic assay of cultured amniotic fluid cells and electrophoretic analysis of glycosaminoglycans (GAG) in amniotic fluid (Mossman & Patrick 1982).

## MPS I-H (Hurler* syndrome)

MPS I-H or the Hurler syndrome is probably the best known of the MPS. It has been estimated that the prevalence in British Columbia is of the order of 1 in 100 000 (Lowry & Renwick 1971). The disorder has been encountered in many different ethnic groups and it is evident that the gene is widely distributed.

By the second year of life, infants with MPS I-H have a coarse facies, with a wide mouth, thick eyebrows and a protruberant tongue. The joints, particularly those of the hands, are stiff. The thorax is deformed and the aortic and mitral valves often become incompetent. Growth is retarded and intellectual function is impaired. Hepatosplenomegaly and corneal clouding develop in early childhood and death usually occurs by the end of the first

---

* Gertrud Hurler (née Zach), 1889–1965, was a German paediatrician who practised in Neuhausen for more than 45 years.

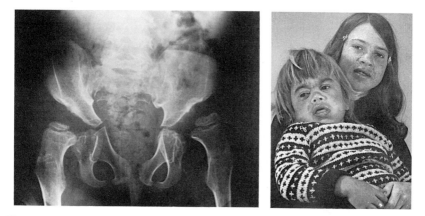

**Fig. 11.4** (left) MPS I: the acetabulae are dysplastic, the femoral capital epiphyses are misshapen and the femoral necks have a valgus deformity.

**Fig. 11.5** (right) MPS I: a 2-year-old girl with the typical facies of MPS I. Her sister and the other members of the kindred are normal.

decade. Radiographic changes of dysostosis multiplex of moderate severity are present in infancy. Urinary excretion of dermatan and heparan sulphate is excessive. The basic abnormality is defective activity of the enzyme α-L-iduronidase.

## Genetics

In a comprehensive review of family data, Jervis (1950) showed that inheritance of the Hurler syndrome was autosomal recessive. Subsequently, Danes & Bearn (1966) observed metachromasia in cultured fibroblasts from six sets of obligatory heterozygote parents of affected children. Omura et al (1976) developed a method of homozygote and heterozygote detection based upon estimation of α-L-iduronidase activity in leucocytes.

The possibility that MPS I-H could be diagnosed antenatally was raised by Nadler (1968) and Fratantoni et al (1969), when they demonstrated metachromasia in amniotic fluid cells from pregnancies in which affected infants were produced at term. An alternative method of antenatal diagnosis by the estimation of amniotic fluid glycosaminoglycan (GAG) content was developed by Matalon et al (1970) and Crawfurd et al (1973). Results obtained in these early days were not consistent but the method of Whiteman (1973) as practised by Mossman & Patrick (1982) has rendered the technique rapid and reliable. Currently at-risk pregnancies are monitored by enzyme assay and radiosulphate uptake in cultured amniotic fluid cells, in conjunction with estimation of ratios of various GAG in the amniotic fluid.

Kleijer et al (1983) monitored 40 at-risk pregnancies and detected 13 affected fetuses while a series of 40 at-risk pregnancies tested by Carey et al (1983) included eight with the Hurler syndrome. These were aborted and the diagnosis was confirmed by analysis of fetal tissues or cultured fetal fibroblasts. Rodeck et al (1983) reported exclusion of MPS I-H in a potentially affected fetus by means of enzymatic assay in fetal blood obtained at fetoscopy, following failure of amniotic cell culture.

MPS I-S (Scheie* syndrome)

MPS I-S, the Scheie syndrome, was initially designated 'MPS V'. However, when it was shown that activity of the enzyme α-L-iduronidase was defective in this condition, as well as in MPS I-H, the classification was adjusted and the two disorders were grouped together. MPS I-S is apparently rare, but in view of the relatively inconspicuous clinical features, it is likely that the condition not infrequently remains undiagnosed.

Individuals with Scheie syndrome have relatively normal height, intelligence and life span. Corneal clouding, rigidity of the digits, aortic incompetence and a tendency to develop the carpal tunnel syndrome are the major clinical problems. Radiographic changes are of minor degree. Urinary and enzymatic abnormalities are identical to those found in MPS I-H.

*Genetics*

Although there is doubt concerning the accuracy of the diagnosis in a number of early case descriptions, there are indisputable reports of affected sibs with normal parents (Scheie et al 1962, McKusick et al 1965). It is evident that MPS I-S is inherited as an autosomal recessive.

Activity of the enzyme α-L-iduronidase is defective in both MPS I-H and MPSI-S (Weismann & Neufeld 1970). It is therefore likely that the determinant genes are allelic. McKusick et al (1972) suggested that an individual with one MPS I-H gene and one MPS I-S gene would have stigmata which were intermediate between those of MPS I-H and MPS I-S. These authors described a group of seven patients with phenotypic features of this nature and postulated that they represented mixed heterozygotes or 'genetic compounds'. The results of fibroblast and enzymatic studies were consistent with this hypothesis and the condition was designated MPS I-H/S, the Hurler-Scheie compound. This condition was subsequently diagnosed in early adulthood in two Japanese brothers, who had dwarfism, Hurler-like facies and normal intelligence (Kajii et al 1974).

The concept of allelism of the Hurler, Scheie and Hurler-Scheie compounds was supported by complementation analysis of fibroblasts from

---

* Harold Scheie was professor of ophthalmology at the University of Pennsylvania during the period 1946–1975. His distinguished career and academic productivity were formally recognised when his name was given to a new eye institute at his university.

affected persons, using a heterokaryon enrichment procedure (Mueller et al 1984). Further complexities remain, as not all patients with alpha-L-iduronidase deficiency have the Hurler or Scheie phenotypes (Roubicek et al 1985). It can be anticipated that these problems will be resolved in the foreseeable future by enzymatic and molecular studies.

### MPS II (Hunter* syndrome)

MPS II, the Hunter syndrome, is X-linked. Two forms of the condition are recognised, the juvenile or severe type (MPS IIA) and the adult or mild type (MPS IIB). The Hunter syndrome is about as common as the Hurler syndrome.

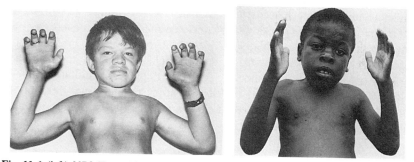

**Fig. 11.6** (left) MPS II: an 18-year-old male with the mild form of the Hunter syndrome. The fingers are fixed in flexion and stature is reduced.

**Fig. 11.7** (right) MPS II: a 10-year-old male with the severe form of the Hunter syndrome. This boy is dwarfed and his mentality is impaired. He attends a special school for the deaf.

Individuals with the severe type of MPS II usually die before adolescence, while those with the mild type survive into middle age. Stature is reduced, intelligence is impaired, the facies are coarse and the hands are clawed. Hepatosplenomegaly and involvement of the cardiac valves are frequent problems. Corneal clouding does not usually develop, but deafness is a common complication. Radiographic abnormalities are similar to those of MPS I-H. Excessive quantities of dermatan and heparan sulphate are excreted in the urine and the enzyme L-iduronosulphate sulphatase is defective in both forms of the condition. Liebaers & Neufeld (1976) have shown that estimation of the activity of this enzyme in serum, lymphocytes or fibroblasts is a relatively simple diagnostic procedure.

---

* Charles Hunter, 1873–1955, was a prominent physician in Winnipeg, Canada. He attained eponymous immortality following a presentation of two brothers with 'gargoylism' at the Royal Society of Medicine, London, during his military service in the First World War.

*Genetics*

The X-linked nature of MPS II was recognised by Noja (1946), when he reported 'a sex-linked type of gargoylism'. Subsequently several pedigrees showing X-linked transmission have been published. Those of Beebe & Formel (1954) and DiFerrante & Nichols (1972) are particularly impressive.

Danes & Bearn (1967) demonstrated that the clinically normal mothers of affected males have metachromasia in about 50% of their cultured fibroblasts. Similarly, female heterozygotes can be detected by the recognition of two types of cells in tissue culture, with regard to the uptake of radioactive sulphate. These findings are consistent with the Lyon hypothesis of random X-chromosome inactivation.

In view of the X-linked mode of inheritance of the Hunter syndrome, female relatives of affected males are potential carriers of the faulty gene. These asymptomatic, heterozygous females were initially recognised by determination of levels of activity of iduronate sulphate sulphatase activity in hair roots (Nwokoro & Neufeld 1979, Archer et al 1981, 1983). Studies of serum or lymphocyte enzymatic activity yield more accurate results and are now in general use (Zlotogora & Bach 1984). The mutation rate is unknown; Archer et al (1983) found that approximately one third of the affected males in Britain represented new mutations, in accordance with the classical genetic hypothesis for a lethal X-linked recessive condition. Conversely, analysis of Ashkenazi Jewish families in Israel (Zlotogora et al 1985) and Black, Indian, Jewish and Afrikaner kindreds in South Africa (Petersen, unpublished data) failed to reveal any new mutations amongst the mothers of males with the disorder. These investigators also observed a deviation in the segregation ratio of affected and unaffected offspring of heterozygous mothers and postulated that pre- or post-zygotic selection favouring the chromosome bearing the Hunter gene might be operative.

A girl infant with the complete clinical phenotype of the Hunter syndrome was documented by Broadhead et al (1986). Cytogenetic studies revealed a partial deletion of the long arm of one X chromosome, and the authors suggested that a Hunter gene on the other X chromosome had been unmasked.

Fetal sexing was the first approach to antenatal diagnosis of MPS II. Niermeijer et al (1976) monitored two pregnancies in which there was a likelihood that the unborn child would have the Hunter syndrome. In each instance, the fetal sex was shown to be female and a normal outcome was successfully predicted. Antenatal diagnosis is now undertaken by measuring enzyme levels in amniotic fluid and cultured amniotic fluid cells together with analysis of GAGs in the cell-free fluid (Liebaers et al 1977, Kleijer et al 1979, Archer et al 1984).

The Hunter syndrome has been diagnosed prenatally following chorionic villus biopsy (Lykkelund et al 1983, Kleijer et al 1984, Harper et al 1984, Pannone et al 1986). This procedure permits determination of the fetal status in the first trimester and it is now the method of choice for obtaining biological material. It can be foreseen that diagnostic molecular techniques will supplement or supplant the existing biochemical methodology.

## MPS III (Sanfilippo syndrome)

Sanfilippo syndrome comprises a heterogeneous group of lysosomal disorders which are the result of deficiencies of enzymes involved in heparan sulphate metabolism. Autosomal recessive inheritance was established following the recognition of affected siblings and parental consanguinity (Sanfilippo et al 1963, Maroteaux et al 1966) and large series of patients have been reviewed by Spranger (1972) and van de Kamp et al (1981). At present four biochemically distinct forms are designated MPS III A-D.

A lucid diagram of the underlying metabolic pathways was published by Petersen (1986). These disorders are clinically similar and differ only in their basic enzymatic abnormalities. Profound mental retardation is the major feature and death often occurs by early adulthood. Stature is relatively normal and there are few systemic ramifications. Some degree of joint rigidity and moderate hirsutism are usually present. Radiographic changes are very mild. Excess heparan sulphate is demonstrable in the urine.

## MPS III-A

In MPS III-A heparan N-sulphatase activity is deficient (Kresse & Neufeld 1972, Matalon & Dorfman 1974). MPS III-A is probably the most common form of the disorder and it has predominated in diagnostic surveys in Canada (Gordon et al 1975), Britain (Whiteman & Young 1977), Germany (McKusick et al 1978) and Holland (van de Kamp et al 1981).

The frequency of MPS III-A is unknown but in view of the phenotypic variability it is possible that patients may remain undiagnosed. It has been suggested that the condition can be recognised by infrared spectroscopy of hair (Lubec et al 1985). This non-invasive method would probably be of value, especially in institutionalised patients.

MPS III-A was successfully diagnosed antenatally by Harper et al (1974) by measurement of enzyme activity in amniotic fluid cells and GAG levels in amniotic fluid. First trimester diagnosis by enzyme analysis of chorionic villi was accomplished by Kleijer et al (1986) and now represents the optimal approach.

## MPS III-B

MPS III-B was separated from MPS III-A by cross-correction studies (Kresse et al 1971). It was then determined that N-Ac-alpha-D-glucosaminidase was defective in this disorder (von Figura & Kresse 1972, O'Brien

1972). Obligatory heterozygotes for MPS III-B have been shown to have a partial reduction in enzymatic activity (von Figura et al 1973, Liem et al 1976). MPS III-B has a wide geographic distribution, although it is less common than type A. In Greece, however, type B predominates (Beratis et al 1986), possibly because of an early founder effect.

The clinical variability of MPS III-B was emphasised by van de Kamp et al (1976) following their studies of six definite and two probable patients in two related consanguineous Dutch sibships. The majority of affected persons die in childhood but some survive into middle age. It is uncertain whether these severe and mild forms of MPS III-B are separate entities. Cell fusion of cultured fibroblasts from persons with these disorders did not result in complementation in heterokaryons and on this basis Ballabio et al (1984) suggested than the mutant genes were allelic.

MPS III-B was diagnosed antenatally by demonstration of defective enzymatic activity in cultured amniotic cells (Mossman et al 1983). Marsh & Fensom (1985) assessed a new flurogenic substrate and suggested that it would be applicable to chorionic villus biopsy prenatal diagnosis in the first trimester.

## MPS III-C

MPS III-C was delineated when Kresse et al (1976) demonstrated metabolic cross-correction in co-cultivation studies with MPS III type A and B fibroblasts from two related Greek patients with the clinical stigmata of the Sanfilippo syndrome. On this evidence the authors proposed that these individuals had a new disease entity, which they designated 'Sanfilippo syndrome type C'. Thereafter, Klein et al (1978) demonstrated deficiency of the enzyme acetyl-CoA: alpha-glucosaminade N-acetyltransferase.

MPS III-C is less common than types A and B and the majority of known cases formed part of a Dutch series published by van de Kamp et al (1981). Affected sisters have been reported from Sweden (Uvebrant 1985) and both affected Black and Indian cases have been diagnosed in South Africa (Petersen 1986). The phenotype is very variable; the former author emphasised that the manifestations could be very subtle, whereas the latter likened the clinical and radiological appearances of some cases to those of Morquio A disease (MPS IV).

## MPS III-D

MPS III-D results from deficiency of N-acetyl-glucosamine-6-sulphate-sulphatase (Kresse et al 1980). Two affected Italians were reported by Gatti et al (1982) but otherwise very few affected persons are known.

## MPS IV (Morquio* syndrome)

MPS IV or the Morquio syndrome was described independently in 1929 by Morquio and Brailsford. Although the conjoined eponym was popular for some years, the single designation is now preferred. The term 'Morquio syndrome' still leads to immense semantic confusion, as it is often used haphazardly and erroneously for any syndrome of dwarfism and spinal malalignment. MPS IV is much less common than MPS I or II.

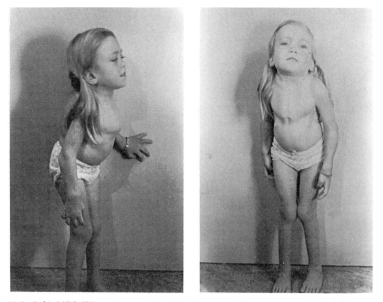

**Fig. 11.8** (left) MPS IV: an 8-year-old girl with a crouching stance and thoracic asymmetry.

**Fig. 11.9** (right) MPS IV: mild facial changes and genu valgum are present. Atlantoaxial dislocation occurred during general anaesthesia for a minor operation. (From Beighton P, Craig J 1973 Journal of Bone and Joint Surgery 55: 478)

Individuals with MPS IV are dwarfed, with a thoracolumbar gibbus, and a protuberant sternum. In distinction to MPS I and II, the joints of the hands are lax, the facies are relatively normal and the intellect is unimpaired. Corneal clouding and deafness develop in later childhood. Involvement of the cardiac valves contributes to cardiorespiratory embarrassment and death usually occurs in early adulthood. Genu valgum and spinal cord compression may necessitate operative intervention. The odontoid process may be hypoplastic, and atlantoaxial subluxation is a potentially lethal hazard during anaesthesia (Beighton & Craig 1973).

* Luis Morquio (1867–1935) was professor of paediatrics in Montevideo, Uruguay. James Brailsford (1888–1961) was a distinguished radiologist in Birmingham, England.

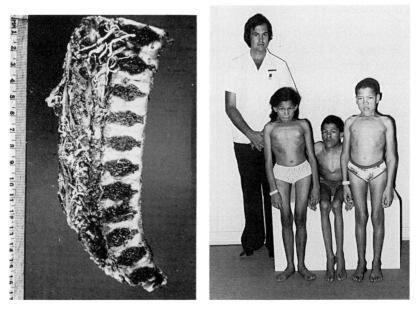

**Fig. 11.10** (left) MPS IV: a vertical section of the spine. Platyspondyly is clearly demonstrated.

**Fig. 11.11** (right) MPS IV-B: siblings with stunted stature and thoracic asymmetry. (From Beck et al 1987 South African Medical Journal, 72: 704–707.)

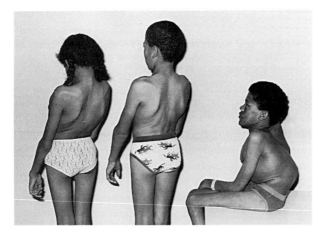

**Fig. 11.12** MPS IV-B: spinal malalignment is the major complication, and cord compression has led to paraplegia in the eldest brother. (From Beck et al 1987 South African Medical Journal, 72: 704–707.)

The radiographic features are those of dysostosis multiplex in severe degree. Excess keratan sulphate is usually present in the urine and inclusion bodies may be demonstrated in leucocytes.

MPS IV-A is the classical form of the condition in which activity of the enzyme N-acetylgalactosamine-6-sulphate sulphatase in cultured fibroblasts is defective (Matalon et al 1974). This disorder is heterogeneous, and severe, intermediate and mild forms are recognised. On a basis of complementation studies, it has been suggested that the causative mutations of these subtypes are allelic (Glössl et al 1981). Some persons with severe manifestations have an additional, probably secondary defect of neuraminidase activity in their fibroblasts (Glössl et al 1984).

MPS IV-B results from deficiency of beta-galactosidase (O'Brien et al 1976, Arbisser et al 1977). The manifestations are variable but generally milder than those of classical MPS IV (Groebe et al 1980, Trojak et al 1980, Van Gemund et al 1983). Spranger (1977) discussed the syndromic status of MPS IV-B and concluded that there was probably considerable heterogeneity. A report of three affected siblings and a review of the biomolecular mechanisms was published by Beck et al (1987).

MPS IV-C comprises a group of patients with the mild manifestations of the Morquio phenotype in whom N-Ac-galactosamine-6-suphate sulphatase and beta-galactosidase activity are normal (Maroteaux et al 1982). At present the basic defect in this disorder is unknown and the syndromic status is uncertain.

*Genetics*

The parents of four affected sibs in the original family reported by Morquio (1929) were consanguineous. Gadbois et al (1973) investigated 48 cases in 27 kindreds in Quebec and found that the ratios of affected and unaffected sibs were consistent with autosomal recessive transmission. Although the Morquio syndrome is uncommon, it has a wide geographical distribution. The author has encountered affected persons in South Africa and MPS IV-A has been recognised in a consanguineous Hutterite community in Canada (Lowry et al 1985). The issue of heterogeneity in the Morquio syndrome has been reviewed by Beck et al (1986).

Prenatal diagnosis by enzymatic estimations of cultured amniotic cells has been accomplished by von Figura et al (1982) and Yuen & Fensom (1985). The latter authors pointed out that prenatal diagnosis using chorionic villi should also be feasible.

## MPS V

MPS V was originally the Scheie syndrome but after allelism with MPS I had been demonstrated, the condition was redesignated 'MPS I-S' in order to avoid nosological confusion. The category MPS V has been left vacant.

MPS VI (Maroteaux-Lamy* syndrome)

MPS VI, the Maroteaux-Lamy syndrome, is one of the least common disorders in the MPS group. Following the original description by Maroteaux et al (1963), fewer than 50 cases have been reported. The intellect is normal, but the clinical and radiographic stigmata are similar to those of the Hurler syndrome, although milder in degree. Excess dermatan sulphate is present in the urine. The leucocytes are packed with granules and cultured fibroblasts show metachromasia. Activity of the enzyme arylsulphatase-B is defective (Beratis et al 1975).

*Genetics*

Autosomal recessive inheritance was established following early reports of affected sibs and parental consanguinity (Spranger et al 1970). As there is disparity in radiographic changes in MPS VI it has been suggested that there might be a mild variant (Spranger et al 1974). Studies of enzymatic activity in cultured fibroblasts supported this contention (O'Brien et al 1974). The concept of heterogeneity was validated following studies of the disorder in a consanguineous German-Acadian (Cajun) kindred in Louisiana, USA. The electrophoretic activity and residual enzymatic activity of arylsulphatase B in this family differed from that of the mutant enzyme in other families (Black et al 1986).

In a review of technical aspects of assay of lysosomal enzymes in amniotic fluid cells and chorionic villi, Gatti et al (1985) concluded that these techniques would be applicable to the antenatal diagnosis of MPS VI.

MPS VII (Beta-glucuronidase deficiency)

MPS VII or β-glucuronidase deficiency was recognised in a Negro infant with mild Hurler-like stigmata, who did not have mucopolysacchariduria. Abnormal accumulation of labelled sulphate and deficient activity of β-glucuronidase was demonstrated in cultured skin fibroblasts from this individual (Sly et al 1973). In keeping with autosomal recessive inheritance, the parents and several sibs were shown to have intermediate levels of enzyme activity. Additional patients with β-glucuronidase deficiency have been reported by Gehler et al (1974) and Beaudet et al (1975). There is some inconsistency in both clinical features and the urinary excretion pattern and it is possible that there are allelic forms of the abnormal gene.

---

* Pierre Maroteaux, born 1926, is the director of the National Centre of Scientific Research, Hôpital des Enfants Malades, Paris.
   Maurice Lamy (1895–1975) was a paediatrician and the first professor of medical genetics at the University of Paris.

About 10 cases have been documented and mild and severe forms have been recognised (Pfeiffer et al 1977). Pregnancy monitoring by measurement of enzymatic activity in cultured amniotic fluid cells was accomplished by Guibaud et al (1979) and Poenaru et al (1982).

## B. MUCOLIPIDOSES AND RELATED DISORDERS
(this section was contributed by Ms E. M. Petersen, Cape Town)

The term 'genetic mucolipidoses' (ML) coined by Spranger & Wiedemann (1970) pertained to disorders such as I-cell disease (ML II), pseudo-Hurler polydystrophy (ML III), mannosidosis, fucosidosis and various conditions now reclassified as sialidoses (ML I and ML IV). Together with those disorders listed below, they all involved defective catabolism of the oligosaccharide portion of glycoproteins and consequently may be grouped together with the MPS as heteroglycanoses. The term 'oligosaccharidoses' has also been suggested (Maroteaux & Humbel 1976) but the currently preferred nomenclature is 'glycoproteinoses' (Strecker 1977).

Phenotypic features resemble a mild form of the Hurler syndrome, while radiographic changes are those of dysostosis multiplex in varying degree (Grossman & Dorst 1973). The lymphocytes contain vacuoles and inclusions are present in cultured fibroblasts. With a few exceptions, mucopolysacchariduria is absent, but abnormal urinary oligosaccharides, glycopeptides or free sialic acid are demonstrable by chromatographic techniques. Accumulation of radiosulphate by cultured fibroblasts occurs in some conditions in this group (Hieber et al 1975). All are associated with a varying degree of mental retardation and all are inherited as autosomal recessives. In view of the rarity of many of these disorders only those likely to be encountered or of special interest are discussed.

*True glycoproteinoses (primary defective glycoprotein catabolism)*
1. Sialidosis (includes former ML I and ML IV)
   a. Sialidosis Types 1 and 2
   b. Mucolipidosis IV
   c. Galactosialidosis
2. Fucosidosis
3. Mannosidosis
4. Aspartylglycosaminuria

Heteroglycanoses with altered glycoprotein catabolism
1. I-cell disease (formerly ML II)
2. Pseudo-Hurler polydystrophy (formerly ML III)
3. GM1 gangliosidoses
4. Sialuria
5. Multiple sulphatase deficiency

## TRUE GLYCOPROTEINOSES

### Sialidosis

Defects in the activity of lysosomal neuraminidase (or sialidase, the terms being interchangeable) give rise to several different clinical entities, namely Sialidosis types 1 and 2, Mucolipidosis IV and Galactosialidosis.

*Sialidosis types 1 and 2*

A primary deficiency of oligosaccharide-specific neuraminidase is the basic defect in both types of sialidosis, but there is a clear clinical differentiation into a normosomic group (Type 1) and a dysmorphic group (Type 2).

Type 1 patients, with the so-called cherry-red spot myoclonus syndrome, present during mid or late childhood. The condition follows a slowly progressive, prolonged course with normal growth, usually normal intellect and neither skeletal nor somatic involvement.

Type 2 patients, formerly classified as ML I, show wide clinical variability with dysmorphism ranging from mildly coarse facies to an appearance which resembles the Hurler syndrome. Radiographic changes are similar to those of the MPS and include beaking of the lumbar vertebrae, broadening of the ribs and thickening of the calvarium (Staalman & Becker 1984). Early cases were reported by Berard et al (1968) and Spranger et al (1968) and the condition has been comprehensively reviewed by Lowden & O'Brien (1977) and Spranger et al (1977).

An extremely severe condition, nephrosialidosis, has also been described (Maroteaux 1978, Aylsworth et al 1980). Confusion in the literature is gradually being reduced as fuller understanding of the underlying biochemical and molecular defects facilitates diagnosis of new cases and reclassification of those previously misdiagnosed.

*Mucolipidosis IV*

Though the original name of this disorder continues to be used the basic defect is now known to be a deficiency of ganglioside-specific neuraminidase (Bach et al 1979, Ben-Yoseph et al 1981). The condition presents in infancy with corneal clouding, athetosis and psychomotor retardation (Berman et al 1974). The ocular abnormalities have been discussed in detail by Riedel et al (1985). The diagnosis, suspected upon recognition of inclusions in conjunctival biopsy specimens or cultured skin fibroblasts, may be confirmed by enzyme assay, which is also the method of choice for prenatal diagnosis. This was successfully accomplished by electron microscopic studies of amniotic fluid cells (Kohn et al 1977, Sekeles et al 1978); the value of this approach was critically reviewed by Kohn et al (1982). The vast majority of documented persons with the condition have been of Ashkenazi Jewish origin.

*Galactosialidosis*

As implied from the name of this disorder, there is a combined deficiency in the activities of both sialidase and beta-galactosidase (Andria et al 1981, Mueller & Shows 1982). Many of the original patients were Japanese (Sakai et al 1982, Sakuraba et al 1983, Loonen et al 1984), although more recent reports indicate that the condition is probably panethnic. The phenotypic features are very variable but in general resemble those of the other hetero-glycanoses and in particular, those of the dysmorphic Sialidosis type 2. Early infantile, late infantile and juvenile forms are recognised, resulting from various mutations of the protective protein for beta-galactosidase. This protein is necessary both to prevent premature degradation of this enzyme and for activation of neuraminidase which functions only when complexed to beta-galactosidase.

## Fucosidosis

Durand et al (1969) reported two sibs who both died in early childhood from a progressive neurological disorder. These authors found excess fucose in the tissues and termed the condition 'fucosidosis'. Deficient activity of the enzyme alpha-fucosidase was demonstrated in the livers of these infants.

A severe infantile form (Type 1) and a chronic, milder form with survival into adulthood (Type 2) are now recognised and both are characterised radiographically by dysostosis multiplex. Because of the heterogeneous biochemical nature of alpha-fucosidase, prenatal diagnosis does not always give clear-cut results (Matsuda et al 1975). The first successful identification of a fetus affected with fucosidosis was achieved by Durand et al (1979).

## Mannosidosis

Ockerman (1967) described a boy with a generalised storage disorder clinically reminiscent of the MPS but without mucopolysacchariduria, in whom tissue alpha-mannosidosis was profoundly deficient. By 1981 over 80 reports had been published and it is evident that the condition is relatively common and panethnic. Prenatal diagnosis is routinely performed in specialised centres. The radiographic features of alpha-mannosidosis have been reviewed by Spranger et al (1976).

Beta-mannosidosis, long recognised in goats, has recently been described in man (Cooper et al 1986, Wenger et al 1986). Precise delineation of the phenotype will depend upon the recognition of further cases but in general it resembles the other heteroglycanoses.

### Aspartylglycosaminuria

Although aspartylglycosaminuria (AGU) was first described in England (Pollitt et al 1968) it is a common cause of mental retardation in Finland, where 138 patients had been identified by 1982. The clinical characteristics of AGU have been analysed in detail by Autio (1972) and the condition was reviewed by Maury (1982).

Diagnosis is based upon recognition of urinary oligosaccharides and enzyme assay of aspartylglucosaminidase, which is deficient in many tissues. Carrier detection is possible and prenatal diagnosis feasible.

## HETEROGLYCANOSES WITH ALTERED GLYCOPROTEIN CATABOLISM

### I-cell disease

The manifestations of I-cell disease (formerly ML II) are reminiscent of a severe form of MPS, though mucopolysacchariduria is absent. Initially, the eponym 'Leroy' was applied to the disorder, following the early case description by Leroy et al (1969), while the alternative term, I-cell disease, pertained to the inclusion bodies which are seen in cultured fibroblasts. These coarse, granular structures are large lysosomes which contain heterogeneous material (Hanai et al 1971, Tondeur et al 1971). Irregular periosteal thickening and widening of the diaphyses of the long bones are important radiographic features (Patriguin et al 1977). Death occurs in early childhood.

I-cell disease is probably less rare than the other disorders in the group and more than 20 cases have been reported from Japan (Okada et al 1985). Affected sibs and parental consanguinity have been recorded and abnormal inclusions have been demonstrated in cultured fibroblasts from obligatory heterozygotes. The diagnosis is routinely established by summation of the clinical and radiographic features, together with the biochemical demonstration of gross imbalance of extracellular to intracellular ratios of many lysosomal hydrolase activities. Affected fetuses in at-risk pregnancies can be identified prenatally by increased activity of several lysosomal enzymes in cell-free amniotic fluid, and correspondingly decreased activity in cultured amniotic fluid cells (Matsuda et al 1975, Aula et al 1975, Owada et al 1980), or by abnormal accumulation of radiosulphate by cultured amniocytes. Prenatal diagnosis of I-cell disease has been discussed by Gehler et al (1976).

The basic molecular defect is deficient N-Ac-glucosaminyl-phosphotransferase (Reitman et al 1981). This enzyme is responsible for attaching a phosphomannosyl marker to the various lysosomal enzymes necessary for post-translational 'targeting' from the ribosomes where they are made to the lysosomes where they function. Absence of this recognition marker causes 're-routing' of the enzymes to the outside of the cell, resulting in the observed 10–20 fold increase in extracellular concentrations and the greatly reduced intracellular levels.

## Pseudo-Hurler polydystrophy

Pseudo-Hurler polydystrophy, formerly designated ML III, is biochemically very similar to I-cell disease and in some cases the mutations have been shown to be allelic. However, somatic cell hybridisation studies have demonstrated at least two complementation groups, thus confirming heterogeneity.

Clinically, pseudo-Hurler polydystrophy is much less severe than I-cell disease and compatible with survival into adulthood. Patients have a coarse facies, short stature and severely stiff joints, sometimes accompanied by mild mental deterioration. With the accumulation of case reports, it has become obvious that the clinical phenotype is very variable and overlaps other disorders in this general category; this has been reviewed in depth by Kelly et al (1975). The radiological features have been discussed by Nolte & Spranger (1976).

## GM1 gangliosidosis

Although classified as a gangliosidosis on the basis of ganglioside accumulation, the deficiency of the heterocatalytic acid beta-galactosidase also results in incomplete catabolism of glycoproteins with resultant storage of oligosaccharides, thus warranting inclusion among the heteroglycanoses.

In addition to the predominant neurodegeneration which is characteristic of the gangliosidoses, clinical features in common with those of the heteroglycanoses include coarsening of the facies, dysarthria, dysostosis multiplex and mild hepatosplenomegaly. Mild mucopolysacchariduria, in addition to the gross oligosacchariduria, is often present. At least three clinical subtypes are recognised, each resulting from a different allelic mutation which affects post-translational processing of the enzyme beta-galactosidase (Hoogeveen et al 1984).

### GM1 gangliosidosis type 1

In generalised gangliosidosis type I, also known as neurovisceral lipidosis, Lauding disease or pseudo-Hurler disease, Hurler-like clinical and radiographic stigmata are present soon after birth. Widespread periosteal reaction in the long bones is an important diagnostic feature. Gangliosides accumulate in the liver and brain, causing failure to thrive and muscular hypotonia with spasticity, decerebrate rigidity and death by the end of the second year.

*GM1 gangliosidosis type 2*

This variant, known as juvenile GM1 gangliosidosis, does not manifest until late infancy when locomotor ataxia is usually the initial symptom. Bone changes are of minor degree but accumulation of gangliosides in the brain leads to progressive degeneration and death by the end of the first decade.

*GM1 gangliosidosis type 3*

This is the adult form of the condition, characterised by mild phenotypic manifestations and long-term survival (O'Brien et al 1976, Stevenson et al 1978, Taylor et al 1980). Presentation is very variable and stigmata commonly include spondyloepiphyseal dysplasia, angiomata and ataxia, with gradual intellectual impairment.

Case descriptions with intermediate phenotypes probably represent compound heterozygotes of the various mutations. Morquio disease type B, also caused by beta-galactosidase deficiency, results from yet another mutation which affects substrate specificity in such a way that severe skeletal dysplasia occurs without neurological impairment (van der Horst et al 1983, Groebe et al 1980). Despite the heterogeneity and complexity of the causative molecular mechanisms, identification of homozygotes and heterozygotes, together with prenatal diagnosis of the former, may be readily achieved by enzyme assay in various tissues.

## Sialuria

As the name implies, this group of disorders is characterised by tissue storage and urinary excretion of free sialic acid (i.e. N-Ac-neuraminic acid or NANA) (Renlund et al 1979, Baumkotter et al 1985). The basic biochemical defect has not yet been precisely identified but is known to result in defective transport of sialic acid out of the lysosomes, where it accumulates as a consequence of the normal catabolism of gangliosides and sialated oligosaccharides by neuraminidase (Renlund et al 1983).

Various severe, early onset forms have been described which are characterised by massive NANA excretion but the best known and clinically most important form is Salla disease. This latter designation is not an eponym but pertains to the geographical region in Finland where most of the initial patients originated. The usual presentation is severe mental retardation with slowly progressive ataxia and mild MPS-like clinical and radiographic features. The condition has been comprehensively reviewed by Renlund et al (1983) and Renlund (1984).

At present there are no tests for heterozygote detection in Salla disease though pedigree analysis shows clearly that the mode of inheritance is autosomal recessive. Prenatal diagnosis was originally attempted by histochemical staining techniques in cultured amniotic fluid cells (Virtanen et al 1980). It has now been successfully achieved by NANA quantitation in amniotic fluid and cultured AF cells from a pregnancy at risk for the severe, infantile form of sialic acid storage (Vamos et al 1986).

## Multiple sulphatase deficiency

Multiple sulphatase deficiency (MSD) has a very variable phenotype, and can closely resemble the MPS group of disorders (Burk et al 1984). The condition usually presents in infancy (Rampini et al 1970) but detection in the neonatal period has been reported (Vamos et al 1981, Burch et al 1986).

The basic defect in MSD is deficient activity of several sulphohydrolases, which leads to the accumulation of sulphated compounds (Austin 1973, Benson & Fensom 1985). Inheritance is autosomal recessive.

## C. SPHINGOLIPIDOSES

### GAUCHER* DISEASE

Gaucher disease is the only member of the sphingolipidosis group of disorders in which bone changes are an important feature. Although not classified amongst the heteroglycanoses, its metabolic background warrants its inclusion in this section.

Three forms of Gaucher disease are recognised: the infantile cerebral type, the juvenile variety and the adult or chronic non-neuropathic type. They are all inherited as autosomal recessives and activity of the enzyme beta-glucosidase is defective in each. However, they differ in their clinical manifestations, and, at a phenotypic level, they are separate conditions.

The infantile and juvenile forms are lethal, due to accumulation of cerebrosides in the brain. Conversely, the adult form is compatible with a relatively normal life span. Splenomegaly is often the presenting feature of this type of Gaucher disease and skeletal complications include osteitis, pathological fractures and degenerative arthropathy (Goldblatt et al 1978). The radiographic features of 17 patients have been reviewed by Myers et al (1975).

*Genetics*

The adult non-neuropathic form of Gaucher disease has a high prevalence in individuals of Ashkenazi Jewish stock. (The Ashkenazim are Jewish people of European derivation, while the Sephardic Jews had their origins in the Mediterranean region. This distinction is of clinical importance, as there are marked differences in the incidence of certain genetic disorders in these two populations.) Fried (1973) ascertained 100 patients with

---

* Phillipe Charles Ernest Gaucher (1854–1918) was professor of dermatology and syphilology of the University of Paris at the turn of the century.

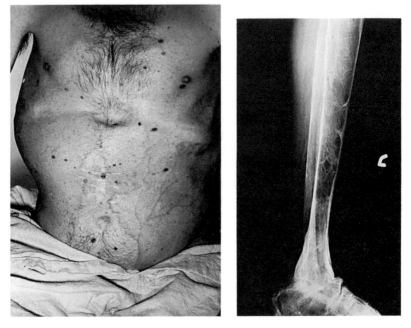

**Fig. 11.13** (left) Gaucher disease: a middle-aged male with massive hepatomegaly, in the terminal stages of the disorder. Splenectomy had been undertaken several years previously.

**Fig. 11.14** (right) Gaucher disease: diffuse rarifaction of the tibia and fibula give the characteristic 'soap bubble' appearance. (From Myers H S et al 1975 British Journal of Radiology 48: 465.)

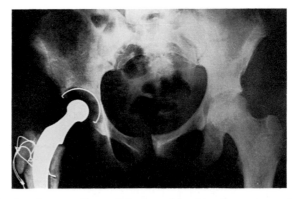

**Fig. 11.15** Gaucher disease: collapse of the femoral head is a frequent complication. Prosthetic replacement has proved to be of value. (Courtesy of Mr S. Sacks, FRCS, Johannesburg.)

Gaucher's disease in Israel, estimated that the minimum prevalence was 1 in 100 000 and, after applying correction for various biases, calculated that the incidence at birth was 1 in 2500. The gene frequency of 0.02 for the

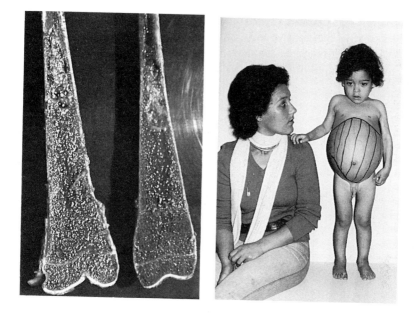

**Fig. 11.16** (left) Gaucher disease: autopsy specimen of the femora showing marrow infiltration and alteration of the normal bone contours.

**Fig. 11.17** (right) Gaucher disease: an affected boy aged 9 years, with massive hepatosplenomegaly. He has the characteristic skeletal changes, but there is no neurological involvement.

Ashkenazim, which is based upon this figure, contrasts dramatically with the probable frequency of less than 0.005 in non-Ashkenazi Jewish populations (Fried 1973). Gaucher disease is particularly common in the Jewish community of South Africa, in which there is a prevalence of at least 1 in 6000, with a carrier rate of between 1 in 20 and 1 in 30 (Beighton & Sacks 1974).

The condition has been identified in non-Jewish populations, including Afrikaners of Dutch stock (Goldblatt & Beighton 1979a) and African negroes (Goldblatt & Beighton 1979b). In view of the occurrence of the disorder in disparate populations, it is possible that non-neuropathic Gaucher disease is heterogeneous. This problem was discussed by Devine et al (1982) in the proceedings of a workshop devoted to the condition. Three distinct forms of the disorder were recognised following comparative kinetic, thermostability and immunotitration investigation of fibroblasts from patients of different ethnic groups (Grabowski et al 1985). Restriction fragment mapping of polymorphisms in the region of the gene for Gaucher disease confirmed that the condition is heterogeneous at the molecular level (Sorge et al 1985).

The heterozygote can be detected by the recognition of diminished activity of β-glucosidase in leucocytes (Beutler & Kuhl 1970). The results are not altogether consistent, and as the laboratory techniques are somewhat laborious, this method is not applicable to population screening. The enzymatic abnormality is expressed in cultured fibroblasts, and Gaucher disease is detectable in the fetus by investigation of cultured amniotic fluid cells.

## REFERENCES

Spranger J 1987 Mini review: inborn errors of complex carbohydrate metabolism. American Journal of Medical Genetics 28: 489–499

### A. Mucopolysaccharidoses
*Preamble*
Akle C, McColl I, Dean M et al 1985 Transplantation of amniotic epithelial membranes in patients with mucopolysaccharidoses. Experimental and Clinical Immunogenetics 2/1: 43–48
Grossman H, Dorst J P 1973 The mucopolysaccharidoses and mucolipidoses. Progress in Pediatric Radiology 4: 495
Huang K-C, Sukegawa K, Orii T 1985 Screening test for urinary glycosaminoglycans and differentiation of various mucopolysaccharidoses. Clinica Chimica Acta 151: 147–156
Hussels I E, Eikman E A, Kenyon K R, McKusick V A 1974 Treatment of mucopolysaccharidoses. Birth Defects: Original Article Series 10/12: 212
Mossman J, Patrick A D 1982 Prenatal diagnosis of mucopolysaccharidosis by two-dimensional electrophoresis of amniotic fluid glycosaminoglycans. Prenatal diagnosis 2: 169–176
Pennock C A, Barnes I C 1976 The mucopolysaccharidoses. Journal of Medical Genetics 13: 169
Spranger J W, Langer L O, Wiedemann H R 1974 Bone Dysplasias. An Atlas of Constitutional Disorders of Skeletal Development. Gustav Fisher Verlag, Stuttgart, p 143

### MPS I-H (Hurler Syndrome)
Carey W J, Hopwood J H, Poulos A et al 1984 Prenatal diagnosis of lysosomal storage diseases: review of experience in 145 patient referrals over a period of 8 years. Medical Journal of Australia 140/5: 203–208
Crawfurd M, Dean M, Hunt D, Johnson D, MacDonald R, Muir H, Payling Wright E, Payling Wright C 1973 Early prenatal diagnosis of Hurler's syndrome with termination of pregnancy and confirmatory findings in the fetus. Journal of Medical Genetics 10: 144
Danes B S, Bearn A G 1966 Hurler's syndrome: effect of retinol (vitamin A alcohol) on cellular mucopolysaccharides in cultured human skin fibroblasts. Journal of Experimental Medicine 124: 912
Fratantoni J C, Neufeld E F, Uhlendorf B W, Jacobson C B 1969 Intrauterine diagnosis of the Hurler and Hunter syndromes. New England Journal of Medicine 280: 686
Jervis G A 1950 Gargoylism: study of 10 cases with emphasis on the formes frustes. Archives of Neurology and Psychiatry 63: 681
Kleijer W J, Thompson E J, Niermeijer M F 1983 Prenatal diagnosis of the Hurler syndrome: Report on 40 pregnancies at risk. Prenatal Diagnosis 3: 179–186
Lowry R B, Renwick S H G 1971 The relative frequency of the Hurler and Hunter syndromes. New England Journal of Medicine 284: 221
Matalon R, Dorfman A, Nadler H L, Jacobson C B 1970 A chemical method for the antenatal diagnosis of mucopolysaccharidoses. Lancet 1: 83
Mossman J, Patrick A D 1982 Prenatal diagnosis of mucopolysacharidosis by two-dimensional electrophoresis of amniotic fluid glycosaminoglycans. Prenatal Diagnosis 2: 169–176
Nadler H L 1968 Antenatal detection of hereditary disorders. Pediatrics 42: 912
Omura K, Higami S, Tada 1976 α-L-Iduronidase activity in leukocytes: diagnosis of homozygotes and heterozygotes of the Hurler syndrome. European Journal of Pediatrics 112/2: 103
Rodeck C H, Tansley L R, Benson P F, Fensom A H, Ellis M 1983 Prenatal exclusion of Hurler's disease by leucocyte alpha-L-iduronidase assay. Prenatal Diagnosis 3: 61–63

Whiteman P 1973 The quantitative determination of glycosaminoglycans in urine with alcian blue 8GX. Biochemical Journal 13: 351–357

**MPS I-S (Scheie syndrome)**

Kajii T, Matsuda I, Ohsawa T, Katsunama H, Ichida Γ, Arashima S 1974 Hurler/Scheie genetic compound (mucopolysaccharidosis IH/IS) in Japanese brothers. Clinical Genetics 6/5: 394

McKusick V A, Kaplan D, Wise D, Hanley W B, Suddarth S B, Sevick M E, Maumenee A E 1965 The genetic mucopolysaccharidoses. Medicine 44: 445

McKusick V A, Howell R R, Hussels I E, Neuffeld E F, Stevenson R 1972 Allelism, non-allelism and genetic compounds among the mucopolysaccharidoses: hypotheses. Lancet 1: 993

Mueller O T, Shows T B, Opitz J M 1984 Apparent allelism of the Hurler, Scheie and Hurler/Scheie syndromes. American Journal of Medical Genetics 18: 547–556.

Roubicek M, Gehler J, Spranger J 1985 The clinical spectrum of alpha-L-iduronidase deficiency. American Journal of Medical Genetics 20: 471–481

Scheie H G, Hambrick G W, Barness L A 1962 A newly recognised forme fruste of Hurler's disease (gargoylism). American Journal of Ophthalmology 53: 753

Wiesmann U, Neufeld E F 1970 Scheie and Hurler syndromes: apparent identity of the biochemical defect. Science 169: 72

**MPS II (Hunter syndrome)**

Archer I M, Harper P S, Wusteman F S 1981 An improved assay for iduronate 2-sulphate sulphatase in serum and its use in the detection of carriers of the Hunter syndrome. Clinica Chimica Acta 112: 107–112

Archer I M, Kingston H M, Harper P S 1984 Prenatal diagnosis of Hunter syndrome. Prenatal Diagnosis 4: 195–200

Archer I M, Young I D, Ress D W, Oladimeji A, Wusteman F S, Harper P S 1983 Carrier detection in Hunter syndrome. American Journal of Medical Genetics 16: 61–69

Beebe R T, Formel P F 1954 Gargoylism: sex-linked transmission in nine males. Archives of Neurology and Psychiatry 63: 681

Broadhead D M, Kurk J M, Burt A J et al 1986 Full expression of Hunter's Disease in a female with an X-chromosome deletion, leading to non-random inactivation. Clinical Genetics 30/5: 392–398

Danes B S, Bearn A G 1967 Hurler's syndrome: a genetic study of clones in cell culture, with particular references to the Lyon hypothesis. Journal of Experimental Medicine 126: 509

DiFerrante N, Nichols B L 1972 A case of the Hunter syndrome with progeny. Johns Hopkins Medical Journal 130: 325

Harper P S, Bamforth S, Rees D, Roberts A, Upadhyaya M 1984 Chorion biopsy for prenatal testing in Hunter's syndrome. Lancet ii: 812–813

Kleijer W J, Mooy P D, Liebaers I, Van de Kamp J J P 1979 Prenatal monitoring for the Hunter syndrome. The heterozygote female fetus. Clinical Genetics 15: 113–117

Kleijer W J, Van Diggelen O P, Janse H C, Galjaard H, Dumez Y, Boue J 1984 First trimester diagnosis of Hunter syndrome on chorionic villi. Lancet ii: 472

Liebaers I, Neufeld E F 1976 Iduronate sulfatase activity in serum, lymphocytes, and fibroblasts: simplified diagnosis of the Hunter syndrome. Pediatric Research 10/7: 733

Liebaers I, Di Natale P, Neufeld E F 1977 Iduronate sulphatase in amniotic fluid: An aid in the prenatal diagnosis of the Hunter syndrome. Journal of Pediatrics 90: 423–425

Lykkelund C, Søndegaard F, Therkelsen A J, Tønnesen T, Rasmussen V, Mikkelsen M, Güttler F, Nyland M H 1983 Feasibility of first trimester prenatal diagnosis of Hunter syndrome. Lancet ii: 1147

Niermeijer M F, Sachs E S, Jahodova M, Tichelaar-Klepper C, Kleijer W J, Galjaard H 1976 Prenatal diagnosis of genetic disorders. Journal of Medical Genetics, 13: 182.

Noja A 1946 A sex-linked type of gargoylism. Acta Pediatrica 33: 267

Nwokoro N, Neufeld E F 1979 Detection of Hunter heterozygotes by enzymatic analysis of hair roots. American Journal of Human Genetics 31: 42–49

Pannone N, Gatti R, Lombardo C, Di Natale P 1986 Prenatal diagnosis of Hunter syndrome using chorionic villi. Prenatal Diagnosis 6: 207–210

Zlotogora J, Bach G 1984 Heterozygote detection in Hunter syndrome. American Journal of Medical Genetics 17: 661–665

Zlotogora J, Schaap T, Zeigler M, Bach G 1985 Hunter syndrome among Ashkenazi Jews in Israel; evidence for prenatal selection favouring the Hunter allele. Human Genetics 71: 329–332

## MPS III (Sanfilippo syndrome)

Ballabio A, Pallini R, di Natale P 1984 Mucopolysaccharidosis III B: Hybridization studies on fibroblasts from a mild case and fibroblasts from severe patients. Clinical Genetics 25: 191–195

Beratis N G, Sklower S L, Wilbur L, Matalon R 1986 Sanfilippo disease in Greece. Clinical Genetics 29: 129–132

Figura K von, Logering M, Mersmann G, Kresse H 1973 Sanfilippo B disease: serum assays for detection of homozygous and heterozygous individuals in three families. Journal of Paediatrics 83: 607

Gatti R, Barrone C, Durand P 1982 Sanfilippo type D disease: Clinical findings in two patients with a new variant of mucopolysaccharidosis III. European Journal of Pediatrics 138: 168–171

Gordon B A, Feleki V, Budreau C H, Tyler L 1975 Defective heparan sulphate metabolism in the Sanfilippo syndrome and assay of the defect in the assessment of the mucopolysaccharidosis patient. Clinical Biochemistry 8: 184–193

Harper P S, Laurence K M, Parkes A, Wusteman F, Kresse H, Figura K von, Ferguson-Smith M, Duncan D, Logan R, Hall F, Whiteman P 1974 Sanfilippo A disease in the fetus. Journal of Medical Genetics 11: 123

Klein U, Kress H, von Figura K 1978 Sanfilippo syndrome type C. Deficiency of acetyl-CoA:alpha-glucosaminide N-acetyltransferase in skin fibroblasts. Proceedings of National Academy of Science 75: 5185–5189

Kleijer W J, Janse H C, Vosters R P L 1986 First trimester diagnosis of mucopolysaccharidosis III A (Sanfilippo disease). New England Journal of Medicine 314: 185–186

Kresse H, Neufeld E F 1972 The Sanfilippo A corrective factor. Purification and mode of action. Journal of Biological Chemistry 247: 2164–2170

Kresse H, Figura K von, Bartsocas C 1976 Clinical and biochemical findings in a family with Sanfilippo disease type C. Clinical Genetics 10/6: 364

Kresse H, Paschke E, von Figura K, Gilberg W, Fuchs W 1980 Sanfilippo disease type D. Deficiency of N-acetylglucosamine-6-sulphate sulphatase required for heparan sulphate degradation. Proceedings of National Academy of Science 77: 6822–6826

Kresse H, Wiessmann U, Cantz M, Hall C W, Neufeld E F 1971 Biochemical heterogeneity of the Sanfilippo syndrome: Preliminary characterisation of two deficient factors. Biochemical and Biophysical Research Communication 42: 892–898

Liem K O, Giesberts M A H, Van de Kamp J J P, Van Pelt J F, Hooghwinkel G J M 1976 Sanfilippo B disease in two related sibships. Biochemical studies in patients, parents and sibs. Clinical Genetics 10/5: 273

Lubec G, Patrick A D, Nauer G, Mossman J 1985 Screening for Sanfilippo disease type A by infrared spectroscopy of hair. Lancet 1: 526–527

Maroteaux P, Frezal J, Tahbaz-Zadeh, Lamy M 1966 Une observation familiale d'oligophrenie polydystrophique. Journal de Génétique Humaine 15: 93

McKusick V, Neufeld E F, Kelly T E 1978 The mucopolysaccharide storage disease. In: Stanbury J B, Wyngaarden J B, Fredrickson D S, Goldstein J L, Brown M S (eds). The Metabolic Basis of Inherited Disease 5th edn. McGraw-Hill, New York p 1282–1307

Marsh J, Fensom A H 1985 4-methylumbelliferyl alpha-N-acetylglucosaminidase activity for diagnosis of Sanfilippo B disease. Clinical Genetics 27: 258–262

Matalon R, Dorfman A 1974 Sanfilippo A syndrome. Sulfamidase deficiency in cultured skin fibroblasts and liver. Journal of Clinical Investigation 54: 907

Mossman J, Young E P, Patrick A D, Fensom A H, Ellis M, Benson P F, Der Kaloustian V M 1983 Prenatal tests for Sanfilippo disease type B in four pregnancies. Prenatal Diagnosis 3: 347–350

O'Brien J S 1972 Sanfilippo syndrome: profound deficiency of alpha-acetylglucosaminidase activity in organs and skin fibroblasts from type B patients. Proceedings of the National Academy of Sciences of the United States of America 69: 1720

Petersen E M 1986 Sanfilippo syndrome type C: The first known case in South Africa. South African Medical Journal 69: 63–68

Sanfilippo S J, Podosin R, Langer L O, Good R A 1963 Mental retardation associated with acid mucopolysacchariduria heparitin (sulfate type). Journal of Paediatrics 63: 837

Spranger J 1972 The systemic mucopolysaccharidoses. Ergebnisse der inneren Medizin und Kinderheilkunde 32: 165

Uvebrant P 1985 Sanfilippo type C syndrome in two sisters. Acta Paediatrica Scandinavica 74: 137–139

Van de Kamp J J P, Niermeyer M F, von Figure K, Giesberts M A H 1981 Genetic heterogeneity and clinical variability in the Sanfilippo syndrome (types A, B and C) Clinical Genetics 20: 152–160

Van de Kamp J J P, Van Pelt J F, Liem K O, Giesberts M A H, Niepoth L T M, Staalman C R 1976 Clinical variability in Sanfilippo B disease: a report on six patients in two related sibships. Clinical Genetics 10/5: 279

Von Figura K, Kresse H 1972 The Sanfilippo B corrective factor. A N-acetyl-alpha-D glucosaminidase. Biochemical and Biophysical Research Communications 48: 262–269

Whiteman P, Young E 1977 The laboratory diagnosis of Sanfilippo disease. Clinica Chimica Acta 76: 139–147

### MPS IV (Morquio syndrome)

Arbisser A I, Donnelly K A, Scott jr C I, DifFerrante N, Singh J, Stevenson R E, Aylesworth A S, Howell R R 1977 Morquio-like syndrome with beta-galactosidase deficiency and normal hexosamine sulfatase activity: Mucopolysaccharidosis IV B. American Journal of Medical Genetics 1: 195–205

Beck M, Beighton P, Petersen E M, Spranger J 1987 Morquio disease type B in three siblings. South African Medical Journal 72: 704–707

Beck M, Glössl J, Grubisic A, Spranger J 1986 Heterogeneity of Morquio disease. Clinical Genetics 29: 325–331

Beighton P, Craig J 1973 Atlanto-axial subluxation in the Morquio syndrome. Journal of Bone and Joint Surgery 55B/3: 478

Brailsford J F 1929 Chondro-osteo-dystrophy: roentgenographic and clinical features of a child with dislocation of vertebrae. American Journal of Surgery 7: 404

Figura J von, van de Kamp J J, Niermeijer M F 1982 Prenatal diagnosis of Morquio's disease type A. Prenatal Diagnosis 2: 67–69

Gadbois P, Moreau J, Laberge C 1973 La maladie de Morquio dans la province de Quebec. L'Union Médicale du Canada 102: 602

Glössl J, Maroteaux P, Di Natale P, Kresse H 1981 Different properties of residual N-acetylgalactosamine-6-sulphate sulphatase in fibroblasts from patients with mild and severe forms of Morquio disease type A. Pediatric Research 15: 976–978

Glössl J, Kresse H, Mendla K, Cantz H, Rosenkranz W 1984 Partial deficiency of glycoprotein neuraminidase in some patients with Morquio disease type A. Pediatric Research 18: 302–305

Groebe H, Krins M, Schmidberger H, von Figura K, Harzer K, Kresse H, Paschke E, Sewell A, Ullrich K 1980 Morquio syndrome (mucopolysaccharidoses IVB) associated with beta-galactosidase deficiency. Report of two cases. American Journal of Human Genetics 32: 258–272

Lowry R B, Snyder F F, Wesenberg R L, Machin G A, Applegarth D A, Morgan K, Carter R J, Toone J R, Holmes T M, Dewar R D 1985 Morquio syndrome (MPS IVA) and hypophosphatasia in a Hutterite kindred. American Journal of Medical Genetics 22: 463–475

McKusick V A 1975 Morquio Syndrome, non-keratosulfate excreting type. Mendelian Inheritance in Man, 4th Edn, Johns Hopkins University Press, Baltimore, p 503

Maroteaux P, Stanescu V, Stanescu R, Kresse H, Hors-Cayla M C 1982 Hétérogeneite des formes frustes de la maladie de Morquio. Archives Française de Pédiatrie 39: 761–765

Matalon R, Arbogast B, Justice P, Brandt I, Dorfman A 1974 Morquio's syndrome: deficiency of a chondroitin sulfate N-acetyle hexosamine sulfate sulfatase. Biochemical and Biophysical Research Communications 61: 759

Morquio L 1929 Sur une forme de dystrophie osseeuse familiale. Bulletins de la Société de Pédiatrie de Paris 27: 145

O'Brien J S, Gugler E, Giedion A, Weissmann R, Herschkowitz N, Meier C, Le Roy J 1976 Spondylo-epiphyseal dysplasia, corneal clouding, normal intelligence and acid beta-galactosidase deficiency. Clinical Genetics 9: 495–504

Spranger J 1977 Beta Galactosidase and the Morquio syndrome. American Journal of Medical Genetics 1: 207–209

Trojak J E, Ho C K, Roesel R A, Levin L S, Kopits S E, Thomas G H, Toma S 1980 Morquio-like syndrome (MPS IVB) associated with deficiency of a beta-galactosidase. Johns Hopkins Medical Journal 146: 75–79

Van Gemund J J, Giesberts M A, Eerdmans R F, Blom W, Kleijer W J 1983 Morquio B disease, spondylo-epiphyseal dysplasia associated with acid beta-galactosidase deficiency. Report of three cases in one family. Human Genetics 64: 50–54

Yuen M, Fensom H 1985 Diagnosis of classical Morquio disease: N-acetyl-galactosamine 6-sulphate sulphatase activity in cultured fibroblasts, leukocytes, amniotic cells and chorionic villi. Journal of Inherited Metabolic Diseases 8: 80–86

**MPS VI (Maroteaux-Lamy syndrome)**

Beratis N G, Turner B M, Weiss R, Hirschhorn K 1975 Arylsulfatase B deficiency in Maroteaux-Lamy syndrome. Cellular studies and carrier identification. Pediatric Research 9: 475

Black S, Pelias M Z, Miller J B, Blitzer M, Shapira E 1986 Maroteaux-Lamy syndrome in a large consanguineous kindred: Biochemical and immunological studies. American Journal of Medical Genetics 25: 273–279

Gatti R, Lombardo C, Filocamo M, Borrone C, Porro E 1985 Comparative study of 15 lysosomal enzymes in chorionic villi and cultured amniotic fluid cells. Prenatal Diagnosis 5: 329–336

Maroteaux P, Leveque B, Marie J, Lamy M 1963 Une nouvelle dysostose avec elimination urinaire de chondroitine sulphate B. Presse Médicale 71: 1849

O'Brien J F, Cantz M, Spranger J 1974 Maroteaux-Lamy disease (mucopolysaccharidosis VI), subtype A; deficiency of N-acetyle galactosamine-4-sulfatase. Biochemical and Biophysical Research Communications 60: 1170

Spranger J, Koch F, McKusick V A, Natzschka J, Wiedemann H R, Zellweger H 1970 Mucopolysaccharidosis VI (Maroteaux-Lamy disease). Helvetica Paediatrica Acta 25: 337

Spranger J W, Langer L O, Wiedemann H R 1974 Mucopolysaccharidosis VI. Bone Dysplasias. Gustav Fischer Verlag, Stuttgart, p 166

**MPS VII (Beta-glucuronidase deficiency)**

Beaudet A L, DiFerrante N M, Ferry G D, Nichols B L, Mullins C W 1975 Variation in phenotypic expression of β-glucuronidase deficiency. Journal of Paediatrics 86: 388

Gehler J, Cantz M, Tolksdorf M, Spranger J 1974 Mucopolysaccharidosis VII: β-glucuronidase deficiency. Humangenetika 23: 149

Guibaud P, Maire I, Gopdon R, Teyssier G, Zabot M T, Mandon G 1979 Mucopolysaccharidosis type VII par deficit en beta-glucuronidase. Etude d'une famille. Journal de Génétique Humaine 27: 29–42

Pfeiffer R A, Kresse H, Baumer N, Sattinger E 1977 Beta-glucuronidase deficiency in a girl with unusual clinical features. European Journal of Pediatrics 126: 155–161

Poenaru L, Castelnau L, Mossman J, Boue J, Dreyfus J C 1982 Prenatal diagnosis of a heterozygote for mucopolysaccharidosis type VII. Prenatal Diagnosis 2: 251–256

Sly W S, Quinton B A, McAllister W H, Rimoin D 1973 β-glucronidase deficiency: report of clinical, radiologic and biochemical features of a new mucopolysaccharidosis. Journal of Paediatrics 82: 249

**B. Mucopolipidoses and related disorders**
*Preamble*

Grossman H, Dorst J P 1973 The mucopolysaccharidoses and mucolipidoses. Progress in Paediatric Radiology 4: 495

Hieber V, Distler J, Jourdian G W, Schmickel R 1975 Accumulation of 35 S-mucopolysaccharides in cultured mucolipidosis cells. Birth Defects: Original Article Series 11/6: 307

Maroteaux P, Humbel R 1976 Les oligosaccharidoses: Un concept nouveau. Archives Françaises de Pédiatrie 33: 641–643

Spranger J, Wiedemann H R 1970 The genetic mucolipidoses: diagnosis and differential diagnosis. Humangenitika 9: 113

Strecker G 1977 Glycoproteins et glycoproteinoses. In: Farriaux J P (ed) Les Oligosaccharidoses. Crouant et Roques, Lille, p 13–30

## True glycoproteinoses
*Sialidosis*

Andria G, Strisciuglio P, Pontarelli G, Sly W S, Dodson W E 1981 Infantile neuraminidase and beta-galactosidase deficiencies with mild clinical courses. Perspectives in Inherited Metabolic Diseases 4: 379–395

Aylsworth A S, Thomas G H, Hood J L, Malouf N, Libert J 1980 A severe infantile sialidosis: Clinical, biochemical and microscopic features. Journal of Pediatrics 96/4: 662–668

Bach G, Zeigler M, Schaap T, Kohn G 1979 Mucolipidosis type IV: Ganglioside sialidase deficiency. Biochemical and Biophysical Research Communications 90: 1341–1347

Ben-Yoseph Y, Hahn L C, Nadler H L 1981 Mucolipidosis Type IV: A Km defect in ganglioside neuraminidase, presented at 6th International Congress of Human Genetics, Jerusalem, Israel

Berard M, Toga M, Bernard R, Dubois R, Mariani R, Hassoun J 1968 Pathological findings in one case of neuronal and mesenchymal storage disease. Its relationship to lipidoses and to mucopolysaccharidoses. Pathology (Europe) 3: 172

Berman E R, Livni N, Shapira E, Merin S, Levij I S 1974 Congenital corneal clouding with abnormal systemic storage bodies: a new variant of Mucolipidosis. Journal of Pediatrics 84: 519–526

Cantz M, Messer H 1979 Oligosaccharide and ganglioside neuraminidase activities of mucolipidosis I (sialidosis) and mucolipidosis II (I-cell disease) fibroblasts. European Journal of Biochemistry 97: 113

Cantz M, Gehler J, Spranger J 1977 Mucolipidosis I: increased sialic acid content and deficiency of an alpha-N-acetylneuraminidase in cultured fibroblasts. Biochemical and Biophysical Research Communications 74: 732

Durand P, Gatti R, Cavalieri S, Borrone C, Tondeur M, Michalski J C, Strecker G 1977 Sialidosis (mucolipidosis I). Helvetica Paediatrica Acta 32: 391

Kelly T E, Bartoshesky L, Harris D J, McCauley R G, Feingold M, Schott G 1981 Mucolipidosis I (Acid neuraminidase deficiency). American Journal of Diseases of Children 135: 703

Kohn G, Sekeles E, Arnon J, Ornoy A 1982 Mucolipidosis IV: Prenatal diagnosis by electron microscopy. Prenatal Diagnosis 2: 301–307

Kohn G, Livni N, Ornoy A, Sekeles E, Beyth Y, Legum C, Bach G, Cohen M M 1977 Prenatal diagnosis of Mucolipidosis IV by electron microscopy. Journal of Pediatrics 90: 62–66

Loonen M C B, Reuser A J J, Visser P, Arts W F M 1984 Combined sialidase (neuraminidase) and beta-galactosidase deficiency. Clinical Genetics 26: 139–149

Lowden J A, O'Brien J S 1979 Sialidosis: A review of human neuraminidase deficiency. American Journal of Human Genetics 31: 1–18

Maroteaux P, Humbell R, Strecker G, Michalski J C, Mande R 1978 Un nouveau type de sialidose avec atteinte rénale: La nephrosialidose. Archives Françaises de Pédiatrie 35: 819

Mueller O T, Shows T B 1982 Human Beta-galactosidase and Alpha-neuraminidase deficient mucolipidosis: genetic complementation analysis of the neuraminidase deficiency. Human Genetics 60: 158–162

Riedel K G, Zwaan J, Kenyon K R et al 1985 Ocular abnormalities in mucolipidosis IV. American Journal of Ophthalmology 99: 125–136

Sakai M, Akago M, Yokoi S, Suzuki Y, Higuchi M 1982 Six adult cases of beta-galactosidase-neuraminidase deficiency in three families. Psychiatry and Neurology, (Japan) 84: 917–938

Sakuraba H, Suzuki Y, Akagi M, Sakai M, Amano N 1983 Beta-galactosidase-neuraminidase deficiency (Galactosialidosis): clinical, pathological and enzymatic studies in a postmortem case. Annals of Neurology 13: 497–503

Sekeles E, Ornoy A, Cohen R, Kohn G 1978 Mucolipidosis IV: Fetal and placental pathology. Monographs in Human Genetics 10: 47–50

Spranger J, Wiedeman N, Tolksdorf M, Graucob E, Caesar R 1968 Lipomucopolysaccharidose. Eine neue Speicherkrankheit. Zeitschrift fur Kinderheilkunde 103: 285

Spranger J, Gehler J, Cantz M 1977 Mucolipidosis 1 — A sialidosis. American Journal of Medical Genetics 1: 21–29

Staalman C R, Becker H D 1984 Mucolipidosis I. Roentgenographic follow-up. Skeletal Radiology 12: 153–161

*Fucosidosis*
Durand P, Borrone C, Della Cella G 1969 Fucosidosis. Journal of Paediatrics 75: 665
Durand P, Gatti R, Borrone C, Costantino G, Cavalieri S, Filocamo M, Romeo G 1979 Detection of carriers and prenatal diagnosis for fucosidosis in Calabria. Human Genetics 51: 195–201
Matsuda I, Arashima S, Oka Y, Mitsuyama T, Ariga S, Ikeuchi T, Ichida T 1975 Prenatal diagnosis of fucosidosis. Clinica Chimica Acta 63: 55–60

*Mannosidosis*
Cooper A, Sardharwalla I B, Roberts M M 1986 Human beta-mannosidase deficiency. New England Journal of Medicine 315: 1231
Ockerman P A 1967 A generalised storage disorder resembling Hurler's syndrome. Lancet 2: 239
Spranger J, Gehler J, Cantz M 1976 The radiographic features of mannosidosis. Radiology 119/2: 401
Wenger D A, Sujansky E, Fennessey P V, Thompson J H 1986 Human beta mannosidase deficiency. New England Journal of Medicine 315: 1201–1205

*Aspartylglycosaminuria*
Autio S 1972 Aspartyglycosaminuria. Analysis of 34 patients. Journal of Mental Deficiency Research Monograph Series 1: 1–39
Maury C P J 1982 Aspartylglycosaminuria: an inborn error of glycoprotein catabolism. Journal of Inherited Metabolic Disease 5: 192–196
Pollitt R J, Jenner F A, Merskey H 1968 Aspartylglycosaminuria: an inborn error of metabolism associated with mental defect. Lancet ii: 253–355

**Heteroglycanoses with altered glycoprotein catabolism**
*I-cell disease*
Aula P, Rapola J, Autio S, Raivio K, Karjalainen O 1975 Prenatal diagnosis and fetal pathology of I cell disease (mucolipidosis type II). Journal of Paediatrics 87/2: 221
Gehler J, Cantz M, Stoeckenius M, Spranger J 1976 Prenatal diagnosis of mucolipidosis II (I cell disease). European Journal of Paediatrics 122/3: 201
Hanai J, Leroy J, O'Brien J S 1971 Ultrastructure of cultured fibroblasts in I-cell disease. American Journal of Diseases of Children 122: 34–38
Leroy J G, Demars R I, Opitz J M 1969 Skeletal dysplasias. The Clinical Delineation of Birth Defects. National Foundation, New York, p 174
Matsuda I, Arashima S, Mitsuyama T, Oka Y, Ikeuchi T, Kaneko Y, Ishikawa M 1975 Prenatal diagnosis of I cell disease. Humangenetika 30/1: 69
Owada M, Nishijao S T, Kitagawa T 1980 Prenatal diagnosis of I-cell disease by measuring altered x-mannosidase activity in amniotic fluid. Journal of International Metabolism Disease 3: 117–121
Okada S, Owada M, Sakiyama T, Yutaka T, Ogawa M 1985 I-cell disease: clinical studies of 21 Japanese cases. Clinical Genetics 28: 207–215
Patriguin H B, Kaplan P, Kind H P, Biedion A 1977 Neonatal mucolipodosis II (I-cell disease): clinical and radiological features in three cases. American Journal of Roentgenology 129: 37–43
Reitman M L, Varki A, Kornveld S 1981 Fibroblasts from patients with I-cell disease and pseudo-Hurler polydystrophy are deficient in UDP-N-acetylglucosamine: glycoprotein-N-acetylglucosaminylphospho-transferase activity. Journal of Clinical Investigation 67: 1574–1579
Tondeur M, Vamos-Hurwitz E, Mockel-Pohl S, Dereunie J P, Cremers N, Loeb J 1971 Clinical, biochemical and structural studies in a case of chondrodystrophy presenting the I-cell phenotype in culture. Journal of Pediatrics 79: 366–378

*Pseudo-Hurler Polydystrophy*
Kelly T E, Thomas G H, Taylor H A, McKusick V A, Sly W S, Glaser J H, Robinson M, Luzzati L, Espiritu C, Feingold M, Bull M J, Ashenhurst E M, Ives E J 1975 Mucolipidosis III (pseudo-Hurler polydystrophy): Clinical and laboratory studies in a series of 12 patients. Johns Hopkins Medical Journal 137: 156–175
Nolte K, Spranger J 1976 Early skeletal changes in mucolipidosis III. Annals of Radiology 19: 151

*GMI Gangliosidoses*

Groebe H, Krins M, Schmidberger H, Von Figura K, Harzer K, Kresse H, Paschke E, Sewell A, Ullrich K 1980 Morquio syndrome (mucopolysaccharidosis IV B) associated with beta-galactosidase deficiency. Report of two cases. American Journal of Human Genetics 32: 258–265

Hoogeveen A T, Graham-Kawashima M, d'Azzo A, Galjaard M 1984 Processing of human beta-galactosidase in GM1-gangliosidosis and Morquio B syndrome. Journal of Biological Chemistry 259/3: 1974–1977

O'Brien J S, Gugler E, Giedion A, Weissman U, Herschkowitz N, Neier C, Leroy J G 1976 Spondyloepiphyseal dysplasia, corneal clouding, normal intelligence and beta-galactosidase deficiency. Clinical Genetics 9: 495–504

Stevenson R E, Taylor H A, Parks S E 1978 Beta-galactosidase deficiency: prolonged survival in three patients following early central nervous system deterioration. Clinical Genetics 13: 305–313

Taylor H A, Stevenson R E, Parks S E 1980 Beta-galactosidase deficiency: studies of two patients with prolonged survival. American Journal of Medical Genetics 5: 235–245

Van der Horst G T J, Kleijer W J, Hoogeveen A T, Huijmans J M G, Blom W, van Diggelen O P 1983 Morquio B syndrome: a primary defect in beta-galactosidase. American Journal of Medical Genetics 16: 261–275


*Sialuria*

Baumkotter J, Cantz M, Mendla K et al 1985 N-acetylneuraminic acid storage disease. Human Genetics 71/2: 155–159

Renlund M 1984 Clinical and laboratory diagnosis of Salla disease in infancy and childhood. Journal of Pediatrics 104: 232–236

Renlund M, Chester M A, Lundblad A, Parkkinen J, Krusius T 1983 Free N-acetylneuraminic acid in tissues in Salla disease and the enzymes involved in its metabolism. European Journal of Biochemistry 130: 39–45

Renlund M, Chester M A, Lundblad A, Aula P, Raivio K O, Autio S, Koskela S L 1979 Increased urinary excretion of free N-acetylneuraminic acid in 13 patients with Salla disease. European Journal of Biochemistry 101: 245–250

Renlund M, Aula P, Raivio O, Autio S, Sainio K, Rapola J, Koskela S L 1983 Salla disease: A new lysosomal storage disorder with disturbed sialic acid metabolism. Neurology 33: 57–66

Vamos E, Liebert J, Elkhazen N, Jaunioux E, Hustin J, Wilkin P, Baumkotter J, Mendla K, Cantz M, Strecker G 1986 Prenatal diagnosis and confirmation of infantile sialic acid storage disease. Prenatal Diagnosis 6: 437–446

Virtanen I, Ekblom P, Laurila P, Nordling S, Raivio K O, Aula P 1980 Characterisation of storage material in cultured fibroblasts by specific lectin binding in lysosomal storage diseases. Pediatric Research 14: 1199–1203


*Multiple sulphatase deficiency*

Austin J U 1973 Studies in metachromatic leukodystrophy. XII Multiple sulfatase deficiency Archives of Neurology 28: 258–264

Benson P F, Fensom A H 1985 Genetic Biochemical Disorders. Oxford University Press, Oxford, p 123–126

Burch M, Fensom A H, Jackson M, Pitts-Tucker T, Congdon P J 1986 Multiple sulphatase deficiency presenting at birth. Clinical Genetics 30: 409–415

Burk R D, Valle D, Thomas G H, Miller C, Moser A, Moser H, Rosenbaum K N 1984 Early manifestations of multiple sulfatase deficiency. Journal of Pediatrics 104: 574–578

Rampini S, Isler W, Baerlocher K, Bischoff A, Ulrich J, Pluss H J 1970 Die Kombination von metachromatischer Leukodystrophie und Mukopolysaccharidoses als selbstandiges Krankheitsbild (Mukosulfatidose). Helvetica Paediatrica Acta 25: 436–446

Vamos E, Liebaers I, Bousard N, Libert J, Perlmutter N 1981 Multiple sulphatase deficiency with early onset. Journal of Inherited Metabolic Diseases 4: 103–104

## C. Sphingolipidoses

### Gaucher disease

Beighton P, Sacks S 1974 Gaucher's disease in Southern Africa. South African Medical Journal 48: 1295

Beutler E, Kuhl W 1970 The diagnosis of the adult type of Gaucher's disease and its carrier state by demonstration of deficiency of beta-glucosidase activity in peripheral blood leukocytes. Journal of Laboratory and Clinical Medicine 76: 747

Devine E A, Beighton P, Petersen E M, Desnick R J 1982 Genetic heterogeneity in Type I Gaucher disease. In: Gaucher Disease: A Century of Delineation and Research. Desnick R J, Gatt S, Grabowski G A (eds) Alan R Liss Inc, New York, p 495–510

Fried K 1973 Population study of chronic Gaucher's disease. Israel Journal of Medical Science 9: 1396

Goldblatt J, Beighton P 1979a Gaucher disease in the Afrikaner population of South Africa. South African Medical Journal 55: 209–210

Goldblatt J, Beighton P 1979b Gaucher disease in South Africa. Journal of Medical Genetics 16: 302–305

Goldblatt J, Sacks S, Beighton P 1978 Orthopaedic aspects of Gaucher disease. Clinical Orthopaedics and Related Research 137: 208–214

Grabowski G A, Goldblatt J, Dinur T, Kruse J, Svennerholm L Gatt S, Desnick R J 1985 Genetic heterogeneity in Gaucher disease: physiokinetic and immunologic studies of the residual enzyme in cultured fibroblasts. American Journal of Medical Genetics 21: 529–549

Myers H S, Cremin B J, Beighton P, Sacks S 1975 Chronic Gaucher's disease: radiological findings in 17 South African cases. British Journal of Radiology 48: 465

Sorge J, Gelbart T, West C et al 1985 Heterogeneity in type I Gaucher disease demonstrated by restriction mapping of the gene. Proceedings of National Acadamy of Science 82: 5442–5445

# 12

# Craniofacial dysostoses

In this group of disorders, the major abnormalities are found in the bones of the skull and face. Categorisation is based upon the pattern of craniofacial changes and the extent of involvement of the extremities and other systems. Gorlin et al (1975) and Cohen (1975) have emphasised that the variations and combinations which may be present in these conditions preclude any simple classification.

1. Craniostenosis
2. Craniofacial dysostosis (Crouzon)
3. Acrocephalosyndactyly (Apert)
4. Acrocephalopolysyndactyly (Carpenter)
   a. Greig cephalopolysyndactyly syndrome
5. Mandibulofacial dysostosis (Treacher Collins)
   a. Nager acrofacial dysostosis
6. Mandibular hypoplasia (Pierre Robin)
7. Cerebrocostomandibular syndrome
8. Mandibuloacral dysplasia
9. Oculomandibulofacial syndrome (Hallermann-Streiff)
10. Craniofrontonasal dysplasia
    a. Acrocallosal syndrome

CRANIOSTENOSIS

Craniostenosis is the result of premature closure of the sutures of the skull. Cranial abnormalities of this type were recognised even in the distant past. For instance, Shou Lao, the Chinese God of Long Life was conventionally depicted with a tower-shaped skull, while Homer mentioned in the Iliad that the lame Thersites had a peaked head.

The abnormal configuration in craniostenosis is determined by the anatomical distribution of the sutures which are involved, the sequence in which the sutures close and by compensatory growth in non-affected regions. For this reason, in the familial forms of the disorder, there is often considerable variation in the manifestations between affected members of

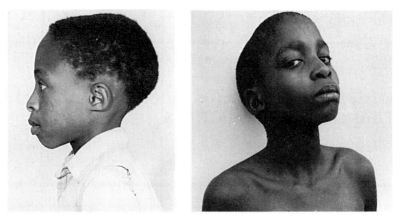

**Fig. 12.1** (left) Craniostenosis: a boy with a long, narrow head (scaphocephaly).

**Fig. 12.2** (right) Craniostenosis: a girl with a pointed skull (acrocephaly).

the same kindred. A number of descriptive designations have been applied to the various forms of skull configuration which result from craniostenosis.

| | |
|---|---|
| Scaphocephaly | — a boat-shaped or long narrow skull |
| Acrocephaly or Oxycephaly | — a sharp or pointed skull |
| Turricephaly | — a tower-shaped, high, broad and short head |
| Plagiocephaly | — asymmetry of the skull |
| Kleeblattschädel | — a trilobal or clover leaf skull (see thanatophoric dwarfism, Ch. 3). |

Microcephaly and hydrocephaly are not included in this account as these abnormalities are secondary to alteration in the dimensions of the contents of the cranium and are not primary disturbances of growth of the bones of the skull. Craniostenosis may be associated with various skeletal and visceral abnormalities in a number of genetic syndromes. It can also result from intra-uterine or perinatal trauma or infection. The pathogenesis of the craniostenoses has been reviewed in detail by Lepintre & Renier (1975).

The clinical importance of craniostenosis lies in the fact that growth and development of the brain may be impaired. Proptosis and visual disturbance are common complications and the facies may become grotesque. Radiographically, alteration in the shape of the cranium and obliteration of the sutures is apparent. Compensatory widening of non-affected sutures may occur, and 'digital' markings are often evident in the calvarium. Advances in surgical technology have had considerable impact upon the management of patients with conditions of this type (Archer et al 1974, Giuffre et al 1975).

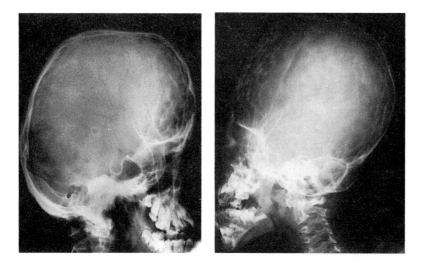

**Fig. 12.3** (left) Craniostenosis: a short high skull (turricephaly).

**Fig. 12.4** (right) Craniostenosis: the sutures are obliterated and digital markings are evident on the calvarium.

*Genetics*

The actual prevalence is unknown but isolated craniostenosis is probably relatively common. Gordon (1959) found six cases of craniostenosis when he undertook a survey of 600 African Negro children. The incidence of non-syndromic craniostenosis in a newborn population was 0.6 per 1000 live births (Shuper et al 1985). Craniostenosis does not always have a genetic basis. Indeed, in a series of 159 patients in whom craniostenosis had been treated surgically, only about 5% had affected kin (Shillito & Matson 1968). A comparable figure was obtained in a large scale family survey undertaken by Carter et al (1982).

Generation to generation transmission of craniostenosis, consistent with dominant inheritance, has been reported by Murphy (1953), Gordon (1959), Bell et al (1961), and Nance & Engel (1967). Several of these reports were presented under the designation 'sarcocephaly'. However, there is often intrafamilial disparity in the anatomical abnormality, and the actual shape of the skull is probably not of great diagnostic importance.

Hunter & Rudd (1976) investigated 214 children with sagittal synostosis and concluded that inheritance was multifactorial. The occurrence of synostosis of the lambdoid suture in four siblings was documented by Lohse et al (1985). Three kindreds in which premature closure of the coronal suture was inherited as an autosomal dominant were reported by Kosnik et al (1975). The existence of an autosomal recessive form of craniostenosis is indicated by reports of affected sibs with normal parents (Duguid 1929,

Gaudier et al 1967, Armandares 1970). Further evidence was provided by the recognition of many examples in an inbred religious isolate, the Amish of Ohio, USA (Cross & Opitz 1969).

A cytogenetic basis for craniosynostosis was proposed by Motegi et al (1985) when they identified an interstitial deletion of the short arm of chromosome 7 in an affected male infant. By means of deletion mapping the critical segment was localised to the mid portion of band 7p 21.

The parents of a child with uncomplicated craniostenosis may well seek counselling concerning the possibility of recurrence. In the absence of any genetic clues, such as affected kin, parental consanguinity or advanced paternal age, it is probably reasonable to give a relatively low risk estimate.

## Craniostenosis — skeletal dysplasia syndromes

In addition to the uncomplicated inherited forms there are a number of atypical combinations of craniostenosis and abnormalities of the extremities. These conditions probably represent rare genetic entities.

1. Gerold (1959) reported a brother and sister with turricephaly and radial aplasia.

2. Lowry (1972) described two brothers with craniostenosis and congenital absence of the fibula. Their unaffected parents were consanguineous.

3. Armendares et al (1975) studied three brothers with craniostenosis, short stature and retinitis pigmentosa, and speculated that inheritance of this new entity was probably autosomal recessive.

4. Sensenbrenner et al (1975) described a brother and sister with premature closure of the sagittal suture, abnormal facies and hair and shortening of the fibula and phalanges.

5. Persons with trigonocephaly have a triangular shaped head, due to frontal constriction and a wide bitemporal diameter. Trigonocephaly can occur in isolation, in certain chromosomal disorders and as part of the arrhinencephaly-holoprosencephaly malformation complex (Currarino & Silverman 1960). The characteristic cranial configuration is a feature of the Opitz C syndrome, in which mental retardation, a typical facies, stunted stature and polysyndactyly are additional manifestations (Opitz et al 1969, Antley et al 1981, Sargent et al 1985). Inheritance is autosomal recessive.

A mother and son with trigonocephaly, normal mentality and minor digital malformations were documented by Hunter et al (1976) and an X-linked variety of trigonocephaly with developmental delay was reported by Say & Meyer (1981). Another autosomal dominant form was proposed by Frydman et al (1984) when they recognised six affected persons in three generations of a Yemenite family.

6. Jackson et al (1976) studied a large Armish kindred with craniostenosis, midfacial hypoplasia and foot abnormalities. This condition is apparently a private syndrome, which is inherited as an autosomal dominant. It

seems to be a separate disorder from the autosomal recessive form of simple craniostenosis, which is also present in the Amish.

## CRANIOFACIAL DYSOSTOSIS (Crouzon* syndrome)

The main stigmata of the Crouzon syndrome or Crouzon craniofacial dysostosis are craniostenosis with midfacial hypoplasia, hypertelorism, proptosis and nasal beaking. Prominence of the eyes produces a 'frog face' appearance. Dental malocclusion and deafness are common concomitants. Although patients resemble each other, there is great diversity in the severity of the disorder, even in members of the same kindred. Radiographic changes include obliteration of sutures, digital impressions on the calvarium, shallowness of the orbits and small paranasal sinuses. Unusual associations of uncertain significance include acanthosis nigricans (Reddy et al 1985, Suslak et al 1985) and a haemangiomatous anomaly of the skull (Alpers & Edwards 1985).

In severe cases the facial appearance may be grotesque but intelligence is normal and psychosocial problems can arise (Cohen et al 1985). Operative intervention can produce considerable improvement but as such procedures

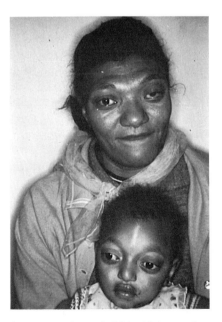

**Fig. 12.5** Crouzon syndrome: a mother and child with proptosis, nasal beaking and hypertelorism. The extreme variability of the phenotype is of great importance in genetic counselling. (Courtesy of Dr S. Zieff, Cape Town.)

* Octave Crouzon (1874–1938) was a neurologist at the Saltpêtrière Hospital, Paris.

are lengthy and complex, comprehensive pre-operative assessment is necessary.

### Genetics

The familial occurrence was mentioned in the initial case description of Crouzon (1912). Since that time, there have been many reports which are compatible with autosomal dominant inheritance. Kindreds with transmission through several generations were investigated by Shiller (1959) and Vulliamy & Normandale (1966). Appreciation of the potential variability of the stigmata is of importance in family studies and genetic counselling.

Franceschetti (1953), using the designation 'pseudo-Crouzon disease', described a disorder in which craniostenosis with prominent calvarial markings was associated with a normal facies. Dolivo & Gillieron (1955) reported a kindred in which members of four generations had the same condition. Inheritance is probably autosomal dominant.

### ACROCEPHALOSYNDACTYLY (Apert* syndrome)

Apert (1906), using the term 'acrocephalosyndactyly', described the association of craniostenosis and severe syndactyly. In a review of 54 cases, Blank (1960) distinguished 'typical' and 'atypical' forms of the disorder. The atypical forms were subsequently divided into a number of separate entities under the eponyms Vogt, Chotzen, Pfeiffer and Summitt (vide infra). Spranger et al (1974) and McKusick (1975) have also applied numerical designations to these conditions. However, there is overlap in these alternative classifications and some controversy concerning the assignment of certain patients to particular categories. Nevertheless, there is general agreement that the typical form or the 'Apert syndrome' should be designated acrocephalosyndactyly type I.

The major clinical features of the Apert syndrome are a high forehead, flat occiput, mid-facial hypoplasia and fusion of the second, third and fourth fingers and toes. Involvement of the first digits is variable. Other inconsistent abnormalities include mild shortening of the upper limbs, limitation of movement of large joints and cardiac, gastro-intestinal and renal anomalies. Ocular problems are not uncommon (Krueger & Ide 1974). Many patients are mentally defective. The occurrence of a frontal encephalocele in an infant with Apert syndrome has been recorded (Waterson et al 1985).

---

* Eugene Apert (1868–1940) was a senior paediatrician at the Hôpital des Enfants Malades, Paris. He had a special interest in congenital deformities and genetic disease.

*Genetics*

More than 200 cases of the Apert syndrome have been reported and an incidence of 1 in 160 000 newborn has been estimated (Blank 1960). The overwhelming majority of patients have been sporadic, with an equal sex frequency. Blank (1960) noted that the mean paternal age was increased. No sets of affected sibs have been encountered and there is no increased parental consanguinity. For these reasons, the evidence points to dominant inheritance, with the sporadic patients representing new mutations. Support for the dominant hypothesis is provided by descriptions of affected mothers

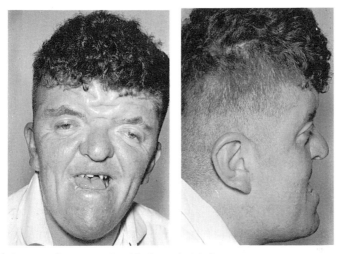

**Fig. 12.6** Apert syndrome: a male with the typical facies.

**Fig. 12.7** (right) Apert syndrome: the shape of the head and the midfacial hypoplasia are characteristic features.

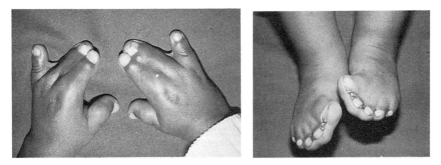

**Fig. 12.8** (left) Apert syndrome: The characteristic mitten appearance of the hands, involvement of the 1st and 5th fingers is often partial.

**Fig. 12.9** (right) Apert syndrome: syndactyly of the toes.

and daughters (Hoover et al 1970, Roberts & Hall 1971, Bergstrom et al 1972). In view of the mental deficiency and the severity of the malformations, it is not surprising that no male patient is known to have procreated.

Dodson et al (1970) found an A-C translocation in an affected female and mentioned three other reports of abnormalities of A-group chromosomes. On the basis of these findings, it is possible that the abnormal gene in the Apert syndrome is located on a chromosome in this group.

Prenatal diagnosis of the Apert syndrome by fetoscopy at the 17th week of a pregnancy of an affected mother has been achieved by Leonard et al (1982).

**Other forms of acrocephalosyndactyly**

There is considerable variation in the stigmata of these uncommon disorders and the classification is neither exact nor complete. The problems involved in accurate diagnosis and categorisation have been discussed by Rochiccioli et al (1974). Attempts have been made to separate these conditions in terms of metabolic parameters based upon sarcosine loading (Minami et al 1975).

There is evidence for syndromic homogeneity, with great variation in phenotypic expression. For instance, in a study of 88 affected persons in a large Amish kindred, Jackson et al (1976) encountered the entire spectrum of acrocephalosyndactyly, excepting the classical Apert syndrome. The problems of syndromic definition of the acrocephalosyndactyly conditions has been reviewed by Bull et al (1979).

1. In the Vogt or Apert-Crouzon syndrome, the facies are said to resemble those of Crouzon syndrome, while involvement of the hands is of lesser degree than in the classical Apert form. All the reported patients have been sporadic. It is possible that these individuals had the Apert syndrome with unusually severe craniofacial manifestations, and that the Vogt syndrome does not exist as an entity in its own right.

2. The Saethre-Chotzen type of acrocephalosyndactyly is characterised by mild cranial changes and soft tissue syndactyly of the second and third fingers and fourth toes. The cranial base and mandibular ramus are reduced in length and the cranial sinuses are undeveloped (Evans & Christiansen 1976). A family in which the Saethre-Chotzen syndrome occurred together with malignancies and decreased in vitro lymphocytic proliferation was reported by McKeen et al (1984). The nature of the pathogenetic inter-relationships is uncertain.

There have been several reports of generation to generation transmission (Chotzen 1932, Bartsocas et al 1970, Pruzanzky et al 1975). Pantke et al (1975) reviewed the findings in six kindreds with 31 cases and concluded that inheritance was autosomal dominant.

3. Pfeiffer (1964) reported a family with mild acrocephaly in association with cutaneous syndactyly of the second and third fingers and marked deviation of broadened thumbs and great toes. Other kindreds with an

autosomal dominant pattern of inheritance have been described by Martsolf et al (1971) and Saldino et al (1972). This Pfeiffer type of acrocephalosyndactyly was probably present in six generations of a family described by Waardenburg et al (1961). In a review, Naveh & Friedman (1976) emphasised that the Pfeiffer syndrome was very rare. A family with transmission through three generations was subsequently reported by Bianchi et al (1985).

## ACROCEPHALOPOLYSYNDACTYLY (Carpenter* syndrome)

Acrocephalopolysyndactyly is distinguished from the acrocephalosyndactyly group of disorders by the presence of extra digits (Temtamy 1966). Initially, two forms were recognised, type I, the Noack syndrome, and type II, the Carpenter syndrome. Subsequently, Pfeiffer (1969) restudied the kindred reported by Noack (1959) and suggested that the condition in question was acrocephalosyndactyly, Pfeiffer type. The same situation arose when Robinow & Sorauf (1975) investigated members of a large family in whom the Noack syndrome had been diagnosed and which conformed with the diagnostic criteria for the Pfeiffer syndrome. The numerical designations have now been discarded and the term 'Carpenter syndrome' has been accepted as being synonymous with 'acrocephalopolysyndactyly'.

Individuals with the Carpenter syndrome have craniostenosis, a high forehead, cutaneous syndactyly, and reduplication of the proximal phalanx of the thumb and great toe. Cardiac abnormalities, mental retardation, obesity and hypogonadism are additional features.

The majority of affected persons have mental retardation but according to Frias et al (1978) this complication may be mitigated by early surgical correction of the craniosynostoses.

### Genetics

Affected sibs have been described by Carpenter (1909), Eaton et al (1974), Temtamy (1966) and Robinson et al (1985) and about 20 cases have now been reported. The author has examined an affected male infant whose sister died following cranial surgery in the neonatal period. The unaffected parents, who were of Afrikaner stock, were consanguineous. Inheritance of the Carpenter syndrome is autosomal recessive. This point is of practical significance, as the phenotype bears some resemblance to that of the Apert syndrome. This latter condition is transmitted as an autosomal dominant,

* George Alfred Carpenter (1859–1910) was a paediatrician at the Queen's Hospital for Children, Hackney, London. He edited several journals and was president of the paediatric section of the Royal Society of Medicine.

and diagnostic accuracy is therefore vital when parents of an affected child request genetic counselling.

Other forms of acrocephalopolysyndactyly have been documented in single families, including an autosomal recessive form (Goodman et al 1979) and a mild autosomal dominant variety (Young & Harper 1982).

### Greig* cephalopolysyndactyly syndrome

The Greig cephalopolysyndactyly syndrome is characterised by hypertelorism, macrocephaly, pre- or postaxial polydactyly of the hands and polysyndactyly of the feet. Following the initial description of an affected mother and daughter (Greig 1926), about 40 cases in 10 families have been recorded. The condition has a wide geographical distribution, including the USA (Duncan et al 1979), Europe (Fryns et al 1981), Canada (Chudley & Houston 1982) and Brazil (Gollop & Fontes 1985). A familial reciprocal translocation between chromosomes 3 and 7 has been observed (Tommerup & Nielsen 1983) but autosomal dominant inheritance is well established.

There are semantic problems with the term 'Greig syndrome' and it has been applied to the combination of hypertelorism and mental retardation and also used for the cranjofrontonasal group of disorders (see p. 302). For these reasons a circumspect approach to published reports is warranted.

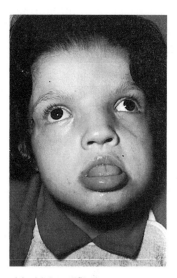

**Fig. 12.10** Greig syndrome: a girl with hypertelorism, macrocephaly and polysyndactyly.

* David Greig (1864–1936) was a surgeon at the Royal Infirmary, Dundee, Scotland. After his retirement he became curator of the Museum at the Royal College of Surgeons of Edinburgh.

## MANDIBULOFACIAL DYSOSTOSIS (Treacher Collins* syndrome)

Treacher Collins (1900) described a patient with an unusual configuration of the eyelids and malar bones. Subsequently, Franceschetti & Zwahlen (1944) and Franceschetti & Klein (1949) reported individuals with the condition, under the title 'mandibulofacial dysostosis'. The features of 200 patients with the disorder were reviewed by Rogers (1964).

The main stigmata of the syndrome are deafness and variable facial abnormalities. The eyes have an anti-Mongoloid slant, the cheek bones are flattened and the mandible is poorly developed. Colobomata are present in the lower eyelids and the external ears are small and malformed. Roberts et al (1975) described the results of a radiocephalometric study in eight cases, and emphasised that patients with the condition tended to resemble each other, in spite of any diversity of their ethnic backgrounds.

The pathogenesis of the Treacher Collins syndrome has been discussed from the embryological point of view by Poswillo (1975). An analagous autosomal recessive condition in the rabbit provides a useful animal model for investigations of developmental mechanisms (Fox & Carey 1979). The surface morphology of the craniofacial skeleton was documented by Kolar et al (1985) following a detailed anthropometric study of 18 affected persons.

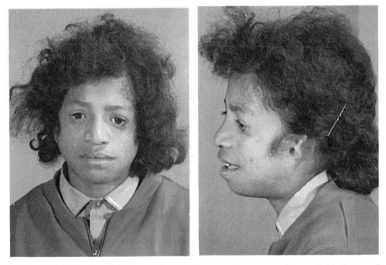

**Fig. 12.11** (left) Treacher Collins syndrome: the eyes have an antimongoloid slant and a coloboma is present in the left lower eyelid. The cheek bones are flattened.

**Fig. 12.12** (right) Treacher Collins syndrome: the external ear is malformed and the lower jaw is hypoplastic. Although this girl was profoundly deaf, a sibling with similar facial features had normal hearing.

* Edward Treacher Collins (1862–1932) was a distinguished ophthalmologist at the Royal Eye Hospital, Moorfields, London.

*Genetics*

Inheritance is autosomal dominant. Kindreds with generation to generation transmission have been reported by several authors, including Franceschetti & Klein (1949), Rovin et al (1964), Rogers (1964) and Fazen et al (1967). Expression of the abnormal gene is notoriously variable. Indeed, in the author's experience, it is not unusual to reach the diagnosis in a mildly affected parent only after a severely affected infant has been produced. The Treacher Collins syndrome is not uncommon, and children with the disorder are encountered in the majority of special schools for the deaf.

Although autosomal dominant inheritance is well established, affected siblings with normal consanguineous parents have been studied in a Hutterite community in Canada (Lowry et al 1985). It is uncertain whether these children had an autosomal recessive form of the disorder, or whether some other process, such as germinal mosaicism, was operative.

It is of practical significance that hearing in the Treacher Collins syndrome can sometimes be improved by middle ear surgery. Equally, the fact that a severe hearing deficiency often compounds the handicap imposed by the abnormal facies is important when a parent with minimal manifestations seeks genetic guidance.

Antenatal diagnosis by fetoscopy was undertaken during the second trimester in four at-risk pregnancies by Nicolaides et al (1984). Facial abnormalities were recognised in two of the fetuses and the diagnosis of the Treacher Collins syndrome was confirmed after termination.

## Nager acrofacial dysostosis

Nager acrofacial dysostosis is characterised by facial malformations which resemble those of the Treacher Collins syndrome, plus abnormalities of the radial side of the upper limb. These include aplasia or hypoplasia of the thumb, radial aplasia and radio-ulnar synostosis. Other variable syndromic components are stunted stature, mild mental retardation, abnormalities of male genitalia (Burton & Nadler 1977) and duplication of the thumbs (Giugliani & Pereira 1984). A sporadic infant with additional abnormalities of the larynx, lungs, ribs and hips was documented by Krauss et al (1985) and another with the tetralogy of Fallot was described by Thompson et al (1985). Severely affected siblings with gross limb defects were recorded by Kawira et al (1984). It is evident that the range of phenotypic manifestations is wider than first thought and their potential severity has important implications for genetic counselling.

The condition was delineated after a report by Bowen & Harley (1974) and an account of 15 cases by Herrmann et al (1975). The total reported cases had risen to 22 when Halal et al (1983) reviewed the literature. It is probable that inheritance is autosomal recessive.

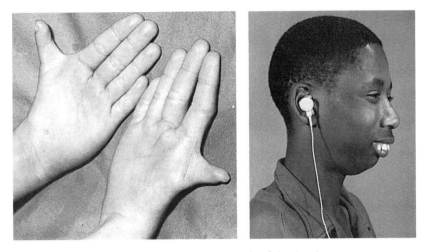

**Fig. 12.13** (left) Nager acrofacial dysostosis: hypoplasia of the thumbs.

**Fig. 12.14** (right) Mandibular hypoplasia: the lower jaw is very underdeveloped. No other members of this boy's large family had similar stigmata.

## MANDIBULAR HYPOPLASIA (Pierre Robin* syndrome)

Mandibular hypoplasia may occur in isolation or as a component of several well defined syndromes. The clinical importance of severe mandibular hypoplasia revolves around airway obstruction in the neonatal period and dental malocclusion and cosmetic appearance in later life.

The term 'Pierre Robin' anomaly or sequence is conventionally applied to the association of mandibular hypoplasia with cleft palate and glossoptosis. In the loose sense, however, this eponym is often used for isolated micrognathia.

### Genetics

Sibs with the Pierre Robin anomaly have been reported by Singh et al (1970), Shah et al (1970) and Bixler & Christian (1971). Several had additional abnormalities and it is possible that they were suffering from unrecognised syndromes. An X-linked disorder in which the Pierre Robin anomaly was associated with congenital heart disease was described by Gorlin et al (1970). In another, a similar condition, which Brude (1984) termed the Catel-Manzke syndrome, mandibular hypoplasia and cleft palate, occur in conjunction with a supernumery ossicle of the index finger.

---

* Pierre Robin (1867–1950) was the doyen of French dental surgeons. He had a profound influence on the development of his speciality.

Affected sisters, one of whom was stillborn, were studied by Dignan et al (1986). The mode of inheritance is unknown. A mother and son with the Pierre Robin anomaly plus hypoplasia of the first and fifth rays of the hands and feet were documented by Robinow et al (1986). Despite these reports the majority of individuals with the isolated abnormality have been sporadic and it must be assumed that the condition is usually non-genetic. Ultrasonic detection of the Pierre Robin anomaly in the third trimester of pregnancy was documented by Pilu et al (1986).

Edwards & Newall (1985) commented that micrognathia and developmental abnormalities of the mandible and palate could be produced experimentally by feeding various drugs to pregnant rodents. If similar mechanisms occur in humans, it is possible that many sporadic cases of the Pierre Robin syndrome may be the result of fetal damage by unknown embryopathic agents.

## CEREBROCOSTOMANDIBULAR SYNDROME

The cerebrocostomandibular syndrome (CCMS) comprises severe micrognathia, rib defects and variable cerebral and skeletal abnormalities. Death in infancy from respiratory complications occurs in about 40% of cases (Smith & Sekar 1985) and psychomotor retardation is present in the majority of survivors (Silverman et al 1980).

The condition was delineated by Smith et al (1966) and the literature has been reviewed by Tachibana et al (1980) and Smith & Sekar (1985). More than 30 cases have now been reported. The majority of affected infants were sporadic (Miller et al 1972, Langer & Herrmann 1974) but siblings with the condition have been documented (McNicholl et al 1970, Hennekam et al 1985) and parental consanguinity has been noted (Clarke & Nguyen 1985). Transmission from an affected mother to her son and daughter (Leroy et al 1981) and from a father to his daughter (Merlob et al 1987) has also been reported. It seems possible that the CCMS is genetically heterogeneous and that there are mild autosomal dominant and severe autosomal recessive forms of the disorder.

## MANDIBULOACRAL DYSPLASIA

In mandibuloacral dysplasia severe underdevelopment of the lower jaw is associated with delayed closure of the cranial sutures, Wormian bones, clavicular hypoplasia, dermal anomalies, acro-osteolysis, stiff joints, alopecia, localised skin atrophy and short stature. Eighteen affected persons in nine families have been recorded (Tenconi et al 1986). Five of the families had Italian origins (Cavallazzi et al 1960, Sensenbrenner & Fiorelli 1971, Zina et al 1981, Pallatta & Morgese 1984). Other kindreds lived in the USA (Young et al 1971, Danks et al 1974) and one was Mexican (Welsh 1975). In a comprehensive review Tenconi et al (1986) analysed the phenotypic spectrum and discussed the apparent Italian aggregation. Several sets of affected siblings with normal parents have been encountered and although there are no reports of parental consanguinity, it is probable that inheritance is autosomal recessive.

## OCULOMANDIBULOFACIAL SYNDROME (Hallermann-Streiff* syndrome)

Reports concerning this disorder were published by Hallermann (1948) and Streiff (1950). The syndrome was finally delineated by Francois (1958), following a review of the literature and the eponym 'Hallermann-Streiff syndrome' is now in general use. An alternative descriptive designation which has found some favour is 'oculomandibulodyscephaly with hypotrichosis'.

Affected individuals have proportionate short stature, and facial features which include frontal bossing, micrognathia, microstomia, microphthalmia, congenital cataracts and a thin pointed nose. The hair is thin and sparse, and the skin is atrophic. Numerous inconsistent concomitants have been described. From the practical point of view, the ocular problems predominate, although airway obstruction sometimes poses problems (Sataloff & Roberts 1984, Friede et al 1985).

### Genetics

The genetic basis of the Hallermann-Streiff syndrome is uncertain. Although more than 60 cases have been reported, the majority have been sporadic. The sex distribution is equal. Dominant inheritance is suggested by the affected father and daughter studied by Guyard et al (1962). Nevertheless, as the father was married to a relative, the genetic situation in this kindred is still unclear. An atypical form of the syndrome in three generations of a kindred was reported by Koliopoulos & Palimeris (1975).

Fraser & Friedmann (1967) favoured the concept of dominant inheritance and suggested that the majority of patients represented new mutations. In view of the stigmata of the condition, fitness to reproduce is probabaly greatly diminished and if the Hallermann-Streiff syndrome is indeed dominant, the likelihood of transmission would be small. The chromosomes were normal in two cases studied by Forsius & de la Chapelle (1964). A pair of distant relatives with the condition, one of whom had a partial duplication of the long arm of chromosome 10, were documented by Schanzlin et al (1980).

---

* Wilhelm Hallermann was born in Dortmund in 1909. He qualified in medicine in Freiburg and became professor of ophthalmology at Gottingen.

Bernado Streiff was born in Genoa, Italy, in 1908 and became professor of ophthalmology at the University of Lausanne, Switzerland.

Although the genetic background of the Hallermann-Streiff syndrome has not been elucidated, for the purposes of genetic counselling it is reasonable to assume that normal parents of an affected child have a low recurrence risk in any further pregnancies.

## CRANIOFRONTONASAL DYSPLASIA

Craniofrontonasal dysplasia is characterised by coronal craniosynostosis and hypertelorism. The nose has a bifid tip and a broad root and a variety of digital abnormalities may be present. Intellect is normal. The manifestations are very variable and range from trivial changes to gross facial malformations.

The nosological situation is confusing as the terms 'median cleft face syndrome' (De Meyer 1967), and 'frontonasal dysplasia' (Sedano et al 1970, Kwee & Lindhout 1983) have been used for the same or a similar disorder. The Greig cephalopolysyndactyly syndrome (p. 296) has also been confused with craniofrontonasal dysplasia, but there is little doubt that these conditions are separate entities. There is also semantic overlap with the acrodysplasia group of conditions (p. 236), especially the condition designated 'acrodysostosis' by Maroteaux & Malamut (1968) and Robinow et al (1971). The majority of persons with this disorder, which has also been termed 'acrofacial dysostosis' would probably now be categorised as having craniofrontonasal dysplasia.

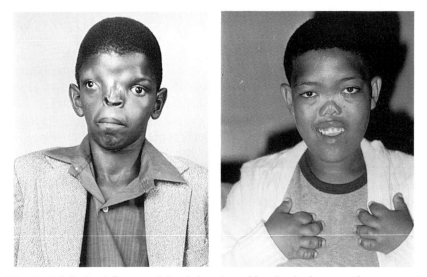

**Fig. 12.16** (left) Craniofrontonasal dysplasia: a boy with a dysplastic nose and gross hypertelorism.

**Fig. 12.17** (right) Acrofacial dysostosis: a boy with a flattened nasal bridge, anteverted nostrils and brachydactyly.

Young & Moore (1984) proposed that the faulty gene in craniofrontonasal dysplasia was lethal in males when fully expressed. In a review of the literature Sax & Flanney (1986) identified eight affected families (Montford 1929, Webster & Deming 1950, Hunter & Rudd 1977, Cohen 1979, Slover & Sujansky 1979, Pruzansky et al 1982, Reynolds et al 1982, Young & Moore 1984). Following a detailed analysis of pedigree data, Sax & Flanney (1986) concluded that the condition did not follow a Mendelian pattern of transmission. They suggested that the disorder was either heterogeneous or caused by a dominant gene which behaves in a non-Mendelian manner.

## Acrocallosal syndrome

Other entities which have been differentiated from this group of disorders include the acrocallosal syndrome in which an unusual facial appearance is associated with pre- and postaxial polydactyly, mental retardation and agenesis of the corpus callosum. Sporadic patients were documented by Schinzel & Schmid (1980), Nelson & Thomson (1982), Sanchis et al (1985) and Legius et al (1985). It became apparent that the mode of inheritance was autosomal recessive when Schinzel & Kaufmann (1986) identified two affected sisters and recognised parental consanguinity in two additional cases.

Maxillonasal dysostosis, or the Binder syndrome, is another similar entity (Ferguson & Thompson 1985). Frontofacionasal dysplasia falls into the same general category but is distinguished by the lack of digital involvement. Affected siblings (Gollop 1981) and parental consanguinity (Gollop et al 1984) have been noted and it is likely that inheritance is autosomal recessive.

REFERENCES

**Preamble**
Cohen M M 1975 An etiologic and nosologic overview of craniosynostosis syndromes. Birth Defects: Original Article Series 11/2: 137
Gorlin R J, Sedano H O, Boggs W S 1975 The face in the diagnosis of dysmorphogenesis. Annales de Pédiatrie 4/3: 10

**Craniostenosis**
Antley R M, Hwang D S, Theopold W et al 1981 Further delineation of the C (trigonocephaly) syndrome. American Journal of Medical Genetics 9: 147–163
Archer D B, Gordon D S, Maguire C J F, Gleadhill C A 1974 Ophthalmic aspects of craniosynostosis. Transactions of the Ophthalmology Society of the United Kingdom 94/1: 172
Armendares S 1970 On the inheritance of craniostenosis. Study of thirteen families. Journal de Génétique Humaine 18: 121
Armendares S, Antillon F, Del Castillo V, Jiminez M 1975 A newly recognised inherited syndrome of dwarfism, craniosynostosis, retinitis pigmentosa and multiple congenital malformations. Birth Defects: Original Article Series 11/5: 49
Bell H S, Clare F B, Wentworth A F 1961 Case reports and technical notes on familial scaphocephaly. Journal of Neurosurgery 18: 239

Carter C O, Till K, Fraser V, Coffey R 1982 A family study of craniosynostosis with probable recognition of a distinct syndrome. Journal of Medical Genetics 194: 280–285

Cross H E, Opitz J 1969 Craniostenosis in the Amish. Journal of Paediatrics 75: 1037

Currarino G, Silverman F N 1960 Orbital hypotelorism, arrhinencephaly and trigonocephaly. Radiology 74: 206–217

Duguid H 1929 An instance of familial scaphocephaly. Journal of Mental Science 75: 704

Frydman M, Kauschansky A, Elian E 1984 Trigonocephaly: A new familial syndrome. American Journal of Medical Genetics 18:

Gaudier B, Laine E, Fontaine G, Castier C, Farriaux J P 1967 Les craniosynostoses (étude de vingt observations). Archives Françaises de Pédiatrie 24: 775

Gerold M 1959 Healing of a fracture in an unusual case of congenital anomaly of the upper extremities. Zeitschrift fur Chirurgie 34: 831

Giuffre R, Scarfo C B, Tomaccini D 1975 Clinical and surgical notes on 78 cases of craniostenosis. Minerva Paediatrica 27/16: 949

Gordon H 1959 Craniostenosis. British Medical Journal 2: 792

Hunter A F, Rudd N L 1976 Craniosynostosis. I. Sagittal synostosis: Its genetics and associated clinical findings in 214 patients. Teratology 14: 185–193

Hunter A G W, Rudd N L, Hoffmann H J 1976 Trigonocephaly and associated minor anomalies in mother and son. Journal of Medical Genetics 13/1: 77

Jackson C E, Weiss L, Renolds W A, Forman T F, Petersen J A 1976 Craniosynostosis, midfacial hypoplasia and foot abnormalities: an autosomal dominant phenotype in a large Amish Kindred. Journal of Paediatrics 38: 963

Keats T E, Smith T H, Sweet D E 1975 Craniofacial dysostosis with fibrous metaphyseal defects. American Journal of Roentgenology, Radium Therapy and Nuclear Medicine 124: 271

Kosnik E J, Gilbert G, Sayers M P 1975 Familial inheritance of coronal craniosynostosis. Developmental Medicine and Child Neurology 17/5: 630

Lepintre J, Renier D 1975 Craniosynostosis. Revue de Pédiatrie 11/8: 433

Lohse D C, Duncan C C, Ment L R 1985 Cranial asymmetry and lambdoid synostosis in a sibship. Neurosurgery 16/6: 836–838

Lowry R B 1972 Congenital absence of the fibula and craniosynostosis in sibs. Journal of Medical Genetics 9: 227

Motegi T, Ohuchi M, Ohtaki C et al 1985 A craniosynostosis in a boy with a del (7) (p15.3p21.3): Assignment by deletion mapping of the critical segment for craniosynostosis to the midportion of 7p21. Human Genetics 71/2: 160–162

Murphy J W 1953 Familial scaphocephaly in father and son. United States Armed Forces Medical Journal 4: 1496

Nance W E, Engel E 1967 Autosomal deletion mapping in man. Science 155: 692

Opitz J M, Johnson R C, McCreadie S R, Smith D W 1969 The C syndrome of multiple congenital anomalies. Birth Defects V(2): 161–166

Sargent C, Burn J, Baraitser M, Pembrey M E 1985 Trigonocephaly and the Opitz C syndrome. Journal of Medical Genetics 22(1): 39–45

Say B, Meyer J 1981 Familial trigonocephaly associated with short stature and developmental delay. American Journal of Diseases of Children 135: 7611–712

Sensenbrenner J A, Dorst J P, Owens R P 1975 New syndrome of skeletal, dental and hair anomalies. Birth Defects: Original Article Series 11/2: 372

Shillito J, Matson D D 1968 Craniostenosis: a review of 159 surgical patients. Paediatrics 41: 829

Shuper A, Merlob P, Grunebaum M, Reisner S H 1985 The incidence of isolated craniosynostosis in the newborn infant. American Journal of Diseases of Children 139: 85–86

**Craniofacial dysostosis (Crouzon syndrome)**

Alpers C E, Edwards M S 1985 Hemangiomatous anomaly of bone in Crouzon's syndrome: Case report. Neurosurgery 16/3: 391–394

Cohen E R, Hesky E M, Bradley W F et al 1985 Life response to Crouzon's disease. Cleft Palate Journal 22: 123–131

Crouzon O 1912 Dysostose cranio-facio héréditaire. Bulletins et Mémoires de la Société Médicale des Hôpitaux de Paris 33: 545

Dolivo G, Gillieron J D 1955 Une famille de pseudo-Crouzon. Confinia Neurologica 15: 114
Franceschetti A 1953 Dysostose crânienne avec calotte cerebriforme (pseudo-Crouzon).
    Confinia Neurologica 13: 161
Reddy B S, Garg Br, Padiyar N V, Krishnaram A S 1985 An unusual association of
    acanthosis nigricans and Crouzon's disease. Journal of Dermatology (Tokyo) 12: 85–90
Shiller J G 1959 Craniofacial dysostosis of Crouzon. A case report and pedigree with
    emphasis on heredity. Paediatrics 23: 107
Suslak L, Glista B, Gertzman G B et al 1985 Crouzon syndrome with periapical cemental
    dysplasia and acanthosis nigricans: The pleiotropic effect of a single gene? Birth Defects:
    Original Article Series 21(2): 127–134
Vulliamy D G, Normandale P A 1966 Cranio-facial dysostosis in a Dorset family. Archives
    of Disease in Childhood 41: 375

**Acrocephalosyndactyly (Apert syndrome)**
Apert M E 1906 Del'acrocephalosyndactylie. Bulletins et Mémoires de la Société des
    Hôpitaux de Paris 23: 1310
Bartsocas C S, Weber A L, Crawford J D 1970 Acrocephalosyndactyly type 3: Chotzen's
    syndrome. Pediatrics 77: 267
Bergstrom L V, Neblett L M, Hemenway W G 1972 Otologic manifestations of
    acrocephalosyndactyly. Archives of Otolaryngology 96: 117
Bianchi E, Arico M, Podesta A F et al 1985 A family with the Saethre-Chotzen syndrome.
    American Journal of Medical Genetics 22: 649–658
Blank C E 1960 Apert's syndrome (a type of acrocephalosyndactyly). Observations on a
    British series of thirty-nine cases. Annals of Human Genetics 24: 151
Bull M J, Escobar V, Bixler D, Antley R M 1979 Phenotype definition and recurrence risk
    in the acrocephalosyndactyly syndromes. Birth Defects 15(5B): 65–74
Chotzen F 1932 Eine eigenartige familiare Entwicklungsstorung (akrocephalosyndaktylie,
    dysostosis craniofacialis und hypertelorismus). Monatsschrift fur Kinderheilkunde 55: 97
Dodson W E, Museles M, Kennedy J L, Al-Aish M 1970 Acrocephalosyndactylia associated
    with a chromosomal translocation. American Journal of Diseases of Children 120: 360
Evans C A, Christiansen R L 1976 Cephalic malformations in Saethre-Chotzen syndrome.
    Radiology 121(2): 399–403
Hoover G H, Flatt A E, Weiss M W 1970 The hand in Apert's syndrome. Journal of Bone
    and Joint Surgery 52A: 878
Jackson C E, Weiss L, Reynolds W A et al 1976 Craniosynostosis, midfacial hypoplasia and
    foot abnormalities: An autosomal dominant phenotype in a large Amish kindred. Journal
    of Pediatrics 88: 963–968
Krueger J L, Ide C H 1974 Acrocephalosyndactyly (Apert's syndrome). Annals of
    Ophthalmology 6/8: 787
Leonard C O, Daikoku N H, Winn K 1982 Prenatal fetoscopic diagnosis of the Apert
    syndrome. American Journal of Medical Genetics 11: 5–9
McKusick V A 1975 Acrocephalosyndactyly type I (typical Apert syndrome). Mendelian
    Inheritance in Man, 4th edn, Johns Hopkins University Press, Baltimore, p 6
McKeen E A, Mulvihill J J, Levine P H et al 1984 The concurrence of Saethre-Chotzen
    syndrome and malignancy in a family with in vitro immune dysfunction. Cancer
    54(12): 2946–2951
Martsolf J T, Cracco J B, Carpenter G G, O'Hara A E 1971 Pfeiffer syndrome: an unusual
    type of acrocephalosyndactyly with broad thumbs and great toes. American Journal of
    Diseases of Children 121: 257
Minami R, Olek K, Wardenbach P 1975 Hypersarcosinemia with craniostenosis
    syndactylism syndrome. Humangenetika 28/2: 167
Naveh Y, Friedman A 1976 Pfeiffer syndrome: report of a family and review of the
    literature. Journal of Medical Genetics 13/4: 277
Pantke O A, Cohen M M, Witkop C J 1975 The Saethre Chotzen syndrome. Birth Defects:
    Original Article Series 11/2: 190
Pfeiffer R A 1964 Dominant erbliche Akrocephalosyndaktylie. Zeitschrift für
    Kinderheilkunde 90: 301
Pruzansky S, Pashayan H, Kreiborg S, Miller M 1975 Roentgencephalometric studies of
    the premature craniofacial synostoses: report of a family with the Saethre-Chotzen
    syndrome. Birth Defects: Original Article Series 11/2: 226

Roberts K B, Hall J B 1971 Apert's acrocephalosyndactyly in mother and daughter. Cleft palate in the mother. Birth Defects: Original Article Series 7/7: 262

Rochiccioli P, Dutau G, Marcou P 1974 Acrocephalosyndactylia. Diagnostic problems with reference to three cases. Journal de Génétique Humaine 22/3: 269

Saldino R M, Steinback H L, Epstein C J 1972 Familial acrocephalosyndactyly (Pfeiffer syndrome). American Journal of Roentgenology, Radium Therapy and Nuclear Medicine 116: 609

Spranger J W, Langer L O, Wiedemann H R 1974 Bone dysplasias, Gustav Fischer Verlag, Stuttgart, p 261

Summitt R L 1969 Recessive acrocephalosyndactyly with normal intelligence. Birth Defects: Original Article Series 5/3: 35

Waardenburg P J, Franceschetti A, Klein D 1961 Genetics and Ophthalmology, Vol 1, Thomas, Springfield, p 301

Waterson J R, Di Pietro M A, Barr M 1985 Apert syndrome with frontonasal encephalocele. American Journal of Medical Genetics 21: 777–783

**Acrocephalopolysyndactyly (Carpenter syndrome)**

Carpenter G 1909 Case of acrocephaly with other congenital malformations. Proceedings of the Royal Society of Medicine 2: 45

Eaton A P, Sommer A, Kontras S B, Sayers M P 1974 Carpenter syndrome — acrocephalopolysyndactyly type II. Birth Defects: Original Article Series 10/9: 249

Frias J L, Felman A H, Rosenbloom A L et al 1978 Normal intelligence in two children with Carpenter syndrome. American Journal of Medical Genetics 2: 191–199

Goodman R M, Sternberg M, Shem-Tov Y et al 1979 Acrocephalopolysyndactyly type IV: A new genetic syndrome in three sibs. Clinical Genetics 15(3): 209–214

Noack M 1959 Ein Beitrag zum Krankheitsbild der Akrozephalosyndaktylie (Apert). Archiv für Kinderheilkunde 160: 168

Pfeiffer R A 1969 Associated deformities of the head and hands. Birth Defects: Original Article Series 5/3: 18

Robinow M, Sorauf T J 1975 Acrocephalopolysyndactyly type Noack in a large kindred. Birth Defects: Original Article Series 11/5: 99

Robinson L K, James H E, Scott J et al 1985 Carpenter syndrome: Natural history and clinical spectrum. American Journal of Medical Genetics 20: 461–469

Temtamy S A 1966 Carpenter's syndrome: acrocephalopolysyndactyly. An autosomal recessive syndrome. Pediatrics 69: 111

Young I D, Harper P S 1982 An unusual form of familial acrocephalosyndactyly. Journal of Medical Genetics 19: 286–288

*Greig syndrome*

Chudley A E, Houston C S 1982 The Greig cephalopolysyndactyly syndrome in a Canadian family. American Journal of Medical Genetics 13: 269–276

Duncan P A, Klein R M, Wilmot P L, Shapiro L R 1979 Greig cephalopolysyndactyly syndrome. American Journal of Diseases of Children 133: 818–821

Fryns J P, Van Noyen G, Van den Berghe H 1981 The Greig polysyndactyly-craniofacial dysmorphism syndrome: Variable expression in a family. European Journal of Pediatrics 136: 217–220

Gollop T R, Fontes L R 1985 The Greig cephalopolysyndactyly syndrome: Report of a family and review of the literature. American Journal of Medical Genetics 22: 59–68

Greig D M 1926 Oxycephaly. Edinburgh Medical Journal 33: 189–218

Tommerup N, Nielsen F 1983 A familial reciprocal translocation t(3:7) (p21.1:p13) associated with the Greig polysyndactyly-craniofacial anomalies syndrome. American Journal of Medical Genetics 16: 313–321

**Mandibulofacial dysostosis (Treacher Collins syndrome)**

Fazen L E, Elmore J, Nadler H L 1967 Mandibulofacial dysostosis (Treacher Collins syndrome). American Journal of Diseases of Children 13: 405

Fox R R, Carey D D 1979 Hereditary macrostomus in the rabbit. A model for Treacher Collins syndrome, one form of mandibulofacial dysostosis. Journal of Heredity 70: 369–372

Franceschetti A, Zwahlen P 1944 Un nouveau syndrome. La dysostose mandibulo-faciale. Bulletin der Schweizerischen Akademie der Medizinischen Wissenschaften 1: 60

Franceschetti A, Klein D 1949 Mandibulofacial dysostosis: new hereditary syndrome. Acta ophthalmologica 27: 143

Herrmann J, Pallister P D, Kaveggia E G, Opitz J M 1975 Acrofacial dysostosis type Nager. Birth Defects: Original Article Series 11/5: 341

Kolar J C, Farkas L G, Munro I R 1985 Surface morphology in Treacher Collins syndrome: An anthropometric study. Cleft Palate Journal 22: 266–274

Lowry R B, Morgan K, Holmes T M et al 1985 Mandibular dysostosis in Hutterite sibs: A possible recessive trait. American Journal of Medical Genetics 22: 501–512

Nicolaides K H, Johansson D, Donnai D, Rodeck C H 1984 Prenatal diagnosis of mandibulofacial dysostosis. Prenatal Diagnosis 4: 201–205

Poswillo D 1975 The pathogenesis of the Treacher Collins syndrome (mandibulofacial dysostosis). British Journal of Oral Surgery 13/1: 1

Roberts F G, Pruzansky S, Aduss H 1975 A radiocephalometric study of mandibulofacial dysostosis in man. Archives of Oral Biology 10/4: 265

Rogers B O 1964 Berry–Treacher Collins syndrome. A review of 200 cases. British Journal of Plastic Surgery 17: 109

Rovin S, Dachi S F, Borenstein D B, Cotter W B 1964 Mandibulofacial dysostosis, a familial study of five generations. Pediatrics 65: 215

Treacher Collins E 1900 Cases with symmetrical congenital notches in the outer part of each lid and defective development of the malar bones. Transactions of the Ophthalmology Society of the United Kingdom 20: 190

*Nager acrofacial dysostosis*

Bowen P, Harley F 1974 Mandibulofacial dysostosis with limb malformations: In: Bergsma D (ed) Limb Malformations. National Foundation March of Dimes, White Plains, p 109–115

Burton B K, Nadler H L 1977 Nager acrofacial dysostosis. Journal of Pediatrics 91: 84–86

Giugliani R, Pereira C H 1984 Nager's acrofacial dysostosis with thumb duplication: Report of a case. Clinical Genetics 26: 228–230

Halal F, Herrmann J, Pallister P D et al 1983 Differential diagnosis of Nager acrofacial dysostosis syndrome. American Journal of Medical Genetics 14: 210–224

Herrmann J, Pallister P D, Kaveggia E G, Opitz J M 1975 Acrofacial dysostosis type Nager. Birth Defects' Original Article Series 11/5: 341

Krauss C M, Hassell L A, Gang D L 1985 Brief clinical report: Anomalies in an infant with Nager acrofacial dysostosis. American Journal of Medical Genetics 21: 761–764

Kawira E L, Weaver D D, Bender H A 1984 Acrofacial dysostosis with severe facial clefting and limb reduction. American Journal of Medical Genetics 17: 641–647

Thompson E, Cadbury R, Baraitser M 1985 The Nager acrofacial dysostosis syndrome with the tetralogy of Fallot. Journal of Medical Genetics 22: 408–410

**Mandibular hypoplasia (Pierre Robin syndrome)**

Bixler D, Christian J 1971 Pierre Robin syndrome occurring in 2 unrelated sibships. The clinical delineation of birth defects vol II. Williams and Wilkins, Baltimore, p 67

Brude E 1984 Pierre Robin sequence and hyperphalangy: A genetic entity. Catel-Manzke syndrome. European Journal of Pediatrics 142: 222–223

Dignan P, Martin L W, Zenni E J 1986 Pierre Robin anomaly with an accessory metacarpal of the index fingers. Clinical Genetics 29: 168–173

Edwards J R, Newall D R 1985 The Pierre Robin syndrome reassessed in the light of recent research. British Journal of Plastic Surgery 38: 339–342

Gorlin R J, Cervenka J, Anderson R C et al 1970 Robin's syndrome. A probably X-linked recessive subvariety exhibiting persistence of left superior vena cava and atrial septal defect. American Journal of Diseases of Children 119: 176–178

Pilu G, Romero R, Reece E A et al 1986 The prenatal diagnosis of Robin anomalad. American Journal of Obstetrics and Gynecology 154/3: 630–632

Robinow M, Johnson G F, Apesos J 1986 Robin sequence and oligodactyly in mother and son. American Journal of Medical Genetics 25: 293–297

Shah C V, Pruzansky S, Harris W S 1970 Cardiac malformations with facial clefts. American Journal of Diseases of Children 114: 238

Singh R P, Jaco N T, Vigna V 1970 Pierre Robin syndrome in siblings. American Journal of Diseases of Children 120: 560


**Cerebrocostomandibular syndrome**

Clarke E A, Nguyen V D 1985 Cerebrocosto-mandibular syndrome with consanguinity. Pediatric Radiology 15: 264–266

Hennekam R C, Beemer F A, Huijbers W A et al 1985 The cerebro-costo-mandibular syndrome: Third report of familial occurrence. Clinical Genetics 28: 118–121

Langer L O, Herrmann J 1974 The cerebrocostomandibular syndrome. Birth Defects: Original Article Series 10/7: 167

Leroy J G, Devos E A, Van den Bulcke L J, Robbe N S 1981 Cerebrocosto-mandibular syndrome with autosomal dominant inheritance. Journal of Pediatrics 99: 441–443

McNicholl B, Egan-Mitchell B, Murray J P 1970 Cerebrocostomandibular syndrome. Archives of Disease in Childhood 45: 421

Merlob P, Schonfeld A, Grunebaum M et al 1987 Autosomal dominant cerebro-costo-mandibular syndrome: Ultrasonographic and clinical findings. American Journal of Medical Genetics 26: 195–202

Miller K E, Allen R P, Davis W S 1972 Rib gap defects with micrognathia. American Journal of Roentgenology, Radium Therapy and Nuclear Medicine 114: 253

Silverman F N, Strefling A M, Stevenson D K, Lazarus J 1980 Cerebro-costo-mandibular syndrome. Journal of Pediatrics 97: 406–416

Smith K G, Sekar K C 1985 Cerebrocostomandibular syndrome: Case report and literature review. Clinical Pediatrics 24: 223–225

Smith D W, Theiler K, Schachenmann G 1966 Rib-gap with micrognathia, malformed tracheal cartilages, and redundant skin: a new pattern of defective development. Journal of Pediatrics 69: 799

Tachibana K, Yamamoto Y, Osaki E, Kuroki Y 1980 Cerebro-costo-mandibular syndrome. A case report and review of the literature. Human Genetics 54: 283–286


**Mandibuloacral dysplasia**

Cavallazzi C, Cremoncini R, Quadri A 1960 Su di un caso didisostosi cleido-cranica. Review of Clinical Pediatrics 65: 313–326

Danks D M, Mayne V, Wettenhall H N, Hall R K 1974 Craniomandibular dermatodysostosis. Birth Defects: Original Article Series X(2): 99–105

Pallatta R, Morgese G 1984 Mandibuloacral dysplasia: A rare progeroid syndrome. Two brothers confirm autosomal recessive inheritance. Clinical Genetics 26: 133–138

Sensenbrenner J A, Fiorelli G 1971 New syndrome manifested by mandibular hypoplasia, acro-osteolysis, stiff joints and cutaneous atrophy in two unrelated boys. In: Bergsma D (ed) The Clinical Delineation of Birth Defects. Orofacial Structures BD:OAS VII(7), Williams & Wilkins, Baltimore, p 291–297

Tenconi R, Miotti F, Miotti A et al 1986 Another Italian family with mandibuloacral dysplasia: Why does it seem more frequent in Italy? American Journal of Medical Genetics 24: 357–364

Welsh O 1975 Study of a family with a new progeroid syndrome. In: Bergsma D (ed) New Chromosomal and Malformation Syndromes, BD:OAS XI(5), p 25–38

Young L W, Radebaugh J F, Rubin P 1971 New syndrome manifested by mandibular hypoplasia, acroosteolysis, stiff joints and cutaneous atrophy in two unrelated boys. In: Bergsma D (ed) The Clinical Delineation of Birth Defects. Orofacial Structures. BD:OAS VII(7): 291–297.

Zina A M, Cravario A, Bundino S 1981 Familial mandibuloacral dysplasia. British Journal of Dermatology 105: 719–723


**Oculomandibulofacial syndrome (Hallermann-Streiff)**

Forsius H, De la Chapelle A 1964 Dyscephalia oculomandibulo-facialis. Two cases in which the chromosomes were studied. Annals of Pediatrics 10: 280

François J 1958 A new syndrome: dyscephalia with bird face and dental anomalies, nanism, hypotrichosis, cutaneous atrophy, microphthalmia and congenital cataract. Archives of Ophthalmology 60: 842

Fraser G R, Friedmann A I 1967 The causes of blindness in childhood. A Study of 776 Children with severe visual handicaps. Johns Hopkins University Press, Baltimore, p 89

Friede H, Lopata M, Fisher E, Rosenthal I M 1985 Cardiorespiratory disease associated with Hallermann-Streiff syndrome. Journal of Craniofacial Genetics 5: Supplement 1: 189–198

Guyard M, Perdriel G, Ceruti F 1962 On two cases of cranial dysostosis with 'bird head'. Bulletins et Mémoires de la Société Francaise d'Ophthalmologie 62: 433

Hallermann W 1948 Vogelgesicht und Cataracta congenita. Klinische Monatsblätter für Augenheilkunde 113: 315

Koliopoulos J, Palimeris G 1975 Atypical Hallermann–Streiff–Francois syndrome in three successive generations. Journal of Pediatrics and Ophthalmology 12/4: 235

Sataloff R T, Roberts B R 1984 Airway management in Hallermann-Streiff syndrome. American Journal of Otolaryngology 5: 64–67

Schanzlin D J, Goldbert D B, Brown S I 1980 Hallermann-Streiff syndrome associated with sclerocornea, aniridia and a chromosomal abnormality. American Journal of Ophthalmology 90: 411–415

Streiff E B 1950 Dysmorphie mandibulo-faciale (tête d'oiseau) et alteration oculaire. Ophthalmologica 120: 79

**Craniofrontonasal dysplasia**

Cohen M M 1979 Craniofrontonasal dysplasia. Birth Defects: Original Article Series XV(5): 85–99

De Meyer W 1967 The median cleft syndrome. Neurology 17: 961–971

Hunter A G, Rudd N L 1977 Craniosynostosis II. Coronal synostosis: Its familial characteristics and associated clinical findings in 109 patients lacking bilateral polysyndactyly or syndactyly. Teratology 15: 301–310

Kwee M L, Lindhout D 1983 Frontonasal dysplasia, coronal craniosynostosis, pre- and postaxial polydactyly and split nails: A new autosomal dominant mutant with reduced penetrance and variable expression? Clinical Genetics 24: 200–205

Maroteaux P, Malamut G 1968 L'acrodysostose. Presse Médicale 76: 2189–2192

Montford T M 1929 Hereditary hypertelorism without mental deficiency. Archives of Disease in Childhood 4: 381–384

Pruzansky S, Costaras M, Rollnick B R 1982 Radiocephalometric findings in a family with craniofrontonasal dysplasia. Birth Defects: Original Article Series XVIII: 121–138

Reynolds J F, Haas R J, Edgerton M T, Kelly T E 1982 Craniofrontonasal dysplasia in a three-generation kindred. Journal of Craniofacial Genetics 2(3): 233–238

Robinow M, Pfeiffer R A, Gorlin R J et al 1971 Acrodysostosis: A syndrome of peripheral dysostosis and nasal hypoplasia. American Journal of Diseases of Children 121: 195–203

Sax C M, Flanney D B 1986 Craniofrontonasal dysplasia: Clinical and genetic analysis. Clinical Genetics 29: 508–515

Sedano H O, Cohen M M Jirasek J, Gorlin R J 1970 Frontonasal dysplasia. Journal of Pediatrics 76: 906

Slover R, Sujansky E 1979 Frontonasal dysplasia with coronal craniosynostosis in three sibs. Birth Defects: Original Article Series XV(5B): 75–83

Webster J P, Deming E G 1950 The surgical treatment of the bifid nose. Plastic and Reconstructive Surgery 6: 1–37

Young I D, Moore J R 1984 Craniofrontonasal dysplasia: A distinct entity with lethality in the male? Clinical Genetics 25: 473–475

*Acrocallosal syndrome*

Ferguson J W, Thompson R P 1985 Maxillonasal dysostosis (Binder syndrome): A review of the literature and case reports. European Journal of Orthodontics 7/2: 145–148

Gollop T R 1981 Fronto-facio-nasal dysostosis. A new autosomal recessive syndrome. American Journal of Medical Genetics 10: 409–412

Gollop T R, Kiota M M, Martins et al 1984 Frontofacionasal dysplasia: Evidence for autosomal recessive inheritance. American Journal of Medical Genetics 19: 301–305

Legius E J, Fryns P, Casaer M et al 1985 Schinzel acrocallosal syndrome: A variant example of the Greig syndrome? Annales Génétique (Paris) 28: 239–240

Nelson M M, Thomson A J 1982 The acrocallosal syndrome. American Journal of Medical Genetics 12: 195–199

Sanchis A L, Cervero A, Martinex A, Valverde C 1985 Duplication of hands and feet, multiple joint dislocations, absence of corpus callosum and hypsarrhythmia: Acrocallosal syndrome? American Journal of Medical Genetics 20: 123–130

Schinzel A, Kaufmann U 1986 The acrocallosal syndrome in sisters. Clinical Genetics 30: 399–405

Schinzel A, Schmid W 1980 Hallux duplication, postaxial polydactyly, absence of corpus callosum, severe mental retardation and additional anomalies in two unrelated patients: A new syndrome. American Journal of Medical Genetics 6: 241–249

# 13

# Vertebral dysostoses

Involvement of the vertebral column is the predominant feature of the disorders in this group. The shoulder girdle and ribs are sometimes malformed and extra-skeletal anomalies may be present.

1. Klippel–Feil syndrome
2. Cervico-oculoacoustic syndrome (Wildervanck)
3. Oculoauriculovertebral syndrome (Goldenhar)
4. Spondylocostal dysostoses
5. Sprengel deformity
6. Miscellaneous vertebral disorders
   a. Cheirolumbar dysostosis
   b. Scheuermann disease

Some of these conditions are relatively common. Nevertheless, they are not clearly defined and there is considerable heterogeneity and overlap among them. Although a few uncommon forms are inherited in a Mendelian fashion, the majority have a multifactorial background.

Single or multiple vertebral abnormalities, without involvement of other regions of the skeleton, are fairly frequent. Various combinations of vertebral maldevelopment may occur, but they all present as spinal malalignment. These isolated vertebral malformations are almost always sporadic and non-genetic. They can be recognised in utero by ultrasonography (Abrams & Filly 1985) and in view of the potentially serious clinical implications, monitoring of subsequent pregnancies by this technique is warranted.

It is becoming increasingly apparent that structural renal abnormalities can occur in conjunction with multiple vertebral anomalies. For instance in a series of 47 patients with different disorders involving maldevelopment of the vertebrae, more than 20% had abnormal kidneys (Bernard et al 1985). In specific conditions such as the Klippel–Feil syndrome (vide infra) the frequency of renal anomalies may be considerably higher. On this basis, renal appraisal is mandatory in any person with vertebral malformations. Non-invasive ultrasonic investigations are now preferred to intravenous pyelography as the initial approach to this situation.

## KLIPPEL–FEIL* SYNDROME

The primary defect in the Klippel–Feil syndrome is fusion of the cervical vertebrae. The syndrome may exist in isolation, or in association with a wide variety of concomitants. It is also a component of a number of specific disorders, such as the Wildervanck syndrome (vide infra).

The neck is short, with a low posterior hair line and a limited range of movement. Torticollis, webbing of the neck, scoliosis and asymmetrical elevation of the scapula are often present. It has been reported that up to 30% of affected persons have some degree of deafness (Palant & Carter 1972). Associated spinal defects include cervical meningomyelocele, syringomyelia and spinal dysraphism (Sherk et al 1974). There is a high risk of neurological complications (Nagib et al 1984, Elster 1984) due in part to spinal stenosis (Prusick et al 1985). Mental deficiency, cleft palate and cardiac malformations are inconsistent features (Schey 1976). Renal abnormalities were present in 30% of a series of 50 patients studied by Hensinger et al (1974), in 64% of 39 patients investigated by Moore et al (1975) and in 52% of 35 persons assessed by Grise et al (1984).

*Genetics*

The birth frequency is approximately 1 in 40 000, with a female preponderance of about 20% (Gorlin et al 1976). This excess of affected girls might be the result of misdiagnosis of the Wildervanck syndrome (cervico-oculo-acoustic syndrome), which is virtually confined to females, and in which fusion of the cervical vertebrae is associated with perceptive deafness and abducens nerve palsy.

The Klippel–Feil syndrome is conventionally subclassified into three types, on a basis of the anatomical distribution of the vertebral abnormalities.

Type I fusion of cervical and upper thoracic vertebrae.
Type II localised fusion of cervical vertebrae.
Type III fusion of cervical and lower thoracic or lumbar vertebrae.

It is by no means proven that this subgrouping reflects true heterogeneity, although Gunderson et al (1967) found a dominant pattern of inheritance in kindreds with the type II abnormality. However, the great majority of cases are sporadic. Dominant inheritance with incomplete expression has been advanced as an explanation of the presence of minor manifestations in relatives. Alternatively, the syndrome may be the result of the action of an unknown environmental agent upon a genetically predisposed fetus. An

---

*Maurice Klippel (1858–1942) was the head of the department of medicine at the Hôspital Tenon, Paris. He published prodigiously and his last article was submitted when he was 84 years of age.

André Feil was Klippel's intern at the time their case report was published.

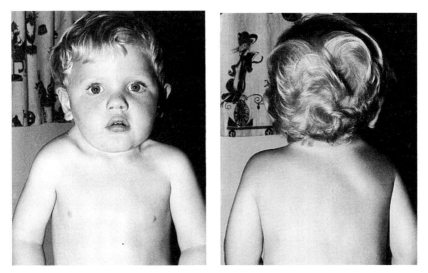

**Fig. 13.1** (left) Klippel–Feil syndrome: the neck is short and movements are restricted.

**Fig. 13.2** (right) Klippel–Feil syndrome: the hairline is low and the right scapula is elevated.

autosomal recessive form of the Klippel–Feil syndrome has been documented by da Silva (1982).

A few non-specific chromosomal abnormalities have been reported. For instance, a 4/14 chromosome translocation was detected in an obese boy with the Klippel–Feil syndrome (Berdel & Burmeister, 1974). The significance of these cytogenetic findings is uncertain.

Following an epidemiological study of 337 patients with congenital vertebral abnormalities, who had presented with scoliosis, Wynne-Davies (1975) suggested that multiple vertebral anomalies might be aetiologically related to spina bifida and anencephaly and estimated that subsequent sibs had a recurrence risk of about 10% for any of these defects. This concept has important implications for pregnancy screening and antenatal diagnosis but it has not achieved general acceptance.

The pathogenesis of the Klippel–Feil syndrome is unknown but Bavinck & Weaver (1986) suggested that interruption of the embryonic blood supply during the sixth week after conception might be responsible. These investigators proposed the term 'subclavian artery supply disruption sequence' for the process which could produce a predictable pattern of birth defects, notably the Poland, Klippel–Feil and Möbius anomalies.

## CERVICO-OCULOACOUSTIC SYNDROME (Wildervanck syndrome)

The cervico-oculoacoustic or Wildervanck syndrome comprises the Klippel–Feil syndrome, with congenital neural or conductive deafness, palsy of the sixth cranial nerve and retraction of the eyeball (Everberg et al 1936, Wildervanck et al 1966). The ocular features, taken alone, are termed the Duane syndrome. Other inconsistent abnormalities include mental retardation, cleft palate, epibulbar dermoids and anomalies of the external ears. Coloboma of the head of the optic nerve has also been recorded (Regenbogen et al 1985). There is considerable phenotypic overlap between the Wildervanck syndrome, the Klippel–Feil syndrome and the oculoauriculovertebral or Goldenhar syndrome (vide infra). Indeed, it is sometimes difficult to assign a patient to a specific category in this group of disorders.

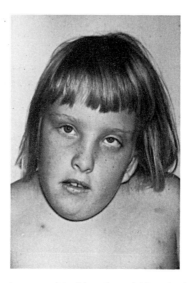

**Fig. 13.3** Wildervanck syndrome: a girl with a short rigid neck, deafness and a palsy of the left sixth cranial nerve.

*Genetics*

There is conjecture and controversy concerning the pathogenesis of the Wildervanck syndrome. As the vast majority of patients are females, it has been proposed that the condition might be an X-linked dominant, with lethality in the hemizygous male (McKusick 1975). An alternative earlier explanation was dominant inheritance with non-penetrance and very variable expression (Kirkham 1970). Konigsmark & Gorlin (1976) suggested that multifactorial inheritance would account for the great diversity of features and the familial tendency. In a comprehensive review of the disorder, Wildervanck (1978) suggested that inheritance was polygenic, with limitation to females.

OCULOAURICULOVERTEBRAL DYSPLASIA (Goldenhar syndrome)

Oculoauriculovertebral dysplasia is also known as the Goldenhar syndrome. The components are facial asymmetry, mandibular hypoplasia, abnormal external ears, epibulbar dermoids, colobomata of the upper eyelids and vertebral abnormalities (Goldenhar 1952). The condition is clinically similar to the Treacher Collins syndrome and the Wildervanck or cervico-oculo-acoustic syndrome. The similarity of the descriptive designations in this group of disorders is a source of confusion, and for this reason, there is obvious advantage in the retention of the eponyms.

The syndromic boundaries of the Goldenhar syndrome are ill-defined and the issue of heterogeneity is unresolved. It is probable that hemifacial microsomia and the Goldenhar syndrome are part of a spectrum of a single entity (Figueroa & Friede, 1985). Unusual manifestations in affected persons, such as tracheo-oesophageal fistula (Mendelberg et al 1985) and anterior encephalocele (Gustavson & Chen 1985) might provide clues to the pathogenesis, which has not yet been elucidated. The phenotypic features of 294 affected persons have been analysed by Rollnick et al (1987).

*Genetics*

The Goldenhar syndrome was inherited as an autosomal dominant in the large kindred studied by Summitt (1969) and dominant transmission has also been proposed by Godel et al (1982) and Regenbogen et al (1982). Conversely, autosomal recessive inheritance was possible in the affected sibs with normal parents reported by Saraux et al (1963) and Krause (1970). This anomalous situation could be the result of heterogeneity or diagnostic confusion.

Many patients have been sporadic, including those mentioned by Labrune & Choulot (1974), Ebbersen & Petersen (1982) and Stoll et al (1984). Monozygotic twins who were discordant for the condition were documented by Papp et al (1974). The Goldenhar syndrome was present in one of three infants produced by in vitro fertilisation (Yovich et al 1985). A deletion of the short arm of chromosome 18 was reported by Buffoni et al (1976). Trisomy 9 mosaicism in an affected girl was documented by Wilson & Barr (1983). Konigsmark & Gorlin (1976) suggested that inheritance of the Goldenhar syndrome was probably multifactorial. In a large scale survey based upon 97 patients with hemifacial microsomia or the Goldenhar syndrome, Rollnick & Kaye (1983) identified the condition in 8% of first degree relatives and documented great phenotypic variability. These authors interpreted their data as indicating syndromic homogeneity and multifactorial inheritance.

Mounoud et al (1975) described a young boy with the Goldenhar syndrome whose mother had accidently ingested a large quantity of vitamin A during the second month of pregnancy. These authors discussed the possible teratogenic effects of vitamin A and extensively reviewed the other aetiological factors which had previously been reported in the syndrome. Fetal exposure to the anticonvulsant drug, primidone, was recorded by Gustavson & Chen (1985).

Four patients with a unilateral blepharo-oculocranial dysplasia, which bore some resemblance to the Goldenhar syndrome, were reported by Lund (1974). There was no evidence to indicate that this particular disorder had a genetic basis.

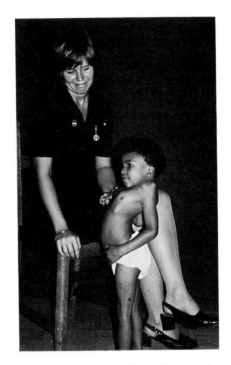

**Fig. 13.4** Spondylocostal dysplasia: an 8-year-old girl with gross abnormalities of the vertebrae and ribs.

## SPONDYLOCOSTAL DYSOSTOSES

The spondylocostal dysostoses or dysplasias (SCD) are a heterogeneous group of disorders in which abnormalities of vertebral segmentation are associated with distortion, fusion or absence of the ribs. Clinically, the trunk is shortened, with thoracic asymmetry and spinal deformity. Spinal cord compression is a potentially dangerous complication. Prognostication is not easy however, as the characteristics of the individual entities within this category have not been clearly defined.

The manifestations of 14 patients with spondylocostal dysplasia were analysed by Kozlowski (1984); 11 of these persons had additional anomalies which were also components of the VATER or VACTERL associations (see Ch. 16). He concluded that spondylocostal dysplasia could represent a heterogeneous genetic entity, either dominant or recessive, or be a part of a number of complex malformation syndromes.

*Genetics*

The classification of costovertebral segmentation anomalies is a matter of controversy and in a number of reports the title 'spondylothoracic dysostosis' (STD) has been employed for autosomal recessive conditions which fit into the diagnostic category of SCD (Moseley & Bonforte 1969, Castroviejo et al 1973). However, Solomon et al (1978) concluded, with some justification, that SCD and STD should be regarded as separate entities. In their review of the literature, these authors were able to identify 18 patients whom they considered to have STD and 17 with SCD.

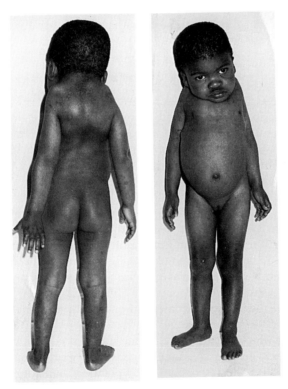

**Fig. 13.5** (left) Spondylocostal dysplasia: the neck is short and rigid. An elder sister is similarly affected, but the parents and her other sibs are normal. Inheritance in this kindred is autosomal recessive. (From Beighton & Horan 1981 Clinical Genetics 19: 23–25.)

**Fig. 13.6** (right) Spondylocostal dysplasia: the thorax is small and malformed.

A benign autosomal dominant form of SCD is recognised (Rimoin et al 1968), but there is disagreement concerning the separate syndromal status of the various autosomal recessive types of SCD. Devos et al (1978) contended that autosomal recessive SCD was a single entity. However, the presence in some individuals of additional manifestations, such as mental deficiency, renal involvement, cardiovascular changes and abnormalities of the extremities, may be indicative of heterogeneity (Pochaczevsky et al 1971, Silengo et al 1978). Nevertheless, there is no doubt that by far the most common form of uncomplicated SCD is inherited as an autosomal recessive trait. Good examples are the sets of siblings reported by Lavy et al (1966), Langer & Moe (1975) and Beighton & Horan (1981). A mildly affected girl born to consanguineous parents highlights the question as to whether or not there are separate mild and severe autosomal recessive forms of uncomplicated SCD (Young & Moore 1984).

Hemivertebrae and rib anomalies were present in an infant born to a woman who had taken lysergic acid diethylamide (LSD) in early pregnancy (Eller & Morton 1970), and in a mother and daughter with 14–15 chromosomal translocations (DeGrouchy et al 1963). It is not known if the spondylocostal malformations in these individuals were causally related to these exogenous or endogenous factors.

## Other forms of spondylocostal dysplasia

1. Under the designation 'cerebrofaciothoracic dysplasia', Castroviejo et al (1975) described two girls and a boy with mental retardation, unusual facies, vertebral anomalies and abnormality of the ribs. The authors considered that this disorder represented a new syndrome, and suggested that inheritance might be autosomal recessive.

2. Perez Comas & Garcia Castro (1974) reviewed the features of 'occipito-facial-cervico-thoracico-abdomino-digital dysplasia'. This anatomical litany is somewhat cumbersome and the original eponymous designation 'Jarcho–Levin syndrome' has considerable merit. In this rare condition, which is lethal in infancy, the ribs are often decreased in number and the thorax has a radiographic crab-like configuration (Jarcho & Levin 1938). This eponymn has been employed in case reports by Manzia et al (1976), Bull & Policastro (1976), Poor et al (1983), Cassidy et al (1984). It is uncertain whether this syndrome is the same condition as autosomal recessive SCD.

A disorder which resembles the Jarcho–Levin syndrome was documented in two brothers from a consanguineous Italian gypsy family by Ventruto & Catani (1986). The authors tabulated points of similarity and difference, and concluded that the disorders were separate entities.

3. Abnormalities of the ribs and vertebrae, together with short metacarpals and mandibular cysts, are a feature of the naevoid basal cell carcinoma syndrome (Koutnik et al 1975). The skeletal abnormalities are oversha-

dowed by the dermal naevi, which are prone to malignant degeneration. There have been over 200 reports of the condition, which is inherited as an autosomal dominant (Gorlin & Goltz 1960, Ferrier & Hinrichs 1967, Lile et al 1968).

4. Sibs with multiple hemivertebrae, with or without rib abnormalities, have been described by Caffey (1967) and Bartsocas et al (1974). Norum (1969) reported affected sibships in an inbred community. The relationship of these disorders and the other spondylocostal dysplasias is uncertain. However, inheritance was evidently autosomal recessive.

5. A syndrome of costovertebral segmentation defect with mesomelia was delineated by Wadia et al (1978). In addition to vertebral and rib abnormalities, these patients had an unusual facies with a triangular mouth, shortening of the forearms and contractures at the elbows. A further case was documented by Aymé & Preus (1986). Inheritance of this rare entity is probably autosomal recessive.

## SPRENGEL* DEFORMITY

Sprengel's shoulder, or congenital elevation of the scapula, may occur as an isolated anomaly. However, it is often accompanied by abnormalities of the vertebrae or ribs and it is a frequent concomitant of the Klippel–Feil syndrome. The deformity may be unilateral or bilateral. The literature concerning congenital abnormalities of the scapula, including the Sprengel deformity, has been reviewed by McClure & Raney (1975).

### Genetics

There have been reports of members of several generations of a kindred having uncomplicated Sprengel deformity (Engel 1943), and it is evident that the condition can be transmitted as a dominant trait. Nevertheless, the majority of cases are sporadic and the disorder does not usually have a genetic basis.

## MISCELLANEOUS VERTEBRAL DISORDERS

### Cheirolumbar dysostosis

Cheirolumbar dysostosis is a skeletal dysplasia in which narrowing of the spinal canal in the lumbar region is associated with brachydactyly. More than 30 cases of this disorder have been documented by Wackenheim (1978, 1985) and the importance of neurological complications has been emphasised. Affected persons have been diagnosed as having the Lawrence-Moon-

---

*Otto Sprengel (1852–1915) graduated at the University of Marburg and became professor of surgery at Braunschweig, Germany.

Biedl syndrome and pseudohypoparathyroidism and it is possible that cheirolumbar dysostosis is a component of these and other disorders rather than a distinct syndromic entity.

## Scheuermann* disease

Scheuermann disease is a common orthopaedic disorder which presents with backache and dorsal kyphosis in young adults (Scheuermann 1921). The course is very variable but in the majority of patients the disorder is comparatively benign. Radiographically, wedging and end plate irregularity are evident in affected vertebral bodies (Deacon et al 1985). The aetiology is unknown, but familial aggregation and generation to generation transmission are well documented (Bjersand 1980, Halal et al 1978). The syndromic boundaries are ill-defined, and it is possible that conditions termed 'Scheuermann disease' are a heterogeneous group of genetic and non-genetic disorders.

*Holgar Scheuermann (1877–1960) was a radiologist in Copenhagen, Denmark.

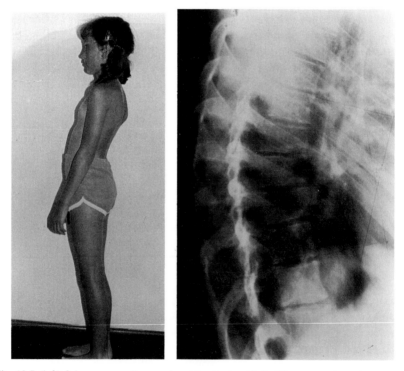

**Fig. 13.7** (left) Scheuermann disease: dorsal kyphosis with insidious onset in early adolescence.

**Fig. 13.8** (right) Scheuermann disease: lateral radiograph of the spine, showing the characteristic anterior vertebral wedging.

## REFERENCES

**Preamble**

Abrams S L, Filly R A 1985 Congenital vertebral malformations: Prenatal diagnosis using ultrasonography. Radiology 155/3: 762

Bernard T N Jr, Burke S W, Johnston C E III, Roberts J M 1985 Congenital spine deformities. A review of 47 cases. Orthopedics 8/6: 777–783

**Klippel–Feil syndrome**

Bavinck J N B, Weaver D D 1986 Subclavian artery supply disruption sequence: Hypothesis of a vascular etiology for Poland, Klippel–Feil, and Möbius anomalies. American Journal of Medical Genetics 23: 903–918

Berdel D, Burmeister W 1974 Pickwickian and Klippel–Feil syndromes in a boy aged 12 yers. Klinische Paediatrie 106/6: 467

Da Silva E O 1982 Autosomal recessive Klippel–Feil syndrome. Journal of Medical Genetics 19(2): 130–134

Elster A D 1984 Quadriplegia after minor trauma in the Klippel–Feil syndrome. A case report and review of the literature. Journal of Bone and Joint Surgery 66(9): 1473–1474

Gorlin R J, Pindborg J J, Cohen M M 1976 Syndromes of the Head and Neck 2nd edn. McGraw-Hill, New York, p 408

Grise P, Lefort J, Dewald M, Mitrofanoff P, Borde J 1984 Urinary malformations in synostoses of the cervical vertebrae (Klippel–Feil syndrome). Apropos of 35 cases. Annales D'Urologie 18(4): 232–235

Gunderson C G, Greenspan R H, Glaser G H, Lubs H A 1967 The Klippel–Feil syndrome: genetic and clinical re-evaluation of cervical fusion. Medicine 46: 491

Hensinger R N, Lang J E, Macewan G D 1974 Klippel–Feil syndrome. A constellation of associated anomalies. Journal of Bone and Joint Surgery 56/6: 1246

Juberg R C, Gershanikk J J 1976 Cervical vertebral fusion (Klippel–Feil) syndrome with consanguineous parents. Journal of Medical Genetics 13(3): 246–249

Moore W B, Matthews T J, Rabinowitz R 1975 Genitourinary anomalies associated with Klippel–Feil syndrome. Journal of Bone and Joint Surgery 57/3: 355

Nagib M G, Maxwell R E, Chou S N 1984 Identification and management of high-risk patients with Klippel–Feil syndrome. Journal of Neurosurgery 61(3): 523–530

Palant D J, Carter B L 1972 Klippel–Feil syndrome and deafness. American Journal of Diseases of Children 123: 218

Prusick V R, Samberg L C, Wesolowski D P 1985 Klippel–Feil syndrome associated with spinal stenosis. A case report. Journal of Bone and Joint Surgery 67(1): 161–164

Sherk H H, Shut L, Chung S 1974 Iniencephalic deformity of the cervical spine with Klippel–Feil anomalies and congenital elevation of the scapula. Report of three cases. Journal of Bone and Joint Surgery 56/6: 1254

Schey W 1976 Vertebral malformations and associated somaticovisceral abnormalities. Clinical Radiology 27: 341

Wynne-Davies R 1975 Congenital vertebral anomalies: aetiology and relationship to spina bifida cystica. Journal of Medical Genetics 12/3: 280

**Cervico-oculoacoustic syndrome (Wildervanck syndrome)**

Everberg G, Ratjen E, Sorensen H 1936 Klippel–Feil's syndrome associated with deafness and retraction of the eyeball. British Journal of Radiology 36: 562

Kirkham T H 1970 Inheritance of Duane's syndrome. British Journal of Ophthalmology 54: 323

Konigsmark B W, Gorlin R J 1976 Klippel–Feil anomolad and abducens paralysis with retracted bulb and sensorineural or conduction deafness. Genetic and Metabolic Deafness. Saunders, Philadelphia, p 188

McKusick V A 1975 Mendelian Inheritance in Man, 4th edn. Johns Hopkins University Press, Baltimore, p 663

Regenbogen L, Godel V 1985 Cervico-oculo-acoustic syndrome. Ophthalmic Paediatrics and Genetics 6/3: 183–187

Wildervanck L S 1978 The cervico-oculo-acusticus syndrome. In: Vinken P J, Bruyn G W, Myranthopoulos N C (eds) Handbook of Clinical Neurology Vol 32. North-Holland, Amsterdam, p 123–130

Wildervanck L S, Hoeksema P E, Penning L 1966 Radiological examination of the inner ear of deaf-mutes presenting the cervico-oculoacusticus syndrome. Acta Otolarynologica 61: 445

## Oculoauriculovertebral dysplasia (Goldenhar syndrome)

Buffoni L, Tarateta A, Aicardi G 1976 Hypophyseal nanism and Goldenhar multiple deformities in a subject with deletion of the short arm of chromosome 18. Minerva Paediatrica 28/12: 716

Ebbersen F, Petersen W 1982 Goldenhar's syndrome: discordance in monozygotic twins and unusual anomalies. Acta Paediatrica Scandinavica 71(4): 685–687

Figueroa A A, Friede H 1985 Craniovertebral malformations in hemifacial microsomia. Journal of Craniofacial Genetics and Developmental Biology 1: 167–178

Godel V, Regenbogen L, Goya V, Goodman R M 1982 Autosomal dominant Goldenhar syndrome. Birth Defects 18(6): 621–628

Goldenhar M 1952 Associations malformatives de l'oeil et de l'orielle. En particulier, le syndrome: dermoide épibulbaire-appendices auriculaires-fistula auris congenita et ses relations avec la dysostose mandibulo-faciale. Journal de Génétique Humaine 1: 243

Gustavon E E, Chen H 1985 Goldenhar syndrome, anterior encephalocele, and aqueductal stenosis following fetal primidone exposure. Teratology 32/1: 13–17

Konigsmark B W, Gorlin R J 1976 Klippel–Feil anomalad and abducens paralysis with retracted bulb and sensorineural or conduction deafness. Genetic and Metabolic Deafness, Saunders, Philadelphia, p 189

Krause U 1970 The syndrome of Goldenhar affecting two siblings. Acta Ophthalmologica 48: 494

Labrune B, Choulot J J 1974 Goldenhar's oculoauriculovertebral syndrome. Semaine des Hôpitaux de Paris 50/39: 713

Lund O E 1974 Blepharo-oculocranial dysplasia: a hitherto unknown syndrome. Medizinische Klinik 69/42: 1715

Mendelberg A, Ariel I, Mogle P, Arad I 1985 Tracheo-oesophageal anomalies in the Goldenhar anomalad. Journal of Medical Genetics 22(2): 149–150

Mounoud R L, Klein D, Weber F 1975 A case of Goldenhar's syndrome: acute maternal vitamin A poisoning during pregnancy. Journal de Génétique Humaine 23/2: 135

Papp Z, Gardo S, Walawska J 1974 Probable monozygotic twins with discordance for Goldenhar syndrome. Clinical Genetics 5/2: 86

Regenbogen L, Godel V, Goya V, Goodman R M 1982 Further evidence for an autosomal dominant form of oculoauriculovertebral dysplasia. Clinical Genetics 21(3): 161–167

Rollnick B R, Kaye C I 1983 Hemifacial microsomia and variants: pedigree data. American Journal of Medical Genetics 15: 233–253

Rollnick B R, Kaye C I, Nagatoshi K, Hauck W, Martin A O 1987 Oculoauriculovertebral dysplasia and variants: phenotypic characteristics of 294 patients. American Journal of Medical Genetics 26: 361–375

Saraux H, Grignon J L, Dhermy P 1963 A propos d'une observation familiale de syndrome de Franceschetti-Goldenhar. Bulletins et Mémoires de la Société Française d'Ophthalmologie 63: 705

Stoll C, Roth M P, Dott B, Bigel P 1984 Discordance for skeletal and cardiac defect in monozygotic twins. Acta Geneticae Medicae et Gemellologiae 33/3: 501–504

Summitt R L 1969 Familial Goldenhar syndrome. The Clinical Delineation of Birth Defects. 5/2. National Foundation, New York, p 106

Wilson G N, Barr M Jr 1983 Trisomy 9 Mosaicism: Another Etiology for the Manifestations of Goldenhar syndrome. Journal of Craniofacial Genetics and Developmental Biology 3: 313–316

Yovich J L, Stanger J D, Grauaug A A, Lunay G G, Hollingsworth P, Mulcahy M T 1985 Fetal abnormality (Goldenhar syndrome) occurring in one of triplet infants derived from in vitro fertilization with possible monozygotic twinning. Journal of In Vitro Fertilization and Embryo Transfer 2(1): 27–32

## Spondylocostal dysostoses

Aymé S, Preus M 1986 Spondylocostal/Spondylothoracic dysostosis: The clinical basis for prognosticating and genetic counseling. American Journal of Medical Genetics 24: 599–606

Bartsocas C S, Kiossoglou K A, Papas C V, Xanthou-Tsingoglou M, Anagnostakis D E Daskalopoulou H D 1974 Costovertebral dysplasia. Birth Defects: Original Article Series 10/9: 221

Beighton P, Horan 1981 Spondylocostal dysostosis in South African sisters. Clinical Genetics 19: 23–25

Bull M J, Policastro A 1976 Spondylothoracic dysplasia (costovertebral dysplasia, Jarcho–Levin syndrome). American Journal of Diseases of Children 130: 513–514

Caffey J P 1967 Normal vertebral column. Pediatric X-ray diagnosis 5th edn. Year Book Medical Press, Chicago, p 1101

Cassidy S B, Herson V, Tibbets J 1984 Natural history of Jarcho–Levin syndrome. Proceedings of the Greenwood Genetics Center 3: 90–91

Castroviejo I P, Rodriguez-Costa, Castillo F 1973 Spondylo-thoracic dysplasia in three sisters. Developmental Medicine and Child Neurology 15: 348–354

Castroviejo I, Santolaya J M, Lopez Martin V 1975 Cerebro-facio-thoracic dysplasia: report of three cases. Developmental Medicine and Child Neurology 17/3: 343

De Grouchy J, Mlynarski J C, Maroteaux P, Lamy M, Deshaies G, Benichou C, Salmon C 1963 Syndrome polydysspondylique par translocation 14–15 et dyschondrostéose chez un même sujet. Segregation familiale. Comptes Rendus Hebadomadaires de Séances de l'Académie 256: 1614

Devos E A, Leroy J G, Braeckman J J, van den Bulcke L J, Langer L O 1978 Spondylocostal dysostosis and urinary tract anomaly: Definition and review of an entity. European Journal of Pediatrics 128: 7–15

Eller J L, Morton J M 1970 Bizarre deformities in offspring of user of lysergic acid diethylamide. New England Journal of Medicine 283: 395

Ferrier P E, Hinrichs W L 1967 Basal-cell carcinoma syndrome. American Journal of Diseases of Children 113: 538

Gorlin R J, Goltz R W 1960 Multiple nevoid basal-cell epithelioma, jaw cysts and bifid rib: a syndrome. New England Journal of Medicine 262: 908

Jarcho S, Levin P M 1938 Hereditary malformation of the vertebral bodies. Bulletin of the Johns Hopkins Hospital 62: 216–226

Koutnik A W, Kolodny S C, Hooker S P, Roche W C 1975 Multiple nevoid basal cell epithelioma, cysts of the jaw and bifid rib syndrome: report of a case. Journal of Oral Surgery 33/9: 686

Kozlowski K 1984 Spondylo-costal dysplasia. A further report — review of 14 cases. Fortschritte auf dem Gebiete der Röntgenstrahlen und der Nuklearmedizin 140/2: 204–209

Lavy N W, Palmer C G, Merrit A D 1966 A syndrome of bizarre vertebral anomalies. Journal of Pediatrics 69: 1121

Langer L O, Moe J H 1975 A recessive form of congenital scoliosis different from spondylothoracic dysplasia. Birth defects: Original Article Series 11/6: 83

Lile H A, Rogers J F, Gerald B 1968 The basal cell nevus syndrome. American Journal of Roentgenology, Radium Therapy and Nuclear Medicine 103: 214

Manzia S, Cortesi M, Grazioli M, Bonacini G C 1976 La sindrome delle anomalie vertebro-costali di Jarcho–Levin. Minerva Pediatrica 28: 2141–2146

Moseley J E, Bonforte R J 1969 Spondylothoracic dysplasia: A syndrome of congenital anomalies. American Journal of Roentgenology 106: 166–169

Norum R A 1969 Costovertebral anomalies with apparent recessive inheritance. The Clinical Delineation of Birth Defects, Vol 4. National Foundation, New York, p 326

Perez Comas A, Garcia Castro J M 1974 Occipitofacial-cervico-thoracic-abdomino-digital dysplasia: Jarcho–Levin syndrome of vertebral anomalies. Journal of Pediatrics 85/3: 388

Pochaczevsky M D, Ratner H, Perles D, Kassner G, Naysan P 1971 Spondylothoracic dysplasia. Radiology 98: 53–58

Poor M A, Alberti O, Griscom N T, Driscoll S G, Holmes L B 1983 Nonskeletal malformations in one of three siblings with Jarcho–Levin syndrome of vertebral anomalies. Journal of Pediatrics 103: 270–272

Rimoin D, Fletcher B D, McKusick V A 1968 Spondylocostal dysplasia: a dominantly inherited form of short-trunked dwarfism. American Journal of Medicine 45: 948

Silengo M C, Cavallaro C, Franceschini P 1978 Recessive spondylocostal dysostosis: Two new cases. Clinical Genetics 13: 289–294

Solomon L, Jimenez R B, Reiner L 1978 Spondylothoracic dysostosis. Archives of Pathological & Laboratory Medicine 102: 210–205

Ventruto V, Catani L 1986 New Syndrome: Progressive scoliosis by unilateral unsegmented fusion bar, foot deformity, joint laxity, congenital inguinal herniae, peculiar face. American Journal of Medical Genetics 25: 429–432

Wadia R S, Shirole D B, Dikshit M S 1978 Recessively inherited costovertebral segmentation defect with mesomelia and peculiar facies (COVESDEM syndrome). A new genetic entity? Journal of Medical Genetics 15: 123–127

Young I D, Moore J R 1984 Spondylocostal dysostosis. Journal of Medical Genetics 21: 68–69

## Sprengel deformity

Engel D 1943 The etiology of the undescended scapula and related syndromes. Journal of Bone and Joint Surgery 25: 613

McClure J G, Raney B R 1975 Anomalies of the scapula. Clinical Orthopaedics and Related Research 110: 22

## Miscellaneous vertebral disorders

*Cheirolumbar dysostosis*

Wackenheim A 1978 Une dysostose cheirolombaire (brachymetacarpophalangie et dysostose sténosante de l'arc vertébral postérieur). Journal of Radiology 59: 563

Wackenheim A 1985 Cheirolumbar dysostosis and constitutional narrowness of the cervical spinal canal. Skeletal Radiology 14: 47–52

*Scheuermann disease*

Bjersand A J 1980 Juvenile kyphosis in identical twins. American Journal of Roentgenology 134: 598–599

Deacon P, Berkin C R, Dickson R A 1985 Combined idiopathic kyphosis and scoliosis. An analysis of the lateral spinal curvatures associated with Scheuermann's disease. Journal of Bone and Joint Surgery 67/2: 189–192

Halal F, Gledhill R B, Fraser F C 1978 Dominant inheritance of Scheuermann's juvenile kyphosis. American Journal of Diseases of Children 132: 1105–1107

Scheuermann H W 1921 Kyphosis dorsalis juvenilis. Zeitschrift für Orthopadie und ihre Grenzgebiete 41: 305

# 14

# Stiff joint syndromes

The following stiff joint syndromes form the subject of this chapter:

1. Arthrogryposis multiplex congenita
2. Rigidity syndromes
   a. Ankylosis-pulmonary hypoplasia (Pena-Shokeir I) syndrome
   b. Cerebro-oculofacio-skeletal (Pena-Shokeir II) syndrome
   c. Goodman camptodactyly syndrome
   d. Emery-Nelson syndrome
   e. Liebenberg syndrome
   f. Mietens syndrome
   g. Kuskokwim disease
   h. Contractural myopathies
   i. Juvenile hyaline fibromatosis
3. Stiff hand-foot syndromes (distal arthrogryposis)
   a. Distal arthrogryposis
   b. Camptodactyly-cleft palate-clubfoot syndrome (Gordon)
   c. Digitotalar dysmorphism
   d. Digital-ulnar drift syndrome
   e. Digital deviation-clubfoot syndrome
   f. Tel-Hashomer camptodactyly syndrome
   g. Dominant metatarsus varus
   h. Trismus-pseudocamptodactyly syndrome
4. Flexed digit syndromes
5. Adducted thumb syndromes
6. Pterygium syndromes
   a. Popliteal pterygium syndrome
   b. Lethal pterygium syndrome

Articular rigidity is also a component of many well defined genetic entities, such as diastrophic dwarfism, the Freeman–Sheldon syndrome and the Schwartz syndrome. These conditions fall into a variety of nosological categories and several have been considered elsewhere.

Restricted joint movement may result from neurological damage, as in meningomyelocele and cerebral palsy, and from rheumatic disease during childhood. Diminution in articular mobility may also be the consequence of dislocation or subluxation, which are common complications in a number of the syndromes in which joint laxity predominates. These 'secondary' forms of joint stiffness enter into the differential diagnosis of any patient with articular rigidity.

Large scale surveys and literature reviews of the stiff joint syndromes have been undertaken by Hall et al (1982, 1983) and Gericke et al (1984).

## ARTHROGRYPOSIS MULTIPLEX CONGENITA

The term 'arthrogryposis multiplex congenita' (AMC) connotes restricted articular mobility which is present at birth. This non-specific designation is used interchangeably with the modern title 'multiple congenital contractures' but in order to avoid confusion, 'AMC' will be retained in this chapter. 'Primary' or 'classical' AMC, which is non-genetic, forms the subject of this section.

'Primary AMC', in the narrow sense of the definition, is a condition in

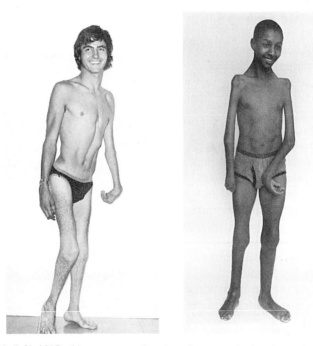

**Fig. 14.1** (left) AMC: this young man has the primary or classical form of the condition. Widespread articular rigidity causes considerable disability.

**Fig. 14.2** (right) AMC: the muscles may be hypoplastic. However, there are no systemic ramifications and general health is good. (From Davidson J, Beighton P 1976 Journal of Bone and Joint Surgery 58: 492.)

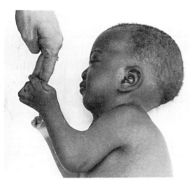

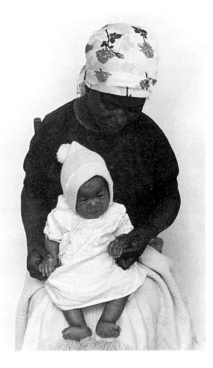

**Fig. 14.3** (above) AMC: soft tissue webbing may be present across the flexor surfaces of rigid joints.

**Fig. 14.4** (right) AMC: articular rigidity is a feature of a number of heritable disorders which must be differentiated from classical AMC for the sake of genetic prognostication. This infant has a familial rigid digit-club foot syndrome.

which infants are born with a limited range of movement or with fixed deformities of multiple joints. Muscular hypoplasia is often a prominent feature, and soft tissue webbing may be present. Due to the rigidity, fractures of the long bones may be sustained during delivery. Congenital dislocation of the hips and talipes equinovarus are common concomitants. AMC is not progressive, but affected individuals are severely disabled. Radiographically, the skeleton is undermineralised and gracile, and the hip joints may be dysplastic.

The basic defect probably lies in the anterior horn cells in the majority of cases (Brown et al 1980). The designation 'amyoplasia' has been applied to this sporadic disorder (Hall et al 1983).

*Genetics*

AMC is not uncommon, and large series of patients have been reviewed by Friedlander et al (1968), Lloyd-Roberts & Lettin (1970), and Hall et al (1976). The overwhelming majority of cases have been sporadic and it is probable that the condition is non-genetic. A maternal age effect was noted by Wynne-Davies & Lloyd-Roberts (1976), in their study of 66 cases. Drachman (1971) suggested that the clinical manifestations might be the consequence of inadequate development of articular and periarticular tissues, due to impairment of fetal movements. Evidence for this attractive hypothesis was provided by experiments in which chick embryos, previously paralysed with curare, were found to have an arthrogrypotic-like condition when they hatched (Drachman & Coulombre, 1962). Further support came from the observations of Jago (1970), who described an infant with AMC born to a mother who had contracted tetanus during early pregnancy, and had been treated with curare. If an environmental determinant does exist, it must be ubiquitous, as cases of AMC have been reported from all parts of the world. In this context, demographic data obtained during an investigation of 26 patients suggests that the condition is increasing in prevalence in South Africa (Davidson & Beighton 1976).

A unifying concept for the mechanism of congenital contractures has been proposed by Swinyard & Bleck (1985). They pointed out that the multiple factors which determine AMC result in loss of muscle mass and imbalance of muscle power at joints. This provokes collagenous thickening of joint capsules, leading to articular fixation.

Chromosomal abnormalities are present in about 10% of sporadic persons with AMC plus mental deficiency or developmental delay (Reed et al 1985). There is no correlation, however, between the nature of the cytogenetic abnormalities and the phenotype.

When familial AMC is encountered, it is virtually certain that the condition represents a distinct genetic entity, which may or may not have been previously delineated. Conversely, sporadic individuals with the classical phenotypic features of primary AMC do not have a heritable disorder and the recurrence risk is therefore very low.

AMC has been diagnosed by ultrasound at the thirtieth week of gestation (Goldberg et al 1986). At present it is not known if early prenatal detection is possible.

## RIGIDITY SYNDROMES

Articular rigidity is present in numerous heritable conditions. In some, non-articular manifestations predominate, while in others, the changes are virtually confined to the joints. Flexion contracture of the digits (camptodactyly) is a feature of many of these disorders.

### Ankylosis-facial anomaly-pulmonary hypoplasia syndrome (Pena-Shokeir* syndrome I)

Pena & Shokeir (1974) described two sisters with camptodactyly, clubfeet, ankylosis of the hips and knees, an unusual facies and pulmonary hypoplasia. The affected infants died in the perinatal period from respiratory

---

* Sergio Pena is the professor of biochemistry at the Federal University of Minas Gerais, Brazil.

Mohamed Shokeir is professor of pediatrics at the University of Saskatchewan, Canada.

problems. Subsequently, Pena & Shokier (1976) encountered a further three cases, including a brother and sister who were the offspring of an incestuous relationship between uncle and niece. The total of reported cases rose to seven when a pair of affected brothers were described by Mease et al (1976).

Although more than 30 cases have now been documented, there is grave doubt concerning syndromic identity (Moerman et al 1983, Lindhout et al 1985). The phenotype can be produced by the Potter sequence, trisomy 18, the lethal pterygium syndrome, and by a lethal autosomal recessive disorder recognised in 16 persons in 10 families in Finland (Herva et al 1985). A similar condition comprising congenital contractures, oedema, hyperkeratosis and intra-uterine growth retardation, has been encountered in endogamous Hutterite and Mennonite families in Canada (Lowry et al 1985).

Pena-Shokeir syndrome may represent an agglomoration of separate entities and it is impossible to quantitate recurrence risks. Prenatal diagnosis by ultrasound has been achieved in the second trimester (MacMilland et al 1985) and after the birth of an affected child surveillance of subsequent pregnancies would be justified.

## Cerebro-oculofacio-skeletal syndrome (Pena-Shokeir syndrome II)

The cerebro-oculofacio-skeletal syndrome (COFS) was delineated by Pena & Shokeir (1974) following their studies of 10 affected infants. The major manifestations were restriction on movements of the knees and elbows, coxa valga, acetabular dysplasia, 'rocker bottom' feet due to vertical talus, clenched hands and camptodactyly. Microcephaly, microphthalmia, micrognathia and large ears were additional features. These infants failed to thrive, and death from respiratory infection took place before the age of three.

Seven of the children were members of a consanguineous American Indian kindred living in Manitoba, Canada. Pena & Shokeir (1974) suggested that the abnormal gene might have been introduced by two French sisters who married into this community in the late nineteenth century. In further reports Preus & Fraser (1974) described an affected infant with consanguineous Italian parents and Lurie et al (1976) mentioned a girl born into an inbred Gipsy kindred.

There is ample evidence that the COFS syndrome is inherited as an autosomal recessive. A similar disorder was encountered by Lowry et al (1971) in a pair of sibs. The hand and foot deformities were less severe in these two infants, and this condition might be a separate entity.

The COFS syndrome is sometimes termed 'Pena-Shokeir syndrome type II' in order to distinguish it from a similar condition described in the same year by the same authors, and now designated 'Pena-Shokeir syndrome type I'.

The COFS and Pena-Shokeir syndromes have some features in common with the Neu-Laxova syndrome, in which joint contractures and syndactyly are associated with microcephaly and other variable anomalies (Neu et al 1971, Laxova et al 1972). About 20 cases of the Neu-Laxova syndrome have been documented and autosomal recessive inheritance is confirmed (Lazjuk et al 1979, Fitch et al 1982, Turkel et al 1983, Shved et al 1985). It has been suggested that the Neu-Laxova and COFS syndromes may be the result of variable phenotypic expression of the same genetic defect (Preus & Fraser 1974). A report of a neonate with features of both disorders lends weight to this hypothesis (Silengo et al 1984).

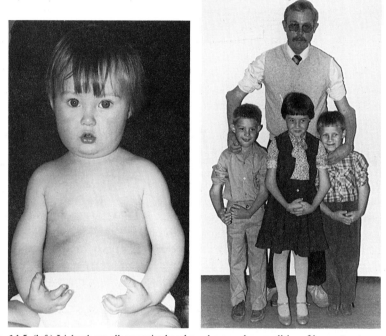

**Fig. 14.5** (left) Liebenberg elbow-wrist-hand syndrome: abnormalities of bone structure produce flexion deformities in the arms and hands. The skeleton is otherwise normal.

**Fig. 14.6** (right) Liebenberg syndrome: a father and three offspring with flexion deformities of the elbows, wrists and hands. This 'private syndrome' is present in five generations of the kindred.

### Goodman⋆ camptodactyly-fibrous tissue hyperplasia-skeletal dysplasia syndrome

Goodman et al (1972) described a young man and his two sisters in whom digital contractures had become evident at the end of the first decade. A Marfanoid habitus, mild thoracic scoliosis, knuckle pads and low intelligence were additional features. The parents, who were of Iranian–Jewish

---

⋆ Richard Goodman is Professor of Human Genetics at Tel Aviv University, Israel.

stock, were first cousins, and it is therefore probable that inheritance was autosomal recessive.

## Emery–Nelson** syndrome

Emery & Nelson (1970) reported a Scottish mother and daughter with clawing of the fingers and toes, extension deformities of the thumbs, mild pes cavus, short stature, a high forehead, depressed nasal bridge, and flattened malar regions. Using the designation 'Emery–Nelson syndrome', Gorlin et al (1976) mentioned that they had also encountered a mother and two sons with the disorder. The mode of inheritance is dominant.

## Liebenberg*** elbow-wrist-hand syndrome

Liebenberg (1973) investigated a South African kindred, of Afrikaner stock, in whom flexion deformities of the elbow, wrist and digits were present in five generations. The restricted joint movements were the consequence of structural abnormalities of the bones of the forearms and hands, and the skeleton was otherwise normal. This disorder is inherited as an autosomal dominant.

## Mietens syndrome

Mietens & Weber (1966) studied four sibs with flexion contractures of the elbows, in association with low intelligence, corneal opacity and nasal narrowing. The tubular bones of the forearms were short and the radial heads were dislocated. Thoracic asymmetry, varus feet and abnormalities in the hip and knee joints were present in some of these children. The parents and two other sibs were normal. There was a suggestion of consanguinity in the kindred, and it is likely that inheritance was autosomal recessive.

## Kuskokwim disease

Kuskokwim disease is an arthrogrypotic-like disorder which is present in a consanguineous Eskimo community in Alaska. Fixed deformities of the large weight bearing joints cause severe disability. The condition is inherited as an autosomal recessive (Petajan et al 1969, Wright & Aase 1969).

---

** Alan Emery is Emeritus Professor of Human Genetics of the University of Edinburgh, Scotland.

Molly Nelson is Senior Lecturer in Human Genetics at the University of Cape Town, South Africa.

*** Fred Liebenberg is a hand surgeon in Pretoria, South Africa.

## Contractural myopathies

Multiple contractures are an early manifestation of several of the infantile neuromyopathies (Emery & Walton 1967, Emery 1971). Distinction between these autosomal recessive disorders and non-genetic primary arthrogryposis is not always an easy matter.

Progressive contractures which are most evident in the ankles and elbows are an important diagnostic marker of Emery-Dreifuss* muscular dystrophy (Emery & Dreifuss 1966). In this X-linked disorder, defective cardiac conduction predisposes to sudden death in adulthood. By means of molecular techniques the faulty gene has been localised to the distal region of the long arm of the X-chromosome (Yates et al 1986, Thomas et al 1986).

## Juvenile hyaline fibromatosis

Joint contractures are a major feature of juvenile hyaline fibromatosis. Other manifestations include gingival hyperplasia, multiple tumours of the face and osteolytic defects. The condition is progressive and affected

* Fritz Dreifuss is Professor of Neurology at the University of Virginia, USA.

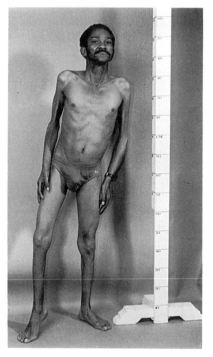

**Fig. 14.7** Emery-Dreifuss syndrome: progressive contractures, maximal at the elbows and ankles. (From Oswald et al 1987 South African Medical Journal 72, 8: 567–570.)

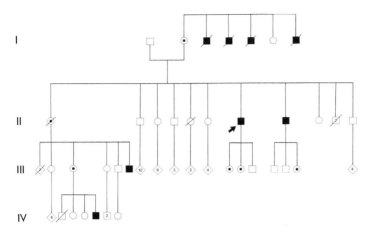

**Fig. 14.8** Emery-Dreifuss syndrome: pedigree, showing X-linked inheritance in an affected kindred. (From Oswald et al 1987 South African Medical Journal 72, 8: 567–570.)

persons can be severely handicapped by the articular changes and disfigured by the skin lesions.

Fayad et al (1987) documented Lebanese siblings with consanguineous parents, and mentioned that a total of 17 cases had been recorded. Other case reports include those of Drescher et al (1967), Kitano et al (1972), Stringer & Hall (1981), Finlay et al (1983). Inheritance is undoubtedly autosomal recessive.

## STIFF HAND-FOOT SYNDROMES

Deformity of the hands and feet, due to rigidity of the joints, is the predominant feature of a group of genetic disorders. Several of these conditions have been reported in only single kindreds and they are therefore regarded as 'private' syndromes. There is little doubt that many others await delineation.

### Distal arthrogryposis

Distal arthrogryposis type I was delineated by Hall et al (1982) in a series of 14 cases, as part of a comprehensive review of congenital contractural syndromes. The main features at birth are tightly clenched fists with medially overlapping fingers and positional foot deformities. Contractures at other joints are variable. The abnormalities tend to improve with the passage of time. Inheritance is autosomal dominant. This disorder is clearly distinguishable from classical arthrogryposis and as it is comparatively mild, it is probably underdiagnosed.

Hall et al (1982) also proposed further forms of the condition which were designated 'distal arthrogryposis type II, subcategories A to E'. These were

**Fig. 14.9** Camptodactyly-cleft palate-clubfoot syndrome: an affected father and daughter.

characterised by additional features, notably spinal and craniofacial abnormalities. Type IIA distal arthrogryposis corresponds to the previously delineated Gordon syndrome, which is outlined below. In type IIB stature is stunted and ptosis, keratoconus and large ears are present. Transmission is autosomal dominant but a sporadic Japanese girl with consanguineous parents raises the possibility of an autosomal recessive form (Tsukhara & Kajii 1984).

Subcategories C to E were documented in very small numbers of patients and syndromic status remains insecure. A mother and affected dizygotic twin offspring had additional manifestations which were suggestive of yet another form of distal arthrogryposis (Kawira & Bender 1985) and an unclassifiable Australian family was recorded by Reiss & Sheffield (1986).

### Camptodactyly-cleft palate-clubfoot syndrome (Gordon* syndrome)

Gordon et al (1969) reported a large Cape Town family of mixed ancestry, in whom flexion deformities of the fingers and talipes equinovarus were variably associated with cleft palate. Since the original report, which concerned members of three generations, further affected individuals have been born into this kindred. The mode of transmission is autosomal dominant.

Other affected families have been reported by Higgins et al (1972), Halal & Fraser (1979) and Robinow & Johnson (1981). In terms of syndromic

* Hymie Gordon is Head of the Division of Medical Genetics at the Mayo Clinic, USA.

delineation, there has been some discussion concerning stature in the Gordon syndrome. It is noteworthy that although short stature has been mentioned as a syndromic component, the physique of the affected members of the original family documented by Gordon et al (1969) is well within the normal range for their population of origin.

### Digitotalar dysmorphism

Digitotalar dysmorphism was recognised in nine members of five generations of a South African kindred of English stock (Sallis & Beighton 1972). These individuals had flexion deformities of the fingers with narrowing of the middle phalanges. The feet had a 'rocker bottom' configuration, due to vertical talus. The author has subsequently encountered digitotalar dysmorphism in a large African Negro family and in a father and son in England. Inheritance is clearly autosomal dominant. A similar condition has been studied in 15 members of four generations of a kindred and reported under the designation 'thumb-clutched hand syndrome' by Canale et al (1976).

The surgical management of the abnormalities in a father and son with the typical stigmata of digitotalar dysmorphism was discussed by Dhaliwal & Myers (1985).

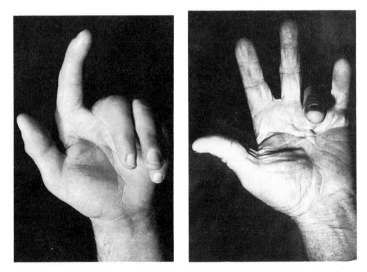

**Fig. 14.10** (left) Digitotalar dysmorphism: flexion deformity of the digits. Vertical talus is the other component of the syndrome. (From Sallis J, Beighton P 1972 Journal of Bone and Joint Surgery 54: 509.)

**Fig. 14.11** (right) Dupuytren contracture: this abnormality can mimic the various flexed digit syndromes.

### Digital-ulnar drift syndrome

Stevenson et al (1975) reported an American kindred with hand abnormalities which resembled those of digitotalar dysmorphism. However, the feet were normal, and the authors concluded that the condition was a distinct autosomal dominant disorder, which they designated 'congenital ulnar drift with webbing and flexion contractures of the fingers.' Powers & Ledbetter (1976) and Fried & Mundel (1976) encountered kindreds with a similar syndrome.

### Digital deviation-clubfoot syndrome

Fisk et al (1974) described a family with dominantly inherited deformities of the extremities. The term 'congenital ulnar deviation of the fingers, with clubfoot deformities' was used in the report, in which it was stated that maldevelopment of the digital extensor tendons was probably an underlying pathogenic factor.

### Tel-Hashomer camptodactyly syndrome

Goodman et al (1972) studied a brother and sister of Jewish Moroccan stock, in whom flexion contractures of the fingers were associated with digital spindling, clubbing of the feet, short stature, thoracic scoliosis, muscular hypoplasia and an unusual facies. Subsequently, Goodman et al (1976) reported two affected sibs from a different family using the designation 'Tel-Hashomer camptodactyly syndrome', and confirmed that inheritance was autosomal recessive.

The condition has been identified in Brazilian siblings born to consanguineous parents (Gollop & Colletto 1984). Fifth finger camptodactyly in the father was interpreted as a possible heterozygous manifestation. Affected siblings have also been documented in a consanguineous Libyan family (Tylki-Szymanska 1986).

### Dominantly inherited metatarsus varus

Metatarsus varus is a form of clubfoot, in which the anterior portion of the extremity is angulated medially. There are various types of this anomaly, which are defined on a basis of their clinical manifestations and response to treatment. As with talipes equinovarus, inheritance is generally multifactorial. Juberg & Touchstone (1974) reported a family in which nine members of four generations had metatarsus varus, together with variable minor anomalies, including fixed extension of digits and other joints. The authors recognised that their pedigree data were consistent with autosomal dominant inheritance. They suggested that metatarsus varus was yet another congenital deformity which displayed heterogeneity, and that it exists in common polygenic and uncommon monogenic forms.

**Fig. 14.12** Trismus-pseudocampodactyly syndrome: a young woman with contractures of her fingers and inability to fully open her mouth.

## Trismus-pseudocamptodactyly syndrome

The trismus-pseudocamptodactyly syndrome is characterised by the association of contractures of the fingers, foot deformities and the inability to fully open the mouth. The condition was delineated by Hecht & Beals (1969) and more than 300 affected persons have now been documented. Phenotypic expression is variable and trismus is not always present (Tsukahara et al 1985). Autosomal dominant inheritance has been firmly established but the locus of the faulty gene is unknown (Robertson et al 1982).

Several of the reported families have been of Dutch stock (de Jong 1971, Ter Haar & van Hoof 1974, Mabry et al 1974). It is uncertain whether there is a genuine preponderance in this population or whether diagnostic awareness has facilitated documentation in this community.

## FLEXED DIGIT SYNDROMES

Flexion deformities of single or multiple digits may occur in isolation or as a component of a large number of malformation syndromes. Gordon et al (1969) listed the conditions in which camptodactyly may be present. This list was greatly expanded when Rozin et al (1984) identified 44 camptodactyly syndromes. In their comprehensive review these authors sub-categorised the disorders in terms of chromosomal anomalies, major central nervous system involvement and multisystem defects. Some of these disorders have been mentioned elsewhere in this chapter and consideration of the majority of conditions is outside the scope of this book.

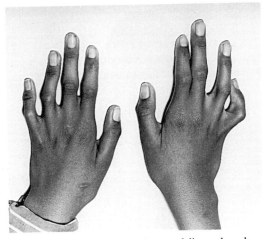

**Fig. 14.13** Digital flexion: the hands of a girl with an undelineated syndrome of digital flexion, deafness and an unusual facies.

The term 'camptodactyly' implies flexion contraction, usually at the proximal interphalangeal joints, while 'clinodactyly' denotes radial curvature, typically of the fifth finger. In practice, it is not always easy to draw a precise distinction between these abnormalities and the designations are often used synonymously. The alternative term 'streblodactyly' pertains to a digit which is twisted or bent, while 'streblomicrodactyly' denotes a shortened, crooked digit. Temtamy & McKusick (1969), in their modification of the classification of dominant uncomplicated brachydactyly syndromes of Bell (1951), listed clinodactyly as 'type A-3' (vide Ch. 15).

Isolated camptodactyly or clinodactyly of the fifth finger is often sporadic. However, there have been several reports of large kindreds with dominant

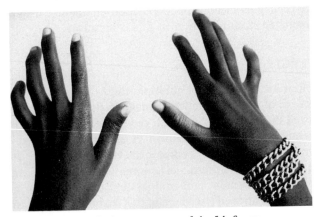

**Fig. 14.14** (left) Clinodactyly: flexion contracture of the 5th fingers.

inheritance of the deformity (Hefner 1941, Spear 1946, Welch & Temtamy 1966). Emphasising the anatomical disparity and the semantic imprecision concerning flexion deformities of the fifth finger, Katz (1970) reported an American kindred with 17 affected individuals in five generations. Of these, seven had incurving of the distal interphalangeal joint, while 10 had streblo-microdactyly. Alluding to the report of Kirner (1927), which concerned flexion and radial bowing of the terminal phalanges of the fifth finger,

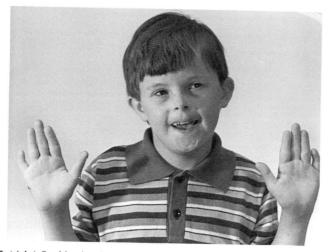

**Fig. 14.15** (right) Streblomicrodactyly: the hands of a child with Down's syndrome. The 5th fingers are short and crooked.

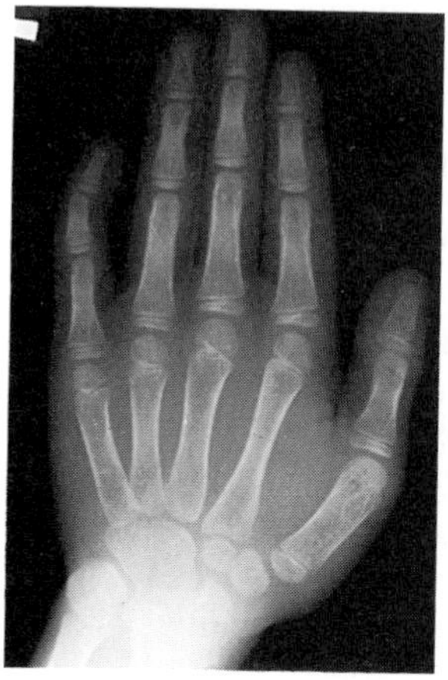

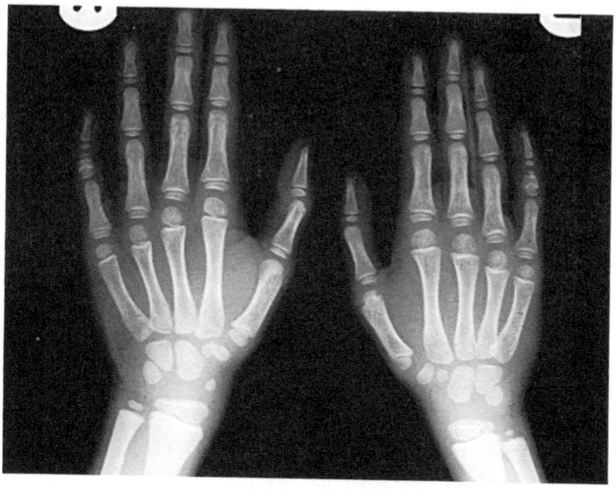

**Fig. 14.16** (left) Clinodactyly: radiograph showing flexion and radial deviation of the 5th finger.

**Fig. 14.17** (right) Clinodactyly: the term 'Kirner deformity' denotes flexion of the 5th finger due to malformation of the middle phalynx. This abnormality is also designated 'brachydactyly type A3'.

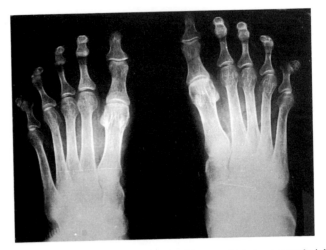

**Fig. 14.18** Transkei foot: lateral deviation of the 5th toe, which is apparently inherited as autosomal dominant in two Xhosa kindreds in Transkei, Southern Africa. (From Schwartz P A et al 1974 South African Medical Journal 48: 961.)

David & Burwood (1972) described nine English kindreds with 18 affected members. The authors discussed the radiological features of the bone changes in the phalanges in 'Kirner's deformity' and concluded that inheritance was autosomal dominant with variable penetrance.

Pronounced lateral deviation of the fifth toes was observed in members of two kindreds during the course of an epidemiological survey of bone and joint disorders in a tribal Xhosa population in Transkei, Southern Africa (Schwartz et al 1974). The condition which was designated 'Transkei foot', was present in successive generations, and it is possible that inheritance was dominant.

Newly delineated flexed digit syndromes are listed below:

1. Ankylosis of the teeth, abnormalities of the jaws and fifth finger clinodactyly. AD (Pelias & Kinnebrew 1985).
2. Lower limb asymmetry, scoliosis and syncamptodactyly of the second and third toes. AD (Halal 1985).
3. Tetralogy of Fallot, characteristic facies, preauricular pits and fifth finger clinodactyly. AD (Jones & Waldman 1985).
4. Digital flexion deformities with aplasia of extensor muscles and polyneuropathy. AR (Hamanishi et al 1986).
5. Familial proportionate short stature with recurrent locking of the fingers. AD (Eng & Strom 1987).

## ADDUCTED THUMB SYNDROMES

Adduction deformity of the thumbs is a prominent feature of a number of genetic syndromes. These have been reviewed by Fitch & Levy (1975). Autosomal dominant entities in this group include the 'clasped thumb' syndrome described by Weckesser et al (1968) and an adducted thumb-mental deficiency syndrome (MASA) reported by Bianchine & Lewis (1974). Adduction contracture of the thumb due to congenital hypoplasia of the muscles of the thenar eminence was observed in eight individuals, including three members of a kindred, by Strauch & Spinner (1976). The authors suggested that inheritance was probably autosomal dominant. Christian et al (1971) described an autosomal recessive adducted thumb syndrome which was present in an inbred Amish community. Joint contractures, craniostenosis and cleft palate were additional features.

A clenched hand, with adduction of the thumb across the palm, may be the presenting feature of trisomy 18 in the newborn. A phenocopy of the disorder, in which the chromosomes are normal, has been reported (Hongre et al 1972). In addition, an adducted thumb is a useful diagnostic feature of X-linked hydrocephalus (Edwards 1961).

## PTERYGIUM SYNDROMES

The pterygium syndromes are a group of rare disorders which share the common feature of webs of skin across joints. In the strict sense, the term 'pterygium' denotes a thin web of skin but it is also applied loosely and incorrectly to the thick fleshy soft tissues which are often present over the flexural aspects of joints which have been immobile in utero. For this reason the arthrogryposis and congenital contractural syndromes have sometimes been mislabelled 'pterygium' to the detriment of nosological clarity.

### Popliteal pterygium syndrome

The popliteal pterygium syndrome is characterised by webs of skin and scanty soft tissues which may extend from the calf to the thigh and severely limit extension of the knee joint (Gorlin et al 1968, Escobar et al 1978). The periphery of the web contains a neurovascular bundle and surgical correction is very difficult. Life span is unimpaired but handicap may be severe. Stature is stunted, frenulae are often present in the vagina and mouth and facial clefts may occur. Autosomal recessive inheritance is well established. This entity and the lethal popliteal pterygium syndrome (vide infra) are not clearly distinguished from each other in the literature.

### Lethal pterygium syndrome

The lethal pterygium syndromes present in the neonate with multiple congenital contractures and they have been subclassified in terms of additional features (Hall et al 1982). The existance of a lethal multiple pterygium syndrome (LMPS) (Gillin & Pryse-Davis 1976, Chen et al 1984,

**Fig. 14.19** Popliteal pterygium syndrome: a girl aged 14 years, in whom ambulation is severely restricted by thin soft tissue webs which extend from her ankles to the back of her thighs.

Isaacson et al 1984) and a lethal popliteal pterygium syndrome (LPPS) (Bartsocas & Papas 1972, Papadia et al 1984) is generally accepted. Hall (1984) proposed further categorisation of the LMPS based upon the presence of spinal fusion (Chen et al 1984) and synostosis of the long bones (van Regemorter 1984). Martin et al (1986) reported three affected siblings with variable manifestations, questioned the validity of subdivision of LMPS and drew attention to the similarity of the phenotype to that of the Pena-Shokeir syndrome (see pp 327 and 328). The situation became even more complex when Mbakop et al (1986) reported an affected male fetus with the additional feature of hydranencephaly. Thereafter Papadia et al (1987) gave a detailed account of a girl with a progressive form of the pterygium syndrome plus nemaline myopathy.

Whatever the outcome of the debate concerning syndromic identity, the evidence points to autosomal recessive inheritance in all the lethal pterygium syndromes. Golbus (1985) recorded failure of antenatal diagnosis by ultrasound and emphasised the problems posed by intrafamilial phenotypic variability.

REFERENCES

**Preamble**
Gericke G S, Hall J G, Nelson M M, Beighton P H 1984 Diagnostic considerations in arthrogryposis syndromes in South Africa. Clinical Genetics 25: 155–162

Hall J G, Reed S D, Driscoll E P 1983 Amyoplasia: A common sporadic condition with congenital contractures. American Journal of Medical Genetics 15: 571–590
Hall J G, Reed S D, Scott C I et al 1982 Three distinct types of X-linked arthrogryposis seen in six families. Clinical Genetics 21:81–97

**Arthrogryposis multiplex congenita**
Bargeton E, Nezelop C, Guran P, Job J C 1961 Étude anatomique d'un case d'arthrogrypose multiple congénitale et familiale. Revista Neurologia 104: 479
Brown L M, Robson M J, Sharrard W J 1980 The pathophysiology of arthrogryposis multiplex congenita neurologica. Journal of Bone and Joint Surgery 62: 291–296
Davidson J, Beighton P 1976 Whence the arthrogrypotics? Journal of Bone and Joint Surgery 58B: 492
Drachman D B, Coulombre A J 1962 Experimental clubfoot and arthrogryposis multiplex congenita. Lancet ii: 523
Drachman D B 1971 The syndrome of arthrogryposis multiplex congenita. Birth Defects: Original Article Series 7/2: 90
Goldberg J D, Chervenak F A, Lipman R A, Berkowitz R L 1986 Antenatal sonographic diagnosis of arthrogryposis multiplex congenita. Prenatal Diagnosis 6: 45–49
Friedlander H L, Westin G W, Wood W L 1968 Arthrogryposis multiplex congenita — review of 45 cases. Journal of Bone and Joint Surgery 50A: 89
Hall J G, Reed S D, Driscoll E P 1983 Amyoplasia: A common sporadic condition with congenital contractures. American Journal of Medical Genetics 15: 571–590
Hall J G, Greene G, Shinkoskey S, McIlvaine R 1976 Arthrogryposis — clinical and genetic heterogeneity. Fifth International Congress of Human Genetics. Mexico, p 78
Jago R H 1970 Arthrogryposis following treatment of maternal tetanus with muscle relaxants. Case report. Archives of Disease in Childhood 45: 277
Lloyd-Roberts G C, Lettin A W F 1970 Arthrogryposis multiplex congenita. Journal of Bone and Joint Surgery 52B: 494
Pena C E, Miller F, Budzilovich G N, Feign I 1968 Arthrogryposis multiplex congenita: report of two cases of a radicular type with familial incidence. Neurology 18: 926
Reed S D, Hall J G, Riccardi V M et al 1985 Chromosomal abnormalities associated with congenital contractures. Clinical Genetics 27: 353–372
Swinyard C A, Bleck E E 1985 The etiology of arthrogryposis (multiple congenital contracture). Clinical Orthopaedics and Related Research 194: 15–29
Wynne-Davies R, Lloyd-Roberts G C 1976 Arthrogryposis multiplex congenita. Archives of Disease in Childhood 51: 618

**Rigidity syndromes**
*Ankylosis-facial anomaly-pulmonary hypoplasia syndrome (Pena-Shokeir I)*
Herva R, Leisti J, Kirkinen P, Seppanen U 1985 A lethal autosomal recessive syndrome of multiple congenital contractures. American Journal of Medical Genetics 20: 431–439
Lindhout D, Hageman G, Beemer F A et al 1985 The Pena-Shokeir syndrome: Report of nine Dutch cases. American Journal of Medical Genetics 21: 655–668
Lowry R B, Machin G A, Morgan K, Mayock D, Marx L 1985 Congenital contractures, edema, hyperkeratosis and intrauterine growth retardation: A fatal syndrome in Hutterite and Mennonite kindreds. American Journal of Medical Genetics 22: 531–543
MacMilland R H, Harbert G M, Davis W D, Kelly T E 1985 Prenatal diagnosis of Pena-Shokeir syndrome type I. American Journal of Medical Genetics 21: 279–284
Mease A D, Yeatman G W, Pettett G, Merenstein G B 1976 A syndrome of ankylosis, facial anomalies and pulmonary hypoplasia secondary to fetal neuromuscular dysfunction. Birth Defects: Original Article Series 12/5: 193
Moerman P H, Frijns J P, Godderis P, Lauwerijns J M 1983 Multiple ankyloses, facial anomalies and pulmonary hypoplasia associated with severe antenatal spinal muscular atrophy. Journal of Pediatrics 103: 238–241
Pena S D J, Shokeir M H K 1974 Syndrome of camptodactyly, multiple ankyloses, facial anomalies, and pulmonary hypoplasia: a lethal condition. Journal of Pediatrics 85: 373
Pena S D J, Shokeir M H K 1976 Syndrome of camptodactyly, multiple ankyloses, facial anomalies and pulmonary hypoplasia — further delineation and evidence for autosomal recessive inheritance. Birth Defects: Original Article Series 13/5: 201

*Cerebro-oculofacio-skeletal (COFS) syndrome (Pena-Shokeir II)*
Fitch N, Resch L, Rochon L 1982 The Neu-Laxova syndrome: Comments on syndrome identification. American Journal of Medical Genetics 134: 445–452
Laxova R, Ohdra P T, Timothy J A D 1972 A further example of a lethal autosomal recessive condition in sibs. Journal of Mental Deficiency and Research 16: 139–143
Lazjuk G I, Lurie I W, Ostrowskaja T I et al 1979 Brief clinical observations: The Neu-Laxova syndrome — A distinct entity. American Journal of Medical Genetics 3: 261–267
Lowry R B, Maclean R, McLean D M, Tischler B 1971 Cataracts, microcephaly, kyphosis and limited joint movement in two siblings: a new syndrome. Journal of Pediatrics 79: 282
Lurie I W, Cherstvoy D E, Lazjuk G I, Nedzued M K, Usoev S S 1976 Further evidence for the autosomal recessive inheritance of the COFS syndrome. Clinical Genetics 10: 343
Neu R L, Kajii T, Gardner L I et al 1971 A lethal syndrome of microcephaly with multiple congenital anomalies in three siblings. Pediatrics 47: 610–612
Pena S D J, Shokeir M H K 1974 Autosomal recessive cerebro-oculofacio-skeletal (COFS) syndrome. Clinical Genetics 5: 285
Preus M, Fraser F C 1974 The cerebro-oculofacio-skeletal syndrome. Clinical Genetics 5: 294
Shved I A, Lazjuk G I, Cherstvoy E D 1985 Elaboration of the phenotypic changes of the upper limbs in the Neu-Laxova syndrome. American Journal of Medical Genetics 20: 1–11
Silengo M C, Davi G, Bianco R et al 1984 The NEU-COFS (cerebro-oculofacio-skeletal) syndrome: Report of a case. Clinical Genetics 25: 201–204
Turkel S B, Ebbin A I, Towner J W 1983 Additional manifestations of the Neu-Laxova syndrome. Journal of Medical Genetics 20: 227–229

*Goodman camptodactyly-fibrous tissue hyperplasia-skeletal dysplasia syndrome*
Goodman R M, Bat-Miriam Katznelson M, Manor E 1972 Camptodactyly: Occurrence in two new genetic syndromes and its relationship to other syndromes. Journal of Medical Genetics 9: 203

*Emery-Nelson syndrome*
Emery A E H, Nelson M M 1970 A familial syndrome of short stature, deformities of the hands and feet, and an unusual facies. Journal of Medical Genetics 7: 379
Gorlin R J, Pindborg J J, Cohen M M 1976 Syndromes of the Head and Neck 2nd edn McGraw-Hill, New York, p 738

*Liebenberg elbow-wrist-hand syndrome*
Liebenberg F 1973 A pedigree with unusual anomalies of the elbows, wrist and hands in five generations. South African Medical Journal 47: 745

*Mietens syndrome*
Mietens C, Weber H 1966 A syndrome characterised by corneal opacity, nystagmus, flexion contractures of the elbows, growth failure and mental retardation. Journal of Pediatrics 69: 624

*Kuskokwim disease*
Petajan J H, Momberger G L, Aase J M, Wright D G 1969 Arthrogryposis syndrome (Kuskokwim disease) in the Eskimo. Journal of the American Medical Association 209: 1481
Wright D G, Aase J 1969 The Kuskokwim syndrome: an inherited form of arthrogryposis in the Alaskan Eskimo. The Clinical Delineation of Birth Defects Vol 3. National Foundation, New York, p 91

*Contractural myopathies*
Emery A E H 1971 The nosology of the spinal muscular atrophies. Journal of Medical Genetics 8: 481
Emery A E H, Dreifuss F E 1966 Unusual type of benign X-linked muscular dystrophy. Journal of Neurology, Neurosurgery and Psychiatry 29: 338

Emery A E H, Walton N 1967 The genetics of muscular dystrophy. Progress in Medical Genetics 5: 116

Oswald H H, Goldblatt J, Horak A R, Beighton P 1987 Lethal cardiac conduction defects in Emery-Dreifuss muscular dystrophy. South African Medical Journal 72: 567–570

Thomas N S, Williams H, Elsas I J et al 1986 Localisation of the gene for Emery-Dreifuss muscular dystrophy to the distal long arm of the X-chromosome. Journal of Medical Genetics 23: 509–515

Yates J R W, Affara N A, Jamieson D M et al 1986 Emery-Dreifuss muscular dystrophy: Localisation to Xq27.3 qter confirmed by linkage to the factor VIII gene. Journal of Medical Genetics 23: 587–590

*Juvenile hyaline fibromatosis*

Drescher E, Woyke S, Markiewicz C, Tegi S 1967 Juvenile fibromatosis in siblings. Journal of Pediatric Surgery 2: 427–430

Fayad M N, Yacoub A, Salman S et al 1987 Juvenile hyaline fibromatosis: Two new patients and review of the literature. American Journal of Medical Genetics 26: 123–131

Finlay A Y, Ferguson S D, Hold P J A 1983 Juvenile hyaline fibromatosis. British Journal of Dermatology 108: 609–616

Kitano Y, Horiki M, Aoki T, Sagami S 1972 Two cases of juvenile hyaline fibromatosis. Archives of Dermatology 106: 877–883

Stringer D A, Hall C M 1981 Juvenile hyaline fibromatosis. British Journal of Radiology 54: 473–478

## Stiff hand-foot syndromes

*Distal arthrogryposis*

Hall J G, Reed S D, Greene G 1982 The distal arthrogryposes: Delineation of new entities — Review and nosologic discussion. American Journal of Medical Genetics 11: 185–239

Kawira E L, Bender H A 1985 Brief clinical report: An unusual distal arthrogryposis. American Journal of Medical Genetics 20:425–429

Reiss J A, Sheffield L J 1986 Distal arthrogryposis type II: A family with varying congenital abnormalities. American Journal of Medical Genetics 24: 255–267

Tsukahara M, Kajii T 1984 Distal arthrogryposis type IIB in a girl: Autosomal recessive inheritance? Japanese Journal of Human Genetics 29: 447–451

*Camptodactyly-cleft palate-clubfoot (Gordon) syndrome*

Gordon H, Davies D, Berman M 1969 Camptodactyly, cleft palate and club foot. Syndrome showing the autosomal dominant pattern of inheritance. Journal of Medical Genetics 6: 266

Halal F, Fraser F C 1979 Camptodactyly, cleft palate and club foot. Journal of Medical Genetics 16: 149–150

Higgins J V, Hackel E, Kapur S 1972 A second family with cleft palate, club feet and camptodactyly. American Journal of Human Genetics 24: 58A

Robinow M, Johnson G F 1981 The Gordon syndrome: Autosomal dominant cleft palate, camptodactyly and club feet. American Journal of Medical Genetics 9: 139–146

*Digitotalar dysmorphism*

Canale S T, Ingram A J, Tipton R E, Wilroy R S 1976 The thumb-clutched hand syndrome. Fifth International Congress of Human Genetics, Mexico.

Dhaliwal A S, Myers T L 1985 Digitotalar dysmorphism. Orthopaedic Review XIV/2: 90–94

Sallis J G, Beighton P 1972 Dominantly inherited digito-talar dysmorphism. Journal of Bone and Joint Surgery 54B: 509

*Digital-ulnar drift syndrome*

Fried K, Mundel G 1976 Absence of distal interphalangeal creases of fingers with flexion limitation. Journal of Medical Genetics 13/2: 127

Powers R C, Ledbetter R H 1976 Congenital flexion and ulnar deviation of the metacarpophalangeal joints of the hand: a case report. Clinical Orthopaedics and Related Research 116: 173

Stevenson R E, Scott C I, Epstein M 1975 Dominantly inherited ulnar drift. Birth Defects: Original Article Series 11/5: 75

*Digital deviation-clubfoot syndrome*
Fisk J R, House J H, Bradford D S 1974 Congenital ulnar deviation of the fingers with clubfoot deformities. Clinical Orthopaedics and Related Research 104: 200

*Tel-Hashomer camptodactyly syndrome*
Gollop T R, Colletto G M 1984 The Tel Hashomer camptodactyly syndrome in a consanguineous Brazilian family. American Journal of Medical Genetics 17: 399–406
Goodman R M, Bat-Miriam Katznelson M, Manor E 1972 Camptodactyly: occurrence in two new genetic syndromes and its relationship to other syndromes. Journal of Medical Genetics 9: 203
Goodman R M, Bat-Miriam Katznelson M, Hertz M, Katznelson A 1976 The Tel-Hashomer camptodactyly syndrome: Journal of Medical Genetics 13: 136–141
Tylki-Szymanska A 1986 Brief clinical report: Three new cases of Tel Hashomer camptodactyly syndrome in one Arabic family. American Journal of Medical Genetics 23: 759–763

*Dominantly inherited metatarsus varus*
Juberg R C, Touchstone W J 1974 Congenital metatarsus varus in four generations. Clinical Genetics 5/2: 127

*Trismus-pseudo camptodactyly syndrome*
De Jong J G Y 1971 A family showing strongly reduced ability to open the mouth and limitation of some movements of the extremities. Humangenetik 13: 210–217
Hecht F, Beals R K 1969 Inability to open the mouth fully: An autosomal dominant phenotype with facultative camptodactyly and short stature. Birth Defects. Original Article Series V(3): 96–98
Mabry C C, Barnett I S, Hutcheson M W, Sorenson H W 1974 Trismus pseudo-camptodactyly syndrome: Dutch-Kentucky syndrome. Journal of Pediatrics 85: 503–508
Robertson R D, Spence M A, Sparkes R S et al 1982 Linkage analysis with the trismus-pseudocamptodactyly syndrome. American Journal of Medical Genetics 12: 115–120
Ter Haar B G A, van Hoof R F 1974 The trismus-pseudocamptodactyly syndrome. Journal of Medical Genetics 11: 41–49
Tsukahara M, Shinozaki F, Kajii T 1985 Trismus-pseudocamptodactyly syndrome in a Japanese family. Clinical Genetics 28: 247–250

**Flexed digit syndromes**
Bell J 1951 On brachydactyly and symphalangism. Treasury of Human Inheritance Vol 5. Cambridge University Press, London, p 1
David T J, Burwood R L 1972 The nature and inheritance of Kirner's deformity. Journal of Medical Genetics 9/4: 430
Eng C E L, Strom C M 1987 New syndrome: familial proportionate short stature, intrauterine growth retardation and recurrent locking of the fingers. American Journal of Medical Genetics 26: 217–220
Gordon H, Davies D, Berman M 1969 Camptodactyly, cleft palate and club foot. Syndrome showing the autosomal dominant pattern of inheritance. Journal of Medical Genetics 6: 266
Halal F 1985 Dominant inheritance of syncamptodactyly of the second and third toes with foot and lower limb asymmetry and scoliosis. American Journal of Medical Genetics 22: 149–156
Hamanishi C, Ueba Y, Tsuji T et al 1986 Congenital aplasia of the extensor muscles of the fingers and thumb associated with generalized polyneuropathy. American Journal of Medical Genetics 24: 247–254
Hefner R A 1941 Crooked little finger (minor streblodactyly). Journal of Heredity 32: 37
Jones M C, Waldman J D 1985 An autosomal dominant syndrome of characteristic facial appearance, preauricular pits, fifth finger clinodactyly and tetralogy of Fallot. American Journal of Medical Genetics 22: 135–141

Katz G 1970 A pedigree with anomalies of the little finger in five generations and 17 individuals. Journal of Bone and Joint Surgery 52A/4: 717

Kirner J 1927 Doppelseitige Verkrummungen des Kleinfingerendgliedes als selbstandiges Krankheitsbild. Fortschritte auf dem Gebiete der Roentgenstrahlen und Nuklearmedizin 36: 804

Rozin M M, Hertz M, Goodman R M 1984 A new syndrome with camptodactyly, joint contractures, facial anomalies and skeletal defects: A case report and review of syndromes with camptodactyly. Clinical Genetics 26: 342–355

Pelias M Z, Kinnebrew M C 1985 Autosomal dominant transmission of ankylosed teeth, abnormalities of the jaws and clinodactyly. Clinical Genetics 27: 496–500

Schwartz P A, Shlugman D, Daynes G, Beighton P 1974 Transkei foot. South African Medical Journal 48: 961

Spear G S 1946 The inheritance of fixed fingers. Journal of Heredity 37: 189

Temtamy S, McKusick V A 1969 Synopsis of hand malformations with particular emphasis on genetic factors. Birth Defects: Original Article Series 5/3: 125

Welch J P, Temtamy S A 1966 Hereditary contracture of the finger (camptodactyly). Journal of Medical Genetics 3: 104

Eng C E L, Strom C M 1987 New syndrome: Familial proportionate short stature, intrauterine growth retardation and recurrent locking of the fingers. American Journal of Medical Genetics 26: 217–220

## Adducted thumb syndromes

Bianchine J, Lewis R 1974 The MASA syndrome: a new heritable mental retardation syndrome. Clinical Genetics 5: 298

Christian J C, Andrews P A, Conneally P M, Muller J 1971 The adducted thumbs syndrome. Clinical Genetics 2: 95

Edwards J H 1961 The syndrome of sex-linked hydrocephalus. Archives of Disease in Childhood 36: 486

Fitch N, Levy E P 1975 Adducted thumb syndromes. Clinical Genetics 8: 190

Hongre J F, Toursei M F, Staquet J P, Farriaux J P, Walbraum R 1972 Phénocopie de la trisomie 18. Annals de Pédiatrie 19: 830

Strauch B, Spinner M 1976 Congenital anomaly of the thumb: absent intrinsics and flexor pollicis longus. Journal of Bone and Joint Surgery 58/1: 115

Weckesser E, Reed J, Heiple K 1968 Congenital clasped thumb (congenital flexion-adduction deformity of the thumb). Journal of Bone and Joint Surgery 50A: 1417

## Pterygium syndromes

*Popliteal pterygium syndrome*

Escobar V, Bixler D, Gleiser S 1978 Multiple pterygium syndrome. American Journal of Diseases of Children 132: 609

Gorlin R, Sedano H, Cervenka J 1968 Popliteal pterygium syndrome. Pediatrics 41(2): 503

*Lethal pterygium syndrome*

Bartsocas C S, Papas C V 1972 Popliteal pterygium syndrome. American Journal of Medical Genetics 9: 222–226

Chen H, Immken L, Lachman R et al 1984 Syndrome of multiple pterygia, camptodactyly, facial anomalies, hypoplastic lungs and heart, cystic hygroma and skeletal anomalies. American Journal of Medical Genetics 17: 809–826

Gillin M E, Pryse-Davis J 1976 Pterygium syndrome: Report of a family. American Journal of Medical Genetics 13: 249–251

Golbus M S 1985 The lethal multiple pterygium syndromes: Is prenatal detection possible? American Journal of Medical Genetics 20: 411–442

Hall J G 1984 Editorial comment: The lethal multiple pterygium syndromes. American Journal of Medical Genetics 17: 803–807

Hall J G, Reed S D, Rosenbaum K N et al 1982 Limb pterygium syndrome: A review and report of 11 patients. American Journal of Medical Genetics 12: 377–409

Isaacson G, Gargus J J, Mahoney M J 1984 Lethal multiple pterygium syndrome in an 18-week fetus with hydrops. American Journal of Medical Genetics 17: 835–839

Martin N J, Hill J B, Cooper D H et al 1986 Lethal multiple pterygium syndrome: Three consecutive cases in one family. American Journal of Medical Genetics 24: 295–305

Mbakop A, Cox J N, Stormann C, Delozier-Blanchet C D 1986 Lethal multiple pterygium syndrome: Report of a new case with hydrencephaly. American Journal of Medical Genetics 25: 575–579

Papadia F, Zimbalatti F, Gentile R C 1984 The Bartsocas-Papas syndrome: Autosomal recessive form of popliteal pterygium syndrome in a male infant. American Journal of Medical Genetics 17: 841–847

Papadia F, Longo N, Serlenga L, Porzio G 1987 Progressive form of multiple pterygium syndrome in association with Nemalin myopathy: Report of a female followed for 12 years. American Journal of Medical Genetics 26: 73–83

Van Regemorter N, Wilkin P, Englert Y et al 1984 Lethal multiple pterygium syndrome. American Journal of Medical Genetics 17: 827–834

# 15

# Digital abnormalities

The nosology and genetics of digital anomalies have been reviewed at length by Temtamy & McKusick (1978) and Temtamy (1982, 1985). The following groups of conditions are discussed in this section:

1. Polydactyly
   a. Preaxial polydactyly
   b. Postaxial polydactyly
2. Syndactyly
3. Symphalangism
4. Brachydactyly
   a. Brachytelophalangy
   b. Other brachydactyly syndromes
5. Ectrodactyly
   a. Ectrodactyly, ectodermal dysplasia, facial cleft syndrome (EEC)
   b. Ectrodactyly with scalp defects
6. Digital dysplasia syndromes
   a. Holt-Oram syndrome
   b. Orofacial-digital syndrome
   c. Oto-palato-digital syndrome
   d. Aase syndrome
   e. Aglossia-adactylia syndrome (Hanhart)
   f. Goltz syndrome (focal dermal hypoplasia)
   g. Poland syndrome

## POLYDACTYLY

The presence of an extra digit is termed 'polydactyly'. Polydactyly is conventionally classified into preaxial and postaxial types. In the former, the thumb and great toe are duplicated, while in the latter, the additional digit is adjacent to the fifth finger and toe. If digital fusion is also present, the term 'polysyndactyly' is employed. In a survey of 10 000 consecutive newborn infants in Turkey, Tuncbilek & Say (1976) found the incidence of all forms of polydactyly to be 2.6 per 1000. Following a study of 13 human embryos with preaxial polydactyly, Yasuda (1975) concluded that the pathogenesis was disordered interaction between the limb ectoderm and mesoderm.

## Preaxial polydactyly

A unilateral extra thumb or hallux is usually non-genetic, but there are several autosomal dominant forms of uncomplicated, bilateral, preaxial digital duplication (Hefner 1940, Swanson & Brown 1962, James & Lamb 1963). Preaxial polydactyly, more severe in the hands than in the feet, was documented in 21 persons in a 5 generation family by Reynolds et al (1984). As with the majority of digital abnormality syndromes, phenotypic expression was very variable. Atasu (1976) observed an extra index finger in members of four generations of a Turkish family.

Preaxial polydactly is a feature of a rare dominantly inherited condition in which syndactyly and bilateral aplasia of the tibia are variable components. Affected kindreds have been reported by Eaton & McKusick (1969) and Pashayan et al (1971). A family in which 15 persons in five generations had varying combinations of preaxial polydactly and partial or complete absence of the tibia was documented by Lamb et al (1983) . The occurrence of potentially severe limb malformations in this autosomal dominant disorder would justify ultrasonic monitoring of at-risk pregnancies.

A triphalangeal thumb contains an additional phalange and resembles a finger. Triphalangeal thumbs are uncommon but when present are usually

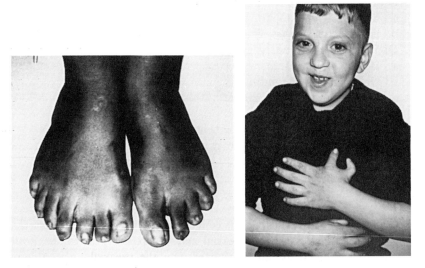

**Fig. 15.1** (left) Polydactyly the right foot has seven toes. This abnormality is rare.

**Fig. 15.2** (right) Preaxial polydactyly: an extra digit lateral to the thumb. (Preaxial implies that the additional digit is anterior to the anatomical axis of the limb.) This isolated abnormality is non-genetic.

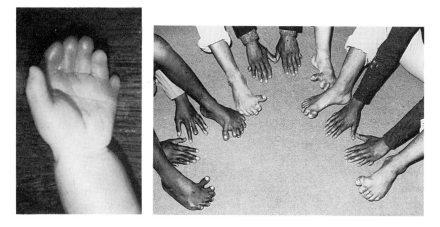

**Fig. 15.3** (left) 'Postaxial polydactyly': an extra digit on the medial side of the hand. This common abnormality is probably inherited as an autosomal dominant with variable phenotypic expression. (Postaxial implies that the additional digit is posterior to the anatomical axis of the limb.)

**Fig. 15.4** (right) Postaxial polydactyly: a family with extra digits on the lateral sides of the hands and feet. Deviation of the great toe is an additional component of this autosomal dominant syndrome.

a component of a genetic syndrome; they are therefore of considerable diagnostic significance. In view of their occasional association with preaxial polydactyly, triphalangeal thumbs are included in this section.

A sporadic female with polydactyly and triphalangeal thumbs in association with upper limb and pectoral dysplasia was described by Temtamy & Dorst (1975). Although no other members of the family had the abnormality, the authors commented that triphalangeal thumbs with or without preaxial polydactyly could be inherited as an autosomal dominant. Say et al (1976) described a mother and three daughters with preaxial polydactyly of the hands and feet, triphalangeal thumbs, brachydactyly, camptodactyly, congenital dislocation of the patella, short stature and low intelligence. The authors listed the conditions in which triphalangeal thumbs were a feature and commented that inheritance of their patient's syndrome was either autosomal dominant or X-linked. Triphalangeal thumbs and duplication of the great toes were inherited as an autosomal dominant trait in five generations of a family documented by Merlob et al (1985).

## Postaxial polydactyly

Postaxial polydactyly is more common than the preaxial type. A minute sixth digit is often present in individuals of African Negro stock. Indeed, Okeahialam (1974) commented that this was the commonest anomaly encountered during a 12 month survey of congenital malformations in Dar-es-Salaam, East Africa. The reason for this high prevalence is unknown, but it has been suggested that inheritance is autosomal dominant, with variable expression (Johnston & Davis, 1953). Large kindreds with autosomal dominant inheritance of extra digits of greatly varying size and distribution have been reported by several authors, including Woolf & Woolf (1970). Autosomal recessive forms of postaxial polydactyly may also exist. Cantu et al (1975) recognised this anomaly in four out of six sibs in a consanguineous Mexican kindred, while Temtamy & Rogers (1975) reported an affected Turkish boy with consanguineous parents. This child had multiple defects, including brachydactyly, craniofacial asymmetry, ear malformations and cardiac anomalies.

Postaxial polydactyly is a component of various potentially lethal conditions such as trisomy 13, asphyxiating thoracic dystrophy and the Saldino–Noonan syndrome. It is also a feature of chondroectodermal dysplasia (Ellis–van Creveld syndrome) and the hydrolethalus syndrome. This latter condition is characterised by the association of hydrocephalus, mandibular hypoplasia, cardiac and pulmonary anomalies (Salonen et al 1981). Inheritance is autosomal recessive.

Postaxial polysyndactyly is found in the Bardet–Biedl syndrome, where it is associated with mental retardation, retinitis pigmentosa and hypogenitalism. Ammann (1970) distinguished this autosomal recessive disorder from the Laurence–Moon syndrome, in which the digits are normal (see Ch. 19). Holt (1975) described two English families with autosomal dominant inheritance of postaxial polydactyly in association with variable shortening of the metacarpals and metatarsals. An autosomal recessive syndrome of postaxial polydactyly, cortical blindness and retarded development was documented in three siblings by Hernandez et al (1985). An autosomal dominant condition in which postaxial polydactyly was associated with progressive myopia was described in nine affected persons in four generations of an Hungarian family (Czeizel & Brooser 1986).

Postaxial hexadactyly was a diagnostic feature in a series of 18 neonates with a syndrome of microcephaly, growth and psychomotor retardation, genital abnormalities, congenital heart disease and cleft palate (Curry et al 1987). The 16 authors of this paper differentiated this condition from the disorder delineated by Smith et al (1964) and proposed the designation 'Smith-Lemli-Opitz syndrome (SLOS) type II'. Further nosological complexity was engendered when Casamassima et al (1987) documented a sibship with features of the SLOS plus Meckel syndrome and cerebellar defects. Following an extensive review of the literature, these authors postulated that the autosomal recessive condition might be an autonomous entity.

Laurence et al (1975) reported a kindred in which two male infants had been born with ulnar polydactyly and broad great toes, together with Hirschsprung's disease and cardiac malformations. A subsequent pregnancy was monitored by fetoscopy at the eighteenth week. The observation that the fetus had normal digits was confirmed when an unaffected boy was delivered. Although leakage of amniotic fluid was troublesome in later pregnancy, there is no doubt that fetoscopy was of real value in this particular case.

## SYNDACTYLY

The term 'syndactyly' pertains to bone or soft tissue union of two or more digits. In severe forms bone fusion of adjacent digits may be complete, while in mild types minimal skin webbing at the digital roots is the only abnormality. Syndactyly is a component of several genetic syndromes and it may also exist in isolation, either as a genetic or non-genetic entity. In general, the genetic forms are bilateral and fairly symmetrical, while the non-genetic varieties are unilateral.

Familial isolated syndactyly has been classified into five numerical categories by Temtamy & McKusick (1978). The majority are inherited as autosomal dominant traits, with variable phenotypic expression.

*Type I.* Zygodactyly comprises partial or complete union of the third and fourth fingers and toes. This is by far the most common form of syndactyly.

*Type II.* Synpolydactyly consists of syndactyly of the third and fourth or fourth and fifth fingers and toes, with a reduplicated digit in the soft tissue web. A three generation Nigerian family with this disorder was reported by Ofodile (1982) and a kindred with 16 affected persons in six generations has been documented by Merlob & Grunebaum (1986). Nonpenetrance resulted in skipped generations in this latter kindred. A recessive form of synpolydactyly of uncertain syndromic status was recorded by Briard & Kaplan (1982).

*Type III.* Ring and fifth finger syndactyly is uncommon in isolation. A significant proportion of affected persons have oculodento-osseous dysplasia (see Ch. 8) or other complex syndromes.

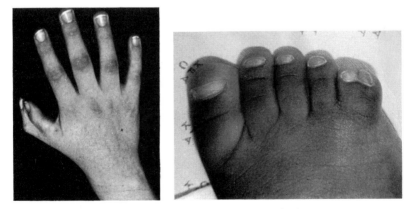

**Fig. 15.5** (left) Preaxial polysyndactyly: the thumb is duplicated and fused.

**Fig. 15.6** (right) Postaxial polysyndactyly: the 5th toe is duplicated and fused.

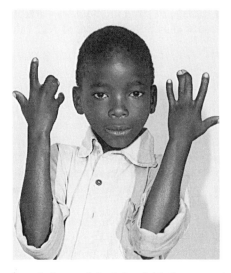

**Fig. 15.7** Syndactyly: the soft tissues of the 3rd and 4th fingers are fused.

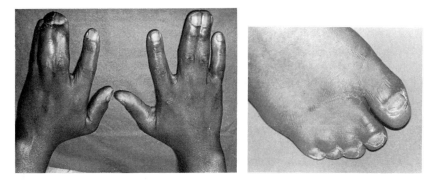

**Fig. 15.8** (left) Syndactyly: fusion of the 3rd and 4th fingers.

**Fig. 15.9** (right) Syndactyly: fusion of the lateral four toes.

*Type IV.* Syndactyly of all digits is a component of the Apert syndrome, while partial digital union occurs in other disorders involving craniostenosis (see Ch. 12). Isolated total syndactyly is a rare autosomal dominant trait.

*Type V.* Syndactyly with metacarpal and metatarsal synostosis resembles the split hand malformation. A mother and three affected children were documented by Robinow et al (1982); as with other conditions in this category, phenotypic expression was variable.

Syndactyly of various types is a component of several well established genetic disorders, such as sclerosteosis, the acrocephalosyndactylies and focal dermal hypoplasia. These conditions are reviewed elsewhere in this book. Other rare inherited disorders in which syndactyly is associated with non-skeletal abnormalities include the following:

1. Fitch et al (1976) reported a pair of half brothers with syndactyly and various cranial, facial, oral and skeletal abnormalities. As the mother had minor deviation deformities of her digits, the authors postulated that inheritance could be either autosomal dominant with variable penetrance or X-linked, with manifestations in the female heterozygote.

2. Goldberg & Pashayan (1976) described a kindred in which 10 individuals in three generations had syndactyly of the hallux and adjacent toes, a small sixth digit on the ulnar side of the hands and an abnormal configuration of the earlobes. This condition seems to be a 'private syndrome' in which inheritance is autosomal dominant, with variable phenotypic expression of the gene.

3. Brazilian siblings with campto-brachy-polysyndactyly, fibular hypoplasia, foot anomalies, mental retardation, stunted stature and an abnormal facies were documented by Richieri-Costa et al (1985). The unaffected parents were consanguineous and autosomal recessive inheritance is probable.

## SYMPHALANGISM

Symphalangism is far less common than syndactyly. In symphalangism, interphalangeal joints are absent, so that movements of the digits are restricted. Unlike syndactyly, there is no fusion between neighbouring digits. One form of symphalangism is inherited as an autosomal dominant. Drawing upon historical sources, Elkington & Huntsman (1967) described how this abnormality had been present in several generations of the aristocratic Talbot family in Britain.

A syndrome of symphalangism and coalition of the small bones of the wrist and ankle is transmitted as an autosomal dominant. Affected kindreds have been reported by Austin (1951) and Strasburger et al (1965). Symphal-

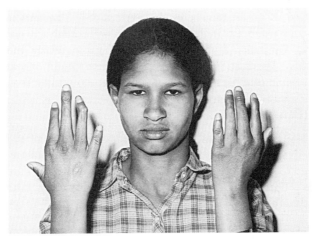

**Fig. 15.10** Symphalangism: the fingers are rigid, due to bony union of the proximal and middle phalanges.

angism with metacarpophalangeal fusions and elbow abnormalities was described by Kassner et al (1976). This condition was present in three generations of a kindred, and inheritance was evidently autosomal dominant. Walbaum et al (1976) reported the association of brachydactyly and symphalangism, which was probably inherited as an autosomal recessive.

Symphalangism of the proximal digits in association with a cylindrical nose, fusion in the wrist and ankles, subluxation of the radial heads and progressive conductive deafness has been recorded by Maroteaux et al (1972), Herrmann (1974) and Pierson et al (1982). This condition was recognised in 28 members of a Brazilian kindred by Da-Silva et al (1984) and autosomal dominant inheritance was confirmed. Deafness, due to ankylosis of the stapes, together with proximal symphalangism, was observed in a father and his two sons by Cremers et al (1985). This disorder is presumably an autosomal dominant trait. A further case was documented in detail by Hurvitz et al (1985) and the designation 'facio-audio-symphalangism syndrome' was proposed.

As reports accumulate, it is becoming evident that symphalangism frequently occurs in conjunction with synostoses, especially in the tarsus and carpus. It is possible that involvement of these latter sites might have been overlooked in some of the earlier studies and the status of isolated symphalangism is therefore questionable. Equally, it is uncertain whether the symphalangism-synostosis syndromes are homogeneous or heterogeneous. These disorders are given further consideration in Chapter 16.

## BRACHYDACTYLY

The term 'brachydactyly' pertains to shortening of a single or multiple digits, due to bone maldevelopment. There are many types of brachydactyly, some of which are components of syndromes or which exist in conjunction with other defects in the extremities. The various forms of dominantly inherited isolated brachydactyly were analysed and listed by Bell (1951). Temtamy & McKusick (1978) subsequently modified and extended this classification. A number of these conditions are 'private' syndromes, while others are relatively common. An outline is given below:

### Type A-1 (Farabee)

The main features are digital shortening, which is maximal in the middle phalanges, dysplasia of the femoral heads and small stature. Haws & McKusick (1963) restudied the kindred first reported by Farabee at the turn of the century, and confirmed that the condition was inherited as an autosomal dominant.

### Type A-2 (Mohr–Wreidt)

In this disorder the middle phalanges of the second finger and toe are short-

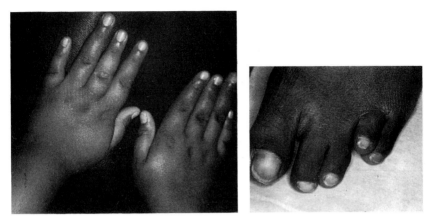

**Fig. 15.11** (left) Brachydactyly: apparent shortening of a finger is often the result of reduction in length of the corresponding metarcarpal. In this patient, the 3rd metacarpals are short, while the phalanges are normal.

**Fig. 15.12** (right) Brachydactyly: the 4th metatarsal is short. This relatively common abnormality may occur in isolation or as a non-specific component of a number of genetic syndromes.

ened, with radial deviation of the terminal phalanx. Mohr & Wreidt (1919) reported a kindred in which two affected parents had produced a daughter with total adactylia. It is possible that this child was homozygous for the abnormal gene.

## Type A-3

A shortened and distorted middle phalanx leads to flexion and incurving of the fifth finger. This is clinodactyly or the Kirner anomaly, as discussed in Chapter 14.

## Type A-4

In this type of brachydactyly the middle phalanges of the second and fifth fingers are shortened. In addition, the middle phalanges of the lateral four toes are sometimes absent.

## Type A-5

Bass (1968) described 13 members of four generations of a large kindred who had absence of the middle phalanges, dysplasia of the nails and duplication of the terminal phalanx of the thumb.

## Type B

The major characteristics are shortening of the middle phalanges, with hypoplasia of the terminal phalanges and dystrophy of the nails of the second to fifth digits, broadening or partial duplication of the terminal phalanx of the thumb and minor cutaneous syndactyly. MacArthur & McCullough (1932) reported this abnormality in three generations of a kindred, under the designation 'apical dystrophy'. Zavala et al (1975) described generation to generation transmission of brachydactyly type B, together with symphalangism, in a Mexican family. A family with an autosomal dominant syndrome comprising type B brachydactyly in association with finger-like thumbs, nail dysplasia and variable ectrodactyly was documented by Kumar & Levick (1986).

## Type C

In this form of brachydactyly, shortening and deformity of the proximal and middle phalanges of the second and third fingers predominates. Haws (1963) identified the abnormality in 86 members of a Mormon kindred. Rimoin et al (1975) encountered type C brachydactyly together with limited flexion of the distal interphalangeal joints in 10 members of five generations of an American Negro family.

## Type D (stub thumb)

Shortening and broadening of the terminal phalanges of the first digit is a common abnormality which is sometimes alluded to as 'murderer's thumb'. By coincidence, and perhaps not inappropriately, J. W. Jailer was a co-author of a paper in which a family with this anomaly was described (Sayles & Jailer 1934). Stub thumb has been studied in kindreds from different ethnic groups by Goodman et al (1965) and the familial nature of the abnormality has been discussed by Davies (1975). The Israeli group retained their interest in stub thumbs and two decades after the original

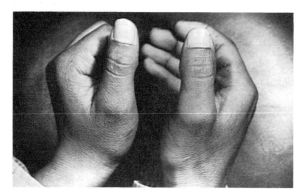

**Fig. 15.13** Brachydactyly type D, 'murderer's thumbs'. The distal phalynx is short and broad.

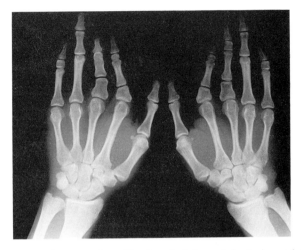

**Fig. 15.14** Brachydactyly: unusual combination of phalangeal shortening and reduplication. The delta configuration of the proximal phalanx of the index finger is characteristic of brachydactyly type C.

publication documented the abnormality in six members of three generations of a family (Goodman et al 1984).

Broad thumbs are also a component of several syndromes, including those bearing the eponyms Larsen, Carpenter and Rubinstein-Taybi. More than 200 cases of this latter disorder have been documented, the vast majority being sporadic. A variety, which is seemingly transmitted as a

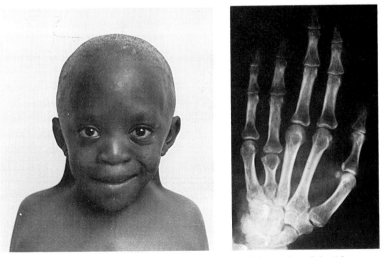

**Fig. 15.15** (left) Turner syndrome: webbing of the neck and shortening of the 4th metacarpal are often present in this chromosomal disorder.

**Fig. 15.16** (right) Turner syndrome: the 4th metacarpal is short.

dominant trait, was recorded by Cotsirilos et al (1987). An extensive review of the thumb in congenital malformation syndromes was published by Poznanski et al (1971).

## Type E

The commonest brachydactyly syndrome involves shortening of the meta-carpals and metatarsals. There is intra and interfamilial variation in the distribution and severity of this abnormality. The condition is hetero-geneous, and kindreds with recognisably different forms of the disorder have been described by several authors, including McKusick & Milch (1964), Cuevas-Sosa & Garcia-Segur (1971) and Gnamey et al (1975).

Shortening of one or more metacarpals and metatarsals, usually the fourth and fifth, is a fairly common sporadic anomaly. A short fourth meta-carpal is also encountered in females with an XO chromosome constitution (Turner's syndrome). All the metacarpals may be shortened in pseudohypoparathyroidism.

Newcombe & Keats (1969) reported two large kindreds with autosomal dominant inheritance of peripheral dysostosis and emphasised that although this condition resembles type E brachydactyly, the disorders are actually different entities. According to these authors, peripheral dysostosis can be distinguished by the involvement of the phalanges and the radiographic appearance of cone-shaped epiphyses in these bones.

## Brachytelophalangy

Brachytelophalangy, or shortening of the terminal phalanges, may occur in isolation, or in certain dwarfing skeletal conditions, notably some forms of multiple epiphyseal dysplasia. It is also a component of the frontonasal dysplasia group of disorders, which are outlined in Chapter 12.

An autosomal dominant syndrome comprising hypoplasia of the terminal phalanges, brachydactyly of the fifth digit, digitalisation of the thumbs and nail dysplasia, was documented by Cooks et al (1985). A mother and son with brachytelophalangy and a characteristic facies were reported by Hunter et al (1986). The boy also had the Kallman syndrome (hypogonadism and anosmia); the genetic significance of this additional abnormality is uncertain.

Using the designation 'Keutal syndrome' Cormode et al (1986) recorded an unusual disorder in which brachytelephalangism was associated with hearing loss, pulmonary stenosis and abnormal tracheobronchial and auricular calcification. In a review of the literature they identified five additional case reports (Say et al 1973, Walbaum et al 1975, Keutel et al 1982, Fryns et al 1984) and suggested that inheritance was probably auto-somal recessive.

In a study of 51 children born to epileptic mothers who were receiving diphenylhydantoin therapy, Kelly (1984) demonstrated that 30% had distal digital hypoplasia. He alluded to previous observations of this phenomenon and suggested that brachytelephalangy might be a specific marker for fetal exposure to diphenylhydantoin.

### Other brachydactyly syndromes

There have been a number of reports of other brachydactyly syndromes. For instance, dominant inheritance of short digits in association with hypertension has been described (Bilginturan et al 1973). Similarly, a kindred with short adducted thumbs and great toes in four generations was studied by Christian et al (1972). A mother and five offspring with brachy-dactyly, which would now be classified as type B, together with pigmented colobomata of the macula, were investigated by Sorsby (1935). Gorlin & Sedano (1971) described a kindred with type E brachydactyly and multiple impacted teeth, using the term 'cryptodontic brachymetacarpalia'. Villav-erde & Silva (1975) found distal brachyphalangy of the thumbs in more than 3 of 852 mentally retarded individuals. Several had additional features such as abnormalities of the feet, cranium and sella turcica. An autosomal dominant syndrome of mixed B and E type brachydactyly was documented by Pitt & Williams (1985).

## ECTRODACTYLY

Maldevelopment of the central rays of the limb buds may produce longi-tudinal splitting of the extremities. In some instances, the split hand or foot has a 'lobster claw' configuration. The term 'ectrodactyly', which has the Greek connotation of 'aborted' digit, is used loosely to denote abnormalities of this type.

Split hand and foot may be sporadic or inherited as an autosomal dominant. Expression is very variable and skipped generations are not uncommon. Large families with dominant transmission have been reported by MacKenzie & Penrose (1951) and Stevenson & Jennings (1960). The latter authors commented that in several kindreds, affected males had a tendency to produce an excess of affected sons. There is no obvious expla-nation for this anomalous situation.

It is likely that there is an uncommon autosomal recessive form of ectro-dactyly. Mosavy & Vakshuri (1975) described a pair of Iranian sisters with splitting of the hands and feet. Nine sibs and the consanguineous parents were normal. Similarly, Verma et al (1976) reported the abnormality in two related consanguineous parents. There do not seem to be any phenotypical features which distinguish these dominant and recessive forms of ectrodactyly.

David (1972) documented a family in which normal parents produced affected offspring, who then transmitted the malformation to their own chil-

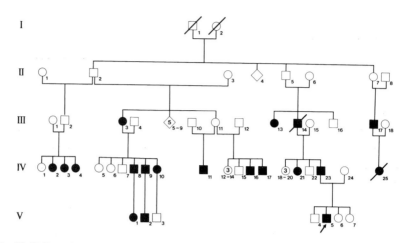

**Fig. 15.17** Ectrodactyly: the pedigree of an affected kindred, showing anomalous transmission. (From Spranger & Schapera 1988, European Journal of Paediatrics, in press.)

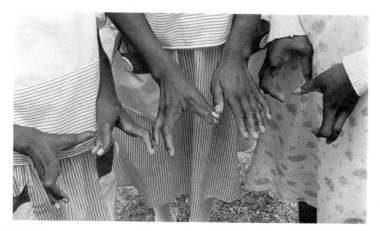

**Fig. 15.18** Ectrodactyly: siblings with variable split-hand malformations. (From Spranger & Schapera 1988 European Journal of Paediatrics, in press.)

dren. This unusual situation was explained on a basis of germinal mosaicism. A similar anomalous kindred was investigated by Spranger & Schapera (1988); three affected siblings with normal parents had produced progeny with the disorder and a total of 19 persons in the family were affected. The phenotype was very variable and the most severely affected person had a unilateral transverse defect of the tibia. Similar tibial involvement had been documented by Majewski et al (1985) in a review of 34 affected persons in six families.

A remarkable form of ectrodactyly is present in the Doma people of the Eastern Zambezi valley. Stories of 'ostrich-footed' men in this remote area

**Fig. 15.19** Ectrodactyly in the Wadoma of Central Africa. (From Viljoen & Beighton 1984 American Journal of Medical Genetics 19: 545–552.)

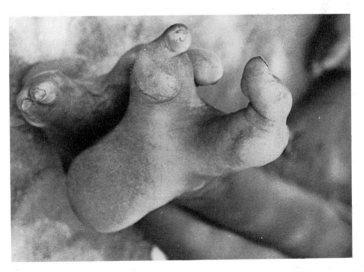

**Fig. 15.20** Ectrodactyly: the feet of a Wadoma family member. (From Viljoen et al 1985 South African Medical Journal 68: 655–658.)

were considered to be mere travellers tales until Gelfand et al (1974) examined an affected male and documented the disorder in detail. The feet were bidactylic, while the hands were normal apart from minimal skin webbing between the third and fourth fingers. A male cousin on the father's side of the family was also known to have the condition. The local population attributed the abnormality to the fact that the individual's mother had imprudently eaten 'two footed' animals during her pregnancy. However, the patient's own family suspected bewitchment. Offering an alternative explanation, the investigators proposed that the condition was an autosomal dominant, with non-penetrance in the father.

The condition was subsequently investigated in six members of three generations of a Talaunda family living in Botswana. This kindred had ancestral links with the Doma and it seems likely that the same mutant gene was responsible for the disorder (Viljoen & Beighton 1984, Viljoen et al 1985).

In view of the existence of sporadic, dominant and recessive forms of the split hand and foot syndrome, the propensity for skipped generations, and the fact that there is considerable phenotypic variability, genetic counselling can be very difficult. The practical implications of this situation have been discussed in detail by Preus & Fraser (1973); ultrasonic monitoring of at-risk pregnancies is an option which warrants consideration.

## Ectrodactyly, ectodermal dysplasia, facial cleft syndrome (EEC)

The association of ectrodactyly, ectodermal dysplasia and cleft lip and palate was reported by Roselli & Gulienetti (1961). Dominant inheritance of 'lobster claw' deformities of the hands and feet together with facial clefts was noted in three kindreds, by Walker & Clodius (1963). Similar features were present in a girl described by Rudiger et al (1970) under the designation 'EEC syndrome'. The manifestations of this syndrome have been reviewed by Bixler et al (1972). Inheritance is apparently autosomal dominant with variable expression and occasional non-penetrance. Penchaszadeh & de Negrotti (1976) emphasised the variability in phenotypic manifestations in affected members of a kindred. Rosenmann et al (1976) described a father and son of Jewish-Polish origin, with ectrodactyly, ectodermal dysplasia and cleft palate. These authors tabulated the features of 18 familial and three sporadic cases of the EEC syndrome and suggested that the condition existed in two forms; one with cleft lip, with or without cleft palate, and the other with cleft palate alone. The ectrodactyly, anodontia and lacrimal duct dysplasia syndrome which was studied in six patients by Pashayan et al (1974) may be yet another separate disorder. Reed et al (1975) described a mother and daughter with similar abnormalities, terming the condition the 'REEDS' syndrome.

Renal dysplasia, possibly secondary to fetal urinary tract obstruction, has been recognised in a number of persons with the EEC syndrome (London et al 1985). The authors emphasised the importance of investigating the renal status of individuals with the disorder.

### Ectrodactyly with scalp defects

Adams & Oliver (1945) reported a kindred in which dominantly inherited scalp defects occurred in conjunction with variable deficiencies of the extremities. In particular, the tibia and fibula were very hypoplastic in one member of this family (see Ch. 16). A further family with variable phenotypic expression and non-penetrance of terminal transverse limb defects and aplasia cutis congenita was documented by Sybert (1985).

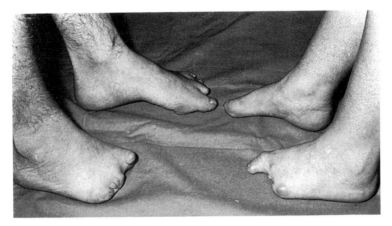

**Fig. 15.21** Ectrodactyly: the feet of a brother and sister, with defects of the toes, fingers and scalp. Members of four generations of the kindred had similar abnormalities. (From Bonafede & Beighton 1979 American Journal of Medical Genetics 3: 35–41.)

A South African kindred, of British stock, has ectrodactyly of the feet associated with minor changes in the fingers and a defect in the scalp (Bonafede & Beighton 1979). Individuals in four generations are affected and it is evident that inheritance is autosomal dominant, with very variable expression. It is uncertain if this disorder is the EEC syndrome without facial clefting, the aplasia cutis congenita-terminal transverse defects syndrome without severe limb anomalies or a separate genetic entity.

## DIGITAL DYSPLASIA SYNDROMES

In view of the semantic problems concerning the term 'ectrodactyly', digital dysplasia syndromes in which the extremities are not usually split are considered in this subsection. In some of these conditions, fused or duplicated digits may be present. In others, extensive bone abnormalities are a feature. It must be emphasised that these disorders have been grouped together for the sake of convenience, and that there is no suggestion that they bear any fundamental genetic or embryological relationship to each other.

### Holt-Oram* syndrome

In the Holt-Oram syndrome, unilateral dysplasia of the thumb is associated with malformations of the heart. Typically, the thumb is a triphalangeal finger-like digit, while the cardiac anomaly is usually an atrial septal defect. Using the designation 'upper limb-cardiovascular syndrome', Lewis et al (1965) pointed out that the thumb may be totally absent and that limb defects, particularly on the radial side, may be extensive. Poznanski et al (1970) drew attention to the presence of extra or abnormal bones in the carpus as an important feature of the Holt-Oram syndrome. The orthopaedic aspects of the syndrome have been reviewed by Letts et al (1976). There have been several reports of generation to generation transmission and inheritance is autosomal dominant, with variable expression of the abnormal gene (Harris & Osborne 1966, Gall et al 1966, Gladstone & Sybert 1982).

Tamari & Goodman (1974) presented an extensive classification of the upper limb-cardiovascular syndromes and discussed the embryologic relationships between the arms and the heart. These authors also described a boy with bifid thumbs, deafness and septal defects, in whom a small extra chromosome was identified. Other cytogenetic abnormalities which have been reported in the Holt-Oram syndrome include a partial deletion of the long arm of a B group chromosome in several members of a large kindred (Rybak et al 1971) and a balanced translocation (6q−, 7p+) in an affected female (Ferrier et al 1975). It is not known whether these chromosomal anomalies are genuine components of the Holt-Oram syndrome. It has been suggested that the underlying pathogenic mechanism might be a segmental defect in the nerve supply to the embryonic cervical segments (Smith et al 1979) or vascular hypoplasia in the fetal upper limb (Hoyme et al 1983).

As the phenotypic manifestations of this autosomal dominant disorder are very variable, decisions concerning the genetic management of a potentially affected fetus in utero can be very difficult. However, in view of the potentially serious cardiac defects, antenatal surveillance certainly warrants consideration. Muller et al (1985) successfully monitored two pregnancies of an affected mother by means of high resolution real-time ultrasonography. In both instances defects of the fetal hands and heart were detected. The outcome of the first pregnancy was premature labour and neonatal death from respiratory distress, while the second was terminated at 18

---

* Mary Holt is a consultant cardiologist at Croydon Hospital, London.
Samuel Oram is senior physician at Kings College Hospital, London and censor of the Royal College of Physicians.

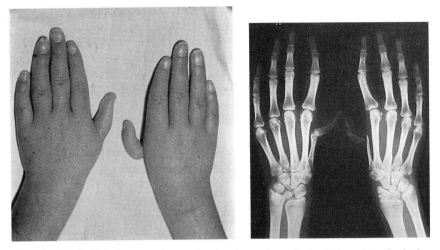

**Fig. 15.22** (left) Holt-Oram syndrome: the thumbs are hypoplastic. This patient also had a ventricular septal defect.

**Fig. 15.23** (right) Holt-Oram syndrome: radiograph showing hypoplasia of the bones of the thumbs.

weeks gestation. In both instances the diagnosis of the Holt-Oram syndrome was confirmed at autopsy.

### Orofacial-digital syndrome

Papillon-Léage & Psaume (1954) described eight females with facial clefts, abnormalities of the mouth and teeth and hand malformations. The disorder was designated the orofacial-digital syndrome (OFD). Digital changes were very variable, but included brachydactyly, syndactyly, clinodactyly, preaxial polydactyly and duplication of the cuneiform bones and hallux. Gorlin & Psaume (1962) subsequently reported 22 cases, all of whom were female, and suggested that the condition was inherited as an X-linked dominant, with lethality in the male. A further large kindred with 15 affected females in four generations was described by Doege et al (1964). Melnick & Shields (1975) reported an additional family, reviewed published pedigrees and confirmed the X-linked dominant mode of transmission. An inconsistent abnormality of chromosome I has been reported by Reuss et al (1962). It is not known if this cytogenetic anomaly has any direct relationship to the syndrome.

Rimoin & Edgerton (1967) suggested that the OFD syndrome might be heterogeneous and proposed that the condition should be subdivided into type I and II, with the eponym 'Mohr' pertaining to the second type. Whelan et al (1975) disagreed, pointing out that there was considerable phenotypic overlap and contended that the two forms of the disorder should be grouped together as one syndrome.

Following an extensive review of the literature Anneren et al (1984) concluded that OFO I and II (Mohr) were indeed separate entities; this view is now generally accepted. More than 150 cases of OFD I have been reported and the concept of X-linked dominant inheritance with male lethality is secure. About 25 instances of OFD II (Mohr) are known; inheritance is autosomal recessive. Haumont & Pelc (1983) have suggested that OFD II (Mohr) might be heterogeneous but this issue remains unresolved.

A form of OFD with the additional feature of cerebromacular degeneration was observed in two sisters by Sugarman et al (1971). McKusick (1983) mentioned that his group had encountered three sibs with the same disorder, proposed the designation 'OFD II' and awarded the asterisk of syndromic identity in the 6th edition of his catalogue of Mendelian disorders.

The nosological situation is complex; Temtamy et al (1975) documented a newborn Philippino girl with phenotypic features which were consistent with a diagnosis of a severe form of the Mohr syndrome or a mild form of the Majewski syndrome and postulated that these two conditions might represent variable expression of the same genetic disorder. Burn et al (1984) reported two Pakistani sisters with similar features. The unaffected parents were consanguineous and it is likely that inheritance is autosomal recessive. The designation 'OFD IV' was proposed for this entity.

Antenatal diagnosis of OFD II (Mohr)* was accomplished by Iaccarino et al (1985) by ultrasonographic recognition of polydactyly and a structural cardiac defect. In view of the autosomal recessive mode of inheritance and the severe complications of this disorder, surveillance of potentially affected pregnancies is indicated.

## Oto-palato-digital syndrome

The oto-palato-digital syndrome type I (OPD I) is characterised by shortening and broadening of the terminal phalanges, which is maximal in the thumbs and great toes, in association with mild mental retardation, stunted stature, conductive deafness, cleft palate, partial anodontia and a characteristic facies (Taybi 1962, Dudding et al 1967, Gorlin et al 1973). Fusions of the hamate and capitate bones, accessory ossification centres at the base of the second metatarsals, limited elbow extension and medial tibial bowing are additional variable skeletal features. Inheritance is probably X-linked, with significant but variable phenotypic expression in heterozygous females (Salinas et al 1979).

* Otto Mohr (1886–1967) was a distinguished geneticist at the University of Oslo. He became Dean of the medical faculty and President of the Norwegian Academy of Science.

OPD II is a similar X-linked disorder, in which skeletal changes are more severe. In particular, the cranial vault is under-ossified, the long bones are sclerotic and the clavicles, ribs, vertebral bodies and pelvis are dysplastic (Fitch et al 1976, Kozlowski et al 1977, Andre et al 1981, Fitch et al 1983). The latter authors suggested that in view of the phenotypic similarities between OPD I and OPD II, the causative mutant genes might be allelic. Following a description of an Italian family in which persons in three generations were affected, Pazzaglia & Beluffi (1986) made the alternative suggestion that the condition might be autosomal dominant with sex limitation of clinical manifestations.

Two severely affected male infants, who died in the neonatal period, were documented by Brewster et al (1985). In keeping with the X-linked mode of transmission, the obligate heterozygote mother had downwardly slanting palpebral fissures.

## Aase syndrome

In the Aase syndrome, triphalangeal thumbs and minor abnormalities of the radius are associated with hypoplastic anaemia and variable cardiac defects. Aase & Smith (1969) reported a pair of brothers with the condition and additional affected boys were described by Murphy & Lubin (1972) and Jones & Thompson (1973). Females with the disorder have been documented by Van Weel-Sipmann et al (1977) and Higginbottom et al (1978). Triphalangeal thumbs also occur in association with congenital anaemias and the syndromic identity of the Aase syndrome is uncertain (Alter 1978).

## Aglossia-adactylia syndrome (Hanhart*)

In the aglossia-adactylia syndrome, defects of the extremities are associated with hypoplasia of the tongue. The skeletal changes are often asymmetrical and range from shortening of a finger or toe to absence of a limb. Mandibular hypoplasia, missing teeth and an aberrant lingual frenulum are common concomitants.

There have been problems with the nosology of the aglossia-adactylia syndrome. In an early report Hanhart (1950) used the term 'peromelia with micrognathia'. This disorder was initially thought to be separate from the aglossia-adactylia syndrome but Herrmann et al (1976) advanced arguments in favour of syndromic unity and proposed the eponymous designation 'Hanhart syndrome'. Goodman & Gorlin (1977) were in agreement with the concept of homogeneity and suggested that an appropriate descriptive title would be the 'oromandibular limb hypogenesis syndrome'.

In a review of the literature Hall (1971) pointed out that all 14 reported cases had been sporadic and that there was no indication of a genetic aeti-

---

* Ernst Hanhart (1891–1973) was a pioneer in medical genetics in Zurich, Switzerland.

ology. Nevin et al (1975) contended that new dominant mutation could not be ruled out. Tuncbilek et al (1977) recognised the disorder in consanguineous Turkish families and proposed that inheritance was autosomal recessive. Dellagrammaticas et al (1982) supported this proposition with a report of two unrelated Greek gypsies with the condition, one of whom had consanguineous parents. The fact remains that all affected persons have been sporadic and the mode of inheritance is still uncertain.

Bokesoy et al (1983) documented a girl with the condition who had been exposed in utero to an antihistaminic drug, meclizine, which had been prescribed for her mother for a dermal mycosis. In this instance cause and effect is unproven but the possibility of an environmental component in the pathogenesis cannot be discarded.

### Goltz* syndrome (focal dermal hypoplasia)

Digital anomalies are a prominent feature of this disorder. Syndactyly of the third and fourth fingers is frequently present, but other digital abnormalities include polydactyly, brachydactyly and adactyly. The clavicles and ribs may be hypoplastic and vertebral malformations are not uncommon. Radiographically, longitudinal striations are evident in the tubular bones. Areas of dermal atrophy and pigmentation, dystrophy of the nails, colobomata of the eyes and papillomata of the mouth are other components of the syndrome. The condition has been reviewed by Goltz et al (1970) and further cases have been documented by Toro-Sola et al (1975), Feinberg & Menter (1976) and Happle & Lenz (1977).

The majority of affected individuals have been sporadic and there has been an overwhelming female preponderance. Transmission through four generations was noted by Goltz et al (1962) and an excessive number of spontaneous abortions in affected kindreds was documented by these authors. These observations are in keeping with X-linked dominant inheritance with lethality in the male.

### Poland** syndrome

The association of unilateral cutaneous synbrachydactyly with ipsilateral absence of the pectoralis minor muscle and the sternal portion of the pectoralis major constitutes the Poland syndrome. Rib defects, together with absence of part or all the hand, are occasional components. The

---

* Robert Goltz was head of dermatology at the University of Minnesota prior to his retirement in 1985.

** Alfred Poland (1822–1872) was a young demonstrator of anatomy when he published the account of the syndrome which bears his name. He became senior ophthalmologist at Guy's Hospital, London.

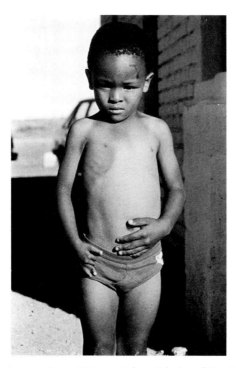

**Fig.15.24** Poland syndrome: a boy with hypoplasia and fusion of the digits of his right hand, together with absence of the sternal head of the ipsilateral pectoralis major muscle.

features of 43 cases were reviewed by Ireland et al (1976) and the occurrence of dextrocardia was documented by Bosch-Banyeras et al (1984). It has been suggested that the underlying mechanism concerns impairment of the blood supply to the developing fetal structures (Bavinck & Weaver (1986).

The Poland syndrome is relatively common and most cases are sporadic. However, the condition was recognised in successive generations by Trier (1965). In an investigation of possible environmental determinants David (1976) noted that several mothers of affected children had made unsuccessful attempts to abort their pregnancies. Inheritance is probably multifactorial.

REFERENCES

Temtamy S A 1982 Classification of hand malformations as isolated defects: An overview. Journal of Human Genetics 30/4: 281–290
Temtamy S A 1985 The Genetics of Hand Malformations: Updated. Congenital Anomalies 25/1: 73–92
Temtamy S, McKusick V A 1978 The Genetics of Hand Malformations. Birth Defects: Original Article Series XIV/3: National Foundation, March of Dimes. Liss, New York

### Polydactyly

Tuncbilek E, Say B 1976 Polydactyly in Turkey. Fifth International Congress of Human Genetics, Mexico
Yasuda M 1975 Pathogenesis of preaxial polydactyly of the hand in human embryos. Journal of Embryology and Experimental Morphology 33: 745

*Preaxial polydactyly*
Atasu M 1976 Hereditary index finger polydactyly. Journal of Medical Genetics 13/6: 469
Eaton G O, McKusick V A 1969 A seemingly unique polydactyly-syndactyly syndrome in four persons in three generations. Birth Defects: Original Article Series 5/3: 221
Hefner R A 1940 Hereditary polydactyly. Journal of Heredity 31: 25
James J I P, Lamb D W 1963 Congenital abnormalities of the limbs. Practitioner 191: 159
Lamb D W, Wynne-Davies R, Whitmore J M 1983 Five-fingered hand associated with partial or complete tibial absence and pre-axial polydactyly. A kindred of 15 affected individuals in five generations. Journal of Bone and Joint Surgery 65: 60–63
Merlob P, Grunebaum M, Reisner S H 1985 Familial opposable triphalangeal thumbs associated with duplication of the big toes. Journal of Medical Genetics 22/1: 78
Pashayan H, Fraser F C, McIntyre J M, Dunbar J S 1971 Bilateral aplasia of the tibia, polydactyly and absent thumb in father and daughter. Journal of Bone and Joint Surgery 53B: 495
Reynolds J F, Sommer A, Kelly T E 1984 Preaxial polydactyly type 4: Variability in a large kindred. Clinical Genetics 25: 267–272
Say B, Field E, Coldwell J G, Warnberg L, Atasu M 1976 Polydactyly with triphalangeal thumbs, brachydactyly, camptodactyly, congenital dislocation of the patellas, short stature and borderline intelligence. Birth Defects: Original Article Series 12/5: 279
Swanson A B, Brown K S 1962 Hereditary triphalangeal thumb. Journal of Heredity 53: 259
Temtamy S A, Dorst J P 1975 Polydactyly of triphalangeal thumbs with upper limb and pectoral dysplasia. Birth Defects: Original Article Series 11/5: 340

*Postaxial polydactyly*
Ammann F 1970 Investigations cliniques et génétiques sur le syndrome de Bardet-Biedl en Suisse. Journal de Génétique Humaine 18: 1
Cantu J M, Del Castillo V, Cortes R, Urrusti J 1975 Autosomal recessive postaxial polydactyly: report of a family. Birth Defects: Original Article Series 11/5: 19
Casamassima A C, Mamunes P, Gladstone I M et al 1987 A new syndrome with features of the Smith-Lemli-Opitz and Meckel-Gruber syndromes in a sibship with cerebellar defects. American Journal of Medical Genetics 26: 321–336
Curry C J R, Carey J C, Holland J S et al 1987 Smith-Lemli-Opitz syndrome Type II. American Journal of Medical Genetics 26: 45–57
Czeizel A, Brooser G 1986 A postaxial polydactyly and progressive myopia syndrome of autosomal dominant origin. Clinical Genetics 30: 406–408
Hernandez A, Garcia-Esquivel L, Reynoso C et al 1985 Cortical blindness, growth and psychomotor retardation and postaxial polydactyly: A probably distinct autosomal recessive syndrome. Clinical Genetics 28: 251–254
Holt S 1975 Polydactyly and brachymetapody in two English families. Journal of Medical Genetics 12: 355
Johnston O, Davis R W 1953 On the inheritance of hand and foot anomalies in six families. American Journal of Human Genetics 5: 356
Laurence K M, Prosser R, Rocker I, Pearson J F, Richards C 1975 Hirschsprung's disease associated with congenital heart malformation, broad big toes and ulnar polydactyly in sibs: a case for foetoscopy. Journal of Medical Genetics 12: 334
Okeahialam T C 1974 The pattern of congenital malformations observed in Dar-es-Salaam. East African Medical Journal 51/1: 101
Salonen R, Herva R, Norio R 1981 The hydrolethalus syndrome: Delineation of a 'new' lethal malformation syndrome based on 28 patients. Clinical Genetics 19: 321
Smith D W, Lemli L, Opitz J M 1964 A newly recognized syndrome of multiple congenital anomalies. Journal of Pediatrics 64: 210–217

Teintamy S A, Rogers J G 1975 A new postaxial polydactyly syndrome? In: New Chromosomal and Malformation Syndromes, BD:OAS. White Plains: The National Foundation 11(5): 344

Woolf C M, Woolf R M 1970 A genetic study of polydactyly in Utah. American Journal of Human Genetics 22: 75

## Syndactyly

Briard M L, Kaplan J 1982 Recessive form of polysyndactyly. Journal of Human Genetics 5: 439–444

Fitch N, Jequier S, Papageorgiou A 1976 A familial syndrome of cranial, facial, oral and limb anomalies. Clinical Genetics 10: 226

Goldberg M J, Pashayan H M 1976 Hallux syndactyly — ulnar polydactyly — abnormal ear lobes: a new syndrome. Birth Defects: Original Article Series 12/5: 255

Merlob P, Grunebaum M 1986 Type II syndactyly or synpolydactyly. Journal of Medical Genetics 23: 237–241

Ofodile F A 1982 Synpolydactyly in three generations of a Nigerian family. East African Medical Journal 59: 835–839

Richieri-Costa A, Colletto G M, Gollop T R, Masiero D 1985 A previously undescribed autosomal recessive multiple congenital anomalies/mental retardation syndrome. American Journal of Medical Genetics 20: 631–638

Robinow M, Johnson G F, Broock G J 1982 Syndactyly type V. American Journal of Medical Genetics 11: 475–482

Temtamy S, McKusick V A 1978 The Genetics of Hand Malformations. Birth Defects: Original Article Series XIV/3. National Foundation, March of Dimes, Liss, New York

## Symphalangism

Austin F H 1951 Symphalangism and related fusions of tarsal bones. Radiology 56: 882

Cremers C, Theunissen E, Kuijpers W 1985 Proximal symphalangia and stapes ankylosis. Archives of Otolaryngology 111: 765–767

Da-Silva E O, Filho S M, de Albuquerque S C 1984 Multiple synostosis syndrome: Study of a large Brazilian kindred. American Journal of Medical Genetics 18: 237–247

Elkington S G, Huntsman R G 1967 The Talbot fingers. A study in symphalangism. British Medical Journal 1: 407

Herrmann J 1974 Symphalangism and brachydactyly syndrome: Report of WL symphalangism-brachydactyly syndrome. Birth Defects: Original Article Series X/5: 23–53

Hurvitz S A, Goodman R M, Hertz M et al 1985 The facio-audio-symphalangism syndrome: Report of a case and review of the literature. Clinical Genetics 28: 61–68

Kassner E G, Katz I, Qazi Q H 1976 Symphalangism with metacarpophalangeal fusions and elbow abnormalities. Pediatric Radiology 4/2: 103

Maroteaux P, Bouvet J P, Briard M L 1972 La maladie des synostoses multiples. La Nouvelle Presse Médicale 1(45): 3041–3047

Pierson M, Tridon P, Wayoff M, Umana L 1982 Symphalangism and multiple synostoses disease: Presentation of two families. Journal of Human Genetics 30/4: 351–358

Strasburger A K, Hawkins M R, Eldridge R, Hargreave R L, McKusick V A 1965 Symphalangism. Genetic and clinical aspects. Bulletin of the Johns Hopkins Hospital 117: 108

Walbaum R, Hazard C, Cordier B 1976 Brachydactylia with symphalangism, probably autosomal recessive. Human Genetics 33/2: 189

## Brachydactyly

Bass H N 1968 Familial absence of middle phalanges with nail dysplasia: a new syndrome. Pediatrics 42: 318

Bell J 1951 On brachydactyly and symphalangism. Treasury of Human Inheritance 5: 1

Cotsirilis P, Taylor J, Matalon R 1987 Dominant inheritance of a syndrome similar to Rubinstein-Taybi. American Journal of Medical Genetics 26: 85–93

Cuevas-Sosa A, Garcia-Segur F 1971 Brachydactyly with absence of middle phalanges and hypoplastic nails. Journal of Bone and Joint Surgery 53B: 101

Davies A B 1975 Stub Thumbs. Journal of Medical Genetics 12/4: 414

Gnamey D, Walbaum R, Fossati P, Prouvost J M 1975 Hereditary brachydactyly type E. Report on a family. Pédiatrie 30/2: 153

Goodman R M, Adam A, Sheba C 1965 A genetic study of stub thumbs among various ethnic groups in Israel. Journal of Medical Genetics 2: 116

Goodman R, Feinstein A, Hertz M 1984 Stub thumbs in Israel revisited. Journal of Medical Genetics 21: 460–462

Haws D V 1963 Inherited brachydactyly and hypoplasia of the bones of the extremities. Annals of Human Genetics 26: 201

Haws D V, McKusick V A 1963 Farabee's brachydactylous kindred revisited. Bulletin of the Johns Hopkins Hospital 113: 20

Kumar D, Levick R K 1986 Autosomal dominant onychodystrophy and anonychia with type B brachydctyly and ectrodactyly. Clinical Genetics 30: 219–225

MacArthur J W, McCullough E 1932 Apical dystrophy as inherited defect of hands and feet. Human Biology 4: 179

McKusick V A, Milch R A 1964 The clinical behaviour of genetic disease: selected aspects. Clinical Orthopaedics and Related Research 33: 22

Mohr O L, Wriedt C 1919 A new type of hereditary brachyphalangy, Publ. 295 Carnegie Institute, Washington, p 5

Newcombe D S, Keats T E 1969 Roentgenographic manifestations of hereditary peripheral dysostosis. American Journal of Roentgenology 106: 178

Poznanski A K, Garn S M, Holt J F 1971 The thumb in the congenital malformation syndromes. Radiology 100/1: 115

Rimoin D L, Hollister D W, Lachman R S 1975 Type C brachydactyly with limited flexion of distal interphalangeal joints. Birth Defects: Original Article Series 11/5: 9

Robinson G C, Wood B J, Miller J R, Baillie J 1968 Hereditary brachydactyly and hip disease. Unusual radiological and dermatoglyphic findings in a kindred. Journal of Pediatrics 72: 539

Sayles L P, Jailer J W 1934 Four generations of short thumbs. Journal of Heredity 25: 377

Temtamy S, McKusick V A 1969 Synopsis of hand malformation with particular emphasis on genetic factors. Birth Defects: Original Article Series 5/3: 125

Temtamy S A, McKusick V A 1978 Genetics of Hand Malformations. Birth Defects: Original Article Series XIV(3) Liss, New York

Zavala C, Hernandez Ortiz J, Lisker R 1975 Brachydactyly type B and symphalangism in different members of a Mexican family. Annales de Génétique 18/2: 131

*Brachytelophalangy*

Cooks R G, Hertz M, Katznelson M B M, Goodman R M 1985 A new nail dysplasia syndrome with onychonychia and absence and/or hypoplasia of distal phalanges. Clinical Genetics 27: 85–91

Cormode E J, Dawson M, Lowry R B 1986 Keutel syndrome: Clinical report and literature review. Americal Journal of Medical Genetics 24: 289–294

Fryns J P, van Fleteren A, Mattelaer P, van den Berghe H 1984 Calcification of cartilages, brachytelephalangy and peripheral pulmonary stenosis: Confirmation of the Keutel syndrome. European Journal of Pediatrics 142: 201–203

Hunter A G W, Feldman W, Sauve G 1986 A characteristic craniofacial appearance and brachytelephalangy in a mother and son with Kallman syndrome in the son. Proceedings of Greenwood Genetic Center 4: 128

Kelly T E 1984 Teratogenicity of anticonvulsant drugs. III: Radiographic hand analysis of children exposed in utero to diphenylhydantoin. American Journal of Medical Genetics 19: 445–450

Keutel J, Jorgensen G, Gabriel P 1982 A new autosomal recessive syndrome: Peripheral pulmonary stenoses, brachytelephalangism, neural hearing loss and abnormal cartilage calcification/ossification. In: Bergsma D (ed) The Clinical Delineation of Birth Defects: The cardiovascular system. Williams and Wilkins, Baltimore. Birth Defects: Original Article Series VIII(5): 60–68

Say B, Balci S, Pirnar T, Israel R, Atasu M 1973 Unusual calcium deposition in cartilage associated with short stature and peculiar facial features: A case report. Pediatric Radiology 1: 127–129

Walbaum R, Boniface L, Tonnel et al 1975 Le syndrome de Keutel. Annales de Pédiatrie 51: 461

*Other brachydactyly syndromes*
Bilginturan N, Zileli S, Karacadag S, Pirnar T 1973 Hereditary brachydactyly associated with hypertension. Journal of Medical Genetics 10: 253
Christian J C, Cho K S, Franken E A, Thompson B H 1972 Dominant preaxial brachydactyly with hallux varus and thumb abduction. American Journal of Human Genetics 24: 694
Gorlin R J, Sedano H O 1971 Cryptodontic brachymetacarpalia. Clinical Delineation of Birth Defects, XI Williams and Wilkins, Baltimore, p 200
Pitt P, Williams I 1985 A new brachydactyly syndrome with similarities to Julia Bell types B & E. Journal of Medical Genetics 22: 202–204
Sorsby A 1935 Congenital coloboma of the macula, together with an account of the familial occurrence of bilateral macular coloboma in association with apical dystrophy of hands and feet. British Journal of Ophthalmology 19: 65
Villaverde M M, Silva J A 1975 Distal brachyphalangy of the thumb in mental retardation. Journal of Medical Genetics 12/4: 401

**Ectrodactyly**
David T J 1972 Dominant ectrodactyly and possible germinal mosaicism. Journal of Medical Genetics 9: 316–320
Gelfand M, Roberts C J, Roberts R S 1974 A two-toed man from the Doma People of the Zambezi valley. Rhodesian History 5: 93
Gemme G, Bonioli E, Ruffa G, Grosso P 1976 EEC syndrome: description of two cases in the same family. Minerva Pediatrica 28/1: 36
Mackenzie H J, Penrose L S 1951 Two pedigrees of ectrodactyly. Clinical Genetics 9: 347
Majewski F, Kuster W, Ter Haar B, Goecke T 1985 Aplasia of tibia with split-hand/split-foot deformities. Report of six families with 35 cases. Human Genetics 70/2: 136–147
Mosavy S H, Vakshuri P 1975 Split hands and feet. South African Medical Journal 49: 1842
Preus M, Fraser F C 1973 The lobster-claw defect with ectodermal defects, cleft lip-palate, tear duct anomaly and renal anomalies. Clinical Genetics 4: 369
Spranger M, Schapera J 1988 Anomalous inheritance in a kindred with the split hand-split foot syndrome. European Journal of Pediatrics (In press)
Stevenson A C, Jennings L M 1960 Ectrodactyly — evidence in favour of a disturbed segregation in the offspring of affected males. Annals of Human Genetics 24: 89
Verma I C, Joseph R, Bhargava S, Mehta S 1976 Split hand and split foot deformity inherited as an autosomal recessive trait. Clinical Genetics 9/1: 8
Viljoen D L, Beighton P 1984 The split-hand and split-foot anomaly in a Central African Negro population. American Journal of Medical Genetics 19: 545–552
Viljoen D, Farrell H M, Brossy J J et al 1985 Ectrodactyly in Central Africa. South African Medical Journal 68: 655–658

*Ectrodactyly, ectodermal dysplasia, facial cleft syndrome*
Bixler D, Spivack J, Bennett J, Christian J C 1972 The ectrodactyly-ectodermal dysplasia-clefting (EEC) syndrome. Clinical Genetics 3: 43
London R, Heredia R M, Israel J 1985 Urinary tract involvement in EEC syndrome. American Journal of Diseases of Children 139: 1191–1193
Pashayan H M, Pruzansky S, Solomon L 1974 The EEC syndrome. Report of six patients. Birth Defects: Original Article Series 10/7: 105
Penchaszadeh V B, de Negrotti T C 1976 Ectrodactyly-ectodermal dysplasia-clefting (EEC) syndrome: dominant inheritance and variable expression. Journal of Medical Genetics 13/4: 281
Reed W B, Brown A C, Sugarman G I, Schlesinger L 1975 The REEDS syndrome. Birth Defects: Original Article Series 11/5: 61
Roselli D, Gulienetti R 1961 Ectodermal dysplasia. British Journal of Plastic Surgery 14: 190

Rosenmann A, Shapira T, Cohen M M 1976 Ectrodactyly, ectodermal dysplasia and cleft palate (EEC syndrome). Report of a family and review of the literature. Clinical Genetics 9: 347

Rudiger R A, Haase W, Passarge E 1970 Association of ectrodactyly, ectodermal dysplasia, with cleft lip-palate. American Journal of Diseases of Children 120: 160

Walker J C, Clodius L 1963 The syndromes of cleft lip, cleft palate and lobster claw deformities of hands and feet. Plastic and Reconstructive Surgery and the Transplantation Bulletin 32: 627

*Ectrodactyly with scalp defects*

Adams F H, Oliver C P 1945 Hereditary deformities in man due to arrested development. Journal of Heredity 36: 3

Bonafede P, Beighton P 1979 Autosomal dominant inheritance of scalp defects with ectrodactyly. American Journal of Medical Genetics 3: 35–41

Sybert V P 1985 Aplasia cutis congenita: a report of 12 families and a review of the literature. Pediatric Dermatology 3: 1–4

**Digital dysplasia syndromes**
*Holt-Oram syndrome*

Ferrier P E, Friedli B, Ferrier S 1975 The heart and hand syndrome (Holt-Oram). Semaine des Hôpitaux de Paris 51/39: 727

Gall J C Jr, Stern A M, Cohen M M, Adams M S, Davidson R T 1966 Holt-Oram syndrome: clinical and genetic study of a large family. American Journal of Human Genetics 18: 187

Gladstone I, Sybert V P 1982 The Holt-Oram syndrome: Penetrance of the gene and lack of maternal effect. Clinical Genetics 21: 98–103

Harris L C, Osborne W P 1966 Congenital absence or hypoplasia of the radius with ventricular septal defect: ventriculo-radial dysplasia. Journal of Pediatrics 68: 265

Hoyme H E, Van Allen M I, Jones K L 1983 The vascular pathogenesis of some sporadically occurring limb defects. Seminars in Perinatology 7: 299–306

Letts R M, Chudley A E, Cumming G, Shokier M H 1976 The upper limb cardiovascular syndrome (Holt-Oram syndrome). Clinical Orthopaedics and Related Research 116: 149

Lewis K B, Bruce R A, Baum D, Motulsky A G 1965 The upper limb-cardiovascular syndrome. An autosomal dominant genetic effect on embryogenesis. Journal of the American Medical Association 193: 1080

Muller L M, De Jong G, Van Heerden K M 1985 The antenatal ultrasonographic detection of the Holt-Oram syndrome, South African Medical Journal 68: 313–315

Poznanski A K, Gall J C Jr, Stern A M 1970 Skeletal manifestations of the Holt-Oram syndrome. Radiology 94: 45

Rybak M, Kozlowski K, Kleczkowska A, Lewandowska J, Sokolowski J, Soltysik-Wilk E 1971 Holt-Oram syndrome associated with ectromelia and chromosomal aberrations. American Journal of Diseases of Children 121: 490

Smith A T, Sack G H, Taylor G J 1979 Holt-Oram syndrome. Journal of Pediatrics 95: 538–543

Tamari I, Goodman R M 1974 Upper limb-cardiovascular syndromes: a description of two new disorders with a classification. Chest 65: 632

*Orofacial-digital syndrome*

Anneren G, Arvidson B, Gustavson K H et al 1984 Oro-facial-digital syndromes I and II: Radiological methods for diagnosis and the clinical variations. Clinical Genetics 26: 178–186

Burn J, Dezateux C, Hall C M, Baraitser M 1984 Orofacialdigital syndrome with mesomelic limb shortening. Journal of Medical Genetics 21: 189–192

Doege T C, Thuline H C, Priest J H, Norby D E, Bryant J S 1964 Studies of a family with the oral-facial-digital syndrome. New England Journal of Medicine 271: 1073

Gorlin R J, Psaume J 1962 Orodigito-facial sysostosis: a new syndrome. Journal of Pediatrics 61: 520

Haumont D, Pelc S 1983 The Mohr syndrome: Are there two variants? Clinical Genetics 24: 41–46

Iaccarino M, Lonardo F, Guigliano M, Della Bruna M 1985 Prenatal diagnosis of Mohr syndrome by ultrasonography. Prenatal Diagnosis 5: 415–418

McKusick V A 1983 Mendelian Inheritance in Man 6th edn. Johns Hopkins University Press, p 831

Melnick M, Shields E D, 1975 Orofaciodigital syndrome, type I: A phenotypic and genetic analysis. Oral Surgery 40/5: 599

Papillon-Léage, Psaume J 1954 Une malformation héréditaire de la muqueuse buccale, brides et freins. Revue de Stomatologie 55: 209

Reuss A L, Pruzansky S, Lis E F 1962 The oral-facial-digital syndrome: A multiple congenital condition of females with associated chromosomal abnormalities. Pediatrics 29: 985–995

Rimoin D L, Edgerton M T 1967 Genetic and clinical heterogeneity in the oral-facial-digital syndromes. Journal of Pediatrics 71: 94

Sugarman G I, Katakia M, Menkes J 1971 See-saw winking in a familial oral-facial-digital syndrome. Clinical Genetics 2: 248–254

Temtamy S A, Levin L S, Miller J D et al 1975 Severe Mohr syndrome or mild Majewski syndrome? Birth Defects: Original Article Series 11/5: 342

Whelan D T, Feldman W, Dost I 1975 The oro-facial-digital syndrome. Clinical Genetics 8: 205


*Oto-palato-digital syndrome*

Andre M, Cigneron J, Didier F 1981 Abnormal facies, cleft palate and generalized dysostosis: A lethal X-linked syndrome. Journal of Pediatrics 98: 747–752

Brewster T G, Lachman R S, Kushner D C et al 1985 Oto-palato-digital syndrome, type II: An X-linked skeletal dysplasia. American Journal of Medical Genetics 20: 249–254

Dudding B, Gorlin R L, Langer L O 1967 The oto-palato-digital syndrome. American Journal of Diseases of Children 113: 214–221

Fitch N, Jequier S, Gorlin R 1983 The oto-palato-digital syndrome, proposed type II. American Journal of Medical Genetics 15: 655–664

Fitch N, Jequier S, Papageorgiou A 1976 A familial syndrome of cranial, facial, oral and limb anomalies. Clinical Genetics 10: 226–231

Gorlin R, Poznanski A, Hendon I 1973 The oto-palato-digital (OPD) syndrome in females. Oral Surgery 35: 218–224

Kozlowski K, Turner G, Scougall J, Harrington J 1977 Oto-palato-digital syndrome with severe X-ray changes in two half brothers. Pediatric Radiology 6: 97–102

Pazzaglia U E, Beluffi G 1986 Oto-palato-digital syndrome in four generations of a large family. Clinical Genetics 30: 338–344

Salinas C, Jorgensen R, Lorenzo R 1979 Variable expression in otopalatodigital syndrome, cleft palate in female. Birth Defects: Original Article Series XV(5b): 329–345

Taybi H 1962 Generalized skeletal dysplasia with multiple anomalies. American Journal of Roentgenology 88: 450–457


*Aase syndrome*

Aase J M, Smith D W 1969 Congenital anemia and triphalangeal thumbs: A new syndrome. Journal of Pediatrics 74: 417

Alter B P 1978 Thumbs and anemia. Pediatrics 62: 613–614

Higginbottom M C, Jones K L, Kung F H et al 1978 The Aase syndrome in a female infant. Journal of Medical Genetics 15: 484–485

Jones B, Thompson H 1973 Triphalangeal thumbs associated with hypoplastic anemia. Pediatrics 52: 609

Murphy S, Lubin B 1972 Triphalangeal thumbs and congenital erythroid hypoplasia: Report of a case with unusual features. Journal of Pediatrics 81: 987

Van Weel-Sipmann M, van de Kamp J J P, de Koning J 1977 A female patient with Aase syndrome. Journal of Pediatrics 91: 753–755


*Aglossia-adactylia syndrome (Hanhart)*

Bokesoy I, Aksuyek C, Deniz E 1983 Oromandibular limb hypogenesis/Hanhart's syndrome: Possible drug influence on the malformation. Clinical Genetics 24: 47–49

Dellagrammaticas H, Tzaki M, Kapiki A et al 1982 Hanhart syndrome: Possibility of autosomal recessive inheritance. Skeletal Dysplasias. Liss, New York, p 299–304

Goodman R, Gorlin R J 1977 Atlas of the Face in Genetic Disorders 2nd edn. Mosby, St Louis

Hall B D 1971 Aglossia-adactylia. A case report, review of the literature and classification of closely related entities. Birth Defects: Original Article Series 7/7: 233

Hanhart E 1950 Uber die Kombination von Peromelie mit Mikrognathie: Ein neues Syndrome beim Menschen, entsprechend der Akroteriasis congenita von Wreidt und Mohr beim Rinde. Archives Julius Stift. Vererbungsforschritte 25: 531–540

Herrmann J, Pallister R D, Gilbert E F et al 1976 Studies of malformation syndromes of man XXXXI B: Nosologic studies in the Hanhart and the Möbius syndrome. European Journal of Pediatrics 122: 19–55

Nevin N C, Burrows D, Allen C, Kernohan D C 1975 Aglossia adactylia syndrome. Journal of Medical Genetics 12/1: 89

Tuncbilek E, Yallin C, Atasu M 1977 Aglossia-adactylia syndrome. Clinical Genetics 11: 421–423

*Goltz syndrome*

Feinberg A, Menter M A 1976 Focal dermal hypoplasia (Goltz syndrome) in a male: Case report. South African Medical Journal 50: 554–555

Goltz R W, Peterson W C Jr, Gorlin R J, Ravits H G 1962 Focal dermal hypoplasia. Archives of Dermatology 86: 708

Goltz R W, Henderson R R, Hitch J M, Ott J E 1970 Focal dermal hypoplasia syndrome. A review of the literature and report of two cases. Archives of Dermatology 101: 1

Happle R, Lenz W 1977 Striation of bones in focal dermal hypoplasia: manifestation of functional mosaicism? British Journal of Dermatology 96: 133–138

Toro-Sola M A, Kistenmacher M L, Punnett H H, DiGeorge A M 1975 Focal dermal hypoplasia syndrome in a male. Clinical Genetics 7: 325–327

*Poland syndrome*

Bavinck J N, Weaver D D 1986 Subclavian artery supply distruption sequence: Hypothesis of a vascular etiology for Poland, Klippel-Feil and Mobius anomalies. American Journal of Medical Genetics 23: 903–918

Bosch-Banyeras J M, Zuasnabar A, Puig A et al 1984 Poland-Mobius syndrome associated with dextrocardia. Journal of Medical Genetics 21: 70–71

David T J 1976 Nature and etiology of the Poland anomaly. New England Journal of Medicine 287: 487–489

Ireland D C R, Takayama N, Flatt A E 1976 Poland's syndrome: A review of 43 cases. Journal of Bone and Joint Surgery 58: 52–58

Trier W C 1965 Complete breast absence. Case report and review of the literature. Plastic and Reconstructive Surgery 36: 431–439

# 16

# Limb dysplasias and synostoses

Major epidemiological studies of limb malformations have been undertaken in Denmark by Birch-Jensen (1949), in Brazil by Freire-Maia (1969), in West Germany and England by Henkel & Willert (1969), in Scotland by Rogala et al (1974) and in Hungary (Czeizel et al 1983). In each of these surveys it was shown that congenital absence of a limb or segment of a limb is usually non-genetic. However, there have been instances of kindreds with aggregation of mixed limb malformations (Falek et al 1968, Hecht & Scott 1981, Stanley et al 1984). Limb reduction is also a feature of a few unusual syndromes which have a proven or possible genetic basis. Finally, synostosis or abnormal bone union is sometimes found in conjunction with limb deficiencies. For this reason, the synostoses are grouped with the limb dysplasia syndromes in this chapter.

1. Non-specific limb reduction
   a. Phocomelia
   b. Amelia
   c. Hemimelia
2. Limb reduction syndromes
   a. Roberts syndrome
   b. Acheiropodia
   c. Proximal focal femoral dysplasia
   d. Tibial deficiency syndromes
   e. Aplasia cutis congenita-terminal transverse limb defect syndrome
   f. Fanconi-pancytopenia syndrome
   g. Radial aplasia-thrombocytopenia syndrome
   h. Radial ray abnormalities (miscellaneous)
   i  Other genetic limb reduction syndromes
3. Synostosis syndromes
   a. Radio-ulnar synostosis
   b. Synostosis of the tarsus
   c. Synostosis of the carpus
   d. Multiple synostoses
   e. Hand-foot-genital syndromes

In a comprehensive review, Henkel & Willert (1969) discussed the classification of limb defects and defined 'dysmelia' as 'malformations in which
there is hypoplasia and partial or total aplasia of the tubular bones of the
extremities, ranging from isolated peripheral hypoplasia to complete loss of
the extremity.' Similarly, 'ectromelia' implies maldevelopment of the long
bones of the limbs and their peripheral rays. The following terms also
pertain to limb anomalies:

| | |
|---|---|
| Phocomelia | hypoplasia of the limbs, so that the hands and feet are attached directly to the limb girdles |
| Amelia | absence of a limb or limbs |
| Hemimelia | absence of a longitudinal segment of a limb |
| Acheira | absence of a hand or hands |
| Apodia | absence of a foot or feet |
| Acheiropodia | absence of the hands and feet. |

The pattern of defective development may embrace more than one of
these anatomical types and multiple mixed abnormalities may be present.
In this context, it is relevant that a 'new international terminology for the
classification of congenital limb deficiencies' has been proposed, following
a conference of delegates from specialised child amputee centres (Kay
1975).

The distinction between the terms 'malformation' and 'deformity' is
important; 'malformation' implies abnormality due to faulty development,
whereas 'deformity' denotes alteration in a previously normal structure. For
example, limb shortening in phocomelia is a malformation, while limb
bowing due to fracturing in osteogenesis imperfecta is a deformity.

The embryological mechanisms by which limb malformations may be
produced have been reviewed by Poswillo (1976) and Wolpert (1976). It has
been suggested that localised deficiencies in vascular supply during fetal
development might be responsible for some transverse limb defects and
similar malformations (Bavinck & Weaver 1986).

A combination of ingenuity and persistence has permitted some individuals to overcome the disability imposed by major limb defects. For
instance, Carl Unthan, who was born without arms, achieved international
recognition as a concert virtuoso, playing his violin with his toes. Similarly,
Eli Bowen, who lacked legs, was a noted acrobat. The non-genetic nature
of these abnormalities is exemplified by the fact that both these remarkable
men had many normal sibs and offspring.

## NON-SPECIFIC LIMB REDUCTION

### Phocomelia

In phocomelia, the malformed extremities articulate directly with the trunk.
This configuration has been likened to the flippers of a seal, hence the

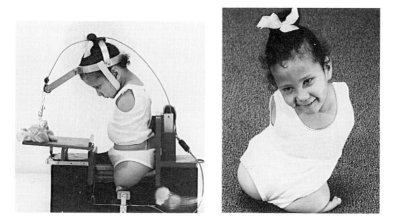

**Fig. 16.1** (left) Limb reduction: technology has much to offer the disabled individual.

**Fig. 16.2** (right) Limb reduction: this girl was born without limbs. A vestigial foot is attached to the left hip.

designation 'phocomelia'. If all four limbs are involved the abnormality is termed 'tetraphocomelia'. The malformation is rare and usually non-genetic. Thalidomide taken during early pregnancy was responsible for the 'epidemic' of phocomelia which occurred in 1961–2. This tragedy is often quoted as the classic example of the embryopathic action of an environmental agent. Di Battista et al (1975) reported the birth of a phocomelic

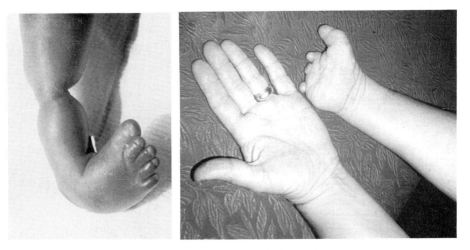

**Fig. 16.3** (left) Limb reduction: constriction of the soft tissues of the shin, due to amniotic bands. If bands of this type cause intrauterine amputation, the stump is smooth.

**Fig. 16.4** (right) Limb reduction: the presence of rudimentary digits indicates that the abnormality of the hand is the result of defective development rather than intrauterine amputation.

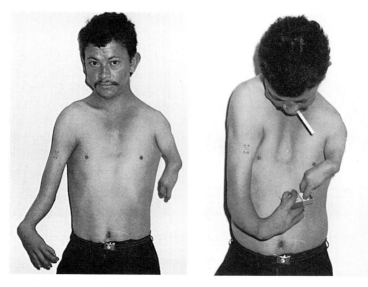

**Fig. 16.5** (left) Limb reduction: the bones of both arms are hypoplastic.

**Fig. 16.6** (right) Limb reduction: in spite of his handicap, this individual is surprisingly dexterous.

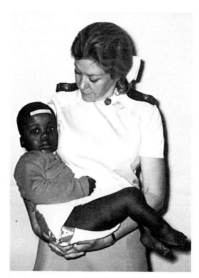

**Fig. 16.7** (left) Limb reduction: hypoplasia of the right tibia and fibula. It is unlikely that this unilateral malformation has a genetic basis.

infant to a mother who had been treated with an anti-inflammatory drug for a 4 day period in early pregnancy. The authors emphasised the import-ance of avoiding potentially toxic agents during the first trimester.

## Amelia (total or partial)

Congenital absence of a portion of the arm, hand, leg or foot is a relatively common abnormality. These defects, which are usually transverse and unilateral, occur in approximately 5 per 10 000 newborns (Cadas et al 1978, van Regemorter et al 1984, Khrouf et al 1986). The vast majority are non-genetic but a few instances of familial recurrence have been documented (Pauli et al 1985). The chromosomes are usually normal in persons with anomalies of this type but there have been a few reports of the association of limb and cytogenetic anomalies (Werchsel & Luzzatti 1965, Bofinger et al 1973, Pfeiffer & Sautelmann 1977). Limb malformations are especially common in trisomy 18 (Kajii et al 1964, Voorhess et al 1964, Christianson & Nelson 1984).

The relationship between oral contraceptives and congenital limb defects was investigated by Drance (1975). No firm conclusions were reached but the author drew attention to the VACTERL (vertebral, anal, cardiac, tracheal, oesophageal, renal and limb) group of anomalies, and mentioned a possible association with maternal ingestion of progesterone or oestrogen during pregnancy. Although more than 40 patients with this acronymic association have been documented, the range of potential manifestations is still uncertain (Evans et al 1985). The term 'VATER' association is also in use and there is some confusion as to whether VACTERL and VATER are the same condition. In any event, apart from a report of affected brothers (Auchterlonie & White 1982) there is no evidence of a genetic aetiology and recurrence risks are low.

Limb reduction has been observed in infants of mothers who have used LSD (Assemarry et al 1970, Blanc et al 1971). Nevertheless, there is no firm evidence that this hallucinogen is, in fact, teratogenic (Long 1972). Absence of the left leg at the knee was recorded in the daughter of a mother who took pyridoxine (vitamin B6) during pregnancy (Gardner et al 1985). The authors commented upon the similarity of the mode of activity of this therapeutic agent to that of thalidomide and speculated that there might be a causal relationship with the limb defect. Maternal alcoholism has recently been implicated as a cause of amelia and other fetal limb malformations (Pauli & Feldman 1986).

Intrauterine amputation by amniotic bands is thought to be a common cause of amelia. It has been estimated that this abnormality has an incidence of between 1 in 5000 and 1 in 10 000 births (Fisher & Cremin 1976). The presence of grooves around limbs, or of partial amputation, is evidence for the action of this process. The distal end of a limb which has amputated in this way is usually smooth. Conversely, if development has been defective, rudimentary digits may be present. In either event, for the purpose of genetic counselling, it is reasonable to offer a very low risk for recurrence. Ultrasonic surveillance of subsequent pregnanices represents an additional safeguard.

The mechanism of amniotic band formation is uncertain. Initial concepts implicated an intrinsic defect in germplasma (Streeter 1930). Subsequently, a process involving rupture of the amnion and the formation of mesodermal bands was favoured (Torpin 1968). The role of this mechanical process in the pathogenesis of birth defects is widely accepted (Baker & Rudolf 1971, Jones et al 1974, Beyth et al 1977) but remains unproven. Reports are accumulating of affected individuals with additional malformations which cannot readily be explained on a basis of amniotic bands alone (Herva & Karkinen-Jaaskelainen 1984). These observations could indicate that the bands might be a component of an underlying fetal defect, rather than the prime causative factor (Hunter & Carpenter 1986).

## Hemimelia

Hemimelia, or longitudinal limb deficiency, is the result of defective development of an embryonic ray. The commonest abnormality of this type in the upper limb is hypoplasia or aplasia of the radius, with corresponding anomalies in the hand. These defects are usually unilateral and non-genetic. However, radial hypoplasia is also a component of a few heritable syndromes (vide infra). Defects of the ulna and the medial side of the hand are much less common, the most important disease association being the de Lange syndrome (see Ch. 18). A review of 65 patients with various categories of ulnar ray deficiency was published by Swanson et al (1984) and a classification based upon clinical manifestations was proposed.

Tibial hemimelia has been encountered in a pair of sisters (Russell 1975). Other members of the kindred had hand and foot deformities which could have represented varying phenotypic manifestations of a single dominant gene. Barrie (1976) described two infants with fibular aplasia and limb reduction defects whose mothers had used copper-containing intrauterine contraceptive devices. The author speculated that there may have been a causal association between the contraceptives and the fetal maldevelopment.

## LIMB REDUCTION SYNDROMES

### Roberts* syndrome

The Roberts syndrome, or pseudophocomelia, is characterised by dysmelia, which ranges from complete phocomelia to minor degrees of deficiency of limb segments, in association with cleft lip and palate. An unusual facies, sparse hair and a variety of visceral abnormalities may also be present. Still-birth is common and the health of the survivors is precarious.

---

* John B. Roberts (1854–1924) had a distinguished career as a plastic surgeon in Philadelphia, USA.

The original case report of Roberts (1919) concerned three affected sibs, born to consanguineous Italian parents. Freeman et al (1974) reviewed the literature and confirmed that the syndrome was inherited as an autosomal recessive. Subsequently, Freeman et al (1975) found chromosomal abnormalities in one patient and further cytogenetic aberrations were recorded by Quazi et al (1979). Zergollern & Hitrec (1976) studied a Yugoslavian family in which four siblings had the Roberts syndrome and affected identical twins were documented by Fryns et al (1980). About 50 cases have now been reported.

Herrmann et al (1969) described a condition which resembled the Roberts syndrome, with the additional feature of mental deficiency. The designation 'SC syndrome' was taken from the surnames of two affected kindreds. Hall & Greenberg (1972) gave details of a child with the disorder and Lenz et al (1975) mentioned three more cases, using the term 'pseudo-thalidomide syndrome.' Several investigators, including Grosse et al (1975) have suggested that the Roberts syndrome and the SC syndrome might be the same entity. Herrmann & Opitz (1977) concurred with this viewpoint and syndromic homogeneity is now widely accepted.

## Acheiropodia

Acheiropodia is a rare disorder in which all components of the limbs are absent, distal to the middle of the forearms and shins. The condition has been recognised in more than 50 individuals in various parts of Brazil (Freire-Maia 1970). The affected kindreds, in which consanguinity is frequent, are known as the 'handless and footless families of Brazil'. Toledo et al (1972) summarised information concerning the manifestations of acheiropodia, pointing out that there were no visceral ramifications, and that neither metabolic nor cytogenetic abnormalities could be detected. Pedigree data is entirely consistent with autosomal recessive inheritance and the mutant gene is fully penetrant in the homozygote (Freire-Maia 1975). It has been estimated that there are a minimum of 25 000 acheiropodia genes in Brazil, all derived from a single mutation (Freire-Maia et al 1975).

More than 22 affected sibships have been identified in Brazil and ongoing investigations have included radiological studies (Freire-Maia et al 1978), pathological aspects (Marcallo et al 1979) and demographic factors (Morton & Barbosa 1981). Experience of the condition gleaned over a 50 year period was reviewed by Freire-Maia (1981).

## Proximal focal femoral dysplasia

Proximal focal femoral dysplasia (PFFD) is an unusual form of femoral maldevelopment. The condition may be unilateral or bilateral. Individuals with the disorder have a remarkable appearance, as their knee joints are in

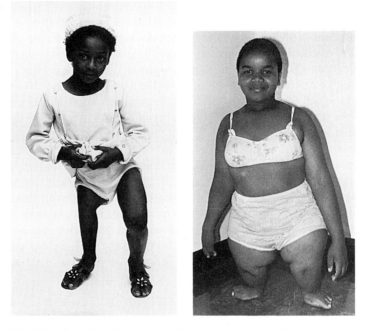

**Fig. 16.8** (left) Proximal focal femoral dysplasia: although the right femur is very hypoplastic, the skeleton is otherwise normal.

**Fig. 16.9** (right) Proximal focal femoral dysplasia: the femora are short and malformed.

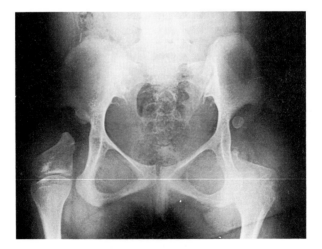

**Fig. 16.10** Proximal focal femoral dysplasia: radiograph showing gross shortening of the femora and a bizarre pelvic configuration. (From Lord & Beighton 1981, Clinical Genetics 20: 267–275.)

close proximity to the pelvic girdle. Nevertheless, they may have a surprising degree of agility. For instance, Hervio Novo achieved fame as a gymnast, despite his bilateral femoral deficiencies (Ferguson 1964). In addition to the femoral abnormalities, distal defects may be present, notably absence of the fibula and malformation of the extremities. The pelvis has a bizarre radiographic appearance, which has not been emphasised in the literature.

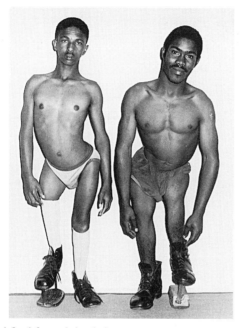

**Fig. 16.11** Proximal focal femoral dysplasia: two young men with unilateral involvement. This condition is much more common than the bilateral type. (From Lord & Beighton 1981 Clinical Genetics 20: 267–275.)

Unilateral focal femoral dysplasia is very much more common than the bilateral type (Hamanish 1980). The cause of unilateral involvement is unknown but there is nothing to indicate a genetic aetiology. Two adult females with the uncommon bilateral form were described and depicted by Bailey & Beighton (1970). One of these persons subsequently married a young man with the same condition but so far there have been no offspring from the union. This patient was documented again by Daentl et al (1975) in a report which drew attention to the facial features of a short nose with a broad tip, a long philtrum, thin upper lip and upward slanting palpebral fissures. The femoral hypoplasia-unusual facies syndrome (FH-UFS) was then delineated and further patients were described by Eastman & Escobar (1978), Gleiser et al (1978) and Hurst & Johnson (1980).

The possibility of dominant inheritance was raised by a report of an affected father and daughter (Lampert 1980) but there have been no other instances of familial aggregation and it seems very unlikely that the disorder is the result of a simple Mendelian mechanism. Following a review of 10 cases identified in the literature, Burck et al (1981) pointed out that cleft palate occurred as a syndromic component only in females and suggested that this feature might represent a diagnostic discriminant for the recognition of heterogeneity.

The syndromic identity of FH-UFS was questioned when a study of nine sporadic persons with bilateral focal femoral dysplasia revealed that only one had the characteristic facial features (Lord & Beighton 1981). In an investigation of 13 affected individuals in the UK, all of whom were sporadic, Burn et al (1984) recognised the characteristic facies in six. Fetal constraint and maternal diabetes were implicated as causative factors in both groups of patients. The recognition of a girl with features of the caudal dysplasia syndrome and FH-UFS led Riedel & Froster-Iskenius (1985) to speculate that both conditions might represent different manifestations of the same disorder.

The severe limb shortening in unilateral and bilateral focal femoral dysplasia renders these disorders suitable candidates for antenatal diagnosis by ultrasound. This has been accomplished at the 33rd week of pregnancy by Graham (1985) following the routine use of limb normograms.

## Tibial deficiency syndromes

Bilateral hypoplasia of the tibia with thickening and bowing of the fibula, in association with polydactyly of the hands and feet, was present in a family studied by Eaton & McKusick (1969). Subsequently, Yujnovsky et al (1974) encountered the same disorder in three generations of a kindred and confirmed that inheritance was autosomal dominant. Pfeiffer & Roeskau (1971) reported a mother and son with similar abnormalities and Pashayan et al (1971) recorded an affected father and daughter. A Mexican family with variable absence of the tibiae, polydactyly and triphalangeal thumbs in four generations was documented by Canun et al (1984). Using the designation 'tibial meromelia' Clark (1975) described a kindred with congenital absence of the tibia. Nine individuals in two generations were affected and the pattern of transmission was consistent with autosomal dominant inheritance. A classification of congenital deficiency of the tibia, based upon orthopaedic assessment of 21 affected children has been proposed by Kalamchi & Dawe (1985).

## Aplasia cutis congenita-terminal transverse limb defect syndrome

Adams & Oliver (1945) reported a family in which terminal transverse defects of the limbs were associated with central skull and scalp deficiencies. Eight individuals in three generations were affected, and inheritance was apparently autosomal dominant. Scribanu & Temtamy (1975) described a second kindred with the condition and emphasised that clinical expression was extremely variable.

Transverse defects of the lower limbs are an occasional component of the ectrodactyly and ectrodactyly-ectodermal dysplasia syndromes, which are discussed in Chapter 15. In view of the considerable intrafamilial phenotypic variability, it is possible that the disorders in which transverse limb defects occur might be homogeneous.

### Fanconi* pancytopenia syndrome

Radial aplasia, absence of the thumb and bone marrow dysfunction constitute the Fanconi pancytopenia syndrome. Small stature and dermal pigmentation are additional features. Haematological problems usually develop in mid-childhood and leukaemia may be a late complication. Bone marrow transplantation from a normal sib produced a dramatic improvement in the blood count of a 15 year old affected boy (Barrett et al 1977).

More than 150 patients have been reported, and there is abundant evidence for autosomal recessive inheritance. In a review of 48 kindreds, Garriga & Crosby (1959) noted that four potentially heterozygous relatives had developed leukaemia. Swift & Hirschhorn (1966) suggested that heterozygotes might have a propensity to the development of malignant tumours. Multiple chromosomal breakages have often been observed in homozygotes but it is not possible to demonstrate consistent levels of induced breakage in heterozygotes (Duckworth-Rysiecki et al 1984).

### Radial aplasia-thrombocytopenia

Defective development of the radius is the main skeletal abnormality in the radial aplasia-thrombocytopenia syndrome. The ulna may also be hypoplastic and a wide variety of anomalies may be present in the extremities and viscera. A significant proportion of infants die from abnormal bleeding. The haematological status of the survivors improves with the passage of time.

The manifestations have been reviewed by Hall et al (1969) under the designation 'thrombocytopenia with absent radius, TAR syndrome'. Autosomal recessive inheritance is well established. Omenn et al (1973) used conventional radiography to investigate a potentially affected fetus. Radii were detected at the 16th gestational week and fetal normality was correctly predicted. Luthy et al (1979) successfully used the same technique to

---

Guido Fanconi* (1892–1979) was Professor of Paediatrics at the University of Zurich, Switzerland.

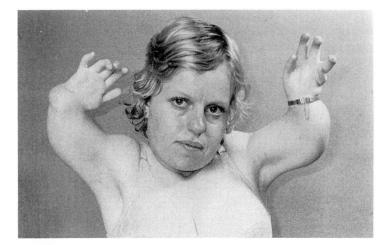

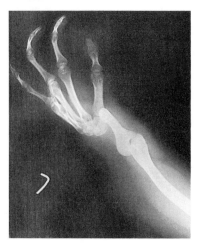

**Fig. 16.12** (above) Radial aplasia —
thrombocytopenia syndrome: a young woman
with abnormalities of the forearms and thumbs.
Thrombocytopenia was troublesome in
childhood but became less severe in later life.

**Fig. 16.13** (left) Radial aplasia —
thrombocytopenia syndrome: radiograph shows
absence of the radius and gross hypoplasia of
the ulna.

recognise two affected and four normal fetuses. Prediction was subsequently achieved by means of ultrasound (Luthy et al 1981). Prenatal assessment of a dizygotic pregnancy presented problems due to oligohydramnios but a combination of radiography, ultrasound and fetoscopy facilitated diagnostic confirmation in both fetuses (Filkins et al 1984).

A variety of the TAR syndrome in which tetraphocomelia is associated with neonatal thrombocytopenia has been recognised in sibs by Pfeiffer & Haenke (1975), Ray et al (1980) and Anyane-Yeboa et al (1985). This disorder seems to be a distinct autosomal recessive entity.

### Radial ray abnormalities, miscellaneous

Bilateral absence of the radius and thumb, in combination with renal abnormalities and malformations of the external ear were encountered in a father and son by Sofer et al (1983). A further pregnancy was monitored by ultrasound and defects of the radius and thumb were detected at the 18th week of gestation (Meizner et al 1986). The diagnosis was confirmed following termination and autopsy.

A familial syndrome of radial ray abnormalities, Duane anomaly of the eye, cervical spine malformations and deafness was documented by Hayes et al (1985). The phenotype in this autosomal dominant disorder was variable and radial ray involvement was manifest as hypoplasia of the thenar eminence.

Radial defects are associated with several chromosomal disorders. Haspeslagh et al (1984) reported a newborn male with severe radial anomalies and a deletion of the short arm of chromosome 4. A female neonate with anomalies of the thumb and first metacarpals was shown by Gardner et al (1984) to have a ring chromosome 1.

## Other genetic limb reduction syndromes

It is likely that other genetic limb reduction syndromes await delineation. For instance, Kucheria, (1976) described an Indian brother and sister with tetraphocomelia, cleft lip and palate and multiple minor abnormalities. The authors contended that the condition was a new syndrome, and that inheritance was autosomal recessive. Similarly, Lazjuk et al (1976) reported amelia, oligodactyly and other malformations in two half-cousins.

Defects of the femur and fibula are associated with malformations of the upper limb in the Femur-Fibula-Ulna complex (FFU) (Kuhne et al 1967, Lenz & Feldmann 1977). This rare anomaly has been recorded in siblings (Zlotogora et al 1983) and the possibility of autosomal recessive inheritance arises. Defects of the fibula and extremities are sometimes a feature of the proximal focal femoral dysplasia (PFFD) syndrome (see p 385) and the distinction between this entity and FFU is not absolute.

Arab siblings with severe deficiencies of the limb bones in association with thoracic dystrophy and an unusual facies were documented by Al-Awadi et al (1985). The parents were consanguineous and autosomal recessive inheritance is possible.

The Weyers syndrome is characterised by deficiency of the ulnar and fibular rays, together with malformation of the kidneys and spleen, digital anomalies and other variable defects (Weyers 1957). Further cases have been reported by Blockey & Lawrie (1963) and prenatal diagnosis by ultrasound at the 19th week of gestation following the previous birth of an affected sibling, has been accomplished by Elejalde et al (1985).

## SYNOSTOSIS SYNDROMES

Synostosis, or bone fusion, is most frequently encountered in the tarsus, carpus and forearms. Coalitions of this type may be isolated anomalies, components of well defined syndromes or discrete genetic entities. Few family studies have been undertaken and there is a paucity of information concerning inheritance of these abnormalities. McCredie (1975) pointed out that congenital fusion is the result of inappropriate organisation of mesenchyme during the 5th week after conception and postulated that the primary defect might lie in the embryonic sensory nerves.

## Radio-ulnar synostosis

Radio-ulnar synostosis occurs as a sporadic defect, as part of non-genetic multiple malformation syndromes and as a simple autosomal dominant trait (Hansen & Anderson 1970). The proximal portions of the radius and ulna are most often involved, with resultant loss of full extension and rotation at the elbow joint.

Following a long-term review of 23 patients Cleary & Omer (1985) concluded that there is usually little impairment of functional ability. The vast majority of affected persons are sporadic; in a series of 33 patients, only one had affected kin. Radio-ulnar synostosis and fusions of the capitate and hamate bones in the carpus was present in two out of 15 children with the fetal alcohol syndrome (Jaffer et al 1981). The clinical manifestations of this common condition can be subtle and the diagnosis warrants consideration in any sporadic person with radio-ulnar fusion.

## Synostosis of the tarsus and carpus

The talus and calcaneum may be united by a fibrous bridge or a bony bar. The only reports of familial occurrence concern a brother and sister (Webster & Roberts 1951) and a mother and her four offsprings (Diamond 1974).

Coalition of the calcaneum and the navicular may present clinically as a painful spastic flat foot. This abnormality was recognised in three males in successive generations of an American kindred by Wray & Herndon (1963). Although inheritance was apparently autosomal dominant in this particular family, it is not known whether all cases of calcaneonavicular synostosis have a similar genetic aetiology.

Kindreds with dominant transmission of talonavicular synostosis have been reported by Rothberg et al (1935) and Boyd (1944). An English family with coalition of the talus and navicular, in association with anomalies of the fifth fingers, has been studied by Challis (1974). Talonavicular coalitions were reviewed by Gill & Sullivan (1985).

Isolated synostosis of the carpus is uncommon, although bone fusion is found in several specific genetic entities and in association with malformations of the digits. Brug (1975) reported a young man with synostosis of the carpus and postaxial symbrachydactyly and mentioned that the anomaly had been present in the male members of the kindred for nearly

200 years. The situation is consistent with Y-linked inheritance, and if these observations are confirmed, this condition would rank with hairy ear lobes in the exclusive category of Y-linked traits!

## Multiple synostoses

Multiple carpal and tarsal synostoses, with phalangeal involvement and radial head subluxation, was encountered in a mother and daughter by Pearlman et al (1964). A similar disorder in two generations of a kindred had previously been reported by Bersani & Samilson (1957). A mother and daughter with fusion of several bones in the carpus and tarsus, in association with asymmetry of the metacarpals and metatarsals, and other variable skeletal changes, were described by Christian et al (1975). The authors suggested that the condition might be inherited as an autosomal dominant. It is possible that these kindreds all had the same genetic disorder, which exhibited variable expression. Equally, they could all have had unique 'private' syndromes. The disorder which was studied in five members of three generations of an Italian family by Ventruto et al (1976) certainly seems to be distinct. These patients had synostoses in the carpus and tarsus, together with symphalangism, brachydactyly, craniostenosis and dysplasia of the hip joints.

Maroteaux et al (1972) introduced the term 'multiple synostoses syndrome' for a condition in which fusions in the carpus and tarsus were associated with symphalangism. The phenotype is variable and it is likely that this disorder has masqueraded under other titles (see Ch. 15, p. 354).

A father and daughter with symphalangism, tarsal and carpal coalitions, short first metacarpals, elbow fusion and conductive deafness were documented by Nixon (1978). The author commented upon the variability of the manifestations in this autosomal dominant syndrome. He suggested that cases reported under diverse titles might be the result of the same mutation and made a plea for nosological simplicity. Other large pedigrees have been published, including a Brazilian family with 28 affected persons in six generations (da-Silva et al 1984).

Drawbert et al (1985) alluded to an earlier report of Furhmann et al (1966) and documented the disorder in 15 members of a five generation family. It is evident that the multiple synostoses syndrome is fairly common and that the detection of symphalangism or bony coalition in the wrist and ankle should raise the suspicion of fusions at other anatomical sites and in other family members.

## Hand-foot-genital syndromes

Stern et al (1970) noted abnormalities in the extremities and genital tracts of 13 members of four generations of a kindred and suggested that the disorder was a new autosomal dominant entity. Terming the condition the 'hand-foot-uterus' syndrome, Poznanski et al (1970) and Giedion & Prader (1976) have given details of the radiographic changes, including trapezium-scaphoid and cuneiform-navicular synostosis and shortening of the first metacarpals and metatarsals. Poznanski et al (1975) used pattern profile analysis to evaluate their patients, and pointed out that genital anomalies were inconsistent, especially in males. They proposed the alternative designation 'hand-foot-genital' syndrome.

Autosomal recessive syndromes in which malformations of the hands and feet are associated with anomalies of the uterus, vagina or urinary tract, include hydrometrocolpos-polydactyly (McKusick 1964) and crypto-phthalmos (Fraser 1962). Lethal autosomal recessive syndromes with similar associations have been recorded by Opitz & Howe (1969) and Halal et al (1980). Disorders in these categories were reviewed by Sarto & Simpson (1978). Halal (1986) documented a French Canadian family in which there was apparent autosomal dominant inheritance of variable ectrodactyly and genital anomalies. Two females with the disorder had successfully borne children, despite having vaginal septa and reduplication of the cervix and uterus. The three affected males, in whom the genital component was expressed as micropenis, are possibly at a greater socio-biological disadvantage and so far none have reproduced.

## Humero-radial synostosis

Humero-radial synostosis may occur in isolation, in conjunction with additional skeletal anomalies, or as a component of multiple malformation syndromes.

Autosomal dominant inheritance of bilateral radio-ulnar synostosis was documented by Birch-Jensen (1949) and Fuhrmann et al (1966) and autosomal recessive transmission by Frostad (1940) and Keutel et al (1970). Sporadic cases were reported by Say et al (1973) and the literature was analysed by Hunter et al (1976) in order to distinguish between the dominant and recessive forms of the disorder.

The potentially lethal association of humero-radial synostosis with craniostenosis, anterior bowing of the femora, medial bowing of the ulnae, midfacial hypoplasia, and cardiac and renal malformations was recorded by Antley & Bixler (1975).

The term 'multisynostotic osteodysgenesis' was introduced by De Lozier et al (1980) and a further report was published by Herva & Seppanen (1985). Inheritance is probably autosomal recessive. Following the birth of an affected child, a successful ultrasonic prenatal diagnosis in a subsequent pregnancy has been accomplished by Savoldelli & Schinzel (1982).

Children with humero-radio-ulnar synostosis, bifurcation of the distal humerus and oligoectrosyndactyly have been recorded by McCredie (1975), Mnaymuch (1978), Gollop & Coates (1983) and Leroy & Speeckaert (1984). These patients were all sporadic and the pathogenesis is uncertain.

REFERENCES

**Preamble**

Bavinck J N, Weaver D D 1986 Subclavian artery supply disruption sequence: Hypothesis of a vascular etiology for Poland, Klippel-Feil and Mobius anomalies. American Journal of Medical Genetics 23: 903–918

Birch-Jensen A 1949 Congenital Deformities of the Upper Extremities. Munksgaard, Copenhagen

Czeizel A, Bod M, Lenz W 1983 Family study of congenital limb reduction abnormalities in Hungary 1975–77. Human Genetics 65/1: 36–45

Falek A, Heath C W, Ebbin A J, McLean W R 1968 Unilateral limb and skin deformities with congenital heart disease in two siblings: A lethal syndrome. Journal of Pediatrics 73: 910–913

Freire-Maia N 1969 Congenital skeletal limb deficiencies — a general view. Birth Defects: Original Article Series 5/3: 7

Hecht J T, Scott C I Jr 1981 Recurrent unilateral hand malformations in siblings. Clinical Genetics 20: 225–228

Henkel L, Willert H G 1969 Dysmelia. Journal of Bone and Joint Surgery 51B: 399

Kay H W 1975 Clinical applications of the new international terminology for the classification of congenital limb deficiencies. Inter-Clinic Information Bulletin 14/3: 1

Poswillo D 1976 Mechanisms and pathogenesis of malformation. British Medical Bulletin 32/1

Rogala E J, Wynne-Davies R, Littlejohn A, Gormley J 1974 Congenital limb anomalies: frequency and aetiological factors. Data from the Edinburgh Register of the Newborn. Journal of Medical Genetics 11/3: 221

Stanley W S, Barr M, Hensinger R et al 1984 Asymmetric skeletal anomalies in siblings. Clinical Genetics 25: 533–537

Wolpert L 1976 Mechanisms of limb development and malformation. British Medical Bulletin 32/1

**Non-specific limb reduction**

*Phocomelia*

Di Battista C, Laudizi L, Tamborino G 1975 Phocomelia and agenesis of the penis in a neonate. Possible teratogenic role of a drug taken by mother during pregnancy. Minerva Pediatrica 27/11: 675

*Amelia*

Assemarry S R, Neu R L, Gardner L I 1970 Deformities in a child whose mother took LSD. Lancet i: 1290

Auchterlonie I A, White M P 1982 Recurrence of the VATER association within a sibship. Clinical Genetics 21/2: 122–124

Baker C J, Rudolf A H 1971 Congenital ring constrictions and intrauterine amputations. American Journal of Diseases of Children 121:

Beyth Y, Perlman M, Ornoy A 1977 Amniogenic bands associated with facial dysplasia and paresis. Journal of Reproductive Medicine 18: 83–86

Blanc W A, Mattison D R, Kane R, Chauhan P 1971 LSD intrauterine amputations and amniotic band syndrome. Lancet ii: 158

Bofinger M K, Dignan P S, Schmidt R E et al 1973 Reduction malformations and chromosome anomalies. American Journal of Diseases of Children 125: 135–143

Cadas C, Trichopoulos D, Papadatos K, Kalapothaki V 1978 Prevalence at birth of congenital malformations in Athens, Greece. International Journal of Epidemiology 7: 251–252

Christianson A L, Nelson M M 1984 Four cases of trisomy 18 syndrome with limb reduction malformations. Journal of Medical Genetics 21/4: 293–297

Drance S M 1975 Oral contraceptives and congenital limb defects. Canadian Medical Association Journal 112/5: 551

Evans J A, Reggin J, Greenberg C 1985 Tracheal agenesis and associated malformations: A comparison with tracheosophageal fistula and the VACTERL association. American Journal of Medical Genetics 21: 21–34

Fisher R M, Cremin B J 1976 Limb defects in the amniotic band syndrome. Paediatric Radiology 5: 24

Gardner L I, Welsh-Sloan J, Cady R B 1985 Phocomelia in infant whose mother took large doses of pyridoxine during pregnancy. Lancet 1/8429: 636

Herva R, Karkinen-Jaaskelainen M 1984 Amniotic adhesion malformation syndrome: Fetal and placental pathology. Teratology 29: 11–19

Hunter A G W, Carpenter B F 1986 Implications of malformations not due to amniotic bands in the amniotic band sequence. American Journal of Medical Genetics 24: 691–700

Jones K L, Smith D W, Hall J G et al 1974 A pattern of craniofacial and limb defects secondary to aberrant tissue bands. Journal of Pediatrics 84: 90–95

Kajii T, Kiyoshi O, Katsuaki I et al 1964 A probable 17–18 trisomy syndrome with phocomelia, exomphalos and agenesis of hemidiaphragm. Archives of Disease in Childhood 39: 519–522

Khrouf N, Spang R, Podgorna T et al 1986 Malformations in 10 000 consecutive births in Tunis. Acta Paediatrica Scandinavica 75: 534–539

Long S 1972 Does LSD induce chromosomal damage and malformations? A review of the literature. Teratology 6: 75

Pauli R M, Lebovitz R M, Meyer R D 1985 Familial recurrence of terminal transverse defects of the arm. Clinical Genetics 27/6: 555–563

Pauli R M, Feldman P F 1986 Major limb malformations following intrauterine exposure to ethanol. Teratology 33: 273–280

Pfeiffer R A, Sautelmann R 1977 Limb anomalies in chromosomal aberrations. Birth Defects Original Article Series 13/2: 319–337

Streeter G L 1930 Focal deficiencies in fetal tissues and their relation to intrauterine amputations. Contribution to Embryology, Carnegie Institute 22: 3–44

Torpin R 1968 Fetal Malformations. Thomas, Springfield, p 3–137

Van Regemorter N, Dodion J, Cruart C et al 1984 Congenital malformations in 10 000 consecutive births at a university hospital: Need for genetic counselling and prenatal diagnosis. Journal of Pediatrics 104: 386–390

Voorhess M L, Aspillaga M J, Gardner L I 1964 Trisomy 18 syndrome with absent radius, varus deformity of hand and rudimentary thumb. Journal of Pediatrics 65: 130–133

Werchsel M E, Luzzatti L 1965 Trisomy 17–18 syndrome with congenital extrahepatic biliary atresia and congenital amputation of the left foot. Journal of Pediatrics 67: 324–327

*Hemimelia*

Barrie H 1976 Congenital malformation associated with intrauterine contraceptive device. British Medical Journal 1/6008: 488

Russell J E 1975 Tibial hemimelia: limb deficiency in siblings. Inter-Clinic Information Bulletin 14/7–8: 15

Swanson A B, Tada K, Yonenobu K 1984 Ulnar ray deficiency: Its various manifestations. Journal of Hand Surgery 9/5: 658–664

**Limb reduction syndromes**
*Roberts syndrome*

Freeman M V R, Williams D W, Schimke N, Temtamy S A 1974 The Roberts syndrome. Clinical Genetics 5: 1

Freeman M V R, Williams D W, Schimke R N et al 1975 The Roberts syndrome. Birth Defects: Original Article Series 11/5: 87

Fryns H, Goddeeris P, Moerman F et al 1980 The tetraphocomelia-cleft palate syndrome in identical twins. Human Genetics 53: 279–281

Grosse F R, Pandel C, Wiedemann H R 1975 The tetraphocomelia-cleft palate syndrome: description of a new case. Humangenetica 28/4: 353

Hall B D, Greenberg M H 1972 Hypomelia-hypotrichosis-facial hemangioma syndrome (pseudothalidomide, SC syndrome, SC phocomelia syndrome). American Journal of Diseases of Children 123: 602

Herrmann J, Opitz J M 1977 The SC phocomelia and the Roberts syndrome: Nosologic aspects. European Journal of Pediatrics 125: 117–134

Herrmann J, Feingold M, Tuffli G A, Opitz J M 1969 A familial dysmorphogenic syndrome of limb deformities, characteristic facial appearance and associated anomalies: the 'pseudothalidomide' or 'SC-syndrome.' In The Clinical Delineation of Birth Defects, III, National Foundation, New York, p 81

Lenz W D, Marquardt E, Weicker H 1975 Pseudothalidomide syndrome. Birth Defects: Original Article Series 11/5: 97

Quazi Q H, Kassner E G, Masakawa et al 1979 The SC phocomelia syndrome: Report of two cases with cytogenetic abnormality. American Journal of Medical Genetics 4: 231

Roberts J B 1919 A child with double cleft of lip and palate, protrusion of the intermaxillary portion of the upper jaw and imperfect development of the bones of the four extremities. Annals of Surgery 70: 252

Zergollern L, Hitrec V 1976 Three siblings with Robert's syndrome. Clinical Genetics 9: 433


*Acheiropodia*

Freire-Maia A 1981 The extraordinary handless and footless families of Brazil. 50 years of acheiropodia. American Journal of Medical Genetics 9: 31–41

Freire-Maia A, Laredo-Filho J, Freire-Maia N 1978 Genetics of acheiropodia ('the handless and footless families of Brazil'): X. Roentgenologic study. American Journal of Medical Genetics 2: 321–330

Freire-Maia A 1970 The handless and footless families of Brazil. Lancet i: 519

Freire-Maia A 1975 Genetics of acheiropodia ('the handless and footless families of Brazil'). VIII. Penetrance and expressivity. Clinical Genetics 7: 98

Freire-Maia A, Li W H, Maruyama T 1975 Genetics of acheiropodia (the handless and footless families of Brazil). VII. Population dynamics. American Journal of Human Genetics 27/5: 665

Marcallo F A, Pilotto R F, Friere-Maia A 1979 Genetics of acheiropodia ('the handless and footless families of Brazil') XI. Pathologic aspects. American Journal of Medical Genetics 4: 287–291

Morton N E, Barbosa C A A 1981 Age, area and acheiropodia. Human Genetics 57: 420–422

Toledo S P A, Saldanha P H, Borelli A, Cintra A B U 1972 Furhter data on acheiropodia. Journal de Génétique Humaine 20/3: 253


*Proximal focal femoral dysplasia*

Bailey J A, Beighton P 1970 Bilateral femoral dysgenesis. Clinical Pediatrics 9: 668

Burck U, Riebel T, Held K R, Stoeckenius M 1981 Bilateral femoral dysgenesis with micrognathia, cleft palate, anomalies of the spine and pelvis and foot deformities. Helvetica Paediatrica Acta 36: 473–482

Burn J, Winter R M, Baraitser M, Hall C M, Fixsen J 1984 The femoral hypoplasia-unusual facies syndrome. Journal of Medical Genetics 21: 331–340

Daentl D L, Smith D W, Scott C I, Hall B D, Gooding C A 1975 Femoral hypoplasia — unusual facies. Journal of Pediatrics 86: 107

Eastman J R, Escobar V 1978 Femoral hypoplasia-unusual facies syndrome: A genetic syndrome? Clinical Genetics 13: 72–76

Ferguson J 1964 Progress of anatomy and surgery during the present century. Lancet ii: 60

Gleiser S, Weaver D D, Escobar V, Nichols G, Escobedo M 1978 Femoral hypoplasia-unusual facies syndrome, from another viewpoint. European Journal of Pediatrics 128: 1–5

Graham M 1985 Congenital short femur: Prenatal sonographic diagnosis. Journal of Ultrasound in Medicine 4/7: 361–363

Hamanish C 1980 Congenital short femur. Journal of Bone and Joint Surgery 62-B: 307–320

Hurst D, Johnson D F 1980 Brief clinical report: Femoral hypoplasia-unusual facies syndrome. American Journal of Medical Genetics 5: 255–258

Lampert R P 1980 Dominant inheritance of femoral hypoplasia-unusual facies syndrome. Clinical Genetics 17: 255–258

Lord J, Beighton P 1981 The femoral hypoplasia-unusual facies syndrome: A genetic entity? Clinical Genetics 20: 267–275

Riedel F, Froster-Iskenius U 1985 Caudal dysplasia and femoral hypoplasia-unusual facies syndrome: Different manifestations of the same disorder? European Journal of Pediatrics 144: 80–82

*Tibial deficiency syndromes*
Canun S, Lomeli R M, Martinez R, Carnevale A 1984 Absent tibiae, triphalangeal thumbs and polydactyly. Description of a family and prenatal diagnosis. Clinical Genetics 25: 182–186
Clark M W 1975 Autosomal dominant inheritance of tibial meromelia: report of a kindred. Journal of Bone and Joint Surgery 57/2: 262
Eaton G O, McKusick V A 1969 A seemingly unique polydactyly-syndactyly syndrome in four persons in three generations. The Clinical Delineation of Birth Defects, III. National Foundation, New York, p 221
Kalamchi A, Dawe R V 1985 Congenital deficiency of the tibia. Journal of Bone and Joint Surgery 67/4: 581–584
Pashayan H, Fraser F C, McIntire J M, Dunbar J S 1971 Bilateral aplasia of the tibiae, polydactyly and absent thumb in father and daughter. Journal of Bone and Joint Surgery 53: 495
Pfeiffer R A, Roeskau M 1971 Agenesie der Tibia, Fibulaverdopplung und spiegelbildliche Polydaktylie (Diplopodie) bei Mutter und Kind. Zeitschrift für Kinderheilkunde 111: 38
Yujnovsky O, Ayala D, Vinvitorio A, Viale H, Sakati N, Nyhan W L 1974 A syndrome of polydactyly-syndactyly and triphalangeal thumbs in three generations. Clinical Genetics 6: 51

*Aplasia cutis congenita terminal-transverse limb defect syndrome*
Adams F H, Oliver C P 1945 Hereditary deformities in man due to arrested development. Journal of Heredity 36: 3
Scribanu N, Temtamy S A 1975 The syndrome of aplasia cutis congenita with terminal transverse defects of limbs. Journal of Pediatrics 87/1: 79

*Fanconi pancytopenia syndrome*
Barrett A J, Brigden W D, Hobbs J R, Hugh-Jones K, Humble J G, James D C O, Retsas S, Rogers T R F, Selwyn S, Sneath P, Watson J G 1977 Successful bone marrow transplant for Fanconi's anaemia. British Medical Journal 1: 420
Duckworth-Rysiecki G, Hulten M, Mann J, Taylor A M R 1984 Clinical and cytogenetic diversity in Fanconi's anaemia. Journal of Medical Genetics 21: 197–203
Garriga S, Crosby W H 1959 The incidence of leukemia in families of patients with hypoplasia of the marrow. Blood 14: 1008
Swift M R, Hirschhorn K 1966 Fanconi's anemia: inherited susceptibility to chromosome breakage in various tissues. Annals of Internal Medicine 65: 496

*Radial aplasia — thrombocytopenia*
Anyane-Yeboa K, Jaramillo S, Nagel C, Grebin B 1985 Brief clinical report: tetraphocomelia in the syndrome of thrombocytopenia with absent radii (TAR syndrome). American Journal of Medical Genetics 20: 571–576
Filkins K, Russo J, Bilinki I et al 1984 Prenatal diagnosis of thrombocytopenia absent radius syndrome using ultrasound and fetoscopy. Prenatal Diagnosis 4: 139–142
Hall J G, Levin J, Kuhn J P, Ottenheimer E J, Van Berkum K A P, McKusick V A 1969 Thrombocytopenia with absent radius (TAR). Medicine 48: 411
Luthy D A, Hall J G, Graham C B 1979 Prenatal diagnosis of thrombocytopenia with absent radii. Clinical Genetics 15: 495–499
Luthy D A, Mack L, Hirsch J, Cheng E 1981 Prenatal ultrasound diagnosis of thrombocytopenia with absent radii. American Journal of Obstetrics and Gynecology 141: 350–351
Omenn G S, Eigley M M, Graham C B, Heinrichs W Le R 1973 Prospects for radiographic intrauterine diagnosis: The syndrome of thrombocytopenia with absent radii. New England Journal of Medicine 288: 777
Pfeiffer R A, Haenke C 1975 The phocomelia thrombocytopenia syndrome. A follow-up report. Humangenetik 26/2: 157
Ray R, Zorn E, Kelly T et al 1980 Lower limb anomalies in the thrombocytopenia absent-radius (TAR) syndrome. American Journal of Medical Genetics 7: 523–528

*Radial ray abnormalities, miscellaneous*
Gardner R J M, Grindley R M, Chewings W E 1984 Ring chromosome 1 associated with radial ray defect. Journal of Medical Genetics 21: 400

Haspeslagh M, Fryns J P, Moerman P H 1984 Severe limb malformations in 4p deletion. Clinical Genetics 25: 353–356

Hayes A, Costa T, Polomeno R C 1985 The Okihiro syndrome of Duane anomaly, radial ray abnormalities and deafness. American Journal of Medical Genetics 22: 273–280

Meizner I, Bar-Ziv J, Barki Y, Abeliovich D 1986 Prenatal ultrasonic diagnosis of radial-ray aplasia and renal anomalies. Prenatal Diagnosis 6: 223–225

Sofer S, Bar-Ziv J, Abeliovich D 1983 Radial ray aplasia and renal anomalies in a father and son. A new syndrome. American Journal of Medical Genetics 14: 151–157

*Other genetic limb reduction syndromes*
Al-Awadi S, Steebi A, Talaatifaret et al 1985 Profound limb deficiency, thoracic dystrophy, unusual facies and normal intelligence: A new syndrome. Journal of Medical Genetics 22: 36–38

Blockey N J, Lawrie J H 1963 An unusual symmetrical distal limb deformity in siblings. Journal of Bone and Joint Surgery 45B: 745–747

Elejalde B R, Elejalde M M, Booth et al 1985 Prenatal diagnosis of Weyers syndrome. American Journal of Medical Genetics 21: 439–444

Kucheria K, Bhargava S K, Bamezai R 1976 A familial tetraphocomelia syndrome involving limb deformities, cleft lip, cleft palate and associated anomalies — a new syndrome. In Fifth International Congress of Human Genetics, Mexico.

Kuhne D, Lenz W, Petersen D et al 1967 Defekt von Femur und Fibula mit Amelia, Peromelia Oder Ulnaren Strahidefekten der Arme: Ein Syndrom. Humangenetik 3: 244–263

Lazjuk G I, Lurie I W, Cherstvoy E D, Ussova Y I 1976 A syndrome of multiple congenital malformations including amelia and oligodactyly occurring in half-cousins. Teratology 13/2: 161

Lenz W, Feldmann U 1977 Unilateral and asymmetric limb defects in man. Delineation of the Femur Fibula Ulna complex. Birth Defects: Original Article Series 13: 269–285

Weyers H 1957 Das Oligodactylie Syndrom des Menschen und seine Parallelmutation bei der Hausmaus. Annales Paediatrica 189: 351–370

Zlotogora J, Rosenmann E, Menashe M et al 1983 The femur, fibula, ulna (FFU) complex in siblings. Clinical Genetics 24: 449–452

**Synostosis syndromes**
McCredie J 1975 Congenital fusion of bones: Radiology, embryology and pathogenesis. Clinical Radiology 26/1: 47

*Radio-ulnar synostosis*
Cleary J E, Omer G E 1985 Congenital proximal radio-ulnar synostosis. Natural history and functional assessment. Journal of Bone and Joint Surgery 67: 539–545

Hansen O H, Anderson N O 1970 Congenital radioulnar synostosis. Report of 37 cases. Acta Orthopaedica Scandinavica 41: 225

Jaffer Z, Nelson M, Beighton P 1981 Bone fusion in the foetal alcohol syndrome. Journal of Bone and Joint Surgery 63: 569–571

*Synostosis of tarsus and carpus*
Boyd H B 1944 Congenital talonavicular synostosis. Journal of Bone and Joint Surgery 26: 682

Brug E 1975 A case of familial hereditary malformation of both hands. Handchirurgie 7/3: 125

Challis J 1974 Hereditary transmission of talonavicular coalition in association with anomaly of the little finger. Journal of Bone and Joint Surgery 56/6: 1273

Diamond L S 1974 Inherited talocalcaneal coalition. Birth Defects: Original Article Series 10/12: 531

Gill P W, Sullivan R W 1985 Talonavicular coalitions. A review and case report. Journal of American Pediatric Medical Association 75/8: 443–445

Rothberg A S, Feldman F W, Schuster T F 1935 Congenital fusion of astragalus and scaphoid: bilateral: inherited. New York State Journal of Medicine 35: 29

Webster F S, Roberts W M 1951 Tarsal anomalies and peroneal spastic flat foot. Journal of the American Medical Association 146: 1099

Wray J B, Herndon C N 1963 Hereditary transmission of congenital coalition of the calcaneus to the navicular. Journal of Bone and Joint Surgery 45A: 365

*Multiple synostoses*

Bersani F A, Samilson R L 1957 Massive familial tarsal synostosis. Journal of Bone and Joint Surgery 39A: 1187

Christian J C, Franken E A, Lindeman J P, Lindseth R E, Reed T, Scott C I Jr 1975 A dominant syndrome of metacarpal and metatarsal asymmetry with tarsal and carpal fusions, syndactyly, articular dysplasia and platyspondyly. Clinical Genetics 8: 75

Da-Silva E O, Filho S M, De Albuquerque S C 1984 Multiple synostosis syndrome: Study of a large Brazilian kindred. Americal Journal of Medical Genetics 18: 237–247

Drawbert J P, Stevens D B, Cadle R G, Hall B D 1985 Tarsal and carpal coalition and symphalangism of the Fuhrmann type. Report of a family. Journal of Bone and Joint Surgery 67: 884–889

Fuhrmann W, Steffens C, Rompe G 1966 Dominant erbliche doppelseitige Dysplasia und Synostose des Ellenbogen-gelenks, mit Symmetrischer Brachymesophalangie und Brachymetakarpie sowie Synostosen im Finger-, Hand- und Fusswurzelzelbereich. Humangenetik 3: 64–75

Maroteaux P, Bouvet J P, Briard M L 1972 La maladie des synostoses multiples. Nouvelle Presse Médicale 1: 3041–3047

Nixon J R 1978 The multiple synostosis syndrome. A plea for simplicity. Clinical Orthopaedics and Related Research 135: 48–51

Pearlman H S, Edkin R E, Warren R F 1964 Familial tarsal and carpal synostosis with radial-head subluxation (Nievergelt's syndrome). Journal of Bone and Joint Surgery 46A: 585

Ventruto V, Girolamo R, Festa B, Romano A, Sebastio G, Sebastio L 1976 Family study of inherited syndrome with multiple congenital deformities: symphalangism, carpal and tarsal fusion, brachydactyly, craniostenosis, strabismus, hip osteochondritis. Journal of Medical Genetics 13/5: 394

*Hand-foot-genital syndromes*

Fraser G R 1962 Our genetic load. A review of some aspects of genetical variation. Annals of Human Genetics 25: 387–315

Giedion A, Prader A 1976 Hand-foot-uterus (HFU) syndrome with hypospadias: the hand-foot-genital (HFG) syndrome. Paediatric Radiology 4/2: 96

Halal F 1986 A new syndrome of severe upper limb hypoplasia and Mullerian duct anomalies. American Journal of Medical Genetics 24: 119–126

Halal F, Desgranges M F, Leduc B et al 1980 Acro-renal-mandibular syndrome. American Journal of Medical Genetics 5: 277–284

McKusick V A, Baver R L, Koop C E, Scott R B 1964 Hydrometrocolpos as a simply inherited malformation. Journal of the American Medical Association 189: 813–816

Opitz J M, Howe J J 1969 The Meckel syndrome. In: Bergsma D (ed) Part II. Malformation Syndromes. Birth Defects, Original Article Series V(2). Liss, New York, p 167–169

Poznanski A K, Kuhns L R, Lapides J, Stern A M 1975 A new family with the hand-foot-genital syndrome: a wider spectrum of the hand-foot-uterus syndrome. Birth Defects: Original Article Series 11/4: 127

Poznanski A K, Stern A M, Gall J C Jr 1970 Radiographic findings in the hand-foot-uterus syndrome (HFUS). Radiology 95:129

Sarto G E, Simpson J L 1978 Abnormalities of the Mullerian and Wolffian duct systems. Birth Defects: Original Article Series XIV(6C)

Stern A M, Gall J C Jr, Perry B L, Stimson C W, Weitkamp L R, Poznanski A K 1970 The hand-foot-uterus syndrome, A new hereditary disorder characterised by hand and foot dysplasia, dermatoglyphic abnormalities, and partial duplication of the female genital tract. Journal of Pediatrics 77: 109

*Humero-radial synostosis*

Antley R M, Bixler D 1975 Trapezoidocephaly, midfacial hypoplasia and cartilage abnormalities with multiple synostosis and skeletal fractures. Birth Defects XI(2): 397–401

Birch-Jensen A 1949 Congenital Deformities of the Upper Extremities. Munksgaard, Copenhagen, p 208

DeLozier C D, Antley R M, Williams R et al 1980 The syndrome of multisynostotic osteodysgenesis with long bone fractures. American Journal of Medical Genetics 7: 391–403

Frostad H 1940 Congenital ankylosis of the elbow joint. Acta Orthopaedica Scandinavica 11: 296–306

Fuhrmann W. Steffens C, Rompe G 1966 Dominant erbliche doppelseitige Dysplasie und Synostose des Ellenbogengelenks. Humangenetik 3: 64–77

Gollop T R, Coates V 1983 Apparent bifurcation of distal humerus with oligoectrosyndactyly. American Journal of Medical Genetics 14: 591–593

Herva R, Seppanen U 1985 Multisynostotic osteodysgenesis. Pediatric Radiology 15: 63–64

Hunter A G, Cox D W, Rudd N L 1976 The genetics of and associated clinical findings in humero-radial synostosis. Clinical Genetics 9: 470–478

Keutel J, Kindermann I, Mockel H 1970 Eine wahrscheinlich autosomal recessive vererbte Skeletmissbildung mit Humeroradialsynostose. Human Genetics 9: 43–53

Leroy J G, Speeckaert M T 1984 Letter to the Editor. Humeroradioulnar synostosis appearing as distal humeral bifurcation in a patient with distal phocomelia of the upper limbs and radial ectrodactyly. American Journal of Medical Genetics 18: 365–368

McCredie J 1975 Congenital fusion of bones: Radiology, embryology and pathogenesis. Clinical Radiology 26: 47–51

Mnaymuch M A 1978 Congenital radio-humeral synostosis: A case report. Clinical Orthopaedics 131: 183–184

Savoldelli G, Schinzel A 1982 Prenatal ultrasound detection of humero-radial synostosis in a case of Antley-Bixler syndrome. Prenatal Diagnosis 2: 219–223

Say B, Balei S, Atsasu M 1973 Humero-radial synostosis. Human Genetics 19: 341–344

# 17

# Connective tissue disorders

Successive editions of McKusick's classical monograph 'Heritable Disorders of Connective Tissue' have brought this group of conditions to the attention of medical science and stimulated investigations into their natural history and pathogenesis. In McKusick's monumental work the interested reader will find a comprehensive account of the manifestations and genetics of these disorders. The skeleton is involved, to some extent, in the majority of them and the following form the subject of this section:

1. Marfan syndrome
   a. Marfanoid hypermobility syndrome
   b. Congenital contractural arachnodactyly
   c. Achard syndrome
   d. Marfanoid habitus with mitral valve prolapse
   e. Other variants of the Marfan syndrome
2. Ehlers-Danlos syndrome
3. Familial articular hypermobility syndromes
4. Fibrodysplasia ossificans progressiva
5. Weill-Marchesani syndrome
6. Homocystinuria
7. Alkaptonuria
8. Tight skin syndromes
   a. Rothmund-Thomson syndrome
   b. Stiff skin syndrome
   c. Syndesmodysplastic dwarfism
   d. Parana hard skin syndrome
9. Loose skin syndromes
   a. Cutis laxa with ligamentous laxity and delayed development
   b. Leprechaunoid syndrome
   c. Wrinkly skin syndrome
10. Disorders of copper transport
    a. Occipital horn syndrome (X-linked cutis laxa)
    b. Menkes syndrome

McKusick has drawn a noteworthy distinction between those conditions in which connective tissues are primarily at fault, as in the Marfan syndrome and the Ehlers-Danlos syndrome, and those in which involvement is secondary to a metabolic abnormality, such as homocystinuria or alkaptonuria. Heterogeneity has been recognised in many of these disorders and with the continuing delineation of new forms, difficulties have arisen concerning syndromic boundaries and nosology (Maroteaux et al 1986). These problems were discussed by a committee of experts at the 1986 International Congress for Human Genetics and the results of their deliberations were promulgated as the Berlin Nomenclature for Inherited Connective Tissue Disorders.

Sporadic patients with poorly defined skeletal and connective tissue abnormalities are frequently encountered and there is little doubt that a large number of rare entities await delineation. Advances in knowledge of the biochemistry and molecular genetics of collagen hold great promise for the ultimate definition and categorisation of conditions of this type.

## MARFAN* SYNDROME

The Marfan syndrome is a well known disorder in which excessive height is associated with a variety of connective tissue abnormalities. Since the original reports at the turn of the century, several hundred cases have been described from all parts of the world. The frequency in the USA is approximately 1 in 10 000 and on this basis, there must be about 23 000 affected persons in that country.

Definitive diagnosis can be a very difficult matter, and in the author's experience, several sporadic patients with indeterminate 'marfanoid' features are seen for every person with the classical syndrome. Diagnostic criteria have been discussed in detail by Pyeritz & McKusick (1979) but the final resolution of this problem awaits the development of biochemical or molecular diagnostic markers.

The National Marfan Foundation of the USA is a lay organisation devoted to the interests of affected persons and their families. Close links are maintained with the medical profession and a regular newsletter is circulated.

### Clinical features

The stigmata of the Marfan syndrome include disproportionately long extremities, a high palate, dislocation of the lens of the eye, thoracic asymmetry, articular hypermobility and a tendency to aneurysm and dissection of the aorta. Robins et al (1971) reviewed the spinal problems in 68 patients

---

*Bernard Marfan (1858–1942) was a founder of French paediatrics.

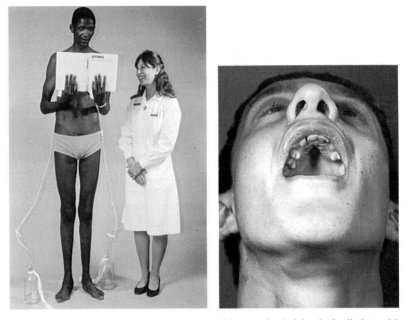

**Fig. 17.1** (left) Marfan syndrome: a young man with excessive height, lanky limbs and long digits. Bilateral spontaneous pneumothorax is being treated by under-water drainage.

**Fig. 17.2** (right) Marfan syndrome: a high arched palate.

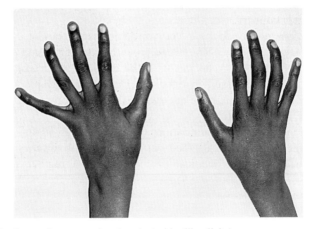

**Fig. 17.3** Marfan syndrome: arachnodactyly (spider-like digits).

and found that more than 50% had significant degrees of scoliosis. Excessive lengthening of the digits, or arachnodactyly, represents one of the cardinal features of the Marfan syndrome. This anomaly is by no means pathognomonic, as arachnodactyly also occurs in isolation and as a component of several other syndromes.

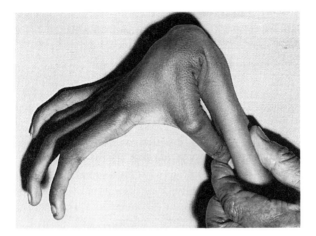

**Fig. 17.4** Marfan syndrome: laxity of the wrist joint.

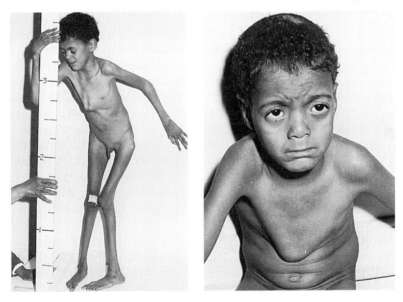

**Fig. 17.5** (left) Marfan syndrome: a boy with severe manifestations.

**Fig. 17.6** (right) Marfan syndrome: marked thoracic deformity.

Ophthalmological involvement has been reviewed in detail by Maumenee (1981) and the cardiovascular features have been analysed by Roberts & Honig (1982). Obstetrical problems have been reviewed by Beighton (1982) and Lind & van Papendrecht (1984), pulmonary involvement by Wood et al (1984) and anaesthetic implications by Verghese (1984). Although the onset of serious cardiac complications is usually delayed until adulthood, death from dissection of the ductus arteriosus has been recorded in infancy (Gillan et al 1984). The intrapartum death of an affected baby with mitral valve involvement has also been documented (Buchanan & Wyatt 1985).

*Genetics*

In Northern Ireland, a minimum prevalence of 1.5 per 100 000 of the population has been calculated (Lynas 1958). Autosomal dominant inheritance is well established and many large pedigrees have been published. The well known variability of expression is typified by the disparity in the manifestations which was observed in affected monozygous twins by Ambani et al (1975). In some families, phenotypic variability can approach virtual nonpenetrance. Homozygosity has been observed in a Yemenite family, where mildly affected parents produced two siblings with very severe stigmata; death occured in infancy (Chemke et al 1984).

Positive diagnosis is sometimes difficult, but if there are unequivocal cases in the kindred, the presence of a limited number of components of the syndrome may permit recognition. The 'metacarpal index', which is derived from measurements of hand radiographs, or a comparison of upper and lower body segment ratios may be helpful, but these objective criteria are by no means infallible. In the sporadic case, it may be impossible to reach a firm decision for the purposes of genetic counselling. A similar problem arises when it is suspected that the newborn child of an individual with the Marfan syndrome might have inherited the disorder. The length of the middle finger, from the tip to the proximal flexion crease, as a percentage of total hand size, is a useful determinant of arachnodactyly in the neonate and it is sometimes possible to resolve the situation in this way (Feingold & Bossert, 1974). Nevertheless, it may be necessary to defer diagnostic judgement until later childhood. Echocardiographic assessment of aortic root dilatation is proving to be a useful diagnostic tool in infants suspected of having the disorder.

Pyeritz & McKusick (1979) estimated that about 15% of patients with the Marfan syndrome represent new mutations. The paternal age effect has been demonstrated in a study of sporadic cases (Murdoch, et al 1972). Lynas & Merrett (1958) undertook linkage studies in kindreds in Northern Ireland using conventional gene markers and further extensive investigations were subsequently carried out in 15 families in the USA by Schleuterman et al (1976). No linkage between the Marfan syndrome locus and the marker loci was demonstrated in either instance. Genotyping for restriction fragment polymorphisms associated with the gene which encodes the pro alpha 2(1) collagen chain in a large family with Marfan syndrome failed to reveal linkage, thus excluding a causative mutation at this locus (Tsipouras et al 1986).

Prenatal diagnosis by ultrasonographic demonstration of excessive limb length was achieved by Koenigsberg et al (1981) at the 24th week of gestation. Following termination, the diagnosis was confirmed by autopsy demonstration of cystic medial necrosis in the aorta. A second 'at risk' fetus was deemed to be normal by ultrasonography; at the age of 21 months, limb length disparity was evident and the child was then thought to be affected. It seems, therefore, that although there is a place for ultrasonography in the monitoring of such pregnancies, diagnosis is not always possible.

Tall stature is a notable feature of the Marfan syndrome, and it is likely that individuals with the condition served in the 'regiment of giants', which was raised by King Fredrich I of Prussia. This military unit is said to have made a significant contribution to the gene pool of the town of Potsdam, where they were based for 50 years. Investigations of the prevalence of the Marfan syndrome in that locality might be of interest!

### Marfanoid hypermobility syndrome

The 'Marfanoid hypermobility syndrome' is characterised by a Marfanoid habitus, extreme joint laxity and extensible skin. A young man with this condition was described by Walker et al (1969). A similar case is recognisable in a previous report by Goodman et al (1965), which was entitled 'the Ehlers–Danlos syndrome occurring with the Marfan syndrome'. The genetic basis of the condition is uncertain and autonomous syndromic status is problematic.

### Congenital contractural arachnodactyly

The clinical manifestations of the congenital contractural arachnodactyly syndrome resemble those of the Marfan syndrome but cardiovascular and ocular problems do not occur. Flexion contractures of the fingers and other joints, severe kyphoscoliosis and small crumpled ears are additional features. Beals & Hecht (1971) encountered the syndrome in two generations of two separate families and mentioned that they could recognise 11 other kindreds in the literature. Male to male transmission had taken place, and inheritance was evidently autosomal dominant. The condition is probably common among the Marfan group of patients and other kindreds have been reported by Lowry & Guichon (1972), MacLeod & Fraser (1973), Steg (1975), Kontras (1975) and Bjerkeim et al (1976).

The presence of mitral valve prolapse in affected persons in three generations of a family with congenital contractural arachnodactyly led Anderson et al (1984) to suggest that the distinction between this disorder and the classical Marfan syndrome might not be clearcut. In a report of four affected families, Arroyo et al (1985) emphasised intra-familial similarity and inter-familial variability.

Passarge (1975) described two unrelated patients with a 'syndrome resembling contractural arachnodactyly'. These individuals were atypical in that

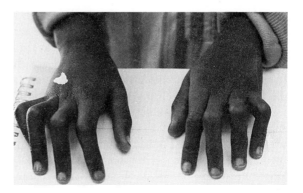

**Fig. 17.7** (left) Congenital contractural arachnodactyly: kyphoscoliosis has been corrected at operation. The limbs are long, the digits are flexed and the pinna of the ear is crumpled.

**Fig. 17.8** (above) Congenital contractural arachnodactyly: fixed deformities of the digits.

the spine was severely involved, while digital contractures were less marked. The author suggested that this condition might represent a new entity. No genetic background was apparent in either kindred. A further report of a severely affected neonate with additional manifestations including duodenal and oesophageal atresia, tracheo-oesophageal fistula and vertebral anomalies raises the possibility that the disorder is heterogeneous (Currarino & Friedman 1986).

### Achard syndrome

The Achard syndrome is characterised by decreased metacarpal width, widespread but mild dysplastic changes in the skeleton and ligamentous laxity. Achard (1902) published his original case report under the designation 'arachnodactyly' and since that time there has been semantic confusion with the Marfan syndrome. In a report of three patients, Duncan (1975) emphasised that the Achard syndrome is clinically distinct from the Marfan syndrome. The mode of genetic transmission is unknown.

### Marfanoid habitus with mitral valve prolapse

Mitral valve prolapse is a common abnormality which may occur in combination with altered bodily proportions and thoracic asymmetry

(Salomon et al 1975, Udoshi et al 1979). Autosomal dominant inheritance of a syndrome of marfanoid habitus and mitral valve prolapse was proposed by Schutte et al (1981). At the mild end of the spectrum the phenotypic manifestations blur with normality, whereas at the severe end, they closely resemble those of the Marfan syndrome. This condition may well be common and in view of the complexity of the collagen molecule, it could be heterogeneous (Beighton 1982). At present, the lack of any biochemical marker makes delineation virtually impossible, but this dilemma may ultimately be resolved at the molecular level.

## Other variants of the Marfan syndrome

The marfanoid habitus is very non-specific and often difficult to distinguish from normality. However, distinctive alterations in bodily proportions have been documented as components of a number of syndromes. For instance, a syndrome of marfanoid habitus, psychomotor retardation, and unusual flat facies was identified in four Mexican siblings by Fragoso & Cantu (1984). The authors proposed that this condition was a new entity which was probably inherited as an autosomal recessive trait.

An X-linked syndrome of mental retardation and a marfanoid habitus was documented by Lujan et al (1984) in four male relatives. The corpus callosum was absent in two of these patients. Hypoplasia of the cerebral cortex was also a feature of a marfanoid infant reported by Tamminga et al (1985).

The Stickler syndrome or hereditary arthro-ophthalmopathy, which has many features in common with Marfan syndrome, is reviewed in Chapter 10. Homocystinuria, an autosomal recessive disorder in which mental retardation is associated with a marfanoid habitus is discussed on page 419 of this chapter.

## EHLERS-DANLOS* SYNDROME

The Ehlers-Danlos syndrome (EDS) is one of the best known inherited disorders of connective tissue, and more than 1000 cases have now been reported. The condition received eponymous recognition at the beginning of the century, but earlier examples can be identified in the literature pertaining to such diverse institutions as the Academy of Leiden and Barnum and Bailey's circus!

The EDS is very heterogeneous and the following varieties have now been recognised:

---

*Eduard Ehlers (1863–1937) was an eminent dermatologist in Copenhagen.
Henri Danlos (1844–1912) was a senior physician and dermatologist in Paris.

| Type | General designation | Subcategories | Inheritance |
|------|---------------------|---------------|-------------|
| EDS I | Gravis type | | AD |
| EDS II | Mitis type | | AD |
| EDS III | Hypermobile type | | AD |
| EDS IV | Vascular type | | Heterogeneous |
| | | IV-A Acrogeric type | AD |
| | | IV-B Acrogeric type | AR |
| | | IV-C Ecchymotic type | AD |
| | | IV-D Others | AD/AR |

(All forms of EDS IV have defective type III collagen except IV-D, AR)

| Type | General designation | Subcategories | Inheritance |
|------|---------------------|---------------|-------------|
| EDS V | X-linked type | | XL |
| EDS VI | Ocular-scoliotic type | | AR |
| | | VI-A Decreased lysyl hydroxylase activity | |
| | | VI-B Normal lysyl hydroxylase activity | |
| EDS VII | Arthrochalasis multiplex congenita | | Heterogeneous |
| | | VII-A Structural defect of pro alpha 1(1) collagen | AD |
| | | VII-B Structural defect of pro alpha 2(1) collagen | AD |
| | | VII-C Procollagen N-Proteinase deficiency? | AR |
| EDS VIII | Periodontosis type | | AD |
| EDS IX | Vacant (formerly occipital horn syndrome, or X-linked cutis laxa, now recategorised) | | — |

| Type | General designation | Subcategories | Inheritance |
|------|--------------------|--------------| ------------|
| EDS X | Fibronectin abnormality | | AR |
| EDS XI | Vacant (formerly familial joint instability, now recategorised) | | — |

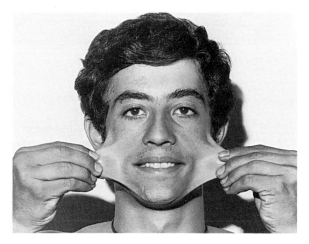

**Fig. 17.9** EDS: hyperextensibility of the skin.

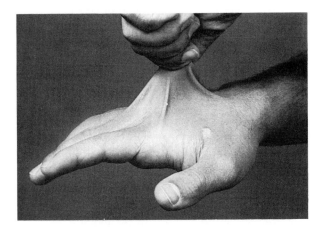

**Fig. 17.10** EDS: cutaneous hyperextensibility.

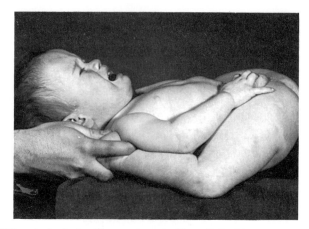

**Fig. 17.11** EDS: articular laxity. Presentation as a 'floppy infant' is not unusual.

*Clinical features*

The major clinical features of the EDS are joint laxity, dermal extensibility and tissue fragility. A bleeding tendency is sometimes present. The skin splits easily and the bony prominences are often covered with distracted papyraceous scars. Hard shotty spheroids can be palpated in the subcutaneous tissues, particularly over the long bones. Hypotonia and articular hypermobility are often the first stigmata to be recognised in early life. Indeed, a professional contortionist with the EDS was heard to lament 'I used to be a floppy infant, but now I'm a loose woman!'

Orthopaedic complications include joint instability, recurrent dislocations, spinal malalignment and pes planus. Serious problems such as dissection of the aorta, rupture of major arteries and spontaneous perforation of the intestine occasionally occur. The phenotypic features of the various forms of the EDS are not clear cut but a summary is given below:

EDS I       Cardinal clinical features in severe degree

EDS II      Cardinal clinical features in mild degree

EDS III     Marked articular hypermobility
            Moderate dermal hyperextensibility
            Minimal scarring

EDS IV      Variable stigmata
            Severe bruising and/or scarring
            Thin skin with prominent venous plexus
            Vascular rupture
            Characteristic facies

EDS V       Cardinal clinical features in moderate degree
            X-linked mode of inheritance

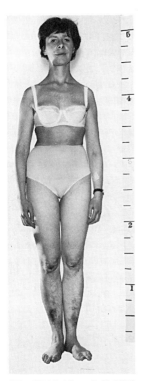

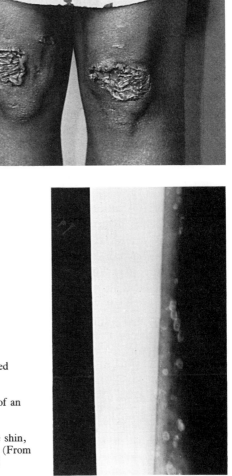

**Fig. 17.12** (above left) EDS: thin pigmented over the knees and shins.

**Fig. 17.13** (above right) EDS: the knees of an affected girl.

**Fig. 17.14** (right) EDS: radiograph of the shin, showing calcified subcutaneous spheroids. (From Beighton P, Lea Thomas M 1969 Clinical Radiology 20: 354.)

| | |
|---|---|
| EDS VI | Cardinal clinical features in severe degree<br>Ocular involvement (microcornea, scleral perforation, retinal detachment)<br>Scoliosis |
| EDS VII | Cardinal clinical features with marked articular hypermobility<br>Short stature and Micrognathia |
| EDS VIII | Cardinal clinical features in moderate degree<br>Aggressive periodontitis, gingival recession, early tooth loss |
| EDS X | Cardinal clinical features but skin texture normal<br>Petechiae<br>Platelet aggregation defect corrected by fibronectin |

Histopathological investigations of collagen in skin biopsy specimens have been undertaken in a search for objective markers of the various forms of the EDS (Black et al 1980, Pierard & Lapier 1983). Although abnormalities have been detected, these have not been clear cut, and laboratory sub-categorisation is now undertaken at the biochemical and molecular levels. This field, which is developing very rapidly, was extensively reviewed by Byers (1983).

The Ehlers-Danlos National Foundation of the USA provides a focus of contact for affected persons and their families. The newsletter 'Loose Connections' has a valuable role in the dissemination of information.

### Genetics

In a study of 100 patients in Southern England, a minimum prevalence of 1 in 150 000 was determined (Beighton et al 1969). The autosomal domi-nant types I, II and III are by far the most common of the various forms of the EDS. They occur in approximately equal frequency, and collectively they constitute over 90% of the total reported cases (Beighton 1970) A severely affected male whose consanguineous parents both had EDS I was documented by Kozlova et al (1984). The likelihood of this individual's homozygosity was supported by the fact that his six offspring had all inherited the disorder. A boy with EDS II and an unbalanced chromosomal translocation (6q:13q) was reported by Scarborough et al (1984). It is not known whether this chromosomal abnormality is aetiologically related to the phenotype.

Despite the rarity of the potentially lethal type IV at least four subtypes have been recognised. Families with autosomal dominant transmission have

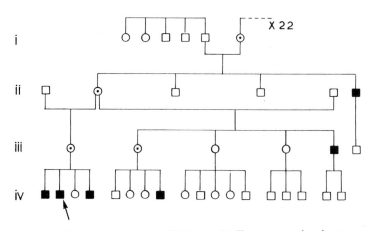

**Fig. 17.15** EDS: X-linked inheritance of EDS type V. Key: □ normal male; ○ normal female; ■ affected male; ⊙ carrier female; / deceased. (From Beighton P 1968 British Medical Journal 53: 409.)

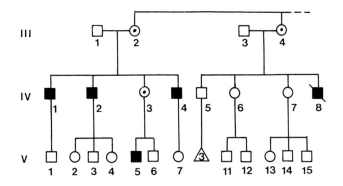

**Fig. 17.16** EDS: the pedigree of the informative branches of the family shown in Figure 17.15, updated after two decades. X-linked inheritance is substantiated. (From Beighton P, Curtis D 1985 Clinical Genetics 27: 472–478.)

been documented by Barabas (1967), Beighton et al (1969) and Pope et al (1977, 1980). The dominant forms, designated EDS IV-A and IV-C, have defective type III pro collagen (Pyeritz & Stolle 1984, Stolle et al 1985). Autosomal recessive inheritance has been proposed in a number of sporadic patients, (EDS IV-B) as their first degree relatives have decreased levels of type III skin collagen (Pope et al 1977). Siblings with the characteristic phenotype, normal type III collagen and unaffected consanguineous parents raise the possibility of a further autosomal recessive form (EDS IV-D) (Sulh et al 1984).

The X-linked EDS type V was initially delineated in two kindreds in England (Beighton 1968). No linkage with the genes for the Xg blood group or colour blindness could be demonstrated in these families. In a follow-up study of one of these families almost two decades later, it was observed that the affected brothers had all produced normal offspring, while their unaffected sister had given birth to a son with the disorder. Greatly to the author's relief, the X-linked mode of inheritance in this family was thus substantiated (Beighton & Curtis 1985). In the other kindred, the two affected brothers had produced male and female offspring with loose joints or extensible skin, but lacking the full phenotype; it is possible that inheritance in this instance is autosomal recessive or that the affected brothers are genetic compounds.

The autosomal recessive background of the uncommon EDS type VI, in which ocular problems predominate, is recognisable in a few early case reports (Beighton 1970). In some persons with this disorder, lysyl hydroxylase is deficient (Krieg et al 1979, Krane 1982), and heterogeneity is possible. Surveillance of a potentially affected pregnancy by enzymatic assay of amniotic fluid cells has been undertaken by Dembure et al (1984).

EDS VII, in which joint laxity is a major feature, is proving to be biochemically heterogeneous. Autosomal dominant and recessive forms have been recognised (Lichtenstein et al 1973) and abnormalities of the pro alpha 1(1) and 2(1) collagen have been detected (Steinmann et al 1980, Cole et al 1986, Eyre et al 1985).

EDS VIII, the periodontosis form, is characterised by chronic inflammation of the gums, in addition to the conventional syndromic stigmata (Nelson & King 1981, Hollister 1982). This disorder is inherited as an autosomal dominant trait.

EDS IX, formerly the occipital horn form of EDS, was also known as X-linked cutis laxa (Byers 1976, Byers et al 1980, Peltonen et al 1983, Satoris et al 1984). This condition is now reclassified as a disorder of copper metabolism and the category EDS IX is vacant.

EDS X, is a mild disorder in which platelet aggregation is defective, possibly on account of a functional abnormality of fibronectin. The condition was recognised in siblings and inheritance is presumably autosomal recessive (Arneson 1980).

EDS XI, formerly the familial joint instability syndrome, is now categorised with the familiar articular hypermobility syndromes.

## FAMILIAL ARTICULAR HYPERMOBILITY SYNDROMES

The familial articular hypermobility syndromes (FAHS) are a group of disorders in which an abnormal range of movements of the joints is the cardinal feature. In some families the condition is benign while in others a wide range of secondary orthopaedic problems may occur. These include subluxations, dislocations, deformity of the feet and malalignment of the spine. The following types of FAHS are recognised:

1. Familial articular hypermobility, uncomplicated type     AD

                                                             AR form?

2. Familial articular hypermobility, dislocating type       AD

The skin and other tissues are not significantly involved in the FAHS and on this basis the Ehlers-Danlos syndrome (EDS) and other articular laxity conditions such as the Larsen syndrome are specifically excluded. This point is of practical importance, as EDS III (hypermobile type) and EDS VII (arthrochalasis multiplex congenita) have phenotypes which are similar to those of the FAHS; the main distinguishing features are some degree of cutaneous hyperextensibility and fragility in the former disorders. Prior to the Berlin conference, the familial articular hypermobility syndromes had been provisionally designated EDS XI, but they are now categorised in their own right and the EDS XI slot remains vacant.

Familial hypermobility was initially recognised by Key (1927) and Sturkie (1941). Thereafter, loose jointedness and multiple dislocations in

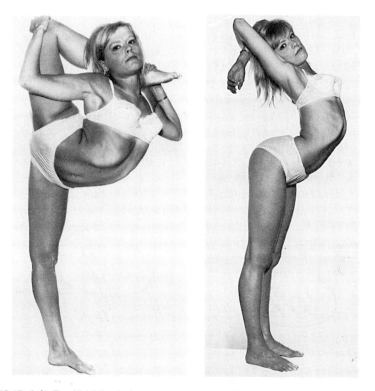

**Fig. 17.17** (left) Familial joint laxity: a young woman with an autosomal dominant form of familial joint laxity. (From Beighton P, Horan F T 1970 Journal of Bone and Joint Surgery 52: 145.)

**Fig. 17.18** (right) Familial joint laxity: the range of movements of all joints is excessive. (From Beighton P, Horan F T 1970 Journal of Bone and Joint Surgery 52: 145.)

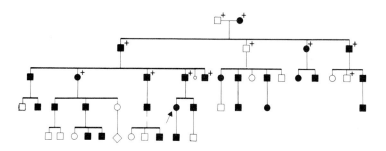

**Fig. 17.19** Familial joint laxity: pedigree showing autosomal dominant inheritance. Key: □ normal male; ○ normal female; ■ affected male; ● affected female; + deceased. (From Beighton P, Horan F T 1970 Journal of Bone and Joint Surgery 52: 145.)

several generations was documented by Hass & Hass (1958) under the designation 'arthrochalasis multiplex congenita'. This term is now applied to EDS VII. The relationship between dislocations and familial joint laxity was further emphasised by Carter & Sweetnam (1958, 1960) and Carter & Wilkinson (1964). Beighton & Horan (1970), used the term 'familial generalised articular hypermobility' in a report of two families with autosomal dominant transmission of loose jointedness. Multiple dislocations occurred in one of these kindreds while in the other, a family of professional contortionists, there were no orthopaedic complications. Horton et al (1980) documented another kindred with autosomal dominant inheritance of joint laxity and dislocations, using the title 'familial joint instability syndrome'.

Two sisters from a consanguineous French-Canadian family were documented by Horan & Beighton (1973). The younger sister had experienced many dislocations but the older was asymptomatic. Inheritance was probably autosomal recessive. The skin was soft and velvety, but neither hyperextensible nor fragile. This phenotype is close to that of EDS VII.

A monograph entitled 'Hypermobility of Joints' contains an account of the articular laxity syndromes, including their clinical implications, biomechanics, basic defects and genetic background (Beighton et al 1983).

## FIBRODYSPLASIA OSSIFICANS PROGRESSIVA

Fibrodysplasia ossificans progressiva (FOP), also known as 'myositis ossificans progressiva,' has been the subject of semantic confusion with other conditions in which soft tissue calcification or ossification takes place. FOP has a long history and at the present time, more than 600 cases have been described. The condition has been extensively reviewed in the monograph 'Soft Tissue Ossification' (Connor 1983).

### Clinical features

The disorder has its onset in childhood. Episodes of spontaneous localised inflammation occur in the soft tissues, particularly in the fascia and tendons. The shoulder girdle, trunk and proximal regions of the limbs are the sites of predilection. These areas subsequently ossify, movements become progressively limited and by early adulthood there is great disability. Nevertheless, general health remains surprisingly good and survival into middle-age is not uncommon. Some degree of microdactyly of the thumb and great toe, due to phalangeal hypoplasia and synostosis, may be present at birth. The radiological features of FOP have been depicted and reviewed by Cremin et al (1982).

Ruderman et al (1974) produced evidence to support the concept that the basic defect might lie in a calcium adenosine-5-triphosphate mediated mechanism of calcification. The diphosphate 'EHDP' may have a therapeutic role in FOP (Smith 1975). However, complications such as de-

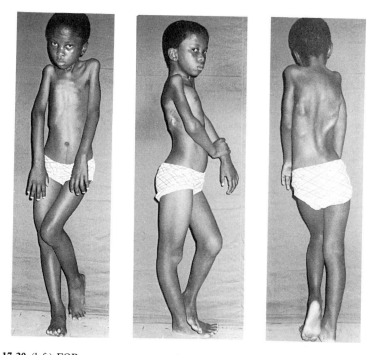

**Fig. 17.20** (left) FOP: movements are restricted, due to extensive soft tissue ossification.

**Fig. 17.21** (centre) FOP: a release operation in the region of the right shoulder precipitated further ossification, and no long-term benefit was obtained.

**Fig. 17.22** (right) FOP: ossified swellings are very obvious over the posterior thorax.

mineralisation and disorganised metaphyseal development have occurred following long-term treatment and the value of EHDP in this condition is doubtful. Computorised tomography is useful in monitoring the response to therapy in FOP (Lindhout et al 1985).

*Genetics*

Fibrodysplasia ossificans progressiva seems to be inherited as an autosomal dominant. A few instances of generation to generation transmission can be recognised in the literature, and there have been two reports of FOP in identical twins (Vastine et al 1948; Eaton et al 1957). Occasionally, phenotypic expression may be limited to microdactyly. However, severely affected individuals rarely reproduce and the majority of patients are probably the result of new dominant mutation. In keeping with this concept, the paternal age effect has been recognised in a study of 23 sporadic cases (Tünte et al 1967).

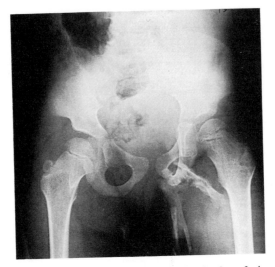

**Fig. 17.23** FOP: radiograph showing extensive ossification in the soft tissue of the left thigh. (From Cremin et al 1982 Clinical Radiology 33: 499–508.)

The majority of reports have emanated from Europe and North America, but FOP has also been encountered in India and in African Negroes (Connor & Beighton 1982).

## WEILL-MARCHESANI* SYNDROME

Individuals with Weill–Marchesani or spherophakia–brachymorphia syndrome have short stature, rigid digits and ectopia lentis. The condition causes little disability and general health, intelligence and life span are normal. The additional syndromic component of aortic stenosis of the subvalvular fibromuscular type was documented by Ferrier et al (1980).

*Genetics*

Affected sibs were mentioned in an early report (Marchesani 1939). The autosomal recessive inheritance of the condition and the fact that the height of heterozygotes is reduced was recognised by Kloepfer & Rosenthal (1955). An anomalous kindred in which a father and two sons had the Weill–Marchesani syndrome was reported by Gorlin et al (1974). As the

*Oswald Marchesani (1900–1952) occupied the Chairs of Ophthalmology at Munster and Hamburg.
George Weill (1924–1937) was Professor of Ophthalmology at the University of Strasbourg.

mother was short, it is possible that she was heterozygous for the condition. If this were so, transmission in this family would be an excellent example of pseudodominant inheritance.

## HOMOCYSTINURIA

Homocystinuria was separated from the Marfan syndrome following a survey of biochemical abnormalities in the urine of mentally retarded individuals in Northern Ireland (Carson & Neill 1962). Once the phenotypic features of the condition had been defined, many other patients were recognised. By means of a world wide questionnaire survey, Mudd et al (1985) obtained information concerning more than 600 affected persons. Homocystinuria is subdivided into pyridoxine sensitive and resistant forms, in terms of response to therapy with vitamin B6. Both are due to cystathimine beta-synthase deficiency and are inherited as autosomal recessive traits.

### Clinical features

The main characteristics of homocystinuria are a marfanoid habitus and dislocation of the lens of the eye. Mental deficiency is present in about 70% of homocystinurics and vascular thrombosis is a frequent complication in adulthood. The skeleton is osteoporotic, the long bones are undermodelled, the epiphyses are widened and vertebral collapse may occur (Schedewie 1973).

### Genetics

Family studies have indicated that homocystinuria is inherited as an autosomal recessive trait (Schimke et al 1965, Carey et al 1968). Variations in phenotypic features, notably mental deficiency, and the inconsistent response to pyridoxine therapy, led Carson & Carré (1969) to postulate that homocystinuria was probably heterogeneous. This fact is now well established, although apart from the two major subtypes the actual extent of heterogeneity is undetermined.

Activity of the enzyme cystathionine synthase is defective in cultured skin fibroblasts from homozygotes and asymptomatic heterozygotes (Uhlendorf & Mudd 1968). As the enzymatic defect is expressed in cultured amniotic fluid cells, Fleisher et al (1974) were able to monitor an 'at risk' pregnancy and successfully predict a normal outcome. Using fetal lymphocytes collected at fetoscopy, Fensom et al (1983) confirmed fetal normality at the 23rd week of gestation.

Boers et al (1985) have raised the intriguing possibility that heterozygotes for homocystinuria might have an increased risk of premature peripheral and cerebral occlusive vascular disease. If substantiated, these observations will have important implications for the understanding of these common medical conditions.

## ALKAPTONURIA

Alkaptonuria is the prototype upon which Garrod based his concept of inborn errors of metabolism. The term 'ochronosis', which is descriptive of dark pigmentation of the connective tissue, is sometimes applied to the condition. However, this designation is not truly synonymous with alkaptonuria as ochronosis can have an acquired basis, such as poisoning with phenol. The title 'homogentisic acid deficiency', which relates to the basic defect, is sometimes used.

### Clinical features

Individuals with alkaptonuria pass urine which is black, or which turns black on standing. The urine can be shown to contain homogentisic acid, which gives a positive reaction to tests for reducing substances. Progressive deposition of black pigment takes place in the connective tissues. By adulthood, this pigmentation is often clinically evident in the pinnae of the ears and in the sclerae. Involvement of the articular cartilages leads to widespread degenerative osteoarthropathy. The spine is often severely affected and patients may be disabled by middle age. Arteriosclerosis, myocardial infarction and involvement of the cardiac valves are common complications. (Levine et al 1978).

Radiographically, degenerative changes are evident in the large joints and the spine. Intra-articular calcified loose bodies may be present and synchondrosis of the pubis is not uncommon. The radiologic manifestations of alkaptonuria have been reviewed by Justesen & Anderson (1984).

### Genetics

Alkaptonuria has a wide geographic distribution and it has been reported in many ethnic groups. In a review of the world literature, O'Brien et al, 1963 were able to identify 520 cases. There is a particularly high prevalence in Dominica (Harrold 1956) and Czechoslovakia (Červeňanský et al 1959, Srsen et al, 1978).

The autosomal recessive inheritance of alkaptonuria was established by Hogben et al (1932) following a formal segregation analysis of pedigree data. A high frequency of parental consanguinity has been recorded (Molony & Kelly 1970). Pedigrees have been published of large kindreds in which inheritance was apparently dominant (Khachadurian & Feisal 1958). However, as with other autosomal recessive disorders which have a particularly high frequency in certain populations, this situation has resulted from marriage between affected homozygotes and clinically normal heterozygotes. It is evident that the alkaptonuria gene is not a recent mutation, as the condition has been recognised radiographically in Egyptian mummies (Simon & Zorab 1961).

## TIGHT SKIN SYNDROMES

This group of very rare disorders share the common feature of skin thickening, to an extent that joint mobility is impaired. In some, the skeleton and other tissues are also involved.

### Rothmund-Thomson* syndrome

Rothmund-Thomson syndrome is characterised by sclerodermatous skin lesions, cataracts, depression of the nasal bridge and alopecia. Bone reabsorption takes place and the peripheral joints become disorganised, presenting a Charcot-like radiographic appearance. Patients may be severely

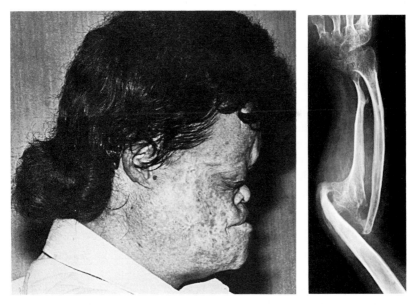

**Fig. 17.24** (left) Rothmund syndrome: a young man with scleroderma of the face and depression of the nasal bridge. Alopecia is hidden by a wig. (From Beighton P 1976 Syndrome Identification 4: 8.)

**Fig. 17.25** (right) Rothmund syndrome: radiographs of the elbow and wrist of the patient depicted in Figure 17.24. The disorganised joints have a Charcot-like appearance.

*August von Rothmund (1830–1906), a man of great jest, was Professor of Ophthalmology and Head of the State Eye Clinic, Munich, Germany.
  Matthew Thomson (1894–1969) was Senior Dermatologist at Kings College Hospital, London.

disabled (Kirkham & Werner 1975, Sivayoham & Ratnaike 1975). Many affected persons have stunted stature, and Kaufmann et al (1986) demonstrated that growth hormone is defective.

The patients documented by Rothmund (1868) had cataracts and depression of the nasal bridge while those reported by Thomson (1936) lacked these features. On this basis, it is possible that these conditions are separate entities, but this remains unproven. In any event, inheritance is autosomal recessive.

### Stiff skin syndrome

Thickening of the skin with restricted joint movements was recognised in a mother and her two offspring by Esterly & McKusick (1971) and termed the 'stiff skin' syndrome. A similar disorder had been previously encountered in a father and his son and daughter by Pichler (1968) and reported under the designation 'hereditary contractures with scleroderma-like changes'. Inheritance was apparently autosomal dominant in each kindred. Transmission through four generations has been recorded (Singer et al 1977). Histopathological studies of skin biopsy specimens reveal large cells which stain metachromatically and which might be indicative of dermal deposition of mucopolysaccharides (Kikuchi et al 1985).

### Syndesmodysplastic dwarfism

Laplane et al (1972) described two brothers with short stature and sclerodermatous skin. The authors considered that the condition was a new entity, which they named 'familial syndesmodysplastic dwarfism'. The genetic background is unknown, but as the affected persons were members of an endogamous Berber community in Algeria, it is possible that inheritance is autosomal recessive.

### Parana hard skin syndrome

Cat et al (1974) described eight individuals in the Parana region of Brazil, in whom widespread dermal thickening appeared in infancy. The condition progressed and the patients gradually become immobilised. Several of them eventually died from respiratory embarrassment. The pedigree data were consistent with autosomal recessive inheritance.

## LOOSE SKIN SYNDROMES

Cutis laxa is a rare disorder which has been the subject of semantic confusion with the Ehlers-Danlos syndrome. In cutis laxa the skin hangs in loose folds and is not fragile, and the joints are not hypermobile. Autosomal dominant and autosomal recessive types of cutis laxa have been recognised

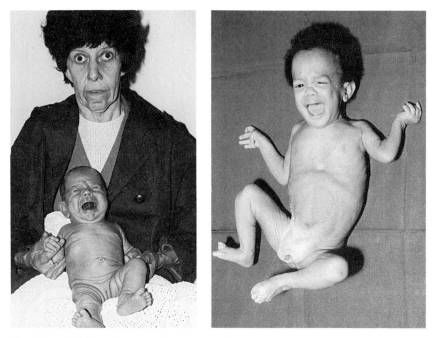

**Fig. 17.26** (left) Cutis laxa: a mother and daughter with the autosomal dominant type of cutis laxa. (From Beighton P 1972 Journal of Medical Genetics 9: 216.)

**Fig. 17.27** (right) Cutis laxa with retarded development and joint hypermobility. This affected boy has large inguinal herniae and redundant skin on the abdomen.

(Beighton 1972). The skeleton is not involved in the classical forms of the disorder but severe bone changes are a feature of at least two cutis laxa variants.

### Cutis laxa with ligamentous laxity and delayed development

Debré et al (1937) reported a prematurely born child who had generalised dermal laxity, genu valgum, abnormality of cranial sutures and loose jointedness in the fingers. No family data was provided in this report. Subsequently, Fittke (1942) described a girl with similar features. Following an extensive study of the girl's kindred, Theopold & Wildhack (1951) published a pedigree showing that the parents were consanguineous and that two of her sisters and a female cousin were also affected. Skeletal changes were severe and included scoliosis, pectus carinatum, congenital dislocation of the hip, wide fontanelles and acrocephaly. Kaye & Fisher (1975) reported an infant with cutis laxa, skeletal abnormalities and genitourinary anomalies; this disorder might be the same entity.

Sakati et al (1983) drew attention to ligamentous laxity and delayed

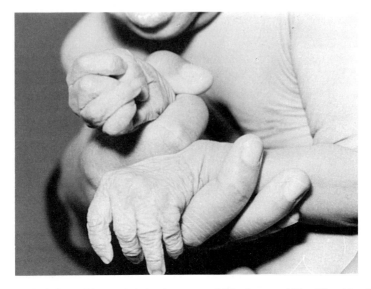

**Fig. 17.28** Cutis laxa with retarded development and joint hypermobility. The skin of the hands is wrinkled and loose.

psychomotor development, as syndromic components and further cases were reported by Rogers & Danks (1985). Four affected siblings were encountered in Saudi Arabi (Karrar et al 1985) and another child with the condition, from the same country, was documented by Allanson et al (1986). There has been a paucity of males amongst the affected children, (Fitzsimmons et al 1985) and it has been suggested that there might be propensity to male lethality (Allanson et al 1986). Accumulating evidence indicates autosomal recessive inheritance and it seems likely that the apparent female preponderance simply reflects ascertainment bias.

### Leprechaunoid syndrome

Donohue & Uchida (1954) were impressed by the pixie-like facial features of two young sisters, who also had marked dermal laxity. These authors quaintly alluded to the condition as 'leprechaunism'. These girls, who had no bone changes, died in infancy. Subsequently, Patterson & Watkins (1962) reported a male infant with gross laxity of the skin of the extremities, in association with radiographic evidence of defective ossification, marked irregularity of the ends of the tubular bones and a low serum alkaline phosphatase. Patterson (1969) described the development of this child, who became cushingoid and died in his seventh year. Dallaire et al (1976) reported three male infants with the same condition from two related consanguineous Italian kindreds. On this evidence, it is likely that inheritance is autosomal recessive. The basic abnormality is an inborn error of high affinity insulin receptor (Elsas et al 1985).

## Wrinkly skin syndrome

Gazit et al (1973) reported two sisters of Iraqi-Jewish stock, who had loose wrinkled skin on the extremities and ventral surfaces, together with kyphosis and winging of the scapula. The parents were consanguineous and it is likely that inheritance was autosomal recessive. This disorder differs from the better known 'Prune belly' syndrome, in which absence of the abdominal muscles is associated with abnormalities of the urinary tract.

## DISORDERS OF COPPER METABOLISM

### Occipital horn syndrome (X-linked cutis laxa)

Using the designation 'X-linked cutis laxa', Byers et al (1976, 1980) described a syndrome in males in four affected families in which mild hyperextensibility of the small joints was associated with skeletal abnormalities and bladder diverticulae. Radiographic studies demonstrated occipital protuberances which appear at puberty and became increasingly prominent. Lysyl oxidase was shown to be deficient.

The condition was initially regarded as a form of the Ehlers-Danlos syndrome and designated EDS IX. Thereafter, Kuivaniemi et al (1982) recognised abnormalities of copper metabolism in two affected brothers; the condition is now termed the occipital horn syndrome and has been reclassified as a disorder of copper transport.

### Menkes syndrome

Menkes syndrome is characterised by lax skin and joints, arterial rupture or thrombosis, osteoporosis, diverticulae of the bladder and ureter and changes in the metaphyses of the long bones. In severely affected infants, multiple fractures and marked cortical thickening may raise suspicion of the battered child syndrome (Wendler & Mutz 1985, Menkes et al 1962, Menkes 1972). The hair is abnormal, psychomotor development is retarded, and death in infancy is usual. Activity of the enzyme lysyl oxidase is defective, due to an inborn error of copper transportation (Danks 1986).

REFERENCES

Maroteaux P, Frëzal J, Cohen-Solal Lola 1986 The differential symptomatology of errors of collagen metabolism: A tentative classification. American Journal of Medical Genetics 24: 219–230

**Marfan syndrome**
Ambani L M, Gelehrter T D, Sheahan D G 1975 Variable expression of Marfan syndrome in monozygotic twins. Clinical Genetics 8/5: 358
Beighton P 1982 Pregnancy in the Marfan syndrome. British Medical Journal 285: 464
Buchanan R, Wyatt G P 1985 Marfan's syndrome presenting as an intrapartum death. Archives of Disease in Childhood 60: 1074–1076

Chemke J, Nisani R, Feigl A, Garty R, Cooper M, Barash Y, Duksin D 1984
Homozygosity for autosomal dominant Marfan syndrome. Journal of Medical Genetics
21: 173–177

Gillan J E, Costigan D C, Keeley F W 1984 Spontaneous dissecting aneurysm of the ductus
arteriosus in an infant with Marfan syndrome. Journal of Pediatrics 105/6: 952–955

Feingold M, Bossert W H 1974 Normal values for selected physical parameters: an aid to
syndrome delineation. Birth Defects: Original Article Series 10/13: 4

Koenigsberg M, Factor S, Cho S, Herskowitz A, Nitoswky H, Morechi R 1981 Fetal
Marfan syndrome. Prenatal ultrasound diagnosis with pathological confirmation of skeletal
and aortic lesions. Prenatal Diagnosis 1: 241–247

Lind J, van Papendrecht HPCMH 1984 Obstetrical complications in a patient with the
Marfan syndrome. European Journal of Obstetrics, Gynaecology and Reproductive
Biology 18: 161–168

Lynas M A 1958 Marfan's syndrome in Northern Ireland; an account of 13 families. Annals
of Human Genetics 22: 289

Lynas M A, Merrett J D 1958 Data on linkage in man; Marfan's syndrome in Northern
Ireland. Annals of Human Genetics 22: 310

Maumenee I H 1981 The eye in the Marfan syndrome. Transactions of the American
Ophthalmological Society 79: 684–733

Murdoch J L, Walker B A, Halpern B L, Kuzma J W, McKusick V A 1972 Life
expectancy and causes of death in the Marfan syndrome. New England Journal of
Medicine 286: 804

Pyeritz R E, McKusick V A 1979 The Marfan syndrome: diagnosis and management. New
England Journal of Medicine 3–0: 772–777

Roberts W C, Honig H S 1982 The spectrum of cardiovascular disease in the Marfan
syndrome: a clinico-morphologic study of 18 necropsy patients and comparison to 151
previously reported necropsy patients. American Heart Journal 104(1): 115–135

Robins P R, Winter R B, Moe J H 1974 Scoliosis in patients with Marfan's syndrome.
Journal of Bone and Joint Surgery 56: 1540

Schleutermann D A, Murdoch J I, Walker B A et al 1976 A linkage study of the Marfan
syndrome. Clinical Genetics 10: 51–53

Tsipouras P, Borresen A, Bamforth S, Harper P S, Berg K 1986 Marfan syndrome:
exclusion of genetic linkage to the COL1A2 gene. Clinical Genetics 30: 428–432

Verghese C 1984 Anaesthesia in Marfan's syndrome. Anaesthesia 39: 917–922

Wood J R, Bellamy D, Child A H, Citron K M 1984 Pulmonary disease in patients with
Marfan syndrome. Thorax 39: 780–784

*Marfanoid hypermobility syndrome*
Goodman R M, Wooley C F, Frazier R L, Covault I 1965 Ehlers-Danlos syndrome together
with the Marfan syndrome. Report of a case with other family members affected. New
England Journal of Medicine 273: 514

Walker B A, Beighton P H, Murdoch J L 1969 The marfanoid hypermobility syndrome.
Annals of Internal Medicine 71: 349

*Congenital contractural arachnodactyly*
Anderson R A, Koch S, Camerini-Otero R D 1984 Cardiovascular findings in congenital
contractural arachnodactyly: report of an affected kindred. American Journal of Medical
Genetics 18: 265–271

Arroyo M A R, Weaver D D, Beals R K 1985 Report of four additional families and review
of literature. Clinical Genetics 27: 570–581

Beals R K, Hecht F 1971 Contractural arachnodactyly, a heritable disorder of connective
tissue. Journal of Bone and Joint Surgery 53A: 987

Bjerkeim I, Skogland L B, Trygstad O 1976 Congenital contractural arachnodactyly. Acta
Orthopaedica Scandinavica 47: 250

Currarino G, Friedman J M 1986 A severe form of congenital contractural arachnodactyly
in two newborn infants. American Journal of Medical Genetics 25: 763–773

Kontras S B 1975 Congenital contractural arachnodactyly. Birth Defects: Original Article
Series 11/6: 63

Lowry R B, Guichon V C 1972 Congenital contractural arachnodactyly: a syndrome
simulating Marfan's syndrome. Canadian Medical Association Journal 107: 531

MacLeod P M, Fraser F C 1973 Congenital contractural arachnodactyly. (A heritable
disorder of connective tissue distinct from Marfan syndrome.) American Journal of
Diseases of Children 126: 810

Passarge E 1975 A syndrome resembling congenital contractural arachnodactyly. Birth Defects: Original Article Series 11/6: 53

Steg N L 1975 Congenital contractural arachnodactyly in a black family. Birth Defects: Original Article Series 11/6: 57

*Achard syndrome*

Achard C 1902 Arachnodactylie. Bulletins et Memoires de la Societe Médicale des Hôpitaux de Paris 19: 834

Duncan P A 1975 The Achard syndrome. Birth Defects: Original Article Series 11/6: 69

*Marfanoid habitus with mitral valve prolapse*

Beighton P 1982 Mitral valve prolapse and a Marfanoid habitus. British Medical Journal 284: 920

Salomon J, Shah P, Heinle R A 1975 Thoracic skeletal abnormalities in idiopathic mitral valve prolapse. American Journal of Cardiology 36: 31–36

Schutte J E, Gaffney F A, Blend L, Blomqvist C G 1981 Distinctive anthropometric characteristics of women with mitral valve prolapse. American Journal of Medicine 71: 533–538

Udoshi M B, Shah A, Fisher V J, Dolgin M 1979 Incidence of mitral valve prolapse in subjects with thoracic skeletal abnormalities — a prospective study. American Heart Journal 97: 303–311

*Other variants of the Marfan syndrome*

Fragoso R, Cantu J M 1984 A new psychomotor retardation syndrome with peculiar facies and marfanoid habitus. Clinical Genetics 25: 187–190

Lujan J E, Carlin M E, Lubs H A 1984 A form of X-linked mental retardation with Marfanoid habitus. American Journal of Medical Genetics 17: 311–322

Tamminga P, Jennekens F G I, Barth P G 1985 An infant with Marfanoid phenotype and congenital contractures associated with ocular and cardiovascular anomalies, cerebral white matter hypoplasia and spinal axonopathy. European Journal of Pediatrics 143: 228–231

*Ehlers-Danlos syndrome*

Arneson M A 1980 A new form of Ehlers-Danlos syndrome. Journal of the American Medical Association 244: 144–147

Barabas A P 1967 Heterogeneity of the Ehlers-Danlos syndrome. British Medical Journal 2: 612

Beighton P 1968 X-linked recessive inheritance in the Ehlers-Danlos syndrome. British Medical Journal 3: 409

Beighton P 1970 The Ehlers-Danlos syndrome. Heinemann, London, p 115

Beighton P 1970 Serious ophthalmological complications in the Ehlers-Danlos syndrome. British Journal of Ophthalmology 54: 263

Beighton P, Curtis D 1985 X-linked Ehlers-Danlos syndrome type V; the next generation. Clinical Genetics 27: 472–478

Beighton P, Price A, Lord J, Dickson E 1969 The variants of the Ehlers-Danlos syndrome. Annals of Rheumatic Disease 28: 228

Black C M, Gathercole L J, Bailey A J, Beighton P 1980 The Ehlers-Danlos syndrome: an analysis of the structure of the collagen fibres of the skin. British Journal of Dermatology 102: 85

Byers P H 1976 Inherited disorders of collagen biosynthesis: Ehlers-Danlos syndrome, the Marfan syndrome, and osteogenesis imperfecta. Clinical Medicine 9/12: 1–41

Byers P H 1983 Inherited disorders of collagen biosynthesis: Ehlers-Danlos syndrome, the Marfan syndrome and osteogenesis imperfecta. Clinical Medicine 1–41

Byers P H, Siegel R C, Holbrook K A, Narayanan A S, Bornstein P, Hall J G 1980 X-linked cutis laxa: defective collagen crosslink formation due to decreased lysyl oxidase activity. New England Journal of Medicine 303: 61

Cole W G, Chan D, Chambers G W, Walker I D, Bateman J F 1986 Delection of 24 amino acids from the pro 1(1) chain of type I procollagen in a patient with the Ehlers-Danlos syndrome type VII. Journal of Biological Chemistry 261: 5496–5503

Dembure P P, Priest J H, Snoddy S C, Elsas L J 1984 Genotyping and prenatal assessment of collagen lysyl hydroxylase deficiency in a family with Ehlers-Danlos syndrome type VI. American Journal of Human Genetics 36: 783–790

Eyre D R, Shapiro F D, Aldridge G F 1985 A heterozygous collagen defect in a variant of the Ehlers-Danlos syndrome type VII. Journal of Biological Chemistry 260: 11322–11329

Hollister D W 1982 Clinical features of Ehlers-Danlos syndrome type VIII and IX in symposium on heritable disorder of connective tissue. American Academy of Orthopaedic Surgery 102–113

Kozlova S I, Prytkov A N Blinnikova O E, Sultanova F A, Bochkova D N 1984 Presumed homozygous Ehlers-Danlos syndrome type I in a highly inbred kindred. American Journal of Medical Genetics 18: 763–767

Krane S M 1982 Hydroxylysine-deficient collagen disease: a form of Ehlers-Danlos syndrome type VI in symposium on heritable disorders of connective tissue. American Academy of Orthopaedic Surgery, eds Akeson, W H et al, Mosby , St Louis, p 61–81

Kreig T, Feldmann U, Kessler W, Muller P K 1979 Biochemical characteristics of Ehlers-Danlos syndrome VI in a family with one affected infant. Human Genetics 46: 41

Lichtenstein J R, Martin G R, Kohn L, Byers P, McKusick V A 1973 Defect in conversion of procollagen to collagen in a form of Ehlers-Danlos syndrome. Science 182: 298

Nelson D L, King R A 1981 Ehlers-Danlos syndrome type VIII. Journal of the American Acadamy of Dermatology 5: 297–303

Peltonen L, Kuivaniemi H, Palotie A, Horn N, Kaitila I, Kivirikko K I 1983 Alterations in copper and collagen metabolism in the Menkes syndrome and a new subtype of the Ehlers-Danlos syndrome. Biochemistry 22: 6165–6163

Pierard F C, Lapier C M 1983 Histopathological aid at the diagnosis of the Ehlers-Danlos syndrome, gravis and mitis types. Journal of Dermatology 22/5: 300–304

Pope F M, Martin G R, McKusick V A 1977 Inheritance of Ehlers-Danlos type IV syndrome. Journal of Medical Genetics 14: 200

Pope F M, Nicholls A C, Jones P M, Wells R S, Lawrence D 1980 EDS IV (acrogeria): new autosomal dominant and recessive types. Journal of the Royal Society of Medicine 73: 180

Pyeritz R E, Stolle C A 1984 Ehlers-Danlos syndrome IV due to a novel defect in type III procollagen. American Journal of Medical Genetics 19: 607–622

Satoris D J, Luzzatti L, Weaver D D, Macfarlane J D, Hollister D W, Parker B R 1984 Type IX Ehlers-Danlos syndrome. Radiology 152: 665–670

Scarbrough P R, Daw J, Carroll A J, Finley S C 1984 An unbalanced (6q:13q) translocation in a male with clinical features of Ehlers-Danlos type II syndrome. Journal of Medical Genetics 21: 226

Steinmann B, Tuderman L, Peltonen L, Martin G R, McKusick V A, Prockop D J 1980 Evidence for a structural mutation of procollagen type I in a patient with the Ehlers-Danlos syndrome type VII. Journal of Biological Chemistry 255: 8887–8893

Stolle C A, Pyeritz R E, Myers J C, Prockop D J 1985 Synthesis of an altered type III procollagen in a patient with type IV Ehlers-Danlos syndrome. Journal of Biological Chemistry 260/3: 1937–1944

Sulh H M B, Steinmann B, Rao V H, Dudin G, Zeid J A, Slim M, Der Kaloustian V M 1984 Ehlers-Danlos syndrome type IV D: an autosomal recessive disorder. Clinical Genetics 25: 278–287

**Familial articular hypermobility syndromes**

Beighton P, Horan F T 1970 Dominant inheritance in familial generalised articular hypermobility. Journal of Bone and Joint Surgery 52: 145–147

Beighton P, Grahame R, Bird H 1983 Hypermobility of joints. Springer-Verlag, Heidelberg

Carter C, Sweetnam R 1958 Familial joint laxity and recurrent dislocations of the patella. Journal of Bone and Joint surgery 40: 664–667

Carter C, Sweetnam R 1960 Recurrent dislocation of the patella and of the shoulder, their association with familial joint laxity. Journal of Bone and Joint surgery 42: 721–727

Carter C, Wilkinson J 1964 Persistent joint laxity and congenital dislocation of the hip. Journal of Bone and Joint Surgery 46: 40–45

Hass J, Hass R 1958 Arthrochalasis multiplex congenita. Journal of Bone and Joint Surgery 40: 663–674

Horan F T, Beighton P. 1973 Recessive inheritance of generalized joint hypermobility. Rheumatology and Rehabilitation 12: 47–49

Horton W A, Collins D L, DeSmet A A, Kennedy J A, Schimke R N 1980 Familial joint instability syndrome. American Journal of Medical Genetics 6: 221–228

Key J A 1927 Hypermobility of joints as a sex-linked hereditary characteristic. Journal of the American Medical Association 88: 1710–1712

Sturkie P D 1941 Hypermobile joints in all descendants for two generations. Journal of Heredity 32: 232–234

### Fibrodysplasia ossificans progressiva

Connor J M 1983 Soft tissue ossification. Springer-Verlag, Heidelberg

Connor J M, Beighton P 1982 Fibrodysplasia ossificans progressiva in South Africa. South African Medical Journal 61: 404–406

Cremin B J, Connor J M, Beighton P 1982 The radiological spectrum of fibrodysplasia ossificans progressiva. Clinical Radiology 33: 499–508

Eaton W L, Conkling W S, Daeschner C W (1957) Early myositis ossificans progressiva occurring in homozygotic twins; a clinical and pathologic study. Journal of Pediatrics 50: 591

Lindhout D, Golding R P, Taets van Amerongen A H M 1985 Fibrodysplasia ossificans progressiva: current concepts and the role of CT in acute changes. Pediatric Radiology 15/3: 211–213

Ruderman R J, Leonard F, Elliot D E 1974 A possible aetiological mechanism for fibrodysplasia ossificans progressiva. Birth Defects: Original Article Series 10/12: 299

Smith R 1975 Myositis ossificans progressiva: a review of current problems. Seminars of Arthritis and Rheumatism 4/4: 369

Tünte W, Becker P E, Knorr G V 1967 Zur Genetik der Myositis ossificans progressiva. Humangenetik 4: 320

Vastine J H, Vastine M E, Oriel A 1948 Myositis ossificans progressiva in homozygotic twins. American Journal of Roentgenology 59: 204

### Weill-Marchesani syndrome

Ferrier S, Nussle D, Friedlei B, Ferrier P E 1980 Le syndrome de Marchesani (spherophakie-brachymorphie). Helvetica Paediatrica Acta 35: 185–198

Gorlin R J, L'Heureux R R, Shapiro I 1974 Weill-Marchesani syndrome in two generations: genetic heterogeneity or pseudodominance? Journal of Pediatric Ophthalmology 11: 139

Kloepfer H W, Rosenthal J W 1955 Possible genetic carriers in the spherophakia-brachymorphia syndrome. American Journal of Human Genetics 7: 398

Marchesani O 1939 Brachydaktylie und angeborene Kugellinse als Systemerkrankung. Klinische Monatsblätter für Augenheilkunde 103: 392

### Homocystinuria

Boers G H J, Smals A G H, Trijbels F J M et al 1985 Heterozygosity for homocystinuria in premature peripheral and cerebral occlusive arterial disease. New England Journal of Medicine 313: 12 709–715

Carey M C, Donovan D E, Fitzgerald O, McAuley F D 1968 Homocystinuria. A clinical and pathological study of nine subjects in six families. American Journal of Medicine 45: 7

Carson N A J, Carré I J 1969 Treatment of homocystinuria with pyridoxine. Archives of Disease in Childhood 44: 387

Carson N A J, Neill D W 1962 Metabolic abnormalities detected in a survey of mentally backward individuals in Northern Ireland. Archives of Disease in Childhood 37: 505

Fensom A H, Benson P F, Crees M J, Ellis M 1983 Prenatal exclusion of homocystinuria (cystathionine B-synthase deficiency) by assay of phytohaemagglutinin-stimulated fetal lymphocytes. Prenatal Diagnosis 3: 127–130

Fleisher L D, Longhi R C, Tallan H H et al 1974 Homocystinuria: investigators of cystathionine synthase in cultured fetal cells and the prenatal determination of genetic status. Journal of Pediatrics 85/5: 677

Mudd S H, Skovby F, Levy H L, Pettigrew K D, Wilcken B, Pyeritz R E, Andria G, Boers G H J, Bromberg I L, Cerone R, Fowler B, Grobe H, Schmidt H, Schweitzer L 1985 The natural history of Homocystinururia due to cystathionine B-synthase deficiency. American Journal of Human Genetics 37: 1–31

Schedewie H 1973 Skeletal findings in homocystinuria: a collaborative study. Pediatric Radiology 1: 12

Schimke R N, McKusick V A, Huang T, Pollack A D 1965 Homocystinuria: studies of 20 families with 38 affected members. Journal of the American Medical Association 193: 711

Uhlendorf B W, Mudd S H 1968 Cystathionine synthase in tissue culture derived from human skin: enzyme defect in homocytinuria. Science 160: 1007

**Alkaptonuria**
Cervenanskỳ J, Sitaj S, Urbánek T 1959 Alkaptonuria and ochronosis. Journal of Bone and Joint Surgery 41: 1169
Harrold A J 1956 Alkaptonuria arthritis. Journal of Bone and Joint Surgery 38: 532
Hogben L, Worrall R L, Zieve I 1932 The genetic basis of alkaptonuria. Proceedings of the Royal Society of Einburgh 52: 264
Justeson P, Andersen P E 1984 Radiologic manifestations in alcaptonuria. Skeletal Radiology 11: 204–208
Khachadurian A, Feisal K 1958 Alkaptonuria; a report of a family with seven cases appearing in four successive generations with metabolic studies in one patient. Journal of Chronic Disease 7: 455
Levine H D, Parisi A F, Holdsworth D E, Cohn L W 1978 Aortic valve replacement for ochronosis of the aortic valve. Chest 74: 466–467
Molony J, Kelly D J 1970 Alkaptonuria, ochronosis, and achronitic arthritis. Journal of the Irish Medical Association 63: 22
O'Brien W M, La Du B N, Bunim J J 1963 Biochemical, pathologic and clinical aspects of alcaptonuria, ochronosis and ochronotic arthropathy; review of world literature. American Journal of Medicine 34: 813
Simon G, Zorab P A 1961 The radiographic changes in alkaptonuric arthritis. British Journal of Radiology 34: 384
Srsen S, Cisarik F, Pasztor L, Harmecko L 1978 Alkaptonuria in the Trencin district of Czechoslovakia. American Journal of Medical Genetics 2: 159–166

**Tight skin syndromes**
*Rothmund-Thomson syndrome*
Kaufman S, Jones M, Culler F L, Lee Jones K 1986 Growth hormone deficiency in the Rothmund-Thomson syndrome. American Journal of Medical Genetics 23: 861–868
Kirkham T H, Werner E B 1975 The ophthalmic manifestations of Rothmund's syndrome. Canadian Journal of Ophthalmolõgy 10/1: 1
Rothmund A 1868 Ueber cataracte in verbindung mit einer eigenthuemlichen hautdegeneration. Graefes Archive for Clinical and Experimental Ophthalmology 14: 159–182
Sivayoham I S S R, Ratnaike V T 1975 Rothmund-Thomson syndrome in an oriental patient.

*Stiff skin syndrome*
Esterly N B, McKusick V A 1971 Stiff skin syndrome. Pediatrics 47: 360
Kikuchi I, Inoue S, Hamada K, Ando H 1985 Stiff skin syndrome. Pediatric Dermatology 3/1: 48–53
Pichler E 1968 Hereditaere Kontrakturen mit sklerodermieartigen Hautveraenderungen. Zeitschrift für Kinderheilkunde 104: 349
Singer H S, Valle D, Rogers J, Thomas G H 1977 The stiff skin syndrome: new genetic biochemical investigations. Birth Defects Original Article Series XIII(3B): 254–255
Thomson M S 1986 Poikiloderma congenitale. British Journal of Dermatology 48: 221

*Syndesmodysplastic dwarfism*
Laplane R, Fontaine J-L, Lagardere B, Sambucy F 1972 Nanisme syndesmodysplasique familial. Une entité morbide nouvelle. Archives Françaises de Pédiatrie (Paris) 29: 831

*Parana hard skin syndrome*
Cat I, Rodrigues-Magdalena N I, Parolin-Marinoni L, Wong M P, Freitas O T, Malfi A, Costa O, Estieves L, Giraldi D J 1974 Parana hard skin syndrome: study of seven families. Lancet i: 215

**Loose skin syndromes**
*Cutis laxa with ligamentous laxity and delayed development*
Allanson J, Austin W, Hecht F 1986 Congenital cutis laxa with retardation of growth and motor development: a recessive disorder of connective tissue with male lethality. Clinical Genetics 29: 133–136

Beighton P 1972 Dominant and recessive forms of Cutis Laxa. Journal of Medical Genetics 9: 216

Debré R, Marie J, Seringe P 1937 'Cutis laxa' avec dystrophies osseuses. Bulletins et Mémoires de la Société Médicale des Hôpitaux de Paris 61: 1038

Fittke H 1942 Über eine ungewöhnliche Form 'multipler Erbabartung' (Chalodermie und dysostose). Zeitschrift für Kinderheilkunde 63: 510

Fitzsimmons J S, Fitzsimmons E M, Guibert P R, Zaldua V, Dodo K L 1985 Variable clinical presentation of cutis laxa. Clinical Genetics 28: 284–295

Karrar Z A, Alarabi K, Adam K A, E I Idrissy A T H 1985 Cutis laxa, intrauterine growth retardation and bilateral dislocation of the hips: a report of four patients. Saudi Medical Journal 6(6): 561–565

Kaye C I, Fisher D E 1975 Cutis laxa and associated anomalies. Birth Defects: Original Article Series 11/2: 130

Rogers J G, Danks D M 1985 Cutis laxa with delayed development. Australian Paediatric Journal 21: 281–283

Sakati NO, Nyhan W L, Shear C S, Kattan H, Akhtar M, Bay C, K L Jones, Schackner L 1983 Syndrome of cutis laxa, ligamentous laxity, and delayed development. Pediatrics 72: 850–856

Theopold W, Wildhack R 1951 Dermatochalasis im Rahmne multipler Abortungen. Monatsschrift für Kinderheilkunde 99: 213

*Leprechaunoid syndrome*

Dallaire L, Cantin M, Melancon S B, Perreault G, Potier M 1976 A syndrome of generalised elastic fiber deficiency with leprechaunoid features: a distinct genetic disease with an autosomal recessive mode of inheritance. Clinical Genetics 10: 1

Donohue W L, Uchida I 1954 Leprechaunism: a euphism for a rare familial disorder. Journal of Pediatrics 45: 505

Elsas L J, Endo F, Strumlauf E, Elders J, Priest J H 1985 Leprechaunism: an inherited defect in a high -affinity insulin receptor. American Journal of Human Genetics 37: 73–88

Patterson J H 1969 Presentation of a patient with leprechaunism. In: Bergsma D (ed) Clinical Delineation of Birth Defects, IV. National Foundation, New York, p 117

Patterson J H, Watkins W L 1962 Leprechaunism in a male infant. Journal of Pediatrics 60: 730

*Wrinkly skin syndrome*

Gazit E, Goodman R M, Bat-Miriam Katznelson M, Rotem Y 1973 The wrinkly skin syndrome: a new heritable disorder of connective tissue. Clinical Genetics 4: 186

**Disorders of copper metabolism**

*Occipital horn syndrome (X-linked cutis laxa)*

Byers P H, Siegel R C, Holbrook K A, Narayanan A S, Bornstein P, Hall J G 1980 X-linked cutis laxa. New England Journal of Medicine 303: 61–65

Byers P H, Narayanan A S, Bornstein P, Hall J G 1976 An X-linked cutis laxa due to deficiency of lysyloxidase. Birth Defects Original Article Series 12(5): 293–298

Kuivaniemi H, Peltonen L, Palotie A, Kaitila I, Kivirikko K I 1982 Abnormal copper metabolism and deficient lysyl oxidase activity in a heritable connective tissue disorder. Journal of Clinical Investigations 69: 730–733

*Menkes syndrome*

Danks D M 1986 Of mice and men, metals and mutations. Journal of Medical Genetics 23: 99–106

Menkes J H 1972 Kinky hair disease. Pediatrics 50: 181–182

Menkes J H, Alter M, Steigleder G K, Weakley D R, Sung J H 1962 A sex-linked recessive disorder with retardation of growth, peculiar hair and focal cerebral and cerebellar degeneration. Pediatrics 29: 764–779

Wendler H, Mutz I 1985 Menkes syndrome with excessive skeletal changes (Germ). Rofo Fortschritte auf dem Gebiete der Rontgenstrahlen und der Nuklearmedizin 143/3: 351–355

# 18

# Primary disturbances of growth

Proportionate or relatively proportionate short stature with low birth weight is a major feature of a number of disorders. Apart from the failure of growth, the skeleton is not usually dysplastic in these conditions. However, as they are frequently confused with other true skeletal dysplasias, they merit consideration in this section.

1. Pituitary dwarfism
   a. Type I: Isolated growth hormone deficiency
   b. Type II: Laron type
   c. Type III: Panhypopituitarism
   d. Other forms of pituitary dwarfism
2. Cornelia de Lange syndrome
3. Seckel syndrome
4. Russell-Silver syndrome
5. Senility syndromes
   a. Progeria
   b. Mandibuloacral dysplasia
   c. Cockayne syndrome
   d. Werner syndrome
6. Bloom syndrome
7. Other disorders of growth
   a. Dubowitz syndrome
   b. 3M syndrome
   c. Nathalie syndrome
   d. KBG syndrome
   e. Oculopalatocerebral dwarfism
   f. Aarskog syndrome
   g. GAPO syndrome
   h. Fetal alcohol syndrome

## PITUITARY DWARFISM

In the early literature, individuals of proportionate short stature were loosely grouped together under the designation 'midget' or as primordial or ateliotic dwarfs. The ateliotic group were conventionally subdivided into 'sexual' and 'asexual' forms. It was subsequently recognised that dysfunction of the pituitary gland was the usual basis of uncomplicated proportionate short stature, and the term 'pituitary dwarfism' gained favour. With the accumulation of genetic and endocrine data, it has become evident that there are several distinct forms of pituitary dwarfism.

Tom Thumb, who was a famous exhibitionist in the circus world, provides an excellent example of pituitary dwarfism. His wife Lavinia Warren was a midget. Her parents and siblings were normal, but as her sister Minnie was also small, it is likely that she had an autosomal recessive form of the disorder.

The term 'pituitary dwarfism' has been replaced by the more specific and less emotive titles 'growth hormone deficiency' or 'familial hypopituitarism'. For the sake of clarity, the old convention is retained in this section, in which a simplified account of these conditions is presented. Delineation is still incomplete, but the following categories are recognised:

### Pituitary dwarfism type I (Isolated growth hormone deficiency)

A high proportion of persons with isolated growth hormone deficiency have a non-genetic disorder (Rona & Tanner 1977) but there are also a number

**Fig. 18.1** Pituitary dwarfism: a Kalahari Bushman with pituitary dwarfism. (From Beighton P 1972 South African Medical Journal 46: 881.)

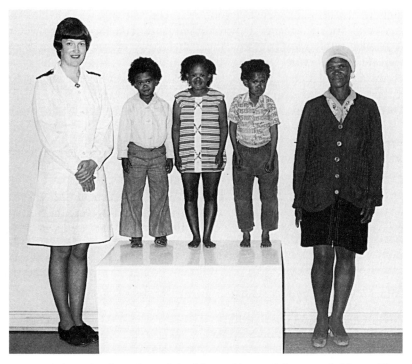

**Fig. 18.2** Pituitary dwarfism: three adult siblings with growth hormone deficiency. The parents are consanguineous and autosomal recessive inheritance is probable.

of uncommon Mendelian forms of this condition. Autosomal dominant and autosomal recessive entities have been delineated and these disorders have been reviewed in detail by Emery & Rimoin (1983).

Autosomal recessive inheritance of isolated growth hormone deficiency was established by Rimoin et al (1966). This disorder has been subcategorised on a molecular basis into type IA, in which there are deletions in the growth hormone gene cluster and type IB, in which deletions are not present in these genes (Phillips et al 1981). In addition it is likely that there is molecular heterogeneity in type IA (Braga et al 1986).

Autosomal dominant growth hormone deficiency was documented in a kindred with several affected generations by Merimee et al (1969). Other similar families have been reported by Sheikholislam & Stempfel (1972), Poskitt & Rayner (1974), van Gelderen & van der Hoog (1981) and Rogol et al (1985).

## Pituitary dwarfism type II (Laron⋆ type)

Laron et al (1966) studied 20 individuals in Israel who had the clinical features of isolated growth hormone deficiency, in the presence of elevated

⋆ Zvi Laron is Professor of Paediatric Endocrinology at the Sackler School of Medicine, Tel Aviv University, Israel.

serum growth hormone levels. These patients were all of Sephardic stock and the pedigrees were consistent with autosomal recessive inheritance. Pertzelan et al (1968) suggested that the growth hormone in this condition might be functionally incompetent although immunoreactive. Alternative explanations would be end-organ unresponsiveness to growth hormone or defective hepatic generation of somatomedin, the intermediate substance (Laron 1974). Jacobs et al (1976) postulated that growth hormone receptors were defective. Non-Jewish cases have been recorded, including Italian brothers (Saldanha & Toledo 1981).

The Pygmies of the Central African rain forests have been shown to have peripheral unresponsiveness to growth hormone (Rimoin et al, 1969). It is likely that this unique ethnic characteristic has a genetic basis, but the mode of inheritance has not been determined. The Bushmen of the Kalahari desert of Southern Africa are also of slight stature. However, their gracile habitus differs from the stocky build of the Pygmies and ethnically these two groups are not closely related. Joffe et al (1971) demonstrated that the Bushmen have normal growth hormone responses to oral glucose, and speculated that, as with the Pygmies, their small stature is the consequence of end-organ unresponsiveness. This issue is complicated by the fact that a Bushman with the classical clinical features of pituitary dwarfism has been encountered (Beighton 1972).

### Pituitary dwarfism type III (Panhypopituitarism)

Individuals with panhypopituitarism have the stigmata of thyroid, adrenal and gonadal dysfunction in addition to proportionate short stature. This condition is far more common that all other forms of pituitary dwarfism. The vast majority of patients are sporadic, with no genetic basis to their disorder. However, in a minority, inheritance is autosomal recessive and affected siblings have been reported in endogamous communities including the Hutterites of Canada (McArthur et al 1985) and the Veglia islanders of the Adriatic Sea (Fraser 1964).

An X-linked form of panhypopituitarism has been documented by Phelan et al (1971), Schimke et al (1971) and Zipf et al (1977).

### Other forms of pituitary dwarfism

Sibs with aplasia of the pituitary gland were described by Steiner & Boggs (1965) and Sadeghi-Hejad & Senior (1974). Two sisters with panhypopituitarism and a minute sella turcica were reported by Ferrier & Stone (1969) and a similar patient, with the additional feature of retinitis pigmentosa, was described by Ozer (1974). Although unproven, it is likely that inheritance is autosomal recessive in these conditions.

Rappaport et al (1976) encountered three children with proportionate short stature and a single central incisor. The authors obtained information concerning three similar children, demonstrated that growth hormone was deficient in two, and termed the condition 'monosuperocentroincisivodontic dwarfism'. No genetic background was apparent in this report.

### Genetic considerations

Genetic counselling is dependent upon accurate recognition of the dominant, recessive and sporadic forms of the disorder. Apart from shortness of stature, individuals with pituitary dwarfism enjoy good health, and it could be argued that genetic distinction is of little practical importance. However, the autobiographical comments of Tom Thumb's midget wife Lavinia Warren are of significance in this context. 'If nature endowed me with any superior personal attraction, it was comparatively small compensation for the inconvenience, trouble and annoyance imposed upon me by my diminutive stature'.

## CORNELIA DE LANGE* SYNDROME

Individuals with the Cornelia de Lange syndrome bear a striking resemblance to each other. The eyebrows are confluent, the anterior and posterior hairlines are low, the nostrils are anteverted and the lips are thin, with down-turned angles. Mental and physical development is retarded. Skeletal abnormalities range from absence of the extremities to minor anomalies such as proximally placed thumbs, fifth finger clinodactyly and limitation of elbow movements. The natural history of the condition has been documented in large series of affected persons by Hawley et al (1985) and Beck & Fenger (1985).

### Genetics

Since Cornelia de Lange's initial description of two affected children in 1933, there have been more than 350 case reports. In their monograph on the syndrome, Berg et al (1970) quoted the incidence in live births at between 1 in 30 000 and 1 in 50 000. The majority of cases have been sporadic, although sibs with the condition have been reported by Falek et al (1966) and Beratis et al (1971). Pashayan et al (1969) analysed genetic data from 54 families and concluded that the recurrence risk to sibs of patients was between 1.5% and 2.2%. Stevenson & Scott (1976) described male twins who were discordant for the syndrome. Garakushansky & Berthier (1976) reported discordant monozygotic female twins and commented that the situation could be explained by postzygotic mutation of a gene of large effect.

---

* Cornelia de Lange (1871–1950) was Professor of Paediatrics at the University of Amsterdam, Holland.

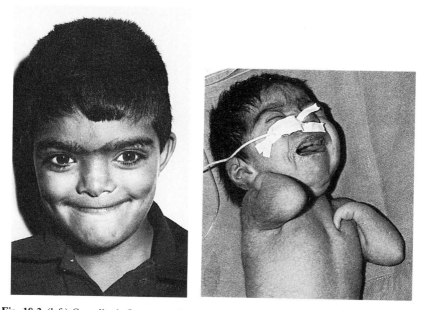

**Fig. 18.3** (left) Cornelia de Lange syndrome: confluent eyebrows, a low anterior hair line and thin lips.

**Fig. 18.4** (right) Cornelia de Lange syndrome: gross malformations of the extremities. (From Begeman G, Duggan R 1976 South African Medical Journal 50: 1475.)

In a report of 11 affected children, Begeman & Duggan (1976) mentioned that minor stigmata were present in the mother of an affected boy. A mother and daughter with the disorder were documented by Leavitt et al (1985) and familial aggregation was reported by Kumar et al (1985). These observations are consistent with autosomal dominant inheritance, with very variable phenotypic expression but this concept does not account for the overwhelming majority of sporadic cases.

There have been many reports of cytogenetic abnormalities in the Cornelia de Lange syndrome. For instance, Craig & Luzzato (1965) found chromosomal anomalies in 11 out of 38 patients. Conversely, the chromosomes were normal in 20 cases investigated by McArthur & Edwards (1967). Although they are inconsistent, the report of cytogenetic abnormalities are too numerous to be ignored. It is possible that there is an inherent predisposition to the development of non-specific chromosomal changes.

For all the above mentioned reasons, the genetic basis of the syndrome remains very uncertain. A statistical analysis of the clinical manifestations of a group of patients was suggestive of heterogeneity (Preus & Fraser 1976) and this may eventually prove to be the case.

## SECKEL* SYNDROME

Following a report of a 'microcephalic midget' (Mann & Russell 1959), Seckel (1960) published a review entitled 'Birdheaded dwarfs'. Subsequently, the eponym 'Seckel' was employed as an alternative designation by Harper et al (1967). The main features of the syndrome are low birth weight, short stature, microcephaly and mental deficiency. The thin pointed face and prominent nose produce an avian appearance. Skeletal abnormalities include scoliosis, absence of ribs, maldevelopment of the radius and fibula, dislocation of the hips, pes planus and talipes equinovarus. Subtle but distinctive radiological findings in the bones of the hands and wrists have been documented by Poznanski et al (1983).

The features of the Seckel syndrome are clearly recognisable in Alfred Chalon's portrait of Carolina Crachani, the Sicilian dwarf, which hangs in the Royal College of Surgeons, London. Carolina died of pulmonary tuberculosis at the age of 8, and her skeleton, which measures less than 55 cm in height, now reposes in the Hunterian Museum of the College. Her condition was attributed to 'maternal impression', as her mother had been bitten on the hand by a monkey during early pregnancy. However, there have been no other similar reports, and it can be confidently assumed that monkey bites are not an important aetiological factor in the Seckel syndrome!

*Genetics*

About 30 cases have now been described. Multiple affected sibs with normal parents have been reported by Black (1961), Harper et al (1967) and McKusick et al (1967). Parental consanguinity has been noted and there is little doubt that the Seckel syndrome is inherited as an autosomal recessive. Antenatal diagnosis by recognition of the characteristic craniofacial features has been documented by Elejalde & Elejalde (1984).

Fitch et al (1970) described a patient with features of the Seckel syndrome, together with premature senility. McKusick (1975) commented that he knew of a similar brother and sister, suggested that inheritance was autosomal recessive and proposed that, in deference to the origins of the first report, this condition should be termed 'Bird-headed dwarfism, Montreal type'.

Majewski et al (1982 a,b,c) gave an in depth review of the Seckel syndrome and defined three subtypes of a similar but separate entity termed 'osteodysplastic primordial dwarfism'. Thompson & Pembrey (1985) drew attention to the importance of differentiating these conditions from the Seckel syndrome and Winter et al (1985) discussed the classification of this heterogeneous group of disorders.

* Helmut Seckel (1900–1961) was Professor of Pediatrics at the University of Chicago School of Medicine.

**Fig. 18.5** (top) Russell-Silver syndrome: the face is triangular and the chin is pointed. Clinodactyly is present in the 5th fingers.

**Fig. 18.6** (right) Russell-Silver syndrome: in this young woman, short stature and disproportion are unusually severe.

## RUSSELL-SILVER* SYNDROME

There has been controversy as to whether the initial case reports of Silver (1953) and Russell (1954) pertained to the same disorder or to separate entities. However, the designation 'Russell-Silver syndrome' is now applied to a dwarfing condition in which lateral asymmetry and low birth weight are associated with a triangular face, clinodactyly of the fifth finger and café-au-lait macules on the trunk. Abnormalities of urinary gonadotrophin excretion and sexual development, which were emphasised by Silver (1964), are inconsistent features. Hypoglycaemia may be a problem during infancy, but general health is otherwise good. Absence of digits and syndactylism has been documented in an isolated case (Keppen & Rennert 1983). The long-term prognosis for stature is very variable but a significant proportion of affected children eventually reach normal height (Angehrn et al 1979, Saal et al 1985).

The Russell-Silver syndrome is probably relatively common and reports of about 200 individuals with the condition have appeared in the literature. Mild cases may remain undiagnosed. In a review of the pattern of growth of 39 affected children, Tanner et al (1975) commented that contrary to earlier reports, no abnormalities of sexual development had occurred.

* Alexander Russell became Professor of Paediatrics at the Hebrew University of Jerusalem in 1966. He is currently the President of the International College of Paediatrics. Henry Silver is Professor of Paediatrics at the University of Colorado, USA.

**Fig. 18.7** Russell-Silver syndrome: an adult female with leg length asymmetry, stunted stature and the characteristic triangular facies.

*Genetics*

The majority of reported cases have been sporadic, but there have been examples of affected monozygous twins (Rimoin 1969) and siblings with consanguineous parents (Fuleihan et al 1971). Familial aggregation has been documented by Gareis et al (1971) and fully reviewed by Escobar et al (1978). A Canadian family of Dutch descent with possible X-linked inheritance of a form of the syndrome was reported by Partington (1986). There is no increased frequency of parental consanguinity, which might indicate autosomal recessive inheritance; equally, neither paternal age nor birth rank point to new mutation of a dominant gene.

The chromosomes have been normal in the majority of patients in whom cytogenetic studies have been undertaken. There has been a single report of trisomy 18 mosaicism (Chauvel et al 1975) and two instances of a 47XXY karyotype (Bianchi et al 1983). A ring 15 chromosome was detected in an affected male infant by Wilson et al (1985); these authors commented that a malformation syndrome with a characteristic facies which resembles the

Russell-Silver Syndrome is related to a deletion in the long arm of chromosome 15.

The pathogenesis of the Russell-Silver syndrome remains uncertain and the issue of heterogeneity is unresolved. For the purpose of genetic counselling it is reasonable to assume that the risks of recurrence are low.

## SENILITY SYNDROMES

### Progeria

Progeria was delineated at the turn of the century by Gilford (1904). However, the condition had been clearly described more than 150 years previously in the obituary of Hopkin Hopkins, a Welsh fairground exhibitionist. Premature aging is the most notable feature. Patients have a senile appearance in childhood and usually die of arteriosclerotic complications, such as myocardial infarction or cerebrovascular accidents, by the end of the second decade. Skeletal changes include osteoporosis, coxa valga and ovoid vertebral bodies. The manifestations of progeria have been extensively reviewed by De Busk (1972).

About 60 patients have now been described. The majority have been sporadic, and the mode of inheritance is uncertain. Progeric Egyptian sisters with consanguineous parents were reported by Gabr et al (1960) and sets of sibs have been described by Rava (1967) and Franklyn (1976). A pair of male monozygotic twins with the condition have been studied by Viegas et al (1974). De Busk (1972) found evidence of parental consanguinity in only 3 out of 19 previously reported cases. A paternal age effect was documented by DeBusk (1972) and Jones et al (1975); Brown (1979) concluded that new dominant mutation was likely. Conversely, in a study of abnormalities of heat labile enzymes, the parents of an affected girl were shown to have intermediate values and autosomal recessive inheritance seemed likely. As affected individuals do not reproduce, evidence to support the former concept will not be forthcoming. It is possible that progeria is heterogeneous.

### Mandibuloacral dysplasia

Atypical progeroid syndromes have been encountered. For instance, Welsh (1975) described a kindred in which two males and two females had an unusual type of premature senility and speculated that this condition might be a new syndrome. The main characteristics were progressive osteolysis of the phalanges, mandible and clavicles with joint rigidity and dystrophic changes in the hair, nails and skin. This disorder, which is evidently inherited as an autosomal recessive trait, had previously been documented and designated 'mandibuloacral dysplasia' by Young et al (1971). A further report of affected Italian brothers was in accordance with this mode of inheritance (Pallotta & Morgese 1984).

## Cockayne* syndrome

The Cockayne syndrome or 'cachectic dwarfism' resembles progeria in that defective growth and progressive loss of adipose tissue produce a senile appearance in late childhood. Mental deficiency, photosensitivity, retinal degeneration and optic atrophy are associated features.

Hydrocephalus and dementia supervene (Brumback et al 1978) and hypertension frequently develops due to renal involvement (Higginbottom et al 1979). Autopsy in an affected person revealed primary degenerative changes in the brain (Soffer et al 1979). Reduction in serum levels of thymic hormone was documented by Bensman et al (1982) but the significance of this observation is uncertain.

About 30 cases have been described, including sibs reported by Cockayne (1946) and Paddison et al (1963) and a severely affected boy born to consanguineous parents (Pfeiffer & Bachmann 1973). Inheritance is autosomal recessive. Prenatal diagnosis by examination of cultured amniotic fluid cells has been accomplished by Lehmann et al (1985).

## Werner** syndrome

The stigmata of the Werner syndrome include premature senility, shortened stature, cataracts and diabetes mellitus. More than 200 cases have been reported, and the accumulated pedigree data are indicative of autosomal recessive inheritance (Epstein et al 1966). The condition seems to be especially common in Japan where 80 affected persons in 42 families have been documented (Goto et al 1981).

## BLOOM SYNDROME

Bloom (1966) described the association of low birth weight, dwarfism, malar hypoplasia and photosensitive telangectasia of the face. Other abnormalities in the Bloom syndrome include dental defects, digital deformities and talipes equinovarus. The immunoglobulins are deficient and leukaemia and other neoplasms are a common complication.

### Genetics

About 50 case reports have now been published and autosomal recessive inheritance is well established. The majority of patients have been of Ashkenazi Jewish stock. As with many genetic disorders which have a high prevalence in the Jewish people, there is a distinctive geographical distri-

---

* Edward Cockayne (1880–1956) was a senior dermatologist at the Hospital for Sick Children, Great Ormond Street, London.

** Otto Werner (1879–1936) was a general practitioner in Eddelak, West Germany.

bution of the abnormal gene. German (1969) has shown that the majority of kindreds with the Bloom syndrome had their origins in the Ukraine region of North-Western Europe.

Excessive chromosome breakage and rearrangement is a consistent feature (Sawitsky et al 1966). Borgeois et al (1975) undertook trypsin banding studies on chromosomes from an affected Jordanian girl and demonstrated that the cytogenic abnormalities were non-random.

Goodman (1975) pointed out that consanguinity was not increased amongst Jewish families with the Bloom syndrome, although several consanguineous non-Jewish kindreds had been reported. The sex ratio is equal in the Jewish cases, but there is a male preponderance in non-Jews. These observations led Goodman (1975) to speculate that the condition might be heterogeneous.

## OTHER DISORDERS OF GROWTH

There are many syndromes in which proportionate short stature is associated with other abnormalities. A full account of these conditions can be found in the classical monograph 'Recognizable patterns of human malformation (Smith 1982). A few of the most interesting and important of these disorders are summarised below:

### Dubowitz* syndrome

Dubowitz (1965) reported two sibs and two unrelated patients with low birth weight and a distinctive facies. Grosse et al (1971) described two more cases and termed the condition 'Dubowitz syndrome'. The stigmata bear some resemblance to the Russell-Silver syndrome but microcephaly, mild mental retardation, and ptosis are additional features. In a review of the syndrome Opitz et al (1973) recognised kindreds with parental consanguinity and multiple affected sibs and concluded that inheritance was autosomal recessive. In a further review Majewski et al (1975) commented that a total of 11 cases had now been described, and drew attention to the similarity of their stigmata to those seen in the fetal alcohol syndrome. Wilroy et al (1978) and Parrish & Wilroy (1980) further refined the phenotype.

### 3M syndrome

Miller et al (1975) reported two sets of sibs with low birth weight dwarfism, a 'hatchet-shaped' cranial configuration, deformity of the thorax and anomalies of the mouth and teeth. The authors termed the condition the '3M

* Victor Dubowitz is Professor of Paediatrics at the Royal Postgraduate Medical School, London.

syndrome' and suggested that, as the unaffected parents in one of the kindreds were consanguineous, the disorder was probably inherited as an autosomal recessive. Spranger et al (1976) compared and contrasted the manifestations of the disorder with those of the Russell-Silver syndrome and concluded that these conditions were separate entities.

### Nathalie syndrome

Using the Christian name of the proband in the title of their paper, Cremers et al (1975) described three sisters and a brother with retarded growth, deafness, cataracts, sexual infantilism and electrocardiographic abnormalities. Mild but widespread skeletal dysplasia was present. Two of the girls developed hip joint problems which resembled Perthes disease, while Scheuermann's disease was diagnosed in the boy. Although the family study did not provide conclusive evidence, it is likely that the Nathalie syndrome was inherited as an autosomal recessive.

### KBG syndrome

The KBG syndrome, named from the initials of the probands, consists of short stature, macrodontia, mental retardation, abnormal vertebrae, short femoral necks and a characteristic facies. Herrmann et al (1975) described the features of seven patients from three kindreds, and concluded that inheritance was autosomal dominant. A large family with a variant form of the KBG syndrome was documented by Parloir et al (1977) and a mother and two daughters with this condition were reported by Fryns & Haspeslagh (1984).

### Oculopalatocerebral dwarfism

Three siblings with microcephaly, mental retardation, persistent hypertrophy of the primary vitreous, cleft palate and dwarfism were documented by Frydman et al (1985). The unaffected parents were consanguineous, and autosomal recessive inheritance is likely.

### Aarskog* syndrome

The Aarskog syndrome or facio-genital dysplasia comprises stunted stature with a characteristic facies and genital anomalies (Aarskog 1970). The disorder is generally regarded as being X-linked (Escobar & Weaver 1978, Berry et al 1980) and female gene carriers often have minor stigmata. Sex-influenced autosomal dominant inheritance has also been proposed (Sugarman et al 1973) and this concept is supported by the recognition of

---

* Dagfinn Aarskog is Professor of Paediatrics at the University of Bergen, Norway.

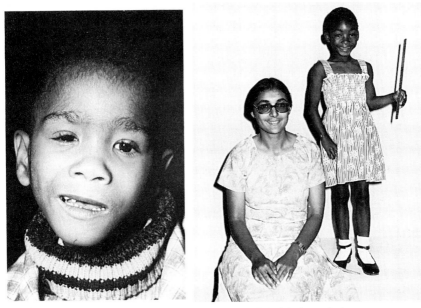

**Fig. 18.8** (left) Fetal Alcohol syndrome: an affected child with the typical blepharophimosis, internal strabismus and a smooth upper lip. His mother was a chronic alcoholic. (From Jaffer et al 1981 Journal of Bone and Joint Surgery 63B: 569–571.)

**Fig. 18.9** (right) Fetal alcohol syndrome: a child with the characteristic facies demonstrating restriction of extension and rotation at the elbow joint due to radio-ulnar synostosis. (From Jaffer et al 1981 Journal of Bone and Joint Surgery 63B: 569–571.)

male to male transmission in a large Dutch family (van de Vooren et al 1983). There is considerable interfamilial variability and it is possible that the Aarskog syndrome is heterogeneous.

## GAPO syndrome

The acronymous GAPO syndrome comprises growth retardation, alopecia, pseudo-anodontia and optic atrophy. In a review of the disorder, Tipton & Gorlin (1984) commented that all five reported patients had a strikingly similar appearance. Parental consanguinity and affected siblings are indicative of autosomal recessive inheritance.

## Fetal alcohol syndrome

It is becoming increasingly evident that maternal alcoholism poses a definite threat to the fetus. The main features of the fetal alcohol syndrome are low birth weight, mental retardation, microcephaly, short palpebral fissures and maxillary hypoplasia (Jones & Smith 1973). Radio-ulnar synostosis is a variable concomitant (Jaffer et al 1981). As the manifestations are not clear-

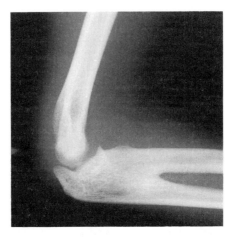

**Fig. 18.10** Fetal alcohol syndrome: radio-ulnar synostosis is sometimes present.

cut, firm diagnosis is not always easy. Nevertheless, it is likely that the condition is fairly common (Jones & Smith 1975). Indeed, in Seattle and Northern France, the incidence exceeds one in 1000 newborns (Clarren & Smith 1978), while in Gothenburg Sweden, the figure may be as high as 1 in 300 (Olegaard et al 1979). The majority of mothers of children with the syndrome have sustained a high alcohol intake throughout pregnancy. However, it is a sobering thought that a single bibulous episode at a crucial stage of fetal development might also be embryopathic!

REFERENCES

**Pituitary dwarfism**
*Pituitary dwarfism type I*
Braga S, Phillips III J A, Joss E, Schwarz H, Zuppinger K 1986 Familial growth hormone deficiency resulting from a 7.6 kb deletion within the growth hormone gene cluster. American Journal of Medical Genetics 25: 443–452
Emery A E H, Rimoin D L 1983 Hereditary forms of growth hormone deficiency. Principles and Practice of Medical Genetics Vol II. Churchill Livingstone, New York, p 1135–1144
Merimee T J, Hall J G, Rimoin D L, McKusick V A 1969 A metabolic and hormonal basis for classifying ateliotic dwarfs. Lancet i: 963
Phillips J A III, Hjelle B L, Seeburg P H, Zachmann M 1981 Molecular basis for familial isolated growth hormone deficiency. Proceedings of the National Academy of Sciences of the United States of America 78: 6372–6375
Poskitt E M, Rayner P H W 1974 Isolated growth hormone deficiency: two families with autosomal dominant inheritance. Archives of Disease in Childhood 49: 55
Rimoin D L, Merimee T J, McKusick V A 1966 Growth hormone deficiency in man: an isolated, recessively inherited defect. Science 152: 1635
Rogol A D, Blizzard R M, Foley T P J R 1985 Growth hormone releasing hormone and growth hormone: Genetic studies in familial growth hormone deficiency. Pediatric Research 19/5: 489–492
Rona R J, Tanner J M 1977 Aetiology of idiopathic growth hormone deficiency in England and Wales. Archives of Disease in Childhood 52: 197–208
Sheikholislam B M, Stempfel R S Jr 1972 Hereditary isolated somatotropin deficiency: effects of human growth hormone administration. Pediatrics 49: 362

van Gelderen H, van der Hoog C E 1981 Familial isolated growth hormone deficiency. Clinical Genetics 20: 173–175

*Pituitary dwarfism type II (Laron type)*
Beighton P 1972 Pituitary dwarfism in a Kalahari Bushman. South African Medical Journal 46: 881
Jacobs L S, Sneid D S, Garland J T, Laron Z, Doughaday W H 1976 Receptor-active growth hormone in Laron dwarfism. Journal of Clinical Endocrinology 42: 403–406
Joffe B I, Jackson W P U, Thomas M E, Toyer M G, Keller P, Pimstone B L, Zamit R 1971 Metabolic responses to oral glucose in the Kalahari Bushmen. British Medical Journal 4: 206
Laron Z 1974 The syndrome of familial dwarfism and high plasma immunoreactive human growth hormone. In: Bergsma D (ed) Clinical Delineation of Birth Defects, Vol XVI, Williams and Wilkins, Baltimore, p 231
Laron Z, Pertzelan A, Mannheimer S 1966 Genetic pituitary dwarfism with high serum concentration of growth hormone. A new inborn error of metabolism? Israel Journal of Medical Science 2: 152
Pertzelan A, Adam A, Laron Z 1968 Genetic aspects of pituitary dwarfism due to absence or biological inactivity of growth hormone. Israel Journal of Medical Science 4: 895
Rimoin D L, Merimee T J, Rabinowitz D, Cavalli Sforza L L, McKusick V A 1969 Peripheral subresponsiveness to human growth hormone in the African pygmies. New England Journal of Medicine 281: 1383
Saldanha P H, Toledo SPA 1981 Familial dwarfism with high IR-GH: report of two affected sibs with genetic and epidemiologic considerations. Human Genetics 59: 367–372

*Pituitary dwarfism type III (Panhypopituitarism)*
Fraser G R 1964 Studies in isolates. Journal de Génétique Humaine 13: 32–46
McArthur R G, Morgan K, Phillips J A III, Bala M, Klassen J 1985 The natural history of familial hypopituitarism. American Journal of Medical Genetics 22: 553–566
Phelan P D, Connelly J, Martin F I R, Wettenhall H N B 1971 X-linked recessive hypopituitarism. Birth Defects: Original Article Series VII(6): 24–27
Schimke R N, Spaulding J J, Hollowell J G 1971 X-linked congenital panhypopituitarism. Birth Defects: Original Article Series VII(6): 21–23
Zipf W B, Kelch R P, Bacon G E 1977 Variable X-linked recessive hypopituitarism with evidence of gonadotropin deficiency in two pre-pubertal males. Clinical Genetics 11: 249–254

*Other forms of pituitary dwarfism*
Ferrier P E, Stone E F Jr 1969 Familial pituitary dwarfism associated with an abnormal sella turcica. Pediatrics 43: 858
Ozer F L 1974 Pituitary dwarfism with retinitis pigmentosa and small sella turcica. In: Bergsma D (ed) Clinical Delineation of Birth Defects Vol XVI. Williams and Wilkins, Baltimore, p 354
Rappaport E B,. Ulstrom R, Gorlin R J 1976 Monosupercentroincisivodontic dwarfism. Birth Defects: Original Article Series 12/5: 243
Sadeghi-Hejad A, Senior B 1974 A familial syndrome of isolated aplasia of the anterior pituitary. Diagnostic studies and treatment in the neonatal period. Journal of Pediatrics 84: 79
Steiner M M, Boggs J D 1965 Absence of pituitary gland, hypothyroidism, hypoadrenalism and hypogonadism in a 17-year-old dwarf. Journal of Clinical Endocrinology 25: 1591

**Cornelia de Lange syndrome**
Beck B, Fenger K 1985 Mortality, pathological findings and causes of death in the de Lange Syndrome. Acta Paediatrica Scandinavica 74: 765–769
Begeman G, Duggan R 1976 The Cornelia de Lange syndrome: A study of nine affected individuals. South African Medical Journal 50: 1475
Beratis N G, Hsu L Y, Hirschhorn K 1971 Familial de Lange syndrome. Report of three cases in a sibship. Clinical Genetics 2: 170
Berg J M, McCreary B D, Ridler M A C, Smith G F 1970 The de Lange syndrome. Institute for Research into Mental Retardation, Monograph No. 2 Pergamon, Oxford

Craig A P, Luzzatto L 1965 Translocation in de Lange's syndrome. Lancet 2: 445
Falek A, Schmidt R, Jervis G A 1966 Familial de Lange syndrome with chromosome abnormalities. Pediatrics 37: 92
Garakushansky G, Berthier C 1976 The de Lange syndrome in one of twins. Journal of Medical Genetics 13/5: 404
Hawley P P, Jackson L G, Kurnit D M 1985 Sixty-four patients with Brachmann-de Lange Syndrome: A survey. American Journal of Medical Genetics 20: 453–459
Kumar D, Blank C E, Griffiths B L 1985 Cornelia de Lange syndrome in several members of the same family. Journal of Medical Genetics 22: 296–300
Leavitt A, Dinno N, Davis C 1985 Cornelia de Lange syndrome in a mother and daughter. Clinical Genetics 28: 157–161
McArthur R G, Edwards J H 1967 de Lange syndrome: report of 20 cases. Canadian Medical Association Journal 96: 1185
Pashayan H, Whelan D, Gittman S, Fraser F C 1969 Variability of the de Lange syndrome: report of 3 cases and genetic analysis of 54 families. Journal of Pediatrics 75: 853
Preus M, Fraser F C 1976 A methodology for establishing a diagnostic index for syndromes of unknown etiology. Clinical Genetics 10/5: 249
Stevenson R E, Scott C I 1976 Discordance for Cornelia de Lange syndrome in twins. Journal of Medical Genetics 13/5: 402

**Seckel syndrome**

Black J 1961 Low birth weight dwarfism. Archives of Disease in Childhood 36: 633
Elejalde M M, Elejalde B R 1984 Visualization of the fetal face by ultrasound. Journal of Craniofacial Genetics and Developmental Biology 4/4: 251–257
Fitch N, Pinsky L, Lachance R C 1970 A form of bird-headed dwarfism with features of premature senility. American Journal of Diseases of Children 120: 260
Harper R G, Orti E, Baker R K 1967 Bird-headed dwarfs (Seckel's syndrome). A familial pattern of developmental, dental, skeletal, genital, and central nervous system anomalies. Journal of Pediatrics 70: 799
Mann T P, Russell A 1959 Study of a microcephalic midget of extreme type. Proceedings of the Royal Society of Medicine 52: 1024
McKusick V A 1975 Bird-headed dwarfism, Montreal type. Mendelian Inheritance in Man 4th Edn, Johns Hopkins University Press, Baltimore, p 366
McKusick V A, Mahloudji M, Abbott M H, Lindenburg R, Kepan D 1967 Seckel's bird-headed dwarfism. New England Journal of Medicine 277: 279
Majewski F, Goecke T 1982a Studies of microcephalic primordial dwarfism I: Approach to a delineation of the Seckel syndrome. American Journal of Medical Genetics 12: 7–21
Majewski F, Ranke M, Schinzel A 1982b Studies of microcephalic primordial dwarfism II: Osteodysplastic Primordial Dwarfism Type II. American Journal of Medical Genetics 12: 13–35
Majewski F, Stoeckenius M, Kemperdick H 1982c Studies of microcephalic primordial dwarfism III: An intrauterine dwarf with platyspondyly and anomalies of pelvis and clavicles — osteodysplastic primordial dwarfism Type III. American Journal of Medical Genetics 12: 37–42
Poznanski A K, Iannaccone G, Pasquino A M, Boscherini B 1983 Radiological findings in the hand in Seckel syndrome (bird-headed dwarfism). Pediatric Radiology 13/1: 19–24
Seckel H P G 1960 Bird-headed Dwarfs. Charles C Thomas, Springfield, p 241
Thompson E, Pembrey M 1985 Seckel syndrome: An overdiagnosed syndrome. Journal of Medical Genetics 22/3: 192–201
Winter R M, Wigglesworth J, Harding B N 1985 Osteodysplastic primordial dwarfism: Report of a further patient with manifestations similar to those seen in patients with types I and III. American Journal of Medical Genetics 21: 569–574

**Russell-Silver syndrome**

Angehrn V, Zachmann M, Prader A 1979 Silver-Russell syndrome. Observations in 20 patients. Helvetica Paediatrica Acta 34: 297–308

Bianchi E, Arico M, Severi F, Pasquali F 1983 Russell-Silver syndrome and XXY syndrome. Pediatrics 71: 220–224

Chauvel P J, Moore C M, Haslam R H A 1975 Trisomy-18 mosaicism with features of Russell-Silver syndrome. Developmental Medicine and Child Neurology 17: 220–224

Escobar V, Gleiser S, Weaver D D 1978 Phenotypic and genetic analysis of the Silver-Russell syndrome. Clinical Genetics 13: 278–288

Fuleihan D S, Vazken B A, Der Kaloustian M, Najjar S S 1971 The Russell-Silver syndrome: report of three siblings. Journal of Pediatrics 78: 654

Gareis F J, Smith D W, Summitt R L 1971 The Russell-Silver syndrome without asymmetry. Journal of Pediatrics 79: 775–781

Keppen L D, Rennert O M 1983 Silver-Russell syndrome with absence of digits and syndactylism of the fingers. Clinical Genetics 24: 453–455

Partington M W 1986 X-linked short stature with skin pigmentation: evidence for heterogeneity of the Russell-Silver syndrome. Clinical Genetics 29: 151–156

Rimoin D L 1969 The Silver syndrome in twins. In: Bergsma D (ed) Clinical Delineation of Birth Defects, Vol II, National Foundation, New York, p 183

Russell A 1954 A syndrome of 'intra-uterine' dwarfism recognisable at birth with craniofacial dysostosis, disproportionately short arms and other anomalies. Proceedings of the Royal Society of Medicine 47: 1040

Saal H M, Pagon R A, Pepin M G 1985 Re-evaluation of Russell-Silver syndrome. Journal of Pediatrics 107/5: 733–737

Silver H K 1953 Syndrome of congenital hemihypertrophy, shortness of stature and elevated urinary gonadotrophins. Pediatrics 12: 368

Silver H K 1964 Asymmetry, short stature, and variations in sexual development. A syndrome of congenital malformations. American Journal of Diseases of Children 107: 495

Tanner J M, Lejarraga H, Cameron N 1975 The natural history of the Silver-Russell syndrome: a longitudinal study of 39 cases. Pediatric Research (Baltimore) 9/8: 611

Wilson G N, Sauder S E, Bush M, Beitins I Z 1985 Phenotypic delineation of ring chromosome 15 and Russell-Silver syndromes. Journal of Medical Genetics 22/3: 233–236

## Senility syndromes
### Progeria

Brown W T 1979 Human mutations affecting ageing — a review. Mechanisms of Ageing and Development 9: 325–336

De Busk F L 1972 The Hutchinson-Gilford progeria syndrome. Journal of Pediatrics 80: 697

Franklyn P P 1976 Progeria in siblings. Clinical Radiology 27/3: 327

Gabr M, Hashem N, Hashem M, Fahmi A, Safouh M 1960 Progeria, a pathologic study. Journal of Pediatrics 57: 70

Gilford H 1904 Progeria: a form of senilisim. Practitioner 73: 188

Goldstein S, Moerman E J 1978 Heat-labile enzymes in circulating erythrocytes of a progeria family. American Journal of Human Genetics 30: 167–173

Jones K L, Smith D W, Harvey M A S, Hall B D, Quan L 1975 Older paternal age and fresh gene mutation: data on additional disorders. Journal of Pediatrics 86: 84–88

Rava G 1967 Su un nucleo familiare di progeria. Minerva Medica 58: 1502

Viegas J, Souza P L R, Salzano F M 1974 Progeria in twins. Journal of Medical Genetics 11/4: 384

### Mandibuloacral dysplasia

Pallotta R, Morgese G 1984 Mandibuloacral dysplasia: a rare progeroid syndrome. Clinical Genetics 26: 133–138

Welsh O 1975 Study of a family with a new progeroid syndrome. Birth Defects: Original Article Series 11/5: 25

Young L W, Radebaugh J F, Rubin P, Sensenbrenner J A, Fiorelli G 1971 New syndrome manifested by mandibular hypoplasia, acroosteolysis, stiff joints and cutaneous atrophy (mandibuloacral dysplasia) in two unrelated boys. Birth Defects Original Article Series VII(7): 291–297

*Cockayne syndrome*

Bensman A, Dardenne M, Bach J F, De Mouillac J V, Lasfargues G 1982 Decrease of thymic hormone serum level in Cockayne syndrome. Pediatric Research 16: 92–94

Brumback R A, Yoder F W, Andrews A D, Peck G L, Robbins J H 1978 Normal pressure hydrocephalus: recognition and relationship to neurological abnormalities in Cockayne's syndrome. Archives of Neurology 35: 337–345

Cockayne E A 1946 Dwarfism with retinal atrophy and deafness. Archives of Disease in Childhood 21: 52

Higginbottom M C, Griswold W R, Jones K L, Vasquez M D, Mendoza S A, Wilson C B 1979 The Cockayne syndrome: an evaluation of hypertension and studies of renal pathology. Pediatrics 64: 929–934

Lehmann A R, Francis A J, Giannelli F 1985 Prenatal diagnosis of Cockayne's syndrome. Lancet 1/8427: 486–488

Paddison R M, Moossy J, Derbes V J, Kloepfer H W 1963 Cockayne's syndrome. A report of five new cases with biochemical, chromosomal, dermatologic, genetic and neuropathologic observations. Dermatologica Tropica et Ecologica Geographica 2: 195

Pfeiffer R A, Bachmann K D 1973 An atypical case of Cockayne's syndrome. Clinical Genetics 4: 28

Soffer D, Grotsky H W, Rapin I, Suzuki K 1979 Cockayne syndrome: unusual neuropathological findings and review of the literature. Annals of Neurology 6(4): 340–348


*Werner syndrome*

Epstein C J, Martin G M, Schultz A L, Motulsky A G 1966 Werner's syndrome. Medicine 45: 177

Goto M, Tanimoto K, Horiuchi Y, Sasazuki T 1981 Family analysis of Werner's syndrome: a survey of 42 Japanese families with a review of the literature. Clinical Genetics 19: 8–15


**Bloom syndrome**

Bloom D 1966 The syndrome of congenital telangiectatic erythema and stunted growth. Journal of Pediatrics 68: 103

Borgeois C A, Claverley M H, Forman L, Polani P E 1975 Bloom's syndrome: a probable new case with cytogenetic findings. Journal of Medical Genetics 12/4: 423

German J 1969 Bloom's syndrome. Genetical and clinical observations in the first 27 patients. American Journal of Human Genetics 21: 196

Goodman R M 1975 Genetic disorders among the Jewish people. Bloom's syndrome. In: Emery A (ed) Modern Trends in Human Genetics 2nd Edn. Butterworth, London, p 285

Sawitsky A, Bloom D, German J 1966 Chromosomal breakage and acute leukaemia in congenital telangiectatic erythema and stunted growth. Annals of Internal Medicine 65: 487


**Other disorders of growth**

Smith D W 1982 Recognisable Patterns of Human Malformation. 3rd ed. Saunders, Philadelphia


*Dubowitz syndrome*

Dubowitz V 1965 Familial low birth weight dwarfism with an unusual facies and a skin eruption. Journal of Medical Genetics 2: 12

Grosse R, Gorlin J, Opitz J M 1971 The Dubowitz syndrome. Zeitschrift für Kinderheilkunde 110: 175

Majewski F, Michaelis R, Moosmann K, Bierich J R 1975 A rare type of low birth weight dwarfism: the Dubowitz syndrome. Zeitschrift für Kinderheilkunde 120/4: 283

Opitz J M, Pfeiffer R A, Herrmann J P R, Kushnick T 1973 Studies of the malformation syndromes of man. XXIV B: the Dubowitz syndrome. Further observations. Zeitschrift für Kinderheilkunde 116: 1

Parrish J M, Wilroy R S Jr 1980 The Dubowitz syndrome: the psychological status of 10 cases at follow-up. American Journal of Medical Genetics 6: 3–8

Wilroy R S Jr, Tipton R E, Summitt R L 1978 The Dubowitz syndrome. American Journal of Medical Genetics 2: 275–284

*3M Syndrome*
Miller J D, McKusick V A, Malvaux P et al 1975 The 3M syndrome: a heritable low birth weight dwarfism. Birth Defects: Original Article Series 11/5: 39
Spranger J, Opitz J M, Nourmand A 1976 A new familial intrauterine growth retardation syndrome the '3-M syndrome'. European Journal of Pediatrics 1;123(2): 115–24

*Nathalie syndrome*
Cremers C W R J, ter Haar B G A, van Rens T J G 1975 The Nathalie syndrome. A new hereditary syndrome. Clinical Genetics 8: 330

*KBG syndrome*
Fryns J P, Haspeslagh M 1984 Mental retardation, short stature, minor skeletal anomalies, craniofacial dysmorphism and macrodontia in two sisters and their mother. Clinical Genetics 26: 69–72
Herrmann J, Pallister P D, Tiddy W, Opitz J M 1975 The KBG syndrome: A syndrome of short stature, characteristic facies, mental retardation, macrodontia and skeletal anomalies. Birth Defects: Original Article Series 11/5: 7
Parloir C, Fryns J P, Deroover J, Lebas E, Goffaux P, Van den Berghe H 1977 Short stature, craniofacial dysmorphism and dento-skeletal abnormalities in a large kindred: a variant of KBG syndrome or a new mental retardation syndrome. Clinical Genetics 12: 263–266

*Oculopalatocerebral dwarfism*
Frydman M, Kauschansky A, Leshem I, Savir H 1985 Oculopalatocerebral dwarfism: A new syndrome. Clinical Genetics 27/4: 414–419

*Aarskog syndrome*
Aarskog D 1970 A familial syndrome of short stature associated with facial dysplasia and genital anomalies. Journal of Pediatrics 77: 856–861
Berry C, Cree J, Mann T 1980 Aarskog's syndrome. Archives of Disease in Childhood 55: 706–710
Escobar V, Weaver D D 1978 Aarskog syndrome: new findings and genetic analysis. Journal of American Medical Association 240: 2638–2644
Sugarmann G I, Rimoin D L, Lachman R S 1973 The facial-digital-genital (Aarskog) syndrome. American Journal of Diseases of Children 126: 248–252
Van de Vooren M J, Niermeijer M F, Hoogeboom A J M 1983 The Aarskog syndrome in a large family, suggestive for autosomal dominant inheritance. Clinical Genetics 24: 439–445

*GAPO syndrome*
Tipton R E, Gorlin R J 1984 Growth retardation, alopecia, pseudo-anodontia, and optic atrophy — The GAPO Syndrome: report of a patient and review of the literature. American Journal of Medical Genetics 19: 209–216

*Fetal alcohol syndrome*
Clarren S K, Smith D W 1979 The fetal alcohol syndrome; a review of the world literature. New England Journal of Medicine 298: 1063
Jaffer Z, Nelson M, Beighton P 1981 Bone fusion in the foetal alcohol syndrome. Journal of Bone and Joint Surgery 63B: 569–571
Jones K L, Smith D W 1973 Recognition of the fetal alcohol syndrome in early infancy. Lancet ii: 999
Jones K L, Smith D W 1975 The fetal alcohol syndrome. Teratology 12: 1
Olegaard R et al 1979 Effects on the child of alcohol abuse during pregnancy. Acta Paediatrica Scandinavica 275: 112

# 19

# Overgrowth syndromes

Conditions in which height is excessive are much less common than those in which growth is impaired. Indeed, there are many forms of dwarfism but few disorders which are characterised by gigantism. Generalised overgrowth disorders and obesity syndromes are reviewed in this chapter. Localised and digital overgrowth syndromes are also outlined.

1. Generalised overgrowth syndromes
   a. Weaver syndrome
   b. Sotos syndrome
   c. Beckwith-Wiedemann syndrome
   d. Miscellaneous infantile overgrowth syndromes
2. Obesity syndromes
   a. Prader-Willi syndrome
   b. Laurence-Moon-Biedl-Bardet syndrome
   c. Cohen syndrome
3. Localised overgrowth syndromes
   a. Klippel-Trenaunay-Weber syndrome
   b. Proteus syndrome
   c. Idiopathic hemihypertrophy
   d. Idiopathic macromelia
4. Digital overgrowth syndromes
   a. Macrodactyly simplex congenita
   b. Macrodystrophia lipomatosa progressiva

## GENERALISED OVERGROWTH SYNDROMES

In clinical practice, it is useful to subcategorise overgrowth syndromes into those presenting in infancy and those which usually become apparent in later childhood. This distinction is arbitrary, as much depends upon the recognition of subtle stigmata, but nevertheless there is practical value in this approach. The infantile gigantism disorders are reviewed in this section.

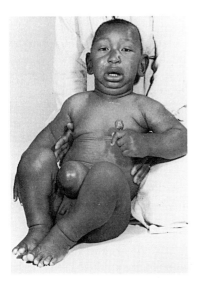

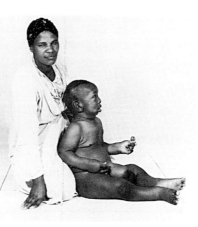

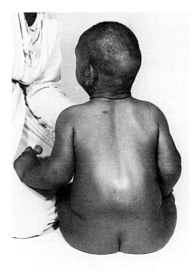

**Fig. 19.1** (top, left) Weaver syndrome: this child weighed 10.20 kg at birth. He has the characteristic large ears, long philtrum and protuberant lower lip. An umbilical hernia and excessive skin folds are evident. (Courtesy of Dr A A Dawood, Durban.)

**Fig. 19.2** (top, right) Weaver syndrome: at the age of 14 months, the boy weighed 30 kg. (Courtesy of Dr A A Dawood, Durban.)

**Fig. 19.3** (right) Weaver syndrome: progressive thoraco-lumbar kyphosis associated with platyspondyly and vertebral wedging. (Courtesy of Dr A A Dawood, Durban.)

Tall stature, which generally becomes evident in childhood rather than infancy, is a manifestation of the following disorders:

Marfan syndrome (see Ch. 17)
Sclerosteosis (see Ch. 8)
Acromegaly
Pituitary gigantism
Klinefelter syndrome (47XXY)
XYY syndrome

The endocrine and chromosomal disorders in this list are outside the scope of this book and will not be considered further.

## Weaver syndrome

The Weaver syndrome is a rare disorder in which accelerated prenatal growth continues in childhood. Affected infants have a broad forehead, wide mouth, large ears, a long philtrum and micrognathia. The skin is loose, finger pads are prominent and umbilical and inguinal hernias are frequent. The digits may be flexed and curved, and the feet are sometimes clubbed. Extension of the elbow and knee joints is limited and the adjacent bones are widened.

Radiographically, bone age is greatly advanced, the vertebrae are dysplastic and flattened and the metaphyses of the long bones are splayed.

During infancy the appetite is insatiable and gigantism becomes increasingly evident, height and weight being over the 97th centile. Mild mental and motor delay is usual. All reported patients have been infants or young children, and the long-term prognosis is unknown.

The syndrome was delineated by Weaver et al (1974) and further cases were reported by Farrell & Hughes (1985) and Dawood (1985). About 10 affected children have been documented, including eight males and two females.

There has been some discussion as to whether the Weaver and the Marshall-Smith syndromes are separate entities. This latter disorder has some features in common with the Weaver syndrome, including advanced skeletal maturation (Marshall 1958). The Marshall-Smith syndrome, of which about 15 cases have been reported, is lethal in infancy; it is possible that the two conditions represent opposite positions of the phenotypic spectrum of a single entity.

Jalaguier et al (1983) argued for homogeneity, while Fitch (1980, 1985) has supported the more attractive concept of separate syndromic identity. Although the majority of infants with the Weaver syndrome have been sporadic, affected siblings have been documented (Roussounis & Crawford 1983) and autosomal recessive inheritance is probable. A 'Weaver-like' syndrome reported by Stoll et al (1985) in a mother and son seems to be a separate entity.

## Sotos syndrome

Sotos syndrome or cerebral gigantism presents at birth with excessive body size and macrocephaly. More than 100 cases have been recorded and the condition is well recognised. During infancy, growth is accelerated, skeletal maturation is advanced and dental eruption is precocious. Affected persons usually have some degree of mental retardation, but general health is good and long-term survival is usual. Occasional concomitants include hepatoma (Sugerman et al 1977), juvenile macular degeneration (Ferrier 1980) and hydronephrosis (Adam et al 1986). Adult stature is normal or tall and the additional features of mandibular prognathism and large hands and feet may lead to confusion with acromegaly.

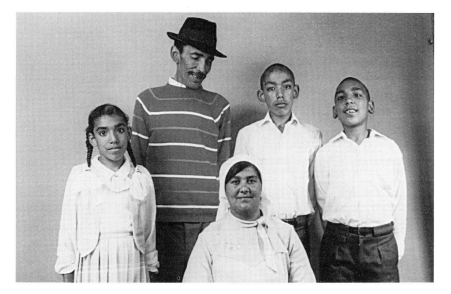

**Fig. 19.4** Sotos syndrome: a father and three offspring with height and head circumference above the 97th percentile. (From Winship I M 1985 Clinical Genetics 28: 243–246.)

Disturbances of plasma somatomedin levels in early childhood have been documented by Wit et al (1985) but there is no other evidence for any endocrinological abnormality in the Sotos syndrome. According to Beemer et al (1986) the phenotype can be mimicked by the Fragile-X syndrome.

After the delineation of the syndrome by Sotos et al (1964), the majority of reported cases were sporadic, and the mode of inheritance remained in doubt. Nevo et al (1974) described affected relatives in a consanguineous kindred and suggested that transmission might be autosomal recessive. Thereafter, families with affected persons in several generations were documented by Zonana et al (1977), Smith et al (1981), Bale et al (1985) and Winship (1985) and autosomal dominant inheritance was established. The issue of possible genetic heterogeneity is unsettled but correlation studies of the metacarpophalangeal pattern profiles of 16 patients were indicative of homogeneity (Butler et al 1985).

### Beckwith-Wiedemann syndrome

The cardinal features of the Beckwith-Wiedemann syndrome are gigantism, macroglossia, omphalocele, and a propensity to episodic hypoglycaemia in the early months of life. The phenotype is variable, and size is not necessarily excessive at the time of birth. However, by the end of the first year, height and weight are usually above the 90th percentile. Macroglossia is inconsistent and may increase or regress. Hyperplastic visceromegaly and hemihypertrophy are other variable features. Horizontal creases on the earlobes are a minor but important diagnostic indicator.

About 10% of reported patients have developed neoplasms, most commonly Wilms tumour and adrenal cortical carcinoma. Intelligence is usually normal, but mental retardation may result from undetected bouts of hypoglycaemia.

Several hundred cases have been reported, of which the majority have been sporadic. There have, however, been several instances of familial aggregation, although no clear-cut Mendelian pattern has emerged. Autosomal recessive transmission was suggested by Wiedemann (1964) and Chemke (1976) and polygenic inheritance was proposed by Wiedemann (1973), Gardner (1973) and Berry et al (1980). The concept of delayed mutation of an unstable premutated allele has attracted some favour (Lubinsky et al 1974, Kosseff et al 1976, Herrmann 1977) but remains unproven. Following a review of the literature, Best & Hoekstra (1981) identified 19 instances of familial recurrence, and concluded that the syndrome was an autosomal dominant trait, with variable phenotypic expression. Further evidence to support this contention was adduced by Niikawa et al (1986) when they identified 18 affected persons in five Japanese families and undertook pedigree analysis of accumulated data from published reports.

Prenatal diagnosis of the Beckwith-Wiedemann syndrome by ultrasonic demonstration of an omphalocele at the 19th week of gestation was accomplished by Winter et al (1986). The pregnancy was monitored because of the previous birth of an affected half sibling. In view of the phenotypic

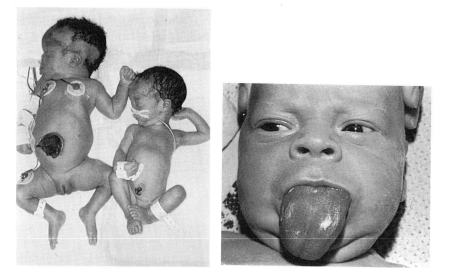

**Fig. 19.5** Beckwith-Wiedemann syndrome: a neonate with macrosomia and omphalocele. The infant on the right is normal. (Courtesy of Dr M Nelson, Cape Town.)

**Fig. 19.6** Beckwith-Wiedemann syndrome: macroglossia. (Courtesy of Dr M Nelson, Cape Town.)

variability, not every affected fetus would be detected by ultrasonography, but in appropriate circumstances, pregnancy surveillance is certainly warranted (Koontz et al 1986).

## Miscellaneous infantile overgrowth syndromes

### X-linked gigantism

A family in which 13 males in five generations had elevated birth weight and length, large hands and a distinctive coarse facies was recorded by Behmel et al (1984). In adulthood affected persons were more than 2 metres in height. The authors suggested that this condition was the same entity as the disorder described by Simpson et al (1973) under the descriptive title 'Bulldog syndrome'.

### Perlman syndrome

Perlman et al (1973, 1975) documented the association of fetal gigantism, renal hamartomas and nephroblastomata in two siblings. Neri et al (1984) reported further cases and additional details of the facial appearance of the original patients were depicted by Perlman (1986). The occurrence in siblings is suggestive of autosomal recessive inheritance.

## OBESITY SYNDROMES

The obesity syndromes, in which weight rather than height is excessive can be confused with the generalised overgrowth disorders. Oedema and nutritional obesity can also mimic these conditions but the presence of additional syndromic manifestations in the obesity syndromes facilitate diagnostic distinction.

## Prader*-Willi syndrome

The Prader-Willi syndrome comprises hypotonia during infancy, obesity which develops in early childhood, mental retardation, small hands and feet, and hypogonadism. Stature is stunted and affected persons have a characteristic facial appearance. The appetite is voracious and obesity may become monumental. The diagnosis of the Prader-Willi syndrome can be suspected from historical photographs of certain fairground exhibitionists, notably Jolly Nellie, Dolly Dimples and Alice from Dallas.

The cause of the Prader-Willi syndrome is unknown, although about 50% of affected persons have deletions involving the long arm of chromosome 15 (Ledbetter et al 1982, Butler & Palmer 1983, Mattei et al 1984). The proportion of patients with and without the deletion has varied from series

*Andrea Prader is Professor of Paediatrics at the University of Zurich, Switzerland

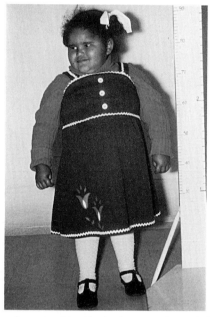

**Fig. 19.7** Prader-Willi syndrome: a girl with obesity, mental retardation, small hands and the characteristic facies.

to series but it is indisputable that the cytogenetic anomaly is both genuine and inconsistent (Bothwell & Merckel 1983). Despite considerable efforts, it has proved impossible to separate these two groups of patients at the phenotypic level. Indeed, detailed anthropometric studies of 38 affected persons yielded similar measurements for those with and without the chromosomal deletion (Butler & Meaney 1987).

## Laurence-Moon-Biedl-Bardet* syndrome

The Laurence-Moon-Biedl-Bardet syndrome (LMBBS) is characterised by stunted stature, marked obesity, mild mental deficiency and postaxial polydactyly. Retinitis pigmentosa, which leads to progressive night blindness and constriction of visual fields compounds the handicap. Severe bow legs is an underemphasised but genuine complication which sometimes arises at adolescence.

*John Zachariah Laurence (1829–1870) was an English ophthalmologist. He founded the Royal Eye Hospital, London.

Richard Charles Moon (1845–1914) was Laurence's assistant. He emigrated to the USA and practised ophthalmology in Philadelphia.

Arthur Biedl (1869–1933) was Professor of Experimental Pathology at the University of Prague and a founder of modern endocrinology.

George Louis Bardet, born in 1885 was a French physician.

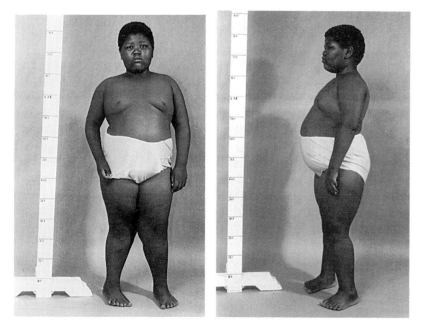

**Fig. 19.8** (left) Laurence-Moon-Biedl-Bardet syndrome: a girl with obesity, mental retardation, retinitis pigmentosa and polydactyly.

**Fig. 19.9** (right) Laurence-Moon-Biedl-Bardet syndrome: preaxial polydactyly of the hands and feet is evident.

The syndromic components are variable and there is unresolved controversy concerning homogeneity or heterogeneity (Ammann 1970, Schachat & Maumenee 1982). Evidence favours the latter viewpoint. There have been numerous instances of affected siblings including those mentioned in the initial report of Laurence & Moon (1866). Whether the LMBBS is one condition or two, there is no doubt that inheritance is autosomal recessive (Klein & Ammann 1969). Polydactyly has been observed in a number of obligate gene carriers but there have been no reports of any systematic studies of other possible phenotypic manifestations in heterozygotes for LMBBS.

### Cohen syndrome

Cohen syndrome is characterised by obesity, mental retardation, stunted stature, prominent upper incisors, a high arched palate and variable ocular abnormalities (Cohen et al 1973, Carey & Hall 1978). In distinction to other conditions in this group, obesity only becomes evident after the age of 5 years (Fryns & van den Berhe 1981). Several sets of affected sibs have been reported, and autosomal recessive inheritance is well established.

## LOCALISED OVERGROWTH SYNDROMES

The hallmark of the localised overgrowth syndromes is asymmetrical involvement of a digit, limb or body segment; in this respect these disorders can be separated from the conditions in which overgrowth is generalised. It must be emphasised, however, that this subdivision is arbitrary and that there is a certain amount of overlap.

The appraisal of asymmetry is not always an easy matter, and it is sometimes difficult to be certain whether an asymmetrical patient has hemihypertrophy or hemiatrophy. For instance, one side of the body may be unusually small in conditions such as the Russell-Silver syndrome (see Ch. 18) and in chromosomal mosaicism. It is also noteworthy that minor degrees of asymmetry are present in normal persons. Indeed, the two halves of the face are not the same size and shape, while discrepancies in limb length are commonplace.

Localised overgrowth can result from trauma, infection, arthritis and neoplasia. It may also be a consequence of fibrous lesions of bone and skeletal disorders such as dysplasia epiphysealis hemimelica, diaphyseal aclasis and the Maffucci syndrome. These secondary forms of localised overgrowth can be readily differentiated from the primary conditions which form the subject of this section.

### Klippel-Trenaunay-Weber* syndrome

The Klippel-Trenaunay-Weber syndrome (KTWS) comprises osseous and soft tissue hypertrophy of one or more limbs together with vascular malformations which include arteriovenous fistulae, haemangiomata, lymphangiomata and varicosities. The trunk is sometimes involved, and digital anomalies and visceral haemangiomata are occasional features.

The condition was first documented by Klippel & Trenaunay (1900) and delineation was continued by Weber (1907, 1918). The natural history of 10 affected persons was analysed by Viljoen et al (1984) and the cutaneous manifestations of 20 individuals with KTWS were reviewed in detail by Viljoen et al (1987). The pathogenesis of the KTWS is unknown but virtually every case has been sporadic and there is no indication of any chromosomal or genetic defect.

As the KTW is non-genetic and non-recurrent, antenatal diagnosis is necessarily serendipitous. The disorder has been detected by ultrasound at the 34th week of gestation by Hatjis et al (1981) while Seoud et al (1984)

---

*Maurice Klippel (1858–1942): see Klippel Feil syndrome, Chapter 13.

Paul Trenaunay, born 1875, was Klippel's junior colleague in Paris when they published their case description in 1900.

Frederick Parkes Weber (1863–1962) was an eminent London physician; he was involved in the delineation of numerous genetic syndromes.

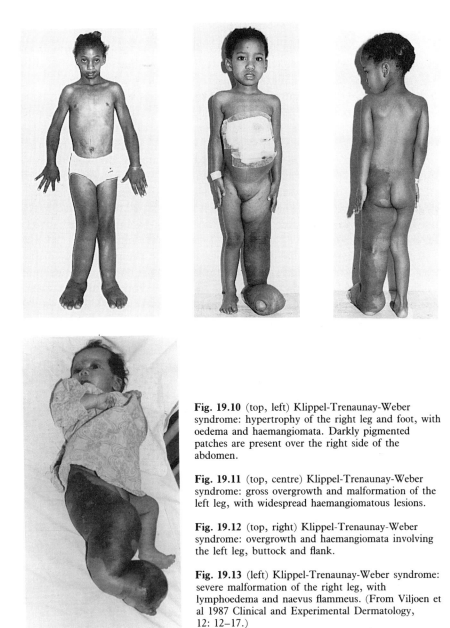

**Fig. 19.10** (top, left) Klippel-Trenaunay-Weber syndrome: hypertrophy of the right leg and foot, with oedema and haemangiomata. Darkly pigmented patches are present over the right side of the abdomen.

**Fig. 19.11** (top, centre) Klippel-Trenaunay-Weber syndrome: gross overgrowth and malformation of the left leg, with widespread haemangiomatous lesions.

**Fig. 19.12** (top, right) Klippel-Trenaunay-Weber syndrome: overgrowth and haemangiomata involving the left leg, buttock and flank.

**Fig. 19.13** (left) Klippel-Trenaunay-Weber syndrome: severe malformation of the right leg, with lymphoedema and naevus flammeus. (From Viljoen et al 1987 Clinical and Experimental Dermatology, 12: 12–17.)

documented the ultrasonographic recognition of a large complex mass with pulsating vascular channels over the anterior chest wall of a fetus at the 17th week of pregnancy. In this instance the diagnosis of the KTWS was established following therapeutic termination.

## Proteus syndrome

The title 'proteus syndrome' pertains to the protean manifestations of an overgrowth disorder which was delineated by Wiedemann et al (1983) in a report of four sporadic children. The stigmata, which are very variable, include asymmetrical gigantism of hands, feet and limbs. The skin of palms and soles is hyperplastic and other cutaneous features are lipomata and linear verrucous epidermal naevi. Macrocephaly and cranial exostoses are frequent concomitants.

Additional cases have been documented by Lezama & Buyse (1984) and Costa et al (1985). The manifestations and natural history of six South African patients, four children and two adults, were recorded by Viljoen et al (1987). At present 15 cases have been recorded; all were sporadic, the sex incidence was equal and there was no known parental consanguinity. So far, the pathogenesis of the proteus syndrome is unknown.

It has been suggested that Joseph Carey Merrick, the grotesquely disfigured 'Elephant Man' of the London Hospital, might have had the proteus syndrome (Tibbles & Cohen 1986). Sir Frederick Treves, surgeon extraordinary to Queen Victoria, rescued this unfortunate individual from his predicament in a fairground peepshow in 1886; Merrick spent his remaining years in sanctuary in the London hospital, where his skeleton still resides in the anatomical museum. A radiographic study of Merrick's skeleton has

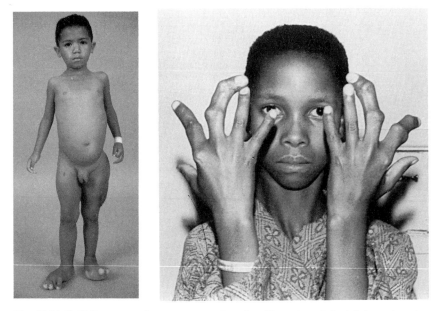

**Fig. 19.14** (left) Proteus syndrome: overgrowth and malformation of the left buttock and both feet. (Courtesy of Dr D Viljoen, Cape Town).

**Fig. 19.15** (right) Proteus syndrome: asymmetrical digital overgrowth. (Courtesy of Dr D Viljoen, Cape Town.)

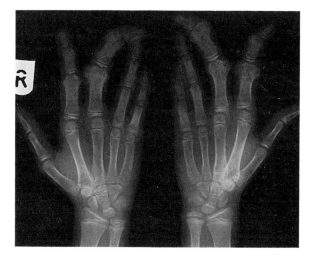

**Fig. 19.16** Proteus syndrome: hand radiograph, showing asymmetrical overgrowth and expansion of the tubular bones. (From Viljoen et al 1987 American Journal of Medical Genetics, 27: 87–97.)

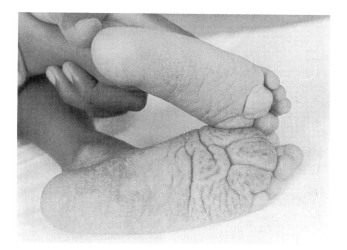

**Fig. 19.17** Proteus syndrome: the left foot is hypertrophic with thickened rugose skin on the sole.

been published by Bean et al (1982) and he has been the subject of a film and two monographs (Montague 1971, Sparks 1980). Although he was initially thought to have neurofibromatosis, his manifestations are more in keeping with those of the proteus syndrome.

## Idiopathic hemihypertrophy

Idiopathic hemihypertrophy is characterised by significant overgrowth of the structures of one side of the body. The overgrowth is present at birth and continues synchronously with normal tissue development. In addition to mechanical and cosmetic problems, a proportion of affected persons have epilepsy and mental retardation (Wakefield & Hines 1933). The recognition of hypertrophy of the oral structures is helpful in diagnosis (Gorlin & Meskin 1962). There is an association with Wilms tumour of the kidney (Miller et al 1964) and possibly with neoplasms of the adrenals and liver (Harris et al 1981); the overall risk of malignancy is probably of the order of 5%.

Idiopathic hemihypertrophy differs from idiopathic macromelia (vide infra) in that the trunk and face are usually involved. The presence of café-au-lait macules may lead to confusion with neurofibromatosis, while cutis mamorata is reminiscent of the Klippel-Trenaunay-Weber syndrome. For these reasons, definitive diagnosis is not always a straightforward matter. Indeed, following an extensive review of the literature concerning hemihypertrophy, it became apparent that many supposedly affected persons actually had the KTWS (Ringrose et al 1965).

About 200 cases of idiopathic hemihypertrophy have been recorded and an estimate of the minimum prevalence can be gained from the fact that nine affected persons were seen during a 12 year period in genetic clinics

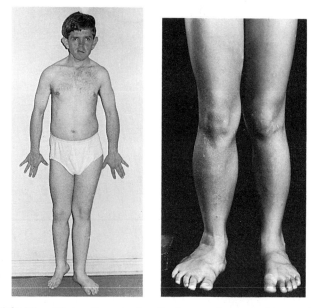

**Fig. 19.18** (left) Idiopathic hemihypertrophy: the face, trunk, arm and leg on the left side are significantly larger than the corresponding structures on the right. (From Viljoen et al 1984 Clinical Genetics 26: 81–86.)

**Fig. 19.19** (right) Idiopathic macromelia: a girl aged 8 years: the right leg and foot are larger than the left. The skin is not involved. (From Pearn et al 1983 South African Medical Journal 64: 905–908.)

and special institutions serving a population of 3 million in the Cape Town area (Viljoen et al 1984). The aetiology is unknown and although there have been a few instances of familial aggregation (Morris & MacGillivray 1955, Fraumeni et al 1967) the condition is probably non-genetic.

## Idiopathic macromelia

Idiopathic macromelia is overgrowth of a limb or limbs without involvement of the trunk or head, and with no cutaneous changes. The disorder is present at birth, and growth is synchronous with development of unaffected parts of the body. Apart from orthopaedic, mechanical and cosmetic considerations, the condition is harmless; in particular, there are no systemic ramifications and no propensity to malignancy (Pearn et al 1983).

In the loose sense, the designation 'idiopathic macromelia' embraces proportionate overgrowth of all components of a hand or foot. However, semantic purists contend, with justification, that manual and pedal gigantism are more accurate terms for these anomalies.

The pathogenesis of idiopathic macromelia is unknown, although it is possible that end organ growth factors may be implicated. There is no evidence to indicate a genetic aetiology.

## DIGITAL OVERGROWTH SYNDROMES

### Macrodactyly simplex congenita

Macrodactyly simplex congenita (MSC) is congenital, non-progressive, primary enlargement of a digit or digits. By definition, the condition is present at birth and size of the involved region, relative to the unaffected digits, does not increase with postnatal growth. In MSC the affected finger

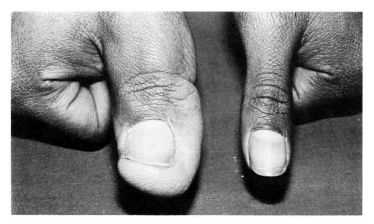

**Fig. 19.20** Macrodactyly simplex congenita: a male aged 22 years with overgrowth of the right thumb. (From Pearn et al 1983 South African Medical Journal 64: 905–908.)

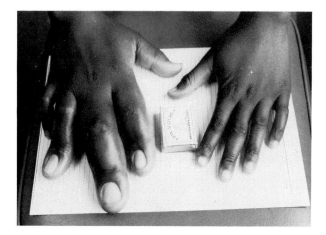

**Fig. 19.21** Macrodactyly simplex congenita: a girl aged 18 years with overgrowth of the digits of the right hand. (Courtesy of J H Davy, FRCS, Harare.)

or toe is increased in length and not simply widened and all tissue components are proportionately involved in the overgrowth.

For the purpose of prognostication, MSC must be differentiated from the macrodystrophica lipomatosa progressiva (MLP) and from disorders such as fibrous dysplasia of bone, haemangiomatosis and neurofibromatosis. Other overgrowth conditions in which digital enlargement may occur, notably Klippel-Trenaunay-Weber syndrome, idiopathic hemihypertrophy, idiopathic macromelia and proteus syndrome can be recognised by their anatomical distribution and syndromic concomitants. Finally, macrodactyly can occur as a component of the ectrodactyly group of malformation syndromes.

Management is centred upon restriction of growth by epiphyseal ablation at the time that the affected digit has reached adult dimensions. The results of these surgical procedures are generally satisfactory.

MSC is a rare disorder and accounts for about 10% of all cases of digital overgrowth. The literature is confusing, as most reports of macrodactyly have included both MSC and MLP (Barsky 1967, Rechnazel 1967, Ofodile & Oluwasanmi 1979, Kumar et al 1985). The nosological status and differential diagnosis of MSC has been reviewed by Temtamy & McKusick (1978) and Pearn et al (1986).

All reported cases of MSC have been sporadic and there is nothing to suggest a conventional genetic aetiology. It is possible that causative mechanisms concern somatic mutation in the limb bubs; these could involve receptor sites for locally acting growth factors, or receptors for chalones which regulate growth by inhibition (Pearn et al 1986). At present, these concepts are purely speculative.

## Macrodystrophia lipomatosa progressiva

Macrodystrophia lipomatosa progressiva (MLP) is an unusual disorder in which proliferation of fibro-fatty connective and neural tissues is associated with progressive digital overgrowth. The condition spreads centripetally to involve the hand or foot and sometimes the arms and legs. Mechanical interference with movement and neurovascular supply may occur and the carpal tunnel syndrome is an occasional complication (Ranawat et al 1968). In contrast to macrodactyly simplex congenita (MSC), the functional and cosmetic results of surgery in MLP are generally unsatisfactory (Kumar & Sundarajan 1987). A comprehensive review of 40 literature reports and five personal cases was published by Yaghmai et al (1976); recent articles have concerned the radiographic and angiographic assessment of affected digits (Goldman & Kaye 1977, Laval-Jeantet et al 1979, Moran et al 1984).

MLS can be distinguished from the other digital overgrowth syndromes by its progressive nature, which leads to increasing disproportion and by the propensity to involvement of adjacent structures. Biopsy of affected soft tissues reveals characteristic histopathological changes.

Reports of MLP have almost always been published in conjunction with accounts of MSC. However, while both disorders are rare, the former is encountered more frequently than the latter (Slavitt et al 1980). The cause is unknown but there is no hint of any genetic aetiology.

REFERENCES

**Generalised overgrowth syndromes**
*Weaver syndrome*
Dawood A A 1985 Weaver's syndrome — primordial excessive growth velocity: a case report. South African Medical Journal 67: 646–648
Farrell S A, Hughes H E 1985 Brief clinical report: Weaver syndrome with pes cavus. American Journal of Medical Genetics 21/4: 737–739
Fitch N 1980 The syndromes of Marshall and Weaver. Journal of Medical Genetics 17: 174–178
Fitch N 1985 Letter to the Editor: update on the Marshall-Smith-Weaver controversy. American Journal of Medical Genetics 20: 559–562
Jalaguier J, Montoya F, Germain M, Bonnet H 1983 Avance de la maturation osséuse et syndrome dysmorphique chez deux germains. Journal of Genetics 31: 385–395
Marshall 1958 Ectodermal dysplasia: report of kindred with ocular abnormalities and hearing defect. American Journal of Ophthalmology 45: 143
Roussounis S, Crawford M 1983 Siblings with the Weaver syndrome. Journal of Pediatrics 102: 595–597
Stoll C, Talon P, Mengus L, Roth M P, Dott B 1985 A Weaver-like syndrome with endocrinological abnormalities in a boy and his mother. Clinical Genetics 28: 255–259
Weaver D D, Graham C B, Thomas I T, Smith D W 1974 A new overgrowth syndrome with accelerated skeletal maturation, unusual facies, and camptodactyly. Journal of Pediatrics 84: 547–552

*Sotos syndrome*
Adam K A R, Al Frayh A R S, Sharma A, Taha S A 1986 Cerebral gigantism with hydronephrosis: a case report. Clinical Genetics 29: 178–180
Bale A E, Drum M A, Parry D M, Mulvihill J J 1985 Familial Sotos syndrome (cerebral gigantism): craniofacial and psychological characteristics. American Journal of Medical Genetics 20: 613–624

Beemer F A, Veenema H, de Pater J M 1986 Cerebral gigantism (Sotos syndrome) in two patients with Fra(X) chromosomes. American Journal of Medical Genetics 23: 221–226

Butler M G, Meaney F J, Kittur S, Hersh J H, Hornstein L 1985 Metacarpophalangeal pattern profile analysis in Sotos syndrome. American Journal of Medical Genetics 20: 625–629

Ferrier P E 1980 Cerebral gigantism with juvenile macular degeneration. Helvetica Paediatrica Acta 35(1): 97–102

Nevo S, Zeltr M, Levy J 1974 Evidence for autosomal recessive inheritance in cerebral gigantism. Journal of Medical Genetics 11: 156–160

Smith A, Farrar J R, Silink M, Judzewitsch R 1981 Investigations in dominant Sotos syndrome. Annales de Génétique 24: 226–228

Sotos J F, Dodge P R, Muirhead D, Cranford J O, Talbot N D 1964 Cerebral gigantism in childhood. New England Journal of Medicine 271: 109–116

Sugerman G I, Henser E T, Reed W B 1977 A case of cerebral gigantism and hepatocarcinoma. American Journal of Diseases of Childhood 131: 631–635

Winship I M 1985 Sotos syndrome — autosomal dominant inheritance substantiated. Clinical Genetics 28: 243–246

Wit J M, Beemer F A, Barth P G 1985 Cerebral gigantism (Sotos syndrome). Compiled data of 22 cases. Analysis of clinical features, growth and plasma somatomedin. European Journal of Pediatrics 144/2: 131–140

Zonana J, Sotos J F, Romshe C A, Fischer R D 1977 Dominant inheritance of cerebral gigantism. Journal of Pediatrics 91: 251–256

*Beckwith-Wiedemann syndrome*

Berry A C, Belton E M, Chantler C 1980 Monozygotic twins discordant for Wiedemann-Beckwith syndrome and the implications for genetic counselling. Journal of Medical Genetics 17: 136–138

Best L G, Hoekstra R E 1981 Wiedemann-Beckwith syndrome: autosomal dominant inheritance in a family. American Journal of Medical Genetics 9: 291–199

Chemke J 1976 Familial macroglossia-omphalocele syndrome. Journal de Génétique Humaine 24: 271–279

Gardner L I 1973 Pseudo-Beckwith-Wiedemann syndrome: Interaction with maternal diabetes. Lancet 2: 911–912

Herrmann J 1977 Clinical aspects of gene expression. Birth Defects Original Article Series XIII (3D): 25–45

Koontz W L, Shaw L A, Lavery J P 1986 Antenatal sonographic appearance of Beckwith-Wiedemann syndrome. Journal of Clinical Ultrasound 14/1: 57–59

Kosseff A L, Herrmann J, Gilbert E F, Viseskul C, Lubinsky M, Opitz J M 1976 Studies of malformation syndromes of man XXIX: the Wiedemann-Beckwith syndrome. Clinical, genetic and pathogenetic studies of 12 cases. European Journal of Pediatrics 123: 139–166

Lubinsky M, Herrmann J, Kosseff A, Opitz J M 1974 Autosomal-dominant sex dependent transmission of the Wiedemann-Beckwith syndrome. Lancet 1: 932

Niikawa N, Ishikiriyama S, Takahashi, Inagawa A, Tonoki H, Ohta Y, Hase N, Kamei T, Kajii T 1986 The Wiedemann-Beckwith syndrome: pedigree studies on five families with evidence for autosomal dominant inheritance with variable expressivity. American Journal of Medical Genetics 24: 41–55

Wiedemann H-R 1964 Complexe malformatif familial avec hernie ombilicale et macroglossie — un 'syndrome nouveau'? Journal de Génétique Humaine 13: 223–232

Wiedemann H-R 1973 EMG syndrome. Lancet 2: 626–627

Winter S C, Curry C J R, Smith J C, Kassel S, Miller L, Andrea J 1986 Prenatal diagnosis of the Beckwith-Wiedemann syndrome. American Journal of Medical Genetics 24: 137–141

*Miscellaneous infantile overgrowth syndromes*
*X-linked gigantism*

Behmel A, Plochl E, Rosenkranz W 1984 A new X-linked dysplasia gigantism syndrome: identical with the Simpson dysplasia syndrome? Human Genetics 67/4: 409–413

Simpson J L, New M, Landey S, German J 1973 A previously unrecognized X-linked syndrome of dysmorphia. Birth Defects Original Article Series XI(2): 18–24

*Perlman syndrome*
Neri G, Martini-Neri N E, Katz B E, Opitz J M 1984 The Perlman syndrome: familial renal dysplasia with Wilms tumor, fetal gigantism and multiple congenital anomalies. American Journal of Medical Genetics 19: 195–207
Perlman M 1986 Perlman syndrome: familial renal dysplasia with Wilms tumor, fetal gigantism, and multiple congenital anomalies. American Journal of Medical Genetics 25: 793–795
Perlman M, Levin M, Wittels B 1975 Syndrome of fetal gigantism, renal hamartomas and nephroblastomatosis with Wilms' tumour. Cancer 35: 1212–1217
Perlman M, Goldberg G M, Bar-Ziv J, Danovitch G 1973 Renal hamartomas and nephroblastomatosis with fetal gigantism: a familial syndrome. Journal of Pediatrics 83: 414–418

**Obesity syndromes**
*Prader-Willi syndrome*
Bothwell R, Merckel L 1983 The Prader-Willi syndrome in Cape Town. South African Medical Journal 63: 883–886
Butler M G, Meaney F J 1987 An anthropometric study of 38 individuals with Prader-Labhart-Willi syndrome. American Journal of Medical Genetics 26: 445–455
Butler M G, Palmer C G 1983 Parental origin of chromosome 15 deletion in Prader-Willi syndrome. Lancet 1: 1285–1286
Ledbetter D H, Mascarello J T, Riccardi V M, Harper V D, Airhart S D, Strobel R J 1982 Chromosome 15 abnormalities and the Prader-Willi syndrome: a follow-up report of 40 cases. American Journal of Human Genetics 34: 278–285
Mattei M G, Souiah N, Mattei J F 1984 Chromosome 15 anomalies and the Prader-Willi syndrome: cytogenetic analysis. Human Genetics 66: 313–334

*Laurence-Moon-Biedl-Bardet syndrome*
Ammann F 1970 Investigations cliniques et génétiques sur le syndrome de Bardet-Biedl en Suisse. Journal de Génétique Humaine 18 (supplement): 1–310
Klein D, Ammann F 1969 The syndrome of Laurence-Moon-Bardet-Biedl and other diseases in Switzerland. Journal of Neurological Sciences 9: 479–513
Laurence J Z, Moon R C 1866 Four cases of retinitis pigmentosa occurring in the same family and accompanied by general imperfection of development. Ophthalmological Review 2: 32–41
Schachat A P, Maumenee I H 1982 The Bardet-Biedl syndrome and related disorders. Archives of Ophthalmology 100: 285–288

*Cohen syndrome*
Carey J C, Hall B D 1978 Confirmation of the Cohen syndrome. Journal of Pediatrics 93: 230–244
Cohen M M Jr, Hall B D, Smith D W, Graham C B, Lampert K J 1973 A new syndrome with hypotonia, obesity, mental deficiency, and facial, oral, ocular and limb anomalies. Journal of Pediatrics 83: 280–284
Fryns J P, van Den Berhe H 1981 The Cohen syndrome. Journal de Génétique Humaine 29/4: 449–453

**Localised overgrowth syndromes**
*Klippel-Trenaunay-Weber syndrome*
Hatjis K G, Philip A G, Anderson G G, Mann L I 1981 The in utero ultrasonographic appearance of Klippel-Trenaunay-Weber syndrome. American Journal of Obstetrics and Gynaecology 139(8): 972–974
Klippel M, Trenaunay P 1900 Du naevus variquex ostéohypertrophique. Archives Generales de Médecine 185: 641–672
Seoud M, Santos-Ramos R, Friedman J M 1984 Early prenatal ultrasonic findings in Klippel-Trenaunay-Weber syndrome. Prenatal Diagnosis 4: 227–230
Viljoen D, Pearn J, Beighton P 1984 On the natural history of the Klippel-Trenaunay-Weber syndrome: a review of 10 cases. Journal of Clinical Dysmorphology 2(4): 2–7

Viljoen D, Saxe N, Pearn J, Beighton P 1987 The cutaneous manifestations of the Klippel-Trenaunay-Weber syndrome. Clinical and Experimental Dermatology 12: 12–17
Weber F P 1907 Angioma formation in connection with hypertrophy of limbs and hemihypertrophy. British Journal of Dermatology 19: 231–235
Weber F P 1918 Haemangiectatic hypertrophy of limbs — congenital phlebarteriectasis and so-called varicose veins. British Journal of Child Diseases 15: 13

*Proteus syndrome*
Bean W B, Felson B, Dolan K D 1982 A nonletter from the editor and a case for all seasons. Seminars in Roentgenology XVII/3: 153–162
Costa T, Fitch N, Azouz E M 1985 Proteus syndrome: report of two cases with pelvic lipomatosis. Pediatrics 76: 984–989
Lezama D B, Buyse M L 1984 The proteus syndrome: the emergence of an entity. Journal of Clinical Dysmorphology 2: 10–13
Montague A 1971 The elephant man: A study in human dignity. London: Dutton
Mucke J, Willgerodt H, Kunzel R, Brock D 1985 Kinderklinik des bereiches medizin der Kerl-Marx-Universitat Leipzig. European Journal of Pediatrics 143/4: 320–323
Sparks C 1980 The elephant man. Ballantine Books. New York
Tibbles J A R, Cohen M M Jr 1986 The proteus syndrome: the Elephant Man diagnosed. British Medical Journal 293: 683–685
Viljoen D L, Nelson M M, de Jongh, Beighton P 1987 Proteus syndrome in Southern Africa, natural history and clinical manifestations in six individuals. American Journal of Medical Genetics 27: 87–97
Wiedemann H R, Gurgio G R, Aldenhoff P, Kunze J, Kaufman H J, Schirg E 1983 The proteus syndrome. European Journal of Pediatrics 140: 5–12

*Idiopathic hemihypertrophy*
Fraumeni J F, Gieser C F, Manning M D 1967 Wilm's tumour and congenital hemihypertrophy: report of five new cases and review of literature. Pediatrics 40: 886–899
Gorlin R J, Meskin L H 1962 Congenital hemihypertrophy. Review of the literature and report of a case with special emphasis on oral manifestations. Journal of Pediatrics 61/6: 870–879
Harris R E, Fuchs E F, Kaempe M J 1981 Medullary sponge kidney and congenital hemihypertrophy: case report and literature review. Urology 126: 676–678
Miller R W, Fraumeni J F, Manning M D 1964 Association of Wilm's tumour with aniridia, hemihypertrophy and other congenital malformations. New England Journal of Medicine 270: 922–924
Morris J V, MacGillivray R C 1955 Mental defect and hemihypertrophy. American Journal of Mental Deficiency 59: 645–650
Ringrose R E, Jabbour J T, Keele D K 1965 Hemihypertrophy. Pediatrics 36: 434–448
Viljoen D, Pearn J, Beighton P 1984 On the natural history of the Klippel-Trenaunay-Weber syndrome: a review of 10 cases. Journal of Clinical Dysmorphology 2(4): 2–7
Wakefield E G, Hines E A 1933 Congenital hemihypertrophy: report of eight cases. American Journal of Medical Science 185: 493–498

*Idiopathic macromelia*
Pearn J, Viljoen D, Beighton P 1983 Limb overgrowth: clinical observations and nosological considerations. South African Medical Journal 64: 905–908

**Digital overgrowth syndromes**
*Macrodactyly simplex congenita*
Barsky A J 1967 Macrodactyly. Journal of Bone and Joint Surgery 49: 1255–1266
Kumar K, Kumar D, Gadegone W M, Kapahtia N K 1985 Macrodactyly of the hand and foot. Orthopedics 9/4: 259–264
Ofodile F A, Oluwasanmi J 1979 Pedal macrodactyly: report of seven cases. East African Medical Journal 56: 283–287

Pearn J, Bloch C E, Nelson M M 1986 Macrodactyly simplex congenita: a case series and considerations of differential diagnosis and aetiology. South African Medical Journal 70: 755–758

Rechnazel K 1967 Megalodactylisms: report of seven cases. Acta Orthopaedica Scandanavica 38: 57–66

Temtamy S A, McKusick V A 1978 Macrodactyly as a part of syndromes, Birth Defects 14: 511–524

*Macrodystrophia lipomatosa progressiva*

Goldman A B, Kaye J J 1977 Macrodystrophia lipomatosa: radiographic diagnosis. American Journal of Roentgenology 128(1): 101–105

Kumar P, Sundarajan M S 1987 Congenital macrodactyly. Saudi Medical Journal 8: 73–76

Laval-Jeantet M, Civatte J, Vadrot D, Katz M, Maarek M, Dassonville M 1979 Macrodystrophia lipomatosa. Angiographic aspects (author's translation). Journal of Radiology 60(10): 653–656

Moran V, Butler F, Colville J 1984 X-ray diagnosis of macrodystrophia lipomatosa. British Journal of Radiology 57(678): 523–525

Ranawat C S, Arora M M, Singh R G 1968 Macrodystrophia lipomatosa with carpal-tunnel syndrome. A case report. Journal of Bone and Joint Surgery 50(6): 1242–1244

Slavitt J A, Hutchinson D L, Brown F L, Mishalanie M A 1980 Macrodystrophia lipomatosa. A case report. Journal of the American Podiatry Association 70(7): 349–352

Yaghmai I, McKowne F, Alizadeh A 1976 Macrodactylia fibrolipomatosis.Southern Medical Journal 69(12): 1565–1568

# Appendix 1. Relevant reviews or monographs

Beighton P, Beighton G 1986 The man behind the syndrome. Springer-Verlag, Heidelberg

Beighton P, Cremin B 1980 Sclerosing bone dysplasias. Springer-Verlag, Heidelberg

Beighton P, Grahame R, Bird H 1983 Hypermobility of joints. Springer-Verlag, Heidelberg

Connor J M 1983 Soft tissue ossification. Springer-Verlag, Berlin

Cremin B, Beighton P 1978 Bone dysplasias of infancy. Springer-Verlag, Heidelberg

Gorlin R J, Pindborg J J, Cohen M Jr 1976 Syndromes of the head and neck 2nd ed. McGraw-Hill, London

Horan F, Beighton P 1982 Orthopaedic problems in inherited disorders. Springer-Verlag, Heidelberg

Kaufmann H J (ed) 1973 Progress in pediatric radiology, intrinsic diseases of bones Vol 4. Karger, Basel

Kozlowski K, Beighton P 1984 Gamut index of skeletal dysplasias. Springer-Verlag, Heidelberg

McKusick V A 1972 Heritable disorders of connective tissue 4th edn. Mosby, St Louis

McKusick V 1986 Mendelian inheritance in man 7th edn, Johns Hopkins University Press, Baltimore

Maroteaux P 1979 Bone diseases of children. Lippincott, Philadelphia

Papadatos C J, Bartsocas C S (eds) 1982 Skeletal dysplasias, Progress in clinical and biological research, vol. 104. Liss, New York

Smith D W 1982 Recognisable patterns of human malformation 3rd edn. Saunders, Philadelphia

Bone dysplasias. Spranger J W, Langer L O, Wiedemann H R 1974 Gustav Fischer Verlag, Stuttgart

Warkany J 1971 Congenital malformations, Year Book Medical Press, Chicago.

Wiedemann H-R, Grosse K-R, Dibbern H 1982 An atlas of characteristic syndromes 2nd edn. Wolfe Medical Publications

Wynne-Davies R, Hall C M, Apley A G 1985 Atlas of Skeletal Dysplasias. Churchill Livingstone, Edinburgh

# Appendix 2. International Nomenclature of Constitutional Diseases of Bone (Revision, May 1983)

OSTEOCHONDRODYSPLASIAS

Abnormalities of cartilage and/or bone growth and development

## Defects of growth of tubular bones and/or spine

*Identifiable at birth:*                                          Transmission

    *usually lethal before or shortly after birth*

| | |
|---|---|
| 1. Achondrogenesis type I (Parenti-Fraccaro) | AR |
| 2. Achondrogenesis type II (Langer-Saldino) | |
| 3. Hypochondrogenesis | |
| 4. Fibrochondrogenesis | AR |
| 5. Thanatophoric dysplasia | |
| 6. Thanatophoric dysplasia with clover-leaf skull | |
| 7. Atelosteogenesis | |
| 8. Short-rib syndrome (with or without polydactyly) | |
|   a. type I (Saldino-Noonan) | AR |
|   b. type II (Majewski) | AR |
|   c. type III (lethal thoracic dysplasia) | AR |

    *usually non-lethal dysplasia*

| | |
|---|---|
| 9. Chondrodysplasia punctata | |
|   a. rhizomelic form autosomal recessive | AR |
|   b. dominant X-linked form | XLD (lethal in males) |
|   c. common mild form (Sheffield) | |
|   Exclude symptomatic stippling (Warfarin, chromosomal aberration . . .) | |
| 10. Campomelic dysplasia | |
| 11. Kyphomelic dysplasia | AR |
| 12. Achondroplasia | AD |
| 13. Diastrophic dysplasia | AR |
| 14. Metatropic dysplasia (several forms) | AR, AD |

15. Chondro-ecto-dermal dysplasia (Ellis-van Creveld)     AR
16. Asphyxiating thoracic dysplasia (Jeune)     AR
17. Spondylo-epiphyseal dysplasia congenita
    a. autosomal dominant form     AD
    b. autosomal recessive form     AR
18. Kniest dysplasia     AD
19. Dyssegmental dysplasia     AR
20. Mesomelic dysplasia
    a. type Nievergelt     AD
    b. type Langer (probable homozygous dyschondrosteosis) AR
    c. type Robinow
    d. type Rheinardt     AD
    e. others
21. Acromesomelic dysplasia     AR
22. Cleido-cranial dysplasia     AD
23. Oto-palato-digital syndrome
    a. type I (Langer)     XLSD
    b. type II (André)     XLR
24. Larsen syndrome     AR, AD
25. Other multiple dislocations syndromes

*Identifiable in later life*

1. Hypochondroplasia     AD
2. Dyschondrosteosis     AD
3. Metaphyseal chondrodysplasia type Jansen     AD
4. Metaphyseal chondrodysplasia type Schmid     AD
5. Metaphyseal chondrodysplasia type McKusick     AR
6. Metaphyseal chondrodysplasia with exocrine pancreatic
    insufficiency and cyclic neutropenia     AR
7. Spondylo-metaphyseal dysplasia
    a. type Kozlowski     AD
    b. other forms
8. Multiple epiphyseal dysplasia
    a. type Fairbank     AD
    b. other forms
9. Multiple epiphyseal dysplasia with early diabetes
    (Wolcott-Rallisson)     AR
10. Arthro-ophthalmopathy (Stickler).     AR
11. Pseudo-achondroplasia
    a. dominant     AD
    b. recessive     AR
12. Spondylo-epiphyseal dysplasia tarda     XLR
13. Progressive pseudo-rheumatoid chondrodyspiasia     AR
14. Spondyloepiphyseal dysplasia, other forms

15. Brachyolmia
   a.  autosomal recessive          AR
   b.  autosomal dominant          AD
16. Dyggve-Melchior-Clausen dysplasia      AR
17. Spondyloepimetaphyseal dysplasia (several forms)
18. Spondyloepimetaphyseal dysplasia with joint laxity    AR
19. Otospondylomegaepiphyseal dysplasia (OSMED)    AR
20. Myotonic chondrodysplasia (Catel-Schwartz-Jampel)    AR
21. Parastremmatic dysplasia         AD
22. Trichorhinophalangeal dysplasia         AD
23. Acrodysplasia with retinitis pigmentosa and nephro-
    pathy (Saldino-Mainzer)          AR

## Disorganised development of cartilage and fibrous components of skeleton

1. Dysplasia epiphyseal hemimelica.
2. Multiple cartilaginous exostoses         AD
3. Acrodysplasia with exostoses (Giedion-Langer)
4. Enchondromatosis (Ollier)
5. Enchondromatosis with hemangioma (Maffucci)
6. Metachondromatosis          AD
7. Spondyloenchondroplasia        AR
8. Osteoglophonic dysplasia
9. Fibrous dysplasia (Jaffe-Lichtenstein)
10. Fibrous dysplasia with skin pigmentation and preco-
    cious puberty (McCune-Albright)
11. Cherubism (familial fibrous dysplasia of the jaws)   AD

## Abnormalities of density of cortical diaphyseal structure and/or metaphyseal modeling

1. Osteogenesis imperfecta (several forms)    AR, AD
2. Juvenile idiopathic osteoporosis
3. Osteoporosis with pseudo-glioma       AR
4. Osteopetrosis
   a.  autosomal recessive lethal      AR
   b.  intermediate recessive        AR
   c.  autosomal dominant         AD
   d.  recessive with tubular acidosis    AR
5. Pycnodysostosis          AR
6. Dominant osteosclerosis type Stanescu    AD
7. Osteomesopycnosis         AD
8. Osteopoikilosis          AD

| | |
|---|---|
| 9. Osteopathia striata | AD |
| 10. Osteopathia striata with cranial sclerosis | AD |
| 11. Melorheostosis | |
| 12. Diaphyseal dysplasia (Camurati-Engelmann) | AD |
| 13. Craniodiaphyseal dysplasia | AR |
| 14. Endosteal hyperostosis | |
|     a. autosomal dominant (Worth) | AD |
|     b. autosomal recessive (van Buchem) | AR |
|     c. autosomal recessive (sclerosteosis) | AR |
| 15. Tubular stenosis (Kenny-Caffey) | AD |
| 16. Pachydermoperiostosis | AD |
| 17. Osteodysplasty (Melnick-Needles) | AD |
| 18. Frontometaphyseal dysplasia | XLR |
| 19. Craniometaphyseal dysplasia | AD, AR |
| 20. Metaphyseal dysplasia (Pyle) | AR |
| 21. Dysosteosclerosis | AR or XLR |
| 22. Osteoectasia with hyperphosphatasia | AR |
| 23. Oculodento-osseous dysplasia | |
|     a. mild type | AD |
|     b. severe type | AR |
| 24. Infantile cortical hyperostosis (Caffey disease, familial type) | AD |

## DYSOSTOSES

Malformation of individual bones, singly or in combination

### Dysostoses with cranial and facial involvement

| | |
|---|---|
| 1. Craniosynostosis (several forms) | |
| 2. Craniofacial dysostosis (Crouzon) | |
| 3. Acrocephalo-syndactyly | |
|     a. type Apert | AD |
|     b. type Chotzen | AD |
|     c. type Pfeiffer | AD |
|     d. other types | |
| 4. Acrocephalo-polysyndactyly (Carpenter and others) | AR |
| 5. Cephalopolysyndactyly (Greig) | AD |
| 6. First and second branchial arch syndromes | |
|     a. mandibulofacial dysostosis (Treacher-Collins, Franceschetti) | AD |
|     b. acrofacial dysostosis (Nager) | |
|     c. oculoauriculovertebral dysostosis (Goldenhar) | AR |
|     d. hemifacial microsomia | |
|     e. others | |
| (probably parts of a large spectrum) | |

7. Oculo-mandibulo-facial syndrome (Hallermann-Streiff-Francois)

## Dysostoses with predominant axial involvement

1. Vertebral segmentation defects (including Klippel-Feil)
2. Cervico-oculo-acoustic syndrome (Wildervanck)
3. Sprengel anomaly
4. Spondylo-costal dysostosis
   a. dominant form                                AD
   b. recessive forms                              AR
5. Oculovertebral syndrome (Weyers)
6. Osteo-onychodysostosis                          AD
7. Cerebrocostomandibular syndrome                 AR

## Dysostoses with predominant involvement of extremities

1. Acheiria
2. Apodia
3. Tetraphocomelia syndrome (Roberts) (SC pseudo    AR
   thalidomide syndrome)
4. Ectrodactyly
   a. isolated
   b. ectrodactyly-ectodermal dysplasia-cleft palate syndrome AD
   c. ectrodactyly with scalp defects              AD
5. Oro-acral syndrome (aglossia syndrome, Hanhart
   syndrome)
6. Familial radio-ulnar synostosis
7. Brachydactyly, types A, B, C, D, E (Bell's classification) AD
8. Symphalangism                                   AD
9. Polydactyly (several forms)
10. Syndactyly (several forms)
11. Polysyndactyly (several forms)
12. Camptodactyly
13. Manzke syndrome
14. Poland syndrome
15. Rubinstein-Taybi syndrome
16. Coffin-Siris syndrome
17. Pancytopenia-dysmelia syndrome (Fanconi)       AR
18. Blackfan-Diamond anemia with thumb anomalies
    (Aase-Syndrome)                                AR
19. Thrombocytopenia-radial-aplasia syndrome       AR
20. Orodigitofacial syndrome
    a. type Papillon-Léage                         XLD (lethal in
                                                   males)
    b. type Mohr                                   AR

21. Cardiomelic syndromes (Holt-Oram and others)    AD
22. Femoral focal deficiency (with or without facial anomalies)
23. Multiple synostoses (includes some forms of symphalangism)    AD
24. Scapulo-iliac dysostosis (Kosenow-Sinios)    AD
25. Hand-foot-genital syndrome    AD
26. Focal dermal hypoplasia (Goltz)    XLD (lethal in males)

## IDIOPATHIC OSTEOLYSES

1. Phalangeal (several forms)
2. Tarso-carpal
   a. including François form and others    AR
   b. with nephropathy    AD
3. Multicentric
   a. Hajdu-Cheney form    AD
   b. Winchester form    AR
   c. Torg form    AR
   d. other forms

## MISCELLANEOUS DISORDERS WITH OSSEOUS INVOLVEMENT

1. Early acceleration of skeletal maturation
   a. Marshall-Smith syndrome
   b. Weaver syndrome
   c. other types
2. Marfan syndrome    AD
3. Congenital contractural arachnodactyly    AD
4. Cerebro-hepato-renal syndrome (Zellweger)
5. Coffin-Lowry syndrome    XLR
6. Cockayne syndrome    AR
7. Fibrodysplasia ossificans progressiva    AD
8. Epidermal nevus syndrome (Solomon)
9. Nevoid basal cell carcinoma syndrome
10. Multiple congenital fibromatosis
11. Neurofibromatosis    AD

## CHROMOSOMAL ABERRATIONS: PRIMARY METABOLIC ABNORMALITIES

### Calcium and/or phosphorus

1. Hypophosphatemic rickets    XLD

2. Vitamin D dependent or pseudo-deficiency rickets
   a. type I with probable deficiency in 25 hydroxy
      vitamin D-1-alpha-hydroxylase                                AR
   b. type II with target-organ resistance                        AR
3. Late rickets (McCance)
4. Idiopathic hypercalciuria
5. Hypophosphatasia (several forms)                               AR
6. Pseudo-hypoparathyroidism (normocalcemic and
   hypocalcemic forms)                                            AD


**Complex carbohydrates**
1. Mucopolysaccharidosis type I (alpha-L-iduronidase
   deficiency)
   a. Hurler form                                                 AR
   b. Scheie form                                                 AR
   c. other forms                                                 AR
2. Mucopolysaccharidosis type II — Hunter
   (sulfoiduronate sulfatase deficiency)                          XLR
3. Mucopolysaccharidosis type III — Sanfilippo
   a. type III A (heparin sulfamidase deficiency)                 AR
   b. type III B (N-acetyl-alpha-glucosaminidase deficiency) AR
   c. type III C (alpha-glucosaminide-N-acetyl transferase
      deficiency)                                                 AR
   d. type III D (N-acetyl-glucosamine-6 sulfate sulfatase
      deficiency)                                                 AR
4. Mucopolysaccharidosis type IV
   a. type IV A — Morquio (N-acetyl-galactosamine-6
      sulfate-sulfatase deficiency)                               AR
   b. type IV B (beta-galactosidase deficiency)                   AR
5. Mucopolysaccharidosis type VI — Maroteaux-Lamy
   (aryl-sulfatase B deficiency)                                  AR
6. Mucopolysaccharidosis type VII (beta-glucuronidase
   deficiency)                                                    AR
7. Aspartylglucosaminuria (Aspartyl-glucosaminidase
   deficiency)                                                    AR
8. Mannosidosis (alpha-mannosidase deficiency)                    AR
9. Fucosidosis (alpha-fucosidase deficiency)                      AR
10. GM1-Gangliosidosis (beta-galactosidase deficiency)
    (several forms)                                               AR
11. Multiple sulfatases deficiency (Austin-Thieffry)              AR
12. Isolated neuraminidase deficiency, several forms.
    Includes:
    a. mucolipidosis I                                            AR
    b. nephrosialidosis                                           AR
    c. cherry red spot-myoclonia syndrome                         AR

13. Phosphotransferase deficiency, several forms.
    a. mucolipidosis II (I cell disease)    AR
    b. mucolipidosis III (pseudo-polydystrophy)    AR
14. Combined neuraminidase beta-galactosidase deficiency    AR
15. Salla disease    AR

## Lipids

1. Niemann-Pick disease (sphingomyelinase deficiency) (several forms)    AR
2. Gaucher disease (beta-glucosidase deficiency) (several types)    AR
3. Farber disease lipogranulomatosis (ceraminidase deficiency)    AR

## Nucleic acids
1. Adenosine-deaminase deficiency and others    AR

## Amino-acids
1. Homocystinuria and others    AR

## Metals
1. Menkes syndrome (Kinky hair syndrome and others)    AR

REFERENCE

Maroteaux et al 1983 Nomenclature des maladies osséuses constitutionnelles (International nomenclature of constitutional diseases of bone).
Ann. Radiol., 26, no 6, 456–462

# Glossary

This short glossary is concerned with genetic, anatomical and orthopaedic terms which are sometimes a source of confusion. Emphasis has been placed upon simple explanation rather than precise definition.

## ANATOMICAL AND ORTHOPAEDIC TERMS

| | |
|---|---|
| **Acromelia** | Shortening in the distal segment of a limb, i.e. hand or foot |
| **Acromesomelia** | Shortening of the middle and distal segment of a limb, i.e. forearm or shin and hand or foot |
| **Amelia** | Absence of a limb |
| **Anomalad** | A malformation with its subsequently derived structural changes, e.g. Klippel–Feil anomalad |
| **Calvarium** | The vault of the skull |
| **Cortex** | The compact outer part of the shaft of a bone |
| **Deformity** | Alteration in shape of a previously normal part, e.g. scoliosis |
| **Diaphysis** | The shaft of a long bone |
| **Dysostosis** | A malformation of an individual bone, either singly or in combination, e.g. mandibulofacial dysostosis |
| **Dysplasia** | A generalised abnormality, e.g. achondroplasia |
| **Ectrodactyly** | Maldevelopment of digits. Used loosely for a split-hand or split-foot malformation |
| **Epiphysis** | Portion of a long bone derived from a centre of ossification distinct from that of the shaft of the bone |
| **Genu valgum** | Knock-knees |
| **Genu varus** | Bowlegs |
| **Gibbus** | Localised sharp backwards angulation of the spine |
| **Hemimelia** | Absence of a segment of a limb |

483

| | |
|---|---|
| **Kyphoscoliosis** | Combined backwards and sideways curvature of the spine |
| **Kyphus** | Backward curvature of the spine |
| **Lordosis** | Forewards curvature of the spine |
| **Madelung deformity** | 'Dinner-fork' deformity of the forearm |
| **Malformation** | Primary structural defect due to failure of normal development, e.g. phocomelia |
| **Medulla** | The inner part or marrow cavity of a bone |
| **Mesomelia** | Shortening in the forearm or shin |
| **Metaphysis** | The region between the epiphysis and the diaphysis of a long bone |
| **Micromelia** | Shortening of all segments of a limb |
| **Phocomelia** | Absence of segments of a limb or limbs, so that 'flipper-like' hands and feet articulate directly with the trunk |
| **Rhizomelia** | Shortening of the proximal segment of a limb |
| **Scoliosis** | Sideways curvature of the spine |
| **Symphalangism** | Union of the bones of the phalanges |
| **Syndactyly** | Union of the digits, either by bone or soft tissues |
| **Synostosis** | Abnormal union of bones |
| **Talipes equinovarus** | Club foot |

## GENETIC TERMS

| | |
|---|---|
| **Allele** | Alternative forms of a gene at the same site or locus on a particular chromosome |
| **Ascertainment** | The tracing of an individual or family with a genetic condition |
| **Autosome** | Any chromosome other than the X or Y sex chromosomes. In every human somatic cell nucleus there are 44 autosomes and two sex chromosomes |
| **Chromosome** | A body made up from DNA which is contained in a cell nucleus, and which is visible by light microscopy in suitable preparations. Each bears tens of thousands of genes |
| **Clone** | Cells which are derived from a single parent cell and which therefore contain identical genetic material |
| **Concordance** | The presence of a particular abnormality or genetic disorder in both members of a pair of twins |

| | |
|---|---|
| **Congenital** | An abnormality which is present at birth. This is not necessarily genetic |
| **Consanguinity** | Relatedness. A consanguineous marriage is one in which the partners have a common ancestor |
| **Dermatoglyphics** | The pattern of skin ridges on the extremities, i.e. finger and palm prints |
| **Dizygous twins** | Twins produced by the union of two sperms with two ova |
| **Embryo** | The product of conception before the end of the 8 week of pregnancy |
| **Expression** | The degree of severity of clinical manifestations of an abnormal gene |
| **Fetus** | The product of conception from the end of the 8th week to the moment of birth |
| **Gene** | A portion of the DNA molecule which determines the structure of a polypeptide chain |
| **Heterozygote** | An individual who possesses a pair of dissimilar genes |
| **Homozygote** | An individual who possesses a pair of similar genes |
| **Karyotype** | A photograph of an individual's chromosomes, arranged according to size and shape |
| **Locus** | The site of a particular gene on a chromosome |
| **Multifactorial inheritance** | Inheritance determined by the interaction of environmental factors and multiple genes |
| **Mutation (gene)** | An alteration in a gene, which is perpetuated in subsequent generations |
| **Penetrance** | The presence or absence of clinical manifestations of an abnormal gene |
| **Phenocopy** | An abnormality produced by an environmental factor which mimics a particular genetic disorder |
| **Phenotype** | Clinical features which reflect the basic genetic constitution |
| **Polygenic** | Determined by multiple genes |
| **Private syndromes** | A syndrome which is apparently confined to a single kindred |
| **Proband** | The first affected individual in a kindred to come to the notice of the investigator |
| **Sex chromosomes** | The two chromosomes which are designated XX in the female and XY in the male |
| **Sib** | Brother or sister |
| **X-linked** | A gene which is situated on an X chromosome |
| **Zygote** | A fertilised ovum. |

# Index